23-40

QMC   622566 1

KV-692-879

DATE DUE FOR RETURN

29. JUN 1984

17. DEC 85

28. FEB 86

2 MAY 1997

19 NOV 1999

3 MAY 2002

Benchmark Papers in Geology

Series Editor: Rhodes W. Fairbridge
Columbia University

Published Volumes

ENVIRONMENTAL GEOMORPHOLOGY AND LANDSCAPE CONSERVATION, VOLUME I: Prior to 1900 / Donald R. Coates
RIVER MORPHOLOGY / Stanley A. Schumm
SPITS AND BARS / Maurice L. Schwartz
TEKTITES / Virgil E. Barnes and Mildred A. Barnes
GEOCHRONOLOGY: Radiometric Dating of Rocks and Minerals / C. T. Harper
SLOPE MORPHOLOGY / Stanley A. Schumm and M. Paul Mosley
MARINE EVAPORITES: Origin, Diagenesis, and Geochemistry / Douglas W. Kirkland and Robert Evans
ENVIRONMENTAL GEOMORPHOLOGY AND LANDSCAPE CONSERVATION, VOLUME III: Non-Urban Regions / Donald R. Coates
BARRIER ISLANDS / Maurice L. Schwartz
GLACIAL ISOSTASY / John T. Andrews
GEOCHEMISTRY OF GERMANIUM / Jon N. Weber
PHILOSOPHY OF GEOHISTORY: 1785–1970 / Claude C. Albritton, Jr.
ENVIRONMENTAL GEOMORPHOLOGY AND LANDSCAPE CONSERVATION, VOLUME II: Urban Areas / Donald R. Coates
GEOCHEMISTRY AND THE ORIGIN OF LIFE / Keith A. Kvenvolden
SEDIMENTARY ROCKS: Concepts and History / Albert V. Carozzi
GEOCHEMISTRY OF WATER / Yasushi Kitano

Additional volumes in preparation

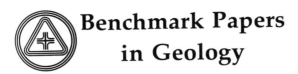

Benchmark Papers in Geology

——— A *BENCHMARK* ® Books Series ———

GEOCHEMISTRY OF WATER

Edited by
YASUSHI KITANO
Nagoya University, Japan

Dowden, Hutchinson & Ross, Inc.
Stroudsburg, Pennsylvania

Distributed by:
HALSTED PRESS *A Division of John Wiley & Sons, Inc.*

Copyright © 1975 by **Dowden, Hutchinson & Ross, Inc.**
Benchmark Papers in Geology, Volume 16
Library of Congress Catalog Card Number: 74-23330
ISBN: 0-470-48950-2

All rights reserved. No part of this book covered by the copyrights hereon may be reproduced or transmitted in any form or by any means—graphic, electronic, or mechanical, including photocopying, recording, taping, or information storage and retrieval systems—without written permission of the publisher.

77 76 75   1 2 3 4 5

Manufactured in the United States of America.

Exclusive Distributor: Halsted Press
a Division of John Wiley & Sons, Inc.

Library of Congress Cataloging in Publication Data

Kitano, Yasushi, 1923-   comp.
   Geochemistry of water.

   (Benchmark papers in geology ; v. 16)
   Bibliography: p.
   Includes indexes.
   1. Water chemistry--Addresses, essays, lectures.
I. Title.
GB697.K57   551.4   74-23330
ISBN 0-470-48950-2

Acknowledgments and Permissions

ACKNOWLEDGMENTS

AMERICAN METEOROLOGICAL SOCIETY—*Journal of Meteorology*
    The Concentration of Chloride, Sodium, Potassium, Calcium, and Sulfate in Rain Water over the United States

GEOLOGICAL SOCIETY OF AMERICA—*Geological Society of America Bulletin*
    Geologic History of Sea Water: An Attempt To State the Problem
    Magmatic, Connate, and Metamorphic Waters

PERMISSIONS

The following articles have been reprinted with the permission of the authors and the copyright holders.

AMERICAN ASSOCIATION FOR THE ADVANCEMENT OF SCIENCE—*Science*
    Isotopic Variations in Meteoric Waters

AMERICAN ASSOCIATION OF PETROLEUM GEOLOGISTS—*Bulletin of the American Association of Petroleum Geologists*
    Evidence of History of Sea Water from Chemistry of Deeper Subsurface Waters of Ancient Basins

AMERICAN CHEMICAL SOCIETY—*Equilibrium Concepts in Natural Water Systems*
    Origin of the Chemical Compositions of Some Springs and Lakes

AMERICAN GEOLOGICAL INSTITUTE—*Geochemistry International*
    The Current Tritium Contents of Natural Waters

AMERICAN GEOPHYSICAL UNION—*Journal of Geophysical Research*
    The Origin of Saline Formation Waters: 1. Isotopic Composition

AMERICAN JOURNAL OF SCIENCE (YALE UNIVERSITY)—*American Journal of Science*
    Chemical Mass Balance Between Rivers and Oceans

ECONOMIC GEOLOGY PUBLISHING COMPANY—*Economic Geology*
    Studies of Fluid Inclusions: I. Low Temperature Application of a Dual-Purpose Freezing and Heating Stage

GEOLOGICAL SOCIETY OF AMERICA—*Petrologic Studies—Volume to Honor A. F. Buddington*
    Model for the Evolution of the Earth's Atmosphere

ISRAEL PROGRAM FOR SCIENTIFIC TRANSLATION LTD.—*Chemistry of the Earth's Crust*
    Migration of Elements Through Phases of the Hydrosphere and Atmosphere
    Principal Geochemical Environments and Processes of the Formation of Hydrothermal Waters in Regions of Recent Volcanic Activity

MICROFORMS INTERNATIONAL MARKETING CORPORATION—*Geochimica et Cosmochimica Acta*
    Abundance of the Elements, Areal Averages and Geochemical Cycles
    Deuterium Content of Natural Water and Other Substances
    The Geologic History of Sea Water: An Attempt To Solve the Problem
    Natural Hydrothermal Systems and Experimental Hot-Water/Rock Interactions

MINERALOGICAL SOCIETY OF AMERICA—*Mineralogical Society of America Special Paper 3*
    Chemical Equilibria and Evolution of Chloride Brines

SPRINGER-VERLAG, INC., NEW YORK—*Hot Brines and Recent Heavy Metal Deposits in the Red Sea*
    Geochemistry and Origin of the Red Sea Brines

THE SWEDISH GEOPHYSICAL SOCIETY—*Tellus*
    Stable Isotopes in Precipitation

SWEDISH SOCIETY OF CHEMISTS AND SWEDISH ASSOCIATION OF CHEMICAL ENGINEERS—*Svensk Kemisk Tidskrift*
    How Has Sea Water Got Its Present Composition?

Series Editor's Preface

The philosophy behind the "Benchmark Papers in Geology" is one of collection, sifting, and rediffusion. Scientific literature today is so vast, so dispersed, and, in the case of old papers, so inaccessible for readers not in the immediate neighborhood of major libraries that much valuable information has been ignored by default. It has become just so difficult, or so time consuming, to search out the key papers in any basic area of research that one can hardly blame a busy man for skimping on some of his "homework."

This series of volumes has been devised, therefore, to make a practical contribution to this critical problem. The geologist, perhaps even more than any other scientist, often suffers from twin difficulties—isolation from central library resources and immensely diffused sources of material. New colleges and industrial libraries simply cannot afford to purchase complete runs of all the world's earth science literature. Specialists simply cannot locate reprints or copies of all their principal reference materials. So it is that we are now making a concerted effort to gather into single volumes the critical material needed to reconstruct the background of any and every major topic of our discipline.

We are interpreting "geology" in its broadest sense: the fundamental science of the planet Earth, its materials, its history, and its dynamics. Because of training and experience in "earthy" materials, we also take in astrogeology, the corresponding aspect of the planetary sciences. Besides the classical core disciplines such as mineralogy, petrology, structure, geomorphology, paleontology, and stratigraphy, we embrace the newer fields of geophysics and geochemistry, applied also to oceanography, geochronology, and paleoecology. We recognize the work of the mining geologists, the petroleum geologists, the hydrologists, the engineering and environmental geologists. Each specialist needs his working library. We are endeavoring to make his task a little easier.

Each volume in the series contains an Introduction prepared by a specialist (the volume editor)—a "state of the art" opening or a summary of the object and content of the volume. The articles, usually some thirty to fifty reproduced either in their entirety or in significant extracts, are selected in an attempt to cover the field, from the key papers of the last century to fairly recent work. Where the original works are in foreign languages, we have endeavored to locate or commission translations. Geologists, because of their global subject, are often acutely aware of the oneness of our world. The selections cannot, therefore, be restricted to any one country, and whenever possible an attempt is made to scan the world literature.

To each article, or group of kindred articles, some sort of "highlight commentary" is usually supplied by the volume editor. This should serve to bring that article into historical perspective and to emphasize its particular role in the growth of the field. References, or citations, wherever possible, will be reproduced in their entirety—for by this means the observant reader can assess the background material available to that particular author, or, if he wishes, he too can double check the earlier sources.

A "benchmark," in surveyor's terminology, is an established point on the ground, recorded on our maps. It is usually anything that is a vantage point, from a modest hill to a mountain peak. From the historical viewpoint, these benchmarks are the bricks of our scientific edifice.

Rhodes W. Fairbridge

Preface

Several thousand papers have been written on the geochemical investigations of natural waters. Most of the interesting papers in this field are lengthy, and indeed the very subject of water geochemistry is too vast to be incorporated into the limited space of a single volume. Accordingly, it was decided to exclude from this volume many noteworthy papers in the following specialities: the chemistry of ocean water, the geochemistry of ice and snow, pollution problems in the hydrosphere, and organic matter in natural waters.

It was not so difficult for me to choose fifty papers from the many published ones. But I did have great difficulty in selecting some twenty papers from the fifty—a reduction necessitated by space limitations. It is my belief, however, that the reader will be led to the omitted papers by means of the many references to them in the articles that are included in the volume.

The twenty-one papers reproduced are arranged under five headings: (1) Origin and Evolution of Natural Water, (2) Isotopes of Natural Waters (D/H, $^{16}O/^{18}O$), (3) Chemical Composition of Rain and River Waters, (4) Chemical Composition of Connate and Thermal Waters, and (5) Chemistry of the Red Sea and the Dead Sea. These papers emphasize the mechanisms of distribution and migration of isotopes and chemical species through the hydrosphere, as well as the chemical systems that control the concentrations and equilibria of chemical species. It is hoped that readers will recognize that the geochemistry of water is a valuable and effective tool in the understanding of the formation and development of the earth's surface.

Cordial acknowledgment is given to Professor Rhodes W. Fairbridge of Columbia University, Series Editor, who has encouraged and led me to undertake the task of editing this book; and to Janet Stroup, also of Columbia, who played an invaluable part in stylistic preparation.

Yasushi Kitano

Contents

Acknowledgments and Permissions	v
Series Editor's Preface	vii
Preface	ix
Contents by Author	xv
Introduction	1

I. ORIGIN AND EVOLUTION OF NATURAL WATER (SEAWATER)

Editor's Comments on Papers 1 Through 6 — 8

1 RUBEY, W. W.: Geologic History of Sea Water: An Attempt To State the Problem — 12
Geol. Soc. America Bull., **62**, 1111–1148 (Sept. 1951)

2 HOLLAND, H. D.: The Geologic History of Sea Water: An Attempt To Solve the Problem — 49
Geochim. Cosmochim. Acta, **36**(6), 637–651 (1972)

3 HOLLAND, H. D.: Model for the Evolution of the Earth's Atmosphere — 64
Petrologic Studies—Volume to Honor A. F. Buddington, A. E. J. Engel, H. L. James, and B. F. Leonard, eds., Geological Society of America, 1962, pp. 447–477

4 SILLÉN, L. G.: How Has Sea Water Got Its Present Composition? — 95
Svensk Kemisk Tidskrift, **75**(4), 161–177 (1963)

5 BARTH, T. F. W.: Abundance of the Elements, Areal Averages and Geochemical Cycles — 112
Geochim. Cosmochim. Acta, **23**(1–2), 1–8 (1961)

6 MACKENZIE, F. T., AND R. M. GARRELS: Chemical Mass Balance Between Rivers and Oceans — 120
Amer. J. Sci., **264**, 507–525 (1966)

II. ISOTOPES OF NATURAL WATERS: ^1H, ^2H, ^3H, ^{16}O, and ^{18}O

Editor's Comments on Papers 7 Through 11 — 140

7 FRIEDMAN, IRVING: Deuterium Content of Natural Water and Other Substances — 144
Geochim. Cosmochim. Acta, **4**(1–2), 89–103 (1953)

8	**CRAIG, HARMON:** Isotopic Variations in Meteoric Waters *Science*, **133**(3465), 1702–1703 (1961)	159
9	**DANSGAARD, W.:** Stable Isotopes in Precipitation *Tellus*, **16**(4), 436–468 (1964)	160
10	**VINOGRADOV, A. P., A. L. DEVIRTS, AND E. I. DOBKINA:** The Current Tritium Contents of Natural Waters *Geochem. Internat.*, **5**(5), 952–966 (1968)	193
11	**CLAYTON, R. N., ET AL.:** The Origin of Saline Formation Waters: 1. Isotopic Composition *J. Geophys. Res.*, **71**(16), 3869–3882 (1966)	208

III. CHEMICAL COMPOSITION OF RAIN AND RIVER WATERS

Editor's Comments on Papers 12, 13, and 14		224
12	**SUGAWARA, KEN:** Migration of Elements Through Phases of the Hydrosphere and Atmosphere *Chemistry of the Earth's Crust*, Vol. 2, A. P. Vinogradov, ed., Israel Program for Scientific Translation Ltd., 1967, pp. 501–510	227
13	**JUNGE, C. E., AND R. T. WERBY:** The Concentration of Chloride, Sodium, Potassium, Calcium, and Sulfate in Rain Water over the United States *J. Meteor.*, **15**(5), 417–425 (1958)	238
14	**GARRELS, R. M., AND F. T. MACKENZIE:** Origin of the Chemical Compositions of Some Springs and Lakes *Equilibrium Concepts in Natural Water Systems*, American Chemical Society, 1967, pp. 222–242	247

IV. CHEMICAL COMPOSITION OF CONNATE AND THERMAL WATERS

Editor's Comments on Papers 15 Through 19		270
15	**WHITE, D. E.:** Magmatic, Connate, and Metamorphic Waters *Geol. Soc. America Bull.*, **68**, 1659–1682 (Dec. 1957)	273
16	**CHAVE, K. E.:** Evidence of History of Sea Water from Chemistry of Deeper Subsurface Waters of Ancient Basins *Bull. Amer. Assoc. Petrol. Geol.*, **44**(3), 357–370 (1960)	297
17	**ROEDDER, E. W.:** Studies of Fluid Inclusions: I. Low Temperature Application of a Dual-Purpose Freezing and Heating Stage *Econ. Geol.*, **57**(7), 1045–1061 (1962)	311

| 18 | IVANOV, V. V.: Principal Geochemical Environments and Processes of the Formation of Hydrothermal Waters in Regions of Recent Volcanic Activity | 328 |

Chemistry of the Earth's Crust, Vol. 2, A. P. Vinogradov, ed., Israel Program for Scientific Translation Ltd., 1967, pp. 260–281

| 19 | ELLIS, A. J., AND W. A. J. MAHON: Natural Hydrothermal Systems and Experimental Hot-Water/Rock Interactions | 350 |

Geochim. Cosmochim. Acta, **28**, 1323–1357 (1964)

V. CHEMISTRY OF THE RED SEA AND THE DEAD SEA

Editor's Comments on Papers 20 and 21   386

| 20 | CRAIG, HARMON: Geochemistry and Origin of the Red Sea Brines | 387 |

Hot Brines and Recent Heavy Metal Deposits in the Red Sea, E. T. Degens and D. A. Ross, eds., Springer-Verlag New York, Inc., 1969, pp. 208–242

| 21 | LERMAN, ABRAHAM: Chemical Equilibria and Evolution of Chloride Brines | 422 |

Mineralog. Soc. Amer. Spec. Paper 3, 291–306 (1970)

Author Citation Index   439
Subject Index   448

Contents by Author

Barth, T. F. W., 112
Chave, K. E., 297
Clayton, R. N., 208
Craig, H., 159, 387
Dansgaard, W., 160
Devirts, A. L., 193
Dobkina, E. I., 193
Ellis, A. J., 350
Friedman, I., 144, 208
Garrels, R. M., 120, 247
Graf, D. L., 208
Holland, H. D., 49, 64
Ivanov, V. V., 328
Junge, C. E., 238

Lerman, A., 422
Mackenzie, F. T., 120, 247
Mahon, W. A. J., 350
Mayeda, T. K., 208
Meents, W. F., 208
Roedder, E. W., 311
Rubey, W. W., 12
Shimp, N. F., 208
Sillén, L. G., 95
Sugawara, K., 227
Vinogradov, A. P., 193
Werby, R. T., 238
White, D. E., 273

Introduction

Water is a relatively unique substance in that it can occur in three phases under the conditions of temperature and atmospheric pressure that exist at the earth's surface. Other substances of light molecular weight occur solely as gases under these surface conditions; water, however, can exist as a gas in the form of steam, as a liquid in the form of water, and as a solid in the form of snow and ice. The volume of water vapor in the earth's atmosphere is small, equivalent at present to the volume of approximately 10 days of precipitation (see Table 1); in other words, vapor in the atmosphere is renewed once in every 10 days through the processes of precipitation and evaporation.

Approximately 97 percent of the earth's surface water is seawater. If the oceans were subject to constant evaporation without renewal through precipitation, it is believed that they would run dry within 4,000 years. It has been suggested, however, that there has been no change in the total volume of seawater in at least the past 3 billion years, which is due to the return of the same volume of water to the oceans directly through precipitation at sea and indirectly through stream flow from land. This indicates that the residence time for water in the sea is approximately 4,000 years.

Water can exist in three phases under normal atmospheric temperature and pressure, owing to the special structure of the water molecule, which has a large dipole moment (1.85×10^{-18} esu). Because of this structure, water has a great potential for dissolving matter.

The separation and concentration of chemical elements in a material occurs during phase transformations such as those which occurred in the mantle when minerals were crystallized in silicate melt at the formation of the earth's crust; what we now refer to as "natural resources" might have been formed by this fractionation. During the distillation of pure water, nonvolatile matters remain in water solution as the solution is changed to vapor; insoluble salts are precipitated and soluble salts remain in solution, when the solution phase is changed to the solid phase. The dissolution and precipitation of chemical elements occurs on the earth's surface through the processes of evaporation, condensation, and fluid transport. Water is able to fractionate chemical elements and isotopes through the processes involved with the phase transformation. And by this fractionation what we refer to as "natural resources" is also formed.

Introduction

Chemical reaction rates in water are very rapid. Organic matter is believed to have originated in the earth's Precambrian seas with the production of nonliving molecules which are essential to the growth of the more complex biological compounds.

Chemical elements and compounds are actively transported by the movement of water, as shown in Fig. 1. The removal, transportation, and deposition of material in water — the "cycle of the hydrosphere" — is the chain of processes most notable in influencing the chemical "image" of the earth's surface; the historical aspects of the geochemistry of water are thus of great importance in determining the formation and development of the earth's morphology.

Although water within the hydrosphere assumes a variety of forms, almost all of these are interdependent within the cycle of phase transformations. Each "type" of water in the hydrosphere (i.e., glacier ice, connate water, rain and river water) has a unique chemical composition. The complex interrelation of their different properties, however, necessitates a complete understanding of the chemical relationships involved in circulation within the hydrosphere.

The disposal by human industry of a great quantity of waste materials into the atmosphere and hydrosphere, where the transportation and deposition of matter are

Table 1. Distribution of water in the hydrosphere (based on U.S. Geological Survey; B. J. Skinner, *Earth Resources,* Prentice-Hall, Inc., Englewood Cliffs, N.J., 1969)

Location	Water volume (liters)	Percentage of total water (%)
Freshwater lake	125×10^{15}	0.009
Saline lakes and inland seas	104×10^{15}	0.008
Average in stream channels	1.1×10^{15}	0.0001
Vadose water, including soil moisture	66.6×10^{15}	0.005
Groundwater within depth of ½ mile	$4,200 \times 10^{15}$	0.31
Groundwater, deep lying	4.200×10^{15}	0.31
Ice caps and glaciers	$29,000 \times 10^{15}$	2.15
Atmosphere	12.9×10^{15}	0.001
World ocean	$1,319,800 \times 10^{15}$	97.2

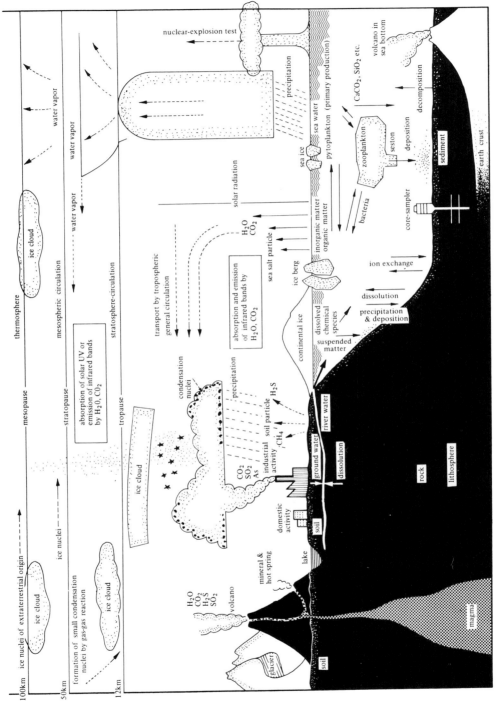

Figure 1.
Movement of water and chemical materials

Introduction

rapid and constant, easily upsets the chemical equilibrium upon which all life is dependent. Man's continued disturbance of this delicate chemical balance may ultimately result in the loss of that unique chain of reactions that supports the biosphere.

Solar radiation penetrating the stratosphere, while generating the thermal energy necessary to cause atmospheric circulation, is not sufficiently strong to counteract tropospheric cooling without the conduction of heat by water during evaporation. "Catastrophic" erosional and meteorological events, such as landslides, mudflows, and typhoons, occur partially as a result of water's tremendous ability to dissolve and its work as transformation and transportation mechanisms. In the disruption of the environmental balance by human activity, water's potential to dissolve and precipitate, and, more importantly, its potential to assist in the process of heat conduction through the atmosphere, may be lost.

The subject of water geochemistry is too vast to be incorporated into a single volume; the studies dealing with the chemistry of ocean water and the effects of pollution on the hydrologic cycle require documentation in a forthcoming volume. Some of the more noteworthy papers in the field of organic chemistry of water are too lengthy to be included here; D. W. Hood has recently edited books entitled *Impingement of Man on the Oceans* (1971) and *Organic Matter in Natural Waters* (1970).

The papers included in this volume have been chosen for their emphasis upon the mechanisms of distribution and migration of chemical species through the hydrosphere and the factors that control these mechanisms.

Geoscientific study of the earth's formation and development has long been dependent upon the study of solid materials, i.e., rocks and minerals. Only recently has the geochemistry of water been recognized as a valuable and effective tool in the understanding of the evolution of the earth's morphology. The great majority of papers written in the field of water geochemistry have only been published in the last 20 years, which is therefore the time period represented by the papers in this volume.

The opening chapter is a compilation of publications that express ideas on the origin and evolution of seawater, and several of these articles discuss the factors that control its chemical composition. The second chapter contains papers dealing with the isotopes of hydrogen and oxygen in hydrospheric water. Recent research has found that the isotopic composition of water might indicate the origin of water on the earth's surface. The progress in research in solution chemistry, which is concerned with the activity and state of chemical species dissolved in natural water, is not given a unique chapter, although several papers in this volume do discuss the application of the results in advanced solution chemistry. Publications that deal with the progress in solution chemistry which might be of interest to the reader include those by Garrels and Christ (1965), Gould (1967), Stumm and Morgan (1970), and Berner (1971). The papers in Part III examine the chemical compositions of river water and precipitation and the factors that control their composition.

Any consideration of the amount of chemical matter concentrated in solution in water must take into account the interaction that may exist between dissolved chemical species and solid material held in suspension. A great deal of study has been devoted to discernment of the equilibrium relation between material held in solution and material in suspension. Such studies should include the measurement of stability constants of ion pairs and complexes, and the examination of equilibrium constants such as solubility constants and exchange ratios. It should be noted that natural phenomena are not always explained simply by a determination of equilibrium relations but very often by an understanding of kinetic processes.

The fourth part contains papers that examine the properties of subsurface and connate waters. The data supplied by these papers indicate that water which has come into contact with sediments changes in chemical composition. Part IV also includes articles that discuss the chemical compositions of thermal and mineral water and the interaction of hot natural waters with rock; certain ideas expressed in these articles imply that the chemical compositions of these waters serves as an indicator for environments and events that occur in subsurface areas.

The two papers in Part V contain information on the Red and Dead seas, which are known to have extremely high salinities. Metal-ore deposits on the Red Sea bottom accumulate in areas of brine concentrations. Dead Sea brine forms during the process of evaporation of saltwater and the simultaneous deposition of large amounts of salt.

In this volume the points under consideration in the geochemical study of natural water are as follows:

1. It is extremely difficult to make an exact determination of trace elements (especially in ppb-order concentration) contained in natural waters.

2. The interaction between dissolved chemical species and suspended particulate matter or sediment should be considered in any discussion of natural water.

3. The dissolved states of chemical elements (ion pair, complexes between inorganic cations and inorganic anions or organic radicals) should be determined by water analysis.

4. Very active movement of chemical elements takes place within the hydrologic cycle. Many complex chemical reactions occur within water environments.

5. Although there are various kinds of water on the earth, almost all are interdependent within the cycle of the hydrosphere. Without a clarification of this series of relationships it is difficult to truly understand the differing chemical properties of each type of natural water.

References

Berner, R. A. (1971). *Principles of Chemical Sedimentology*, McGraw-Hill, New York, 240 pp.
Garrels, R. M., and C. L. Christ (1965). *Solutions, Minerals, and Equilibria*. Harper & Row, New York, 450 pp.
Gould, R. F. (ed.) (1967). *Equilibrium Concepts in Natural Water Systems*, American Chemical Society, Washington, D.C., 344 pp.
Hood, D. W. (ed.) (1970). *Organic Matter in Natural Waters*. Inst. Marine Sci., Univ. Alaska, Occasional Publ. 1, 625 pp.
Hood, D. W. (ed.) (1971). *Impingement of Man on the Oceans*. Wiley-Interscience, New York, 314 pp.
Stumm, W., and J. J. Morgan (1970). *Aquatic Chemistry — An Introduction Emphasizing Chemical Equilibria in Natural Waters*. Wiley-Interscience, New York, 583 pp.

I
The Origin and Evolution of Natural Water (Seawater)

Editor's Comments on Papers 1 Through 6

1 **Rubey:** *Geologic History of Sea Water: An Attempt To State the Problem*

2 **Holland:** *The Geologic History of Sea Water: An Attempt To Solve the Problem*

3 **Holland:** *Model for the Evolution of the Earth's Atmosphere*

4 **Sillén:** *How Has Sea Water Got Its Present Composition?*

5 **Barth:** *Abundance of the Elements, Areal Averages and Geochemical Cycles*

6 **Mackenzie and Garrels:** *Chemical Mass Balance Between Rivers and Oceans*

The origin of the earth's oceans is geologically of historical importance in determining the processes involved in the development of the earth's morphology. There are two approaches to a discussion of the origin and evolution of seawater:
 1. The chemistry of the earth's primeval oceans is reflected in the amount of volatile material present today at the earth's surface. An examination of the change in chemical composition of the primeval ocean water throughout the geologic past gives an indication of the evolution of diagenetic processes involving seawater.
 2. The paleochemical composition of seawater at any one time may be determined by an investigation of the factors that control the chemical compositions of seawater and marine sediments.
 V. M. Goldschmidt (1933) proposed an idea on the geochemical balance of the elements by projecting an outline of the evolution of the chemical composition of seawater during the earth's early history. According to his theory, cations contained in present seawater were produced by the dissolution of rocks in water, and anions have been released, independent of weathering, from hot springs and volcanoes.
 W. W. Rubey's presidential address (Paper 1) was presented to the Geological Society of America at the Washington meetings in November 1950; the address and its published version serve as a point of departure for later studies in the history of ocean water. Rubey pointed out that water is one of several volatile substances on the earth's surface—including Cl, C, N, S, and B—which act as the primary source materials for the atmosphere, hydrosphere, and biosphere. Rubey suggests that only a small portion of the total volume of volatile matter present at the earth's surface today was ever present at any one time in the earth's primeval atmosphere and ocean, and that the hydrosphere and atmosphere have been formed throughout almost the entire geologic past by the continuous escape of volatile matter from the mantle to the earth's surface through hot springs and volcanoes. It is interesting to note the change in approach to the study of the evolution of ocean water from that presented by Rubey in 1950 ("The geologic history of seawater—an attempt to *state* the problem") to the approach adopted by H. D. Holland in 1972, as it is reflected in the title of Holland's presidential address: "The geologic history of sea water—an attempt to *solve* the problem" (Paper 2).
 Both Rubey (1955) and Holland (Paper 3) described the chemistry of the atmosphere and of ocean water by applying the principles of thermodynamic equilibrium to

gaseous volatile matters; an excellent review of this is presented by P. J. Brancazio and A. G. W. Cameron in their book, *The Origin and Evolution of Atmospheres and Oceans* (1964).

Rubey's theory that the atmosphere and the oceans were formed gradually and continuously seems to be acceptable in the light of the "cold origin" explanation for the earth's formation. In considering the "condensation" theory proposed by Turekian and Clark (1969) and Anderson (1972), or the "reduction theory" proposed by Ringwood (1966), however, in which a large fraction of the atmosphere and hydrosphere is presumed to have been already present on the primitive earth's surface (Fanale, 1971), it becomes clear that even now that there are no sufficient explanations for the origin of natural water on earth.

In a determination of the factors that control the chemical composition of seawater at any given time in the earth's past, it is necessary to determine the parameters that presently control its chemical composition. L. G. Sillén (Paper 4) was able to complete his work in this area by making use of equilibrium models. J. R. Kramer (1965) published a paper on the history of seawater that could not be included in this volume because of space limitations; although some parts of his work are not completely accepted and his argument contains certain assumptions that ought to be reexamined carefully, his paper holds important implications for future studies. Kramer assumed that an inorganic ocean is derived from the solution equilibrium of minerals observed commonly on sea bottoms. According to the theory presented in his paper, the chemical composition of seawater is maintained at a given concentration of chloride ions as long as the minerals within the water remain at equilibrium.

Marine sediments deposited in the past 3 billion years seem to contain the minerals noted by both Sillén and Kramer (Ronov, 1964). The concentration of chloride ions in ancient seawater is believed to have been similar to that of present seawater. If Kramer's results are acceptable, and equilibrium has indeed been maintained between minerals and seawater for the past 3 billion years, it can be assumed that the chemical composition of seawater has not changed since that time.

Such a constancy in the chemical composition of seawater would require the removal of dissolved constituents carried to the ocean by streams. Mackenzie and Garrels (Paper 6) calculated the mass of minerals precipitated in the sea for a given period by postulating the chemical reactions involved in this removal of dissolved constituents. Their paper contains valuable information on the deposition and distribution of minerals in the sea; in 1966 this article received the best-paper award from the *Journal of Sedimentary Petrology*.

In 1961, T. F. W. Barth published his presidential address, read to the Geochemical Society in 1960, in which he presents the concept of the use of residence time to analyze the removal rates and processes for elements in the oceans (Paper 5). The concentration of an element at any given time is determined largely by the relationship of input to output during a prior time interval approximately equal to the residence time of the element. All these residence times have turned out to be short compared to the age of the oceans. Livingstone (1963) estimated the age of seawater with regard to the sodium cycle.

Editor's Comments on Papers 1 Through 6

Carbonate sediments record very clearly the history of the paleochemical composition of seawater. Odum (1951) and Lowenstam (1961), by determining the Sr/Ca ratios in calcareous fossils, reported that the chemical composition of seawater has not changed during the last $2-6 \times 10^8$ years. Many papers have been published on the paleochemical composition of seawater from the determination of chemical and physical characteristics of carbonate sediments; in this field R. W. Fairbridge's paper of 1967 is recommended to the reader.

William W. Rubey received his B.A. from Missouri University; two honorary D.Sc. degrees, from Villanova and from Yale; and an LL.D. from California. He has been a professor of geology and geophysics at the University of California (Los Angeles) between 1960 and 1966; from 1966 to the present he has served there as Professor Emeritus. Rubey was the president of the Geological Society of America in 1950, and he received the Penrose Medal in 1963. He has been with the U.S. Geological Survey for 40 years, serving in various capacities from geological aide to research geologist. Rubey was awarded the National Medal of Science by President Johnson in 1966 for his investigations into the origins of seawater.

Heinrich D. Holland was born in 1927 in Mannhein, Germany. He received his B.A. in chemistry in 1946 from Princeton University, and both his M.S. (1948) and his Ph.D. (1952) in geology from Columbia University. Holland was the president of the Geochemical Society from 1970 to 1971; he served as an instructor from 1950 to 1953, an assistant professor from 1953 to 1959, an associate professor from 1959 to 1966, and a full professor between 1966 and 1972 in the Department of Geological and Geophysical Sciences at Princeton University. Holland is presently a professor of geochemistry at Harvard University.

Lars Gunnar Sillén was born July 11, 1916, and died July 23, 1970. During a life-span of 54 years he made numerous important contributions to a wide variety of chemical problems, ranging from X-ray crystallography to solution equilibria and geochemistry, especially the chemistry of seawater, to which he successfully applied equilibrium models in order to understand the present composition of seawater. His scientific career was marked by rapid growth: he gained his doctorate at the age of 24 years, he was instituted as a professor of inorganic chemistry at Chalmers Institute of Technology at 32, and he was a professor at the Royal Institute of Technology at the age of 34. Sillén has actively participated in the production of approximately 500 publications: he was the editor of the section on inorganic chemistry in the publication *Stability Constants*, a useful collection of thermodynamic data from complex formation in solution.

Thomas F. W. Barth was born in Bolsøy, Romsdal, West Norway, May 18, 1899, and died in Oslo March 7, 1971. Barth obtained his doctorate in 1927, and left Oslo at that time for Berlin and later for Leipzig. In 1929 Barth went to Harvard University and from there to the Geophysical Laboratory of the Carnegie Institution in Washington. In 1936 he returned to Oslo as Docent and later professor of crystallography, mineralogy, and petrography. In 1946 Barth was invited to the University of Chicago as a professor of geochemistry; he returned to Oslo in 1949, where he became the director of the Mineralogical and Geological Museum of the University of Oslo. His bibliography includes 200 publications in scientific journals. Barth was the president of the Commission of Geochemistry of the International Union of Pure and Applied Chemistry between 1957 and 1960; he served as the president of the Geochemical Society between 1960 and 1961, and as the president of the International Union of Geological Societies from 1965 to 1969.

Fred T. Mackenzie is presently serving as a professor and chairman of the Department of Geological Sciences at Northwestern University. He studied at Upsala College, at Johns Hopkins University, and he received his doctorate from Lehigh University. Before joining the teaching staff at Northwestern, Mackenzie worked as an exploration geologist for the Shell Oil Company and was a staff geochemist at the Bermuda Biological Research Station. He is coauthor with R. M. Garrels of a textbook, *The Evolution of Sedimentary Rocks*.

References

Anderson, D. L. (1972). The origin and composition of the inner core, outer core and core-mantle boundary. *Trans. Amer. Geophys. Union* (EOS), **53**, 602.

Brancazio, P. J., and A. G. W. Cameron (eds.). (1964). *The Origin and Evolution of Atmospheres and Oceans*. Wiley, New York, 314 pp.

Fairbridge, R. W. (1967). Carbonate rocks and paleoclimatology in the biogeochemical history of the planet, in G. V. Chilingar, H. J. Bissel, and R. W. Fairbridge (eds.), *Carbonate Rocks, Development in Sedimentology*, Vol. 9A, American Elsevier, New York, pp. 399–432.

Fanale, F. P. (1971). Review; a case for catastrophic early degassing of the earth. *Chem. Geol.*, **8**, 79–105.

Goldschmidt, V. M. (1933). Grundlagen der quantitative Geochemie. *Fortsch. Mineral. Krist. Petrog.*, **17**, 112–156.

Kramer, J. R. (1965). History of sea water. Constant temperature–pressure equilibrium models compared to liquid inclusion analyses. *Geochim. Cosmochim. Acta*, **29**, 921–945.

Livingstone, D. A. (1963). The sodium cycle and the age of the ocean. *Geochim. Cosmochim. Acta*, **27**, 1055–1069.

Lowenstam, H. A. (1961). Mineralogy, O^{18}/O^{16} ratios, and strontium and magnesium contents of recent and fossil brachiopods and their bearing on the history of the oceans. *J. Geol.*, **69**, 241–260.

Odum, H. T. (1951). The stability of the world strontium cycle. *Science*, **114**, 407–411.

Ringwood, A. E. (1966). The chemical composition and origin of the earth, in P. M. Hurley (ed.), *Advances in Earth Science*, MIT Press, Cambridge, Mass., pp. 287–353.

Ronov, A. B. (1964). Common tendencies in the chemical evolution of the earth's crust, ocean and atmosphere. *Geochem. Internat.*, **4**, 714–737. Transl. from *Geochimiya*, No. 8, 715–743 (1964).

Rubey, W. W. (1955). Development of the hydrosphere and atmosphere, with special reference to probable composition of the early atmosphere. *Geol. Soc. America Spec. Paper 62*, pp. 631–650.

Turekian, K. K., and S. P. Clark, Jr. (1969). Inhomogeneous accumulation of the earth from the primitive solar nebula. *Earth Planet. Sci. Letters*, **6**, 346–348.

Reprinted from *Geol. Soc. America Bull.*, **62**, 1111–1148 (Sept. 1951)

GEOLOGIC HISTORY OF SEA WATER

An Attempt to State the Problem

(Address of Retiring President of The Geological Society of America)

By William W. Rubey

Abstract

Paleontology and biochemistry together may yield fairly definite information, eventually, about the paleochemistry of sea water and atmosphere. Several less conclusive lines of evidence now available suggest that the composition of both sea water and atmosphere may have varied somewhat during the past; but the geologic record indicates that these variations have probably been within relatively narrow limits. A primary problem is how conditions could have remained so nearly constant for so long.

It is clear, even from inadequate data on the quantities and compositions of ancient sediments, that the more volatile materials—H_2O, CO_2, Cl, N, and S— are much too abundant in the present atmosphere, hydrosphere, and biosphere and in ancient sediments to be explained, like the commoner rock-forming oxides, as the products of rock weathering alone. If the earth were once entirely gaseous or molten, these "excess" volatiles may be residual from a primitive atmosphere. But if so, certain corollaries should follow about the quantity of water dissolved in the molten earth and the expected chemical effects of a highly acid, primitive ocean. These corollaries appear to be contradicted by the geologic record, and doubt is therefore cast on this hypothesis of a dense primitive atmosphere. It seems more probable that only a small fraction of the total "excess" volatiles was ever present at one time in the early atmosphere and ocean.

Carbon plays a significant part in the chemistry of sea water and in the realm of living matter. The amount now buried as carbonates and organic carbon in sedimentary rocks is about 600 times as great as that in today's atmosphere, hydrosphere, and biosphere. If only 1/100 of this buried carbon were suddenly added to the present atmosphere and ocean, many species of marine organisms would probably be exterminated. Furthermore, unless CO_2 is being added continuously to the atmosphere-ocean system from some source other than rock weathering, the present rate of its subtraction by sedimentation would, in only a few million years, cause brucite to take the place of calcite as a common marine sediment. Apparently, the geologic record shows no evidence of such simultaneous extinctions of many species nor such deposits of brucite. Evidently the amount of CO_2 in the atmosphere and ocean has remained relatively constant throughout much of the geologic past. This calls for some source of gradual and continuous supply, over and above that from rock weathering and from the metamorphism of older sedimentary rocks.

A clue to this source is afforded by the relative amounts of the different "excess" volatiles. These are similar to the relative amounts of the same materials in gases escaping from volcanoes, fumaroles, and hot springs and in gases occluded in igneous rocks. Conceivably, therefore, the hydrosphere and atmosphere may have come almost entirely from such plutonic gases. During the crystallization of magmas, volatiles such as H_2O and CO_2 accumulate in the remaining melt and are largely expelled as part of the final fractions. Volcanic eruptions and lava flows have brought volatiles to the earth's surface throughout the geologic past; but intrusive rocks are probably a much more adequate source of the constituents of the atmosphere and hydrosphere. Judged by the thermal springs of the United States, hot springs (carrying only 1 per cent or less of juvenile matter) may be the principal channels by which the "excess" volatiles have escaped from cooling magmas below.

This mechanism fails to account for a continuous supply of volatiles unless it also provides for a continuous generation of new, volatile-rich magmas. Possibly such local magmas form by a continuous process of selective fusion of subcrustal rocks, to a depth of several hundred kilometers below the more mobile areas of the crust. This would imply that the volume of the ocean has grown with time. On this point, geologic evidence permits differences of interpretation; the record admittedly does not prove, but it seems consistent with, an increasing growth of the continental masses and a progressive sinking of oceanic basins. Perhaps something like the following mechanism could account for a continuous escape of volatiles to the earth's surface and a relatively uniform composition of sea water through much of geologic time: (1) selective fusion of lower-melting fractions from deep-seated, nearly anhydrous rocks beneath the unstable continental margins and geosynclines; (2) rise of these selected fractions (as granitic and hydrous magmas) and their slow crystallization nearer the surface; (3) essentially continuous isostatic readjustment between the differentiating continental masses and adjacent ocean basins; and (4) renewed erosion and sedimentation, with resulting instability of continental margins and mountainous areas and a new round of selective fusion below.

CONTENTS

	Page
Introduction	1112
Composition of sea water and atmosphere in the past	1113
Lines of possible evidence	1113
Method of geochemical balances	1114
Possible source of the "excess" volatiles	1117
Nomenclature	1117
Dense primitive atmosphere—solution in the melt	1117
Dense primitive atmosphere—chemical effects	1120
Moderate primitive atmosphere and gradual accumulation of ocean	1123
Significance of carbon dioxide in the atmosphere–ocean system	1124
Inventory of carbon dioxide	1124
Carbon dioxide equilibria in sea water	1125
Effects on organisms	1127
Effects on composition of sediments	1128
Possible limits of variation in the past	1132
Sources of supply	1134
Similarity of "excess" volatiles and magmatic gases	1135
Escape of volatiles during crystallization of magmas	1137
Hot springs as possible channels of escape	1138
Generation of local magmas	1139
Corollaries and tests of suggested mechanism	1140
Isostatic considerations	1140
Other geologic tests	1142
References cited	1143

ILLUSTRATIONS

Figure	Page
1. Carbon dioxide components and pH of sea water with varying P_{CO_2}	1127
2. pH of sea water in relation to total CO_2 of atmosphere-ocean system	1129
3. P_{CO_2} in relation to total CO_2 of atmosphere-ocean system	1133
4. Composition of magmatic gases compared with "excess" volatiles	1136

Introduction

I trust that all of you recognize that the title of this paper is largely a figure of speech. It would be interesting and even, for some inquiries, useful if we knew the history of the earth's sea water and atmosphere. But that history cannot be told until we have solved nearly all other problems of earth history. My title might much better have been "The problem of the source of sea water and its bearing on practically everything else".

Even with this modification, you still may wonder what qualifications I must think I possess to justify undertaking a subject of such global dimensions. I had originally intended to explain that I am neither an oceanographer, a geochemist, nor a Precambrian geologist, and therefore that I have no special qualifications whatever for undertaking this problem. But after discussions with various colleagues—paleontologists, petrologists, geophysicists, structural geologists, and others— I feel somewhat less modest. It seems that the subject I have selected is one in which all geologists, equally, are experts.

My interest in this general problem grew from a paper on which I began working a number of years ago. Trying to test the possibility that the phosphate rock of western Wyoming may have been laid down by direct chemical or biochemical precipitation from sea water, I began searching for some basis on which to estimate the composition of sea water and the carbon dioxide content of the atmosphere in Permian time.

It soon became evident that this question ramifies almost endlessly into nearly every fundamental problem of earth history and far beyond into the foggy borderlands between other scientific disciplines. Eventually I began to realize some of the broader implications of the problem I had tackled. I wish to take this opportunity to review what I think I have learned about the probable history of sea water and to indicate what I think this all means in terms of general earth history.

The principal conclusions of this paper were presented orally, in December 1948 and April 1949, before the Geological Society of Washington under the titles, *The problem of changes in composition of sea water and atmosphere during the geologic past* and *A possible mechanism for the continuous supply of volatiles at the earth's surface*. Since then the main thesis has been given before various groups, in successively revised form and with different points of emphasis.

In preparing this paper I have had the unsuspecting collaboration of nearly every geologist and countless others I have talked with during the past few years. Many of my colleagues of the U. S. Geological Survey and the staff members of the Geophysical Laboratory have been most generous with their criticisms and suggestions. K. J. Murata, James Gilluly, W. T. Pecora, and D. T. Griggs have been especially helpful in leading me across some of my worst gaps of data, logic, and understanding. It is only fair to add, however, that none of these many collaborators are to be held in any way responsible for the pattern of conclusions I have tried to weave from the varied threads of our discussions.

Composition of Sea Water and Atmosphere in the Past

Lines of Possible Evidence

No more than a brief review can be attempted here of the several lines of evidence that afford at least some information about the probable composition of sea water and atmosphere in the past. I cannot refrain, however, from emphasizing the one that seems to me most promising—and most neglected. The intimate dependence of living organisms on the chemical and physical conditions of their environment is so familiar—for example, an adequate supply of oxygen in the air we breathe—that we are inclined to take it all for granted. Were it not for the small amount of ozone in the upper atmosphere, which absorbs most of the deadly ultraviolet radiations, land-living organisms could not survive in direct sunlight (Poole, 1941, p. 346; Giese, 1945, p. 226, 243–246; Allee and others, 1949, p. 74). Yet, if the amount of ozone were much greater than it is, no anti-rachitic or "sunshine" vitamin, essential to the nutrition of most animals, would be produced (Stetson, 1942, p. 18–19). Likewise, many species of modern invertebrates can tolerate only narrow ranges in the salinity of their environment, largely because of the osmotic pressures that are involved (Dakin, 1935; p. 12, 16, 25–27; Rogers, 1938, p. 670; Gunter, 1947). Many organisms have very specific requirements, on the one hand, and tolerances, on the other, for the amounts of dissolved calcium, sodium, potassium, and other elements in the waters in which they live (Rogers, 1938, p. 430, 680–682). It seems not unlikely that the ancestors of some of these modern forms may have been subject to similarly rigid requirements.

Needless to say, it is hazardous to assume that ancient animals and plants had exactly the same chemical and physical requirements as their modern descendants. Conditions may have changed gradually, and the organisms may have modified their requirements by evolutionary adaptations. Yet precisely this same hazard accompanies any effort whatever to interpret the ecology of ancient plant and animal communities. The considerable measure of success that has been attained in paleoecologic interpretations by paleontologists and ecologists, working together and balancing several lines of evidence, shows that, while the problem is difficult, it is not insuperable (Twenhofel, 1936; Vaughan, 1940, p. 457; Ladd, 1944; Lowenstam, 1948, p. 104–114, 140–142; Cloud and Barnes, 1948, p. 31, 58–66).

Paleontologists have given relatively little attention to chemical factors in the environment of ancient organisms. The most noteworthy efforts I know about have been made by a few physiological chemists (Macallum, 1904, p. 561; 1926, p. 317–322, 341–348; Henderson, 1927, especially p. 38–190; Conway, 1942; 1943). Macallum believed that the blood serum and body fluids of many animals reflect closely the composition of sea water at the time when their respective ancestral lines became established. From this hypothesis he deduced a history of the composition of sea water. However, many physiologists do not accept Macallum's main premise. Many of you are familiar with L. J. Henderson's fascinating volume, *The fitness of the environment*, which presents the thesis that life as we know it today would be impossible if the physical and chemical conditions on the earth were much different from those of the present. All in all, it seems likely that the most definite information about the composition of sea water and atmosphere during the past will come, eventually, from the joint efforts of biochemists, ecologists, and paleontologists.

Another possible line of evidence on the composition of ancient sea water has proved much less dependable than was at first hoped. When A. C. Lane (1908, p. 63, 125) defined *connate waters*, it was with the thought that they represent samples of the original water in which a sediment was deposited. As information about these brines has grown, however, it has become increasingly evident that many chemical and physical processes have been modifying these waters since deposition of the enclosing sediments: adsorption, base exchange, dolomitization, evaporation, sulphate reduction, hydration and recrystallization of clay minerals, and other processes (Mills and Wells, 1919, p. 67–68; Newcombe, 1933, p. 189–196; Piper, 1933, p. 82, 89; W. L. Russell, 1933; Crawford, 1940, p. 1221–1222, 1317–1319; Heck, Hare, and Hoskins, 1940; Foster, 1942, p. 846–851). It now seems likely that most so-called connate waters have been so altered in composition that they do not represent at all accurately the original water of deposition.

Two other lines of evidence that afford some information on the probable composition of sea water and atmosphere in the past may be mentioned briefly: (1) Spectroscopic data on the solar and stellar abundances of chemical elements (Goldschmidt, 1938, p. 99–101, 120–121; Brown, 1949b, p. 625–627) and on the composition of the atmospheres of other planets (Wildt, 1942; Kuiper, 1949, p. 309, 326) set limits to the range of permissible speculation. (2) Sedimentary rocks of certain ages appear to have distinctive chemical or mineral characteristics over wide areas. The iron formations of the Precambrian (Eskola, 1932b, p. 39–40, 54; Leith, 1934, p. 161–164), the magnesian limestones of Precambrian and early Paleozoic age (Daly, 1909, p. 163–167), the coal beds of the Carboniferous, and other examples (Rutherford, 1936, p. 1212–1214; Landergren, 1945, p. 26–28; Lane, 1945, p. 396–398) have been mentioned as possibly indicating changes in chemical conditions on the earth with lapse of time. However, either the existence of these supposed worldwide similarities or, where their existence seems clear, this particular interpretation of their significance has been questioned (Clarke, 1924, p. 579; Leith, 1934, p. 161; Pettijohn, 1943, p. 957–960; Conway, 1943, p. 174–179, 200–202; 1945, p. 593–601, 603–604; Bruce, 1945, p. 589–590, 601; Miholić, 1947, p. 719). On the whole, it appears likely that there have been some real changes in the composition of sea water and atmosphere with time but that these changes must have been relatively small.

In the interest of brevity, I may summarize by stating that several lines of evidence seem to indicate some actual changes in the composition of sea water and atmosphere with time. Yet, the more closely one examines the evidence, the more these changes appear to be merely second-order differences. Everything considered, the composition of sea water and atmosphere has varied surprisingly little, at least since early in geologic time. This is certainly no startling conclusion to bring to an audience of geologists. It might even be considered as simply one of the underlying facts of geologic history on which the doctrine of uniformitarianism is based. Yet such a relative constancy of the composition of sea water throughout much of the geologic past has far-reaching implications, and these implications are worth a more careful consideration than they have sometimes received.

Method of Geochemical Balances

The most definite information about the paleochemistry of sea water and atmosphere may come eventually from the biologists, but the best evidence now available seems to lie in a comparison of the composition of rocks that have been weathered and of sediments that have been deposited during the geologic past. This general method has been used by Clarke (1924, p. 31–32), Goldschmidt (1933,

p. 131–133), Kuenen (1946), and others (Leith and Mead, 1915, p. 73; Twenhofel, 1929, p. 395–399) to estimate the total mass of rocks eroded and sediments deposited. Using a modification of this method, I have attempted to bring these earlier estimates up to date. But one must still rely on much the same data used by others; and attempted refinements of the method still give pretty much the same answers as before. Suffice it to say that, when this method is applied to rocks of different ages and to each of the commoner rock-forming elements—silicon, aluminum, iron, calcium, and several others—, the results are in surprisingly good agreement with one another. But for another group it is clear, even with present data, that there is no such agreement. All the constituents of this latter group are much too abundant in the present atmosphere and hydrosphere and in ancient sediments to be accounted for as simply the products of rock weathering—the explanation that fits well enough for the commoner elements. Compared with the commoner rock-forming oxides, the members of this group are all relatively volatile[1]. They are the substances which, in the language of the early chemists, we might call the *distillable spirits* of the earth's solid matter.

I had originally intended to present in this paper a detailed statement of the method by which these estimates have been made. The main paper grew out of all bounds in several directions, however, and these estimates and a fuller statement of the method by which they were derived must be reserved for publication in a separate article. The method of geochemical balances, as it might be called, can be outlined only briefly here.

In any given unit of geologic time, the quantity of material weathered and eroded from (*a*) crystalline rocks and (*b*) previously formed sedimentary rocks, plus (*j*) juvenile matter from volcanic gases, hot springs, etc., equals the quantity of material deposited as sediments on (*c*) continental platforms and (*d*) the deep-sea floor, plus (*s*) matter stored in sea water or escaped into interstellar space.

The results of applying this general equation of geochemical balances are the basis for much of the entire discussion that follows. For some elements, such as Si, Al, and Fe, which are relatively non-

[1] Hence Fenner's convenient term, "volatiles," for the entire group (1926, p. 696–697).

volatile and sparingly soluble at moderate temperatures, the juvenile and stored-in-sea terms are probably negligible; so that for these elements the general equation can be simplified considerably. If we have reasonably good estimates of the chemical composition of crystalline rocks, older sedimentary rocks, new continental deposits, and new deep-sea sediments, we can find for each constituent the relationship between a, the proportion of crystalline rocks to all rocks eroded, and c, the proportion of continental sediments to all sediments deposited. From the percentages of two chemical constituents in each kind of rock eroded and sediment deposited, we can find one expression relating a to c. From percentages of more than two constituents in each rock and sediment, we obtain several simultaneous equations that permit a fairly rigorous test of the chemical compositions we have used.

This is all simple enough in theory, but it is difficult in practice because accurate and representative analyses of sediments are not available. It is encouraging that a National Research Council committee is now being organized to improve available information on the chemical composition of sedimentary rocks. But, even from present information, it is possible to learn something by means of this general equation (or rather the series of equations derived from it) and to obtain results that are not grossly in conflict with other types of evidence.

It turns out that, for average crystalline rocks and sediments of Mesozoic and later ages, the average chemical compositions now available agree surprisingly well with one another for most of the permanent rock-forming oxides—SiO_2, Al_2O_3, total Fe, CaO, Na_2O, K_2O, TiO_2, and P_2O_5. If the average composition of igneous rock in earlier times was somewhat nearer that of plateau basalt, this statement is also true for sedimentary rocks of Paleozoic and even Precambrian age.

This general equation gives only ratios between the several kinds of rocks eroded and sediments deposited. Three independent estimates (based on the amount of sodium now dissolved in the ocean, the quantity of continental sediments remaining uneroded today, and the mean rate of deep-sea sedimentation) accord reasonably well in their results and thus afford means for converting these ratios into quantities of rocks eroded and sediments deposited. When all is done, the final estimates have at least this to recommend them: for the greater part they fall well within the range of estimates that Clarke, Goldschmidt, Kuenen, and others have made by simpler and more direct methods.

The foregoing statements apply to the major rock-making constituents, which are the ones that agree rather well with one another in the general equation of geochemical balances. Collectively they make up from 60 to 95 per cent of all the rocks eroded and sediments deposited. But for the re-

TABLE 1.—ESTIMATED QUANTITIES (in units of 10^{20} grams) of VOLATILE MATERIALS NOW AT OR NEAR THE EARTH'S SURFACE, COMPARED WITH QUANTITIES OF THESE MATERIALS THAT HAVE BEEN SUPPLIED BY THE WEATHERING OF CRYSTALLINE ROCKS

	H_2O	Total C as CO_2	Cl	N	S	H, B, Br, A, F, etc.
In present atmosphere, hydrosphere, and biosphere..	14,600	1.5	276	39	13	1.7
Buried in ancient sedimentary rocks.....	2,100	920	30	4.0	15	15
Total.....	16,700	921	306	43	28	16.7
Supplied by weathering of crystalline rocks..	130	11	5	0.6	6	3.5
"Excess" volatiles unaccounted for by rock weathering	16,600	910	300	42	22	13

maining constituents it is quite a different story. All but two of these latter form relatively volatile compounds at moderate temperatures and pressures; and it is the source of this group of volatile constituents that seems particularly significant in the geologic history of sea water and atmosphere.

Let us look at the way these "volatile spirits" are now distributed on and near the surface of the earth. In Table 1 are summaries of the best estimates I can find or make of the quantities of water, carbon dioxide, chlorine, etc., in today's atmosphere, ocean, fresh water, and organic matter. In the second row are the estimated totals of these constituents that are now buried in sedimentary rocks and that must have been part of the atmosphere and ocean in earlier times. Next we have the sums of the first and second rows, the quantities that must have come from somewhere and whose source should be accounted for. In the fourth row, we have, from the amount of crystalline rocks that must have been eroded to form all the sediments and from the composition of average igneous rocks, the amounts that must have been released by the weathering of crystalline rocks. Finally, in the last row, the differences between the quantities in the third and fourth rows, we have the "excess" volatiles that cannot be accounted for as simply the products of rock weathering.

For the sake of simplicity, C is shown as CO_2 in Table 1. This arbitrary convention oversimplifies the actual situation, for O does not occur in the right amounts to balance C as CO_2 exactly. Estimating directly (in units of 10^{20} g), the amount of O not combined in permanent rock oxides and in water, we find about 12 units in the atmosphere, 26 in the SO_4 of sea water, and small amounts in HCO_3, H_2BO_3, and elsewhere, giving a total of approximately 39 units of O in the present atmosphere, hydrosphere, and biosphere. In sedimentary rocks, estimates indicate about 490 units in the carbonates, 25 in organic matter, 21 in SO_3, 14 in the oxidation of FeO to Fe_2O_3, and 2 in the interstitial water, giving altogether about 550 units of O in the ancient sedimentary rocks. This, together with that in the atmosphere, etc., makes a total of about 589 units. Approximately 8 units of O have probably been released during the weathering of the small amount of CO_2 in crystalline rocks and 75 units from the Na, Mg, Ca, and K now dissolved in sea and interstitial water, which comes to a total of 83 units from rock weathering. This leaves $(589 - 83 =)$ 506 units of O "unaccounted for" and is to be compared with $(\frac{12}{44} \times 910 =)$ 248 units of C similarly "unaccounted for".

If no other complications were involved, these quantities might be explained by the release of 483 units of CO_2 and 271 units of CO from some other unspecified source. But to find a more meaningful balance between the "excess" C and O, one must consider also the probable composition of other gases, besides CO_2 and CO, that may have come from the same unspecified source. Associated with an estimated 68 units of C and 25 units of O in the organic matter in sedimentary rocks, there should be about 9.6 units of H. If this was released originally as H_2O, it would account for 76 of the 506 units of "excess" O in the above calculation. On this assumption, the "excess" of 248 units of C and

(506 − 76 =) 430 units of O would be equivalent to 274 units of CO_2 and 404 units of CO.

It is conceivable, however, that the "excess" Cl, N, S, Br, and F shown in Table 1 were originally released as HCl, NH_3, H_2S, HBr, and HF. If so, the 19 units of H thus required have since lost their identity, presumably by oxidation to H_2O; and this oxidation would have subtracted approximately 150 units of O from the original C-O mixture. On this assumption, the original "excess" of 248 units of C and (430 + 150 =) 580 units of O would have been equivalent to 687 units of CO_2 and 141 units of CO.

In these three hypothetical mixtures of CO_2 and CO, the CO_2 ranges between 40% and 83%. If the original gases also contained significant quantities of methane, CH_4 (Poole, 1941, p. 350–351), the percentage of CO_2 would have been correspondingly higher. It is evident that the proportion of "excess" C originally released as CO_2 cannot be estimated, even approximately, unless the original composition of all the "excess" volatiles is known. For this reason, all C is shown by arbitrary convention in Table 1 as CO_2.

Possible Source of the "Excess" Volatiles

Nomenclature[2]

Only two possible sources of the "excess" volatiles occur to me: Either (a) they are largely or entirely residual from a primitive atmosphere and ocean; or (b) they have largely or entirely risen to the surface from the earth's interior during the course of geologic time.

A few years ago an eminent geologist, in discussing magmas, divided those who have views on the subject into the *pontiffs* and the *soaks*. The classification is not directly applicable to those who have views on the origin of the ocean. But here, also, we have two opposing schools of thought, and, with aplogies to Bowen (1947, p. 264), his classification may readily be modified to fit the occasion. From the very nature of the problem, there can be no anhydrous pontiffs when the origin of the ocean is considered but only soaks of one persuasion or another. Here, we may say, we have the *quick soaks* who prefer to have the wetness of the ocean there at the very beginning, all of it at once; and then there are the *slow soaks* who prefer to increase the liquor gradually by small increments over a much longer period of time. The quantities involved are the same by both courses of action, but the effects are conspicuously different: Taken in small drafts and with a proper regard for timing, an astonishingly large quantity of volatile spirits can be handled by the earth without showing it; but taken hastily, even moderate amounts are almost certain to have noteworthy effects.

Dense Primitive Atmosphere—Solution in the Melt

The two contrasting procedures or viewpoints just mentioned afford a basis for discussing the probable source of these "excess" volatiles. In the opinion of some who have considered the problem—those who are of the "quick-soak" school of thought—, the source of these volatiles can be deduced without too much trouble merely from a consideration of the conditions of the primitive earth. Their argument runs something like this: As the earth was probably once molten throughout, it follows that all or a large part of the water, carbon dioxide, etc., would have been volatilized in a primitive atmosphere. On subsequent cooling, the water vapor would condense into a primitive ocean; and the present hydrosphere and atmosphere are thus residual from this primitive ocean and atmosphere.

Yet, when we examine it more closely, this conclusion is not altogether convincing. To begin with, it is by no means certain that the earth was originally molten (R. T. Chamberlin, 1949, p. 252–253; Latimer, 1950; Slichter, 1950). The inert gases—neon, argon, krypton, and xenon—are from 10^6 to 10^{10} times less abundant on the earth than in the atmospheres of the stars, the lighter gases showing the greater discrepancies. Compounds and elements of the same molecular weights—water, nitrogen, carbon dioxide, carbon monoxide, and oxygen—would have been lost in the same proportions if they too had existed as gases when the earth was formed. The fact that water is present in considerable quantity on the earth today is evidence that much of it was bound in chemical compounds or occluded in solid (that is, in relatively cool) matter when the earth accumulated (Aston, 1924; Jeffreys, 1924; Russell

---

[2] Literally, "name-calling".

and Menzel, 1933, p. 999–1001; Rayleigh, 1939, p. 463; Brown, 1949a; Gibson, 1949, p. 278; Suess, 1949; Jones, 1950, p. 420, 423–424, 428–429). This leaves us with two alternatives: either the earth remained solid from the beginning, except for local melting; or it was first cold, then heated by some process until molten throughout (Urry, 1949, p. 172, 179) and finally solidified again. Whichever one of these two theories may ultimately prevail, the evidence appears distinctly unfavorable to the concept of an atmosphere that ever contained very much of the earth's total volatiles. Instead, this evidence seems to call for some mechanism by which the volatiles were largely retained within the interior of the earth, to the extent of only a few tenths of 1 per cent of the solid matter. This appears entirely possible, whether or not the earth was ever completely molten.

These interpretations are not new, but neither are they as widely accepted as I think they deserve to be. I shall try, therefore, to show why a consideration of the geologic record leads me to these conclusions.

We might start by attempting to follow through, in a semiquantitative way, some of the consequences of a dense primitive atmosphere. If all the water in today's atmosphere and hydrosphere and in ancient sediments were once present in a primitive atmosphere, and if this atmosphere were in contact with molten silicate rock, then some fraction of the water would dissolve in the melt, the amount depending on the vapor pressure, the solubility of water in the melt, and the total quantity of molten material. Using Goranson's data for the solubility of water in a melt of granitic composition, one finds that, at this vapor pressure, the melt would dissolve 2½ per cent by weight of water.

As the deeper-lying rocks probably now contain considerably less than 2½ per cent water, this must be simply a limiting case. If there were then any considerable quantity of molten material and if this were stirred by convection currents, the water that would dissolve in the melt would materially reduce that in the atmosphere until an equilibrium was reached. Making allowance for the water now in deep-lying rocks, calculation shows that a primitive atmosphere in equilibrium with a completely molten earth would contain less than one-tenth the amount of water in the present ocean. As the supposed molten earth cooled and crystallized, the water in the remaining melt and in the atmosphere would be unlikely to remain in equilibrium, for the surface would probably crust over with slag long before the interior had fully crystallized. Thus, when crystallization was completed, the primitive atmosphere, even on this hypothesis, would probably contain at that stage, as it had at the very beginning, much less water than the present ocean.

Goranson's data on the solubility of water in a melt of granitic composition at the temperatures and pressures at which crystallization begins (Goranson, 1932, p. 229–231) appear to be the most nearly applicable to the conditions here postulated. I find no data on the solubility of other gases in silicate melts of the required composition. One would expect, from the work of Morey and Fleischer (1940, p. 1051–1057) on the system CO_2-H_2O-K_2O-SiO_2, that the proportion of CO_2 to H_2O dissolved in the melt would be somewhat higher than that in the vapor phase or atmosphere in equilibrium with it, but not enough higher to modify the results of the following calculation significantly.

Under the assumed conditions, the primitive atmosphere would contain 16.6×10^{20} kg of water vapor. Over the 5.1×10^{18} cm² area of the earth's surface, this quantity of water would exert a pressure of 325 kg/cm². Goranson found that a melt of granitic composition at liquidus temperatures dissolved 3% by weight of water at 400 bars[3] and 2% water at 260 bars (Goranson, 1932, p. 234). Interpolating between these values, it appears that, at a vapor pressure of 325 kg/cm², a melt of this composition would dissolve 2.5% water, and the temperature of crystallization would be about 985°C.

If the rock materials of the earth's crust retained after crystallization, and still contained today, an average of about 2.5% water, the results of this calculation would be of greater significance. Actually they now contain much less than this—only 1.15% in Clarke and Washington's average igneous rock (Clarke, 1924, p. 29). On this hypothesis, over half of the calculated 2.5% water in the melt must have escaped to the surface during or after crystallization. This means that some unspecified quantity of the water in today's atmosphere, ocean, and sedi-

---

[3] 1 bar = 1.01972 kg/cm² (Birch, Schairer, and Spicer, 1942, p. 319.)

mentary rocks was not there earlier and so was not present in the supposed primitive atmosphere. Hence the original vapor pressure would have been less than the 325 kg/cm^2 calculated from today's total of 16.6×10^{20} kg. The amount of this difference depends largely upon the quantity of melt that was involved in the assumed equilibrium.

With present data, the amount of water in the supposed primitive atmosphere can be estimated only within rather broad limits. For purposes of this calculation, the average water content of crystallized rock to a depth of about 40 km may be taken as near 1.15%. This would amount to 6.6×10^{20} kg of water. If this thickness of 40 km of rock were molten, the total water in the primitive atmosphere and melt together would be $(16.6 + 6.6 =) 23.2 \times 10^{20}$ kg. From Goranson's data it may be found that, under conditions of equilibrium, 46% of this total water would be dissolved in the melt; the remainder left in the atmosphere would exert a vapor pressure of 244 kg/cm^2 at the earth's surface. At this pressure the granitic melt would dissolve 1.9% water, and the melting point would be about 1010°C. If it were true that the rocks of the earth's interior now contain 1.15% water to a depth of 300 km, similar reasoning would show that about 85% of the total water would have been dissolved in the melt; the amount remaining in the atmosphere would have had a vapor pressure of 175 kg/cm^2, and the melt would have contained about 1.3% water.

It is extremely doubtful, however, that rocks lying more than a few tens of kilometers below surface contain as much as 1.15% water. For the rocks of the deep interior, direct determination of the water content is, of course, impossible; and there is little on which to base an estimate other than the composition of stony meteorites (Clarke, 1924, p. 42-44; Brown, 1949b, p. 627-629). Meteorites contain few if any hydrous minerals; yet an average of 63 chemical analyses, "of the highest grade obtainable", of stony meteorites gives 0.47% H$_2$O (Merrill, 1930, p. 16-17), and a recent digest of all earlier analyses shows an average content of 0.063% H [equivalent to 0.56% H$_2$O]. (Brown, 1949b, p. 626). Some of this water, in even the most carefully collected and analyzed samples, may be the result of weathering after the meteorites reached the earth. At least some, however, is probably present as original impurities or occlusions, as seems indicated by the quantities of H$_2$, CO$_2$, and other gases in both stony and iron meteorites (R. T. Chamberlin, 1908, p. 26; Farrington, 1915, p. 190-196; Nash and Baxter, 1947, p. 2541-2543). More data on the volatile content of stony meteorites are badly needed; but from present information the best estimate for the siliceous rocks of the earth's interior, from depths of 40 km to 2900 km, is probably about 0.5% H$_2$O.

If the iron meteorites are similarly taken to represent the materials of the earth's iron core, from 2900 km to the center, we may assume the water content there is negligible.

For rocks of the deep interior, one should logically have data on the solubility of water in melts of basaltic or peridotitic composition. However, in comparison with the uncertainties about present water content of these rocks at depth, the lack of data on the solubility of water in a more basic melt is no actual handicap. Increasing pressure at greater depth would increase very considerably the solubility of water in silicate melts (Goranson, 1931, p. 492-494). Under the assumed conditions of a molten earth and with the relatively small quantities of water involved, it is probable that nowhere would the amount of water carried down into the interior by convection currents exceed that which could be dissolved in the deeper-lying melt of different composition. The controlling factor therefore would be the solubility of water in the part of the melt that was in contact with the atmosphere.

For the densities of rocks at various depths to the center of the earth, Bullen's values (1940, p. 246; 1942, p. 28), which are calculated from seismic and astronomic data, may be used.

With these various estimates, it may be calculated that, in an earth molten to the base of the silicate mantle at 2900 km, the total water in the primitive atmosphere and silicate melt would have been $(16.6 + 207 =) 224 \times 10^{20}$ kg. Under conditions of equilibrium, approximately 98% of this total would be dissolved in the melt. The remainder left in the atmosphere would have a vapor pressure of 76 kg/cm^2 at the earth's surface, and the melt would contain about 0.54% by weight of water. In an earth entirely molten to the center of the iron core, approximately 99% of the total water would be dissolved in the melt. The remainder in the atmosphere would have a vapor pressure of 52 kg/cm^2, and the melt would contain 0.37% H$_2$O. This water in the primitive atmosphere, about 2.6×10^{20} kg, would be only 16% of the total in today's atmosphere and ocean and buried in ancient sedimentary rocks. If the estimate used here of 0.5% H$_2$O in the earth's siliceous mantle is too high, the total quantity of water involved in the equilibria would be less, and the remainder in the primitive atmosphere would be even lower than 16% of that today.

For purposes of presentation, this calculation has proceeded through steps of increasing depth of melting from the earth's surface downward to the center of the earth. Actually the hypothesis of an originally molten earth requires the reverse order, starting with a completely molten condition and solidifying from the bottom or, rather, from the base of the silicate mantle, upward (Adams, 1924, p. 467–468). If, at any stage of cooling, the surface crusted over with slag, equilibrium between the water content of the atmosphere and melt could no longer be maintained, and the primitive atmosphere would end up with some earlier water content, which would be much less than 16.6×10^{20} kg.

If, on the other hand, the surface did not crust over before the earth solidified throughout, heat from the deep interior would continue to escape rapidly; and solidification would probably have been completed within a few tens of thousands of years (Jeffreys, 1929, p. 79, 147–148; Slichter, 1941, p. 567, 582–583). With such rapid freezing, it is improbable that equilibrium could have been maintained between the water content of the atmosphere and that of the immense volume of the melt. Thus, on this alternative also, the primitive atmosphere would probably end up with a water content significantly less than 16.6×10^{20} kg.

Dense Primitive Atmosphere—Chemical Effects

A totally different line of approach also indicates that the fraction of present hydrosphere and atmosphere that might be residual from a primitive ocean and atmosphere is probably very small. Let us assume, for purposes of calculation, that all the "excess" volatiles were once in a very hot atmosphere; that, on cooling, the water vapor condensed to liquid and dissolved all the chlorine, fluorine, bromine, etc.; and that the principal gases—carbon dioxide, nitrogen, and hydrogen sulphide—were dissolved in the water according to their solubilities and partial pressures.

This would mean at first a very acid ocean (pH 0.3). This acid water would attack bare rock with which it came in contact, and the primitive ocean would rapidly take into solution increasing quantities of bases from the rocks, until the solubility product of calcite or dolomite was reached. Thereafter, as more dissolved bases were brought in by streams, carbonate would be deposited, and carbon dioxide subtracted from the atmosphere-ocean system.

Eventually conditions would be reached under which primitive life could exist. Certain mosses have been found to tolerate as much as one atmosphere of carbon dioxide (Ewart, 1896, p. 404, 406–407), an amount more than 3000 times that in the atmosphere today. If we take this value as a basis for estimating when life and the photosynthetic production of free oxygen began, we find that sea water then would have a pH of about 7.3; a salinity approximately twice that of the present; and, because of buffer effects, considerably more bicarbonate than chlorine in solution. The bases required to balance the acid radicals, added to those previously deposited as carbonates, would call for the weathering by that time of considerably more igneous rock than appears, from other evidence, to have been weathered in all of geologic time. And, even worse, more than half the original carbon dioxide would still remain in the atmosphere and ocean.

This hypothesis also calls for the deposition of tremendous quantities of carbonate rocks before the carbon dioxide of the atmosphere was lowered to the point where life could first exist. No such excessively large quantities of carbonates are known in the earliest Precambrian sediments. Compilations of measured thicknesses of sedimentary rocks suggest that the percentage of limestone in sediments of Precambrian age is about the same as (or perhaps lower than) that in later sedimentary rocks. Sederholm's estimate (1925, p. 4) indicates a very low proportion of limestones and dolomites in the Precambrian rocks of Finland. Schuchert's (1931, p. 58, 49) figures (25% limestone in the Precambrian, 20% limestone in Cambrian to Pleistocene inclusive) are heavily influenced by estimates, now known to be excessive (Osborne, 1931, p. 27–28), of the thickness of the Grenville series in southeastern Canada and so should be lowered somewhat. Leith's estimates (1934, p. 159–162) show from 1 to 48 (average about 19%) limestone in sections of Precambrian rocks in North America, Africa, and Asia and an average of 22% limestone in sedimentary rocks of later age. There is thus no indication in the geologic record of the tremendous quantities of early Precambrian limestone that would be required under this hypothesis. These and other semiquantitative tests make the hypothesis of a dense primitive atmosphere appear very improbable.

Under the given conditions of temperature, pressure, and composition (essentially no free hydrogen or oxygen), the quantities of the gases, CH_4, NH_3, and SO_2, that would be stable may be taken as negligible, and therefore the "excess" C, N, and S present as CO_2, N_2, and H_2S. If all the Cl, Br, and F were dissolved in 16.6×10^{20} kg water, they would amount to 0.529 acid equivalents per liter. Initially this amount of acid would be balanced almost solely by hydrogen ions, from which a hydrogen-ion concentration of $0.529 = 10^{-0.277}$ or a pH of 0.277 may be calculated.

This acid water would decompose the rocks with which it came in contact, and in so doing it would dissolve some of the bases. As a result, its acidity would decrease, and the chemical effects of dissolved gases would come to play an important part in the over-all balance. Better data are available for estimating the required coefficients at moderate temperatures than at those near the boiling point of water; and 30°C is here chosen as the basis for calculation. At this temperature and in sea water of normal salinity (35°/₀₀ or a "chlorinity" of 19.4 °/₀₀), the solubility coefficients of CO_2, N_2, and H_2S are estimated at 0.026, 0.0009, and 0.077 mols/liter (Harvey, 1945, p. 66; Sverdrup, Johnson, and Fleming, 1946, p. 191); the first and second apparent dissociation constants of CO_2 at 1.11×10^{-6} and 1.27×10^{-9} (Harvey, 1945, p. 64); the two comparable dissociation constants of H_2S at 3.0×10^{-7} and 2.8×10^{-14} (Latimer, 1938, p. 316); the single dissociation constant of H_3BO_3 at 2.26×10^{-9} (Harvey, 1945, p. 62); and the apparent solubility product of $CaCO_3$ at 1.2×10^{-6} (Revelle and Fleming, 1934, p. 2091; Wattenberg, 1936, p. 176; Smith, 1940, p. 182). Equations that express the relationships between various quantities involved in the CO_2 equilibria of sea water (Harvey, 1945. p. 61–68) may, with some patience, be solved by the method of successive approximations.

The proportions of the different bases dissolved when igneous rocks are decomposed by weathering may be estimated roughly from the average composition of waters of the type that drain from areas of igneous rock (75 stream waters lowest in salinity and highest in silica; Clarke, 1924, p. 111, 74–79, 81, 83, 89–91, 95–97, 103–105, 107–108). After correction for "cyclic salts" (Conway, 1942, p. 135–139, 155), this is found to be, in terms of the four principal bases,

Ca	52%
Mg	11%
Na	27%
K	10%
	100%

Other bases, acid radicals, and colloidal aggregates, including significant amounts of silica and iron, in such waters may be ignored for purposes of the present estimate.

An independent estimate of the relative abundance of the principal bases in such stream water may be made by applying Conway's average percentage losses on weathering of igneous rock (Ca —67%, Mg —18%, Na —41%, and K —26%; Conway, 1942, p. 153) to Clarke and Washington's average composition of igneous rocks (Clarke, 1924, p. 29). This gives, as the expected composition of stream water from igneous areas (in per cent),

Ca	53
Mg	8
Na	25
K	14
	100

The subsequent calculations depend primarily upon the proportion of dissolved Ca to total bases, and these two estimates therefore accord closely enough for present purposes.

As decomposition of rocks proceeded, more and more dissolved bases would be carried to the sea until the solubility product of $CaCO_3$ would eventually be reached, and carbonates would begin to precipitate. Using the foregoing data, it may be found that at this stage the concentration of $[Ca^{++}]$ in the sea water would be about 0.15 mols/l, that of $[CO_3^-]$ about 8×10^{-6} mols/l, and the pH 5.1. Approximately 29% of the total CO_2 would then be dissolved in the sea water, 0.6% of the N_2, and 52% of the H_2S. The gases remaining in the atmosphere would have partial pressures of $P_{CO_2} = 12.2$, $P_{N_2} = 1.27$, and $P_{H_2S} = 0.28$. Summing up the quantities of dissolved acids, bases, and atmospheric gases in the sea water would give a "salinity" of about 46 g/kg. (See Table 2, column 2.)[4]

With further rock weathering and transportation of dissolved bases to the sea, carbonates would continue to be deposited and CO_2 subtracted from the atmosphere-ocean system. At the arbitrarily assumed $P_{CO_2} = 1.00$, at which it is conceivable that living organisms could survive, it may be found that 980×10^{20} g $CaCO_3$ would have been precipitated, leaving 53% of the original CO_2 still in the atmosphere and ocean. The concentration of $[Ca^{++}]$ would then have fallen to about 95×10^{-6} mols/l and that of $[CO_3^-]$ would have risen to about 13×10^{-6} mols/l. The pH would be 7.3, and the concentration of Na in the sea water would be about 12.5 g/kg. Approximately 87% of the CO_2 then

---

[4] The full computations would require several pages for adequate presentation; only the results are given here.

remaining in the atmosphere-ocean system would be dissolved in the sea water, 0.5% of the total N_2, and 85% of the H_2S. The gases in the atmosphere would have partial pressures of $P_{CO_2} = 1.00$, $P_{N_2} = 1.07$, and $P_{H_2S} = 0.073$. The solids and gases dissolved in the sea water would give a total "salinity" of about 67 g/kg (Table 2, column 3).

If we apply Conway's percentage losses of bases when igneous rock is weathered and use igneous rock of average composition, it may be found that approximately $17,000 \times 10^{20}$ g of igneous rock must have been weathered to account for the Ca, Mg, Na, and K then dissolved in sea water and previously precipitated as carbonates.

It is of interest to compare the results of these calculations with independent estimates of the quantities of some materials that have been weathered, dissolved, or precipitated during the entire course of geologic history. By methods mentioned briefly in an earlier part of this paper, the writer has estimated the total amount of carbonate rocks of all ages now remaining uneroded at 1500×10^{20} g and the total amount of igneous rock that has been weathered at $11,000 \times 10^{20}$ g (Table 2, columns 3 and 6).

These figures do not accord at all well with those required by the hypothesis of a dense primitive atmosphere. Even after precipitation of 980×10^{20} g $CaCO_3$, 53% of the original CO_2 would still remain in the atmosphere and ocean. To reduce this to its present small amount would mean the weathering of even more than the $17,000 \times 10^{20}$ g of igneous rock required at this stage and the deposition of additional C as $CaCO_3$ and organic carbon. It is thus seen that this assumption of a dense primitive atmosphere requires the weathering of more igneous rock and the deposition of more carbonates than seems likely to have been weathered and deposited in all of geologic time.

The same conclusion may be reached more di-

TABLE 2.—COMPOSITION OF ATMOSPHERE AND SEA WATER UNDER ALTERNATIVE HYPOTHESES OF ORIGIN, COMPARED WITH PRESENT-DAY CONDITIONS

	All "excess" volatiles in primitive atmosphere and ocean (Original P_{CO_2} very high)			Only a fraction of total volatiles in primitive atmosphere and ocean. (Original $P_{CO_2} \leq 1.0$ Life begins early)		Present-day conditions
	Initial stage; before rock weathering	Intermediate stage; $CaCO_3$ begins to precipitate.	Late stage; life begins at $P_{CO_2} = 1.0$	Initial stage; before rock weathering	Intermediate stage; $CaCO_3$ begins to precipitate.	
	(1)	(2)	(3)	(4)	(5)	(6)
Atmosphere (kg/cm²)	14.2	13.8	2.1	<1.1	<1.1	1.0
N_2	9	9	50	7	7	78
CO_2 (% by volume)	89	89	47	90	90	0.03
H_2S	2	2	3	3	3	—
O_2, others	—	—	—	—	tr	22
Ocean ($\times 10^{20}$ g)	16,600	16,600	16,600	<990	<990	14,250
Cl, F, Br	18.3	18.3	18.3	18.3	18.3	19.4
$\Sigma S, \Sigma B$, others	0.8	0.8	1.3	0.1	0.1	2.8
ΣCO_2	14.3	15.8	25.2	<1.1	<1.7	0.1
Ca (g/kg)	—	5.9	tr	—	<5.5	0.4
Mg	—	1.3	5.2	—	<1.2	1.3
Na	—	3.1	12.5	—	<2.9	10.8
K	—	1.2	4.7	—	<1.1	0.4
H	0.5	tr	—	0.5	tr	—
"Salinity" ‰	33.9	46.4	67.2	<20.0	<30.8	35.2
pH	0.3	5.1	7.3	0.3	5.7	8.2
$CaCO_3$ pptd. ($\times 10^{20}$ g)	None	None	980	None	None	1500±
Igneous rock eroded ($\times 10^{20}$ g)	None	4200±	17,000±	None	<240	11,000±

rectly, without using the estimated quantities of igneous rocks eroded and carbonates deposited. The amount of Na now dissolved in sea water is known within relatively narrow limits of uncertainty. Adding any reasonable estimates that one prefers for the Na in ancient salt deposits and the Na dissolved in the interstitial water of sediments, it still comes out that this hypothesis of a dense primitive atmosphere requires the solution of much more Na in sea water, even before life began, than appears, from other evidence, to have been dissolved in all of geologic time.

Results not significantly different from these are obtained if we assume that the first rocks that were decomposed by the supposed highly acid sea water had a composition quite different from that of average igneous rock. Applying Conway's percentage losses to the weathering of average plateau basalt (Daly, 1933, p. 17) gives, as the probable relative abundance of the principal bases in stream water,

Ca	73%
Mg	12%
Na	13%
K	2%
	100%

Plateau basalt makes up only a small fraction of all the various rocks now exposed and undergoing erosion on the continents of the earth. Yet it just happens that present-day streams (Clarke, 1924, p. 119), after correction for "cyclic salts", are carrying these four bases to the ocean in proportions (per cent) similar to those that would be expected in streams from outcrops of plateau basalt:

Ca	73
Mg	11
Na	9
K	7
	100

If stream water of this composition, instead of that from areas of average igneous rock, is used in the above calculations, it may be found that, at the stage when $P_{CO_2} = 1.0$, not quite so much igneous rock needs to have been weathered (13,000 $\times$ 10^{20} instead of 17,000 $\times$ 10^{20} g) but that even more $CaCO_3$ must have been deposited (1500 $\times$ 10^{20} instead of 980 $\times$ 10^{20} g). With these quantities, as with those calculated from the weathering of average igneous rock, the conclusion remains that this hypothesis of a dense original atmosphere seems to require entirely improbable amounts of rock weathering and of carbonate deposition in the early stages of earth history.

Moderate Primitive Atmosphere and Gradual Accumulation of Ocean

Let us contrast this with an alternative hypothesis that seems much more probable. The difficulties just mentioned—with what might be called the "quick-soak" hypothesis—could be avoided if, instead of all the "excess" volatiles, only a small fraction was ever present at any one time in the primitive atmosphere, and if the partial pressure of carbon dioxide never exceeded, let us say, one atmosphere. Following the same reasoning as before about the accumulation of bases in sea water, one finds that—with the alternative "slow-soak" hypothesis—carbonates would begin to precipitate when the atmosphere and ocean contained less than 10 per cent of the total "excess" volatiles. At that stage, the atmosphere would have a total pressure only one-tenth greater than at present; the salinity of the sea water would be nearly the same as that of today; and its pH would be about 5.7. The fourth and fifth columns of Table 2 show some of the conditions for two stages of this alternative hypothesis. With gradual addition of more volatiles, and as free oxygen accumulated after the advent of plant life (Van Hise, 1904, p. 956), conditions would approach closer and closer to those of today (Table 2, last column). Something of this sort seems a much more likely picture of the earth's early atmosphere and ocean.

The preceding calculation started with all the "excess" volatiles in the primitive atmosphere and ocean and followed through the consequences to be expected if the high initial pressure of CO_2 were reduced by deposition of carbonate sediments to some point where primitive forms of life could survive. This alternative hypothesis starts with the assumption that at no time did the "excess" volatiles in the atmosphere and ocean exceed such an amount that the partial pressure of CO_2 was greater than 1.0. This would permit, as seems to be required by geologic and chemical evidence (Rankama, 1948, p. 390–392, 409–414), the existence of primitive forms of life as early as, or earlier than, the deposition of the first carbonate sediments. From this alternative hypothesis (assuming that not enough free oxygen had yet accumulated to oxidize the H_2S to SO_4) and from the previous values for the composition of incoming stream water, it is possible to estimate the conditions that would have prevailed when carbonates first began to precipitate from the sea. It may thus be found that, if P_{CO_2} never exceeded 1.0, then not more than 6% of the total "excess" volatiles could have been present in the early atmosphere and ocean; and the volume of the primitive ocean would therefore have been much smaller than at present.

At this stage the concentration of $[Ca^{++}]$ would be about 0.14 mols/l, that of $[CO_3^-]$ about 8.8×10^{-6} mols/l, and the pH 5.7. Approximately 3.2% of the CO_2 in the system at that time would be dissolved in the small ocean, 0.04% of the N_2, and 6.7% of the H_2S. The gases in the atmosphere would have partial pressures of $P_{CO_2} = 1.0$, $P_{N_2} = 0.077$, and $P_{H_2S} = 0.033$. The solids and gases dissolved in the sea water at this stage would give a "salinity" of about 31 g/kg (Table 2, column 5).

Poole (1941, p. 346–347, 359) has suggested that an initial supply of free oxygen, sufficient to support life and permit photosynthesis by green plants, may have been produced by photochemical dissociation of water vapor in the upper ionized layers of the atmosphere and by subsequent escape of hydrogen. In this connection, it is of interest to note that purple sulfur bacteria and certain other present-day forms can synthesize organic matter from CO_2 in an anaerobic environment that contains some H_2S. It is possible that primitive organisms of this type may have helped prepare the way for the advent of green plants on the earth (Rabinowitch, 1945, p. 4, 82–83, 99–106, 124–125). If, as, here assumed, the partial pressure of CO_2 was once as high as 1.0 atmosphere, even the rain water would have been so highly carbonated that it would leach much Fe and Si from exposed rocks, and the streams would transport significant amounts of these elements in solution (Gruner, 1922, p. 433–436; Moore and Maynard, 1929, p. 276, 293–298, 522–527; Cooper, 1937, p. 307), much as seems to have happened in parts of Precambrian time (Leith, 1934, p. 161–164).

The conditions calculated for stream water from average igneous rocks would remain essentially unchanged if we used stream water from plateau basalt instead. If, however, the partial pressure of CO_2 were less than 1.0, the fraction of total "excess" volatiles in the atmosphere-ocean system would be proportionately less than 6%; and several of the other quantities (Table 2, column 5) would likewise be decreased.

I am sure I need not warn you that the results of these calculations are not to be taken too seriously. They are based on inadequate data and laden with many "ifs and ands". Nevertheless I believe they point rather clearly to the right answer. If so, not

TABLE 3.—INVENTORY OF TOTAL C (AS CO_2) IN ATMOSPHERE, HYDROSPHERE, BIOSPHERE, AND SEDIMENTARY ROCKS

	$\times 10^{18}$ g	
Atmosphere........	2.33	
Ocean and fresh water	130	147
Living organisms and undecayed organic matter....	14.5	
Sedimentary rocks (including interstitial water)....		
Carbonates......	67,000	
Organic C........	25,000	92,000
Coal, oil, etc......	27	

more than a small part of the total "excess" volatiles could be residual from a primitive atmosphere and ocean. The only alternative I can think of is that these volatiles have risen to the surface from the earth's interior.

SIGNIFICANCE OF CARBON DIOXIDE IN THE ATMOSPHERE-OCEAN SYSTEM

Inventory of Carbon Dioxide

Moreover, these volatiles must have risen to the surface gradually and not in a few great bursts. Table 3 is an attempt to bring together an inventory or summary of separate estimates and thus arrive at a figure for the total carbon dioxide and carbon on and near the surface of the earth today. The present atmosphere and ocean contain approximately $1\frac{1}{2} \times 10^{20}$ grams of carbon dioxide—a small part of it in the atmosphere, somewhat more in organic matter, and most of it dissolved in water. It is significant to note that, altogether, this is less than 1 part in 600 of the total carbon dioxide and organic carbon that has, at one time or another, been in circulation in the atmosphere and ocean and is now buried in ancient sedimentary rocks.

The amount of CO_2 in the atmosphere is taken directly from Humphreys (1940, p. 81); that in sea water has been calculated as 129×10^{18} g from data on the solubility and dissociation constants in sea water of $35^0/_{00}$ salinity, at an average temperature of 8°C, and in equilibrium with a partial pressure of 0.0003 atmospheres of CO_2 (Harvey, 1945, p. 59, 62, 64–66). The total quantity of fresh water is here estimated at 335×10^{20} g and, from Clarke's data on average composition of stream and lake waters (1924, p. 119, 138), the amount of dissolved CO_2 is calculated to be 0.9×10^{18} g.

The quantity of CO_2 equivalent to the C in living organisms is estimated at 0.036×10^{18} g in the oceans, using data from Krogh (1934b, p. 433, 436); and at 0.029×10^{18} g on the lands, using data from Krogh (1934a, p. 421, 422), Riley (1944, p. 133), and Rabinowitch (1945, p. 6), and assuming a rough proportionality between the amount of living organic matter and the rate of carbon fixation. Undecayed particulate organic debris is estimated as equivalent to 11.8×10^{18} g CO_2 in the oceans (Krogh, 1934a, p. 422; 1934b, p. 435–436) and, very roughly, as equivalent to 2.6×10^{18} g CO_2 on the land areas, using data from Rabinowitch (1945, p. 6) on the areas of soil types and from Twenhofel (1926, p. 17) on the humus content of soils.

The estimates of carbonates and organic carbon in ancient sediments are those derived from the equations of geochemical balance mentioned above; space does not permit a more complete statement of them here. The estimates for coal, oil, etc., have been taken directly or calculated from the following sources and are here summarized in units of 10^{18} g C: Coal—7.2 (Assoc. for Planning and Regional Reconstruction, Broadsheet No. 10, 1942, p. 4); oil —0.1 (Weeks, 1950, p. 1952); oil shale—0.1; tar sands—0.025; natural gas—0.025 (Hubbert, 1950, p. 174–175); graphite—1.0×10^{-5} (U.S. reserves—Currier and others, 1947, p. 249—multiplied by 10). These total to an amount equivalent to 27.3×10^{18} g CO_2.

Carbon Dioxide Equilibria in Sea Water

If only a small part of this total buried carbon dioxide were suddenly added to today's atmosphere and ocean, it would have profound effects on the chemistry of sea water and on the organisms in the sea. These effects become more evident when we consider the important part played by carbon dioxide in the chemistry of sea water.

Table 4 is presented as a reminder of the composition of sea water and to illustrate several essential points. Sea water varies widely in its total salinity, but scarcely at all in the proportions of its dissolved constituents. The most significant exceptions are the quantities of bicarbonate and carbonate ions, which depend only in part upon salinity but are affected much more by hydrogen-ion concentration and by the amount of carbon dioxide in the atmosphere with which the water is in contact.

Note, in the column to the right, that the positive ions would exceed the negative ions by 2.38 milli-equivalents were it not that the difference between them, the so-called "excess base", is balanced by dissociated ions of carbonic and boric acid. The process of balancing the "excess base" is one of the results of the buffer action of sea water. Note also that, if the amount of sodium in sea water has been increasing continuously through geologic time, as commonly assumed, then some of the acid radicals, such as chlorine, must also have been increasing by the same amounts. Otherwise sea water would have been acid rather than alkaline throughout much the greater part of the past.

The exceedingly delicate balance that prevails between the positive and negative ions in sea water may be appreciated more fully if one considers the consequences of a decrease of only 1 part in 100 of the dissolved Na. For the salinity and temperature shown in Table 4, this would mean reducing the dissolved Na from 480.80 to 475.99 milliequivs/l and the total of dissolved bases from 621.76 to 616.95. If the acid radicals, other than the carbonic and boric acid ions, remain unchanged, then they will exceed the bases; and the "excess base" becomes a negative quantity of $619.38 - 616.95 = 2.43$ milli-equivs/l. It may readily be shown, using the appropriate constants (Harvey, 1945, p. 59, 61–68), that under these conditions the total concentration of carbonic and boric acid ions, $C_{HCO_3^-} + 2C_{CO_3^-} + C_{H_2BO_3^-}$, would be only 5.6×10^{-6} milli-equivs/l, an entirely negligible quantity by comparison. Chemical balance would therefore be maintained by a concentration of $2.43 \times 10^{-3} = 10^{-2.61}$ mols or gram-ions of hydrogen per liter. That is to say, the pH of the sea water would fall from its present average value of 8.17 to 2.61.

The carbon-dioxide equilibria of sea water are numerous and complex, but related to one another in such a way that, for a given temperature, salinity, and partial pressure of carbon dioxide, all the other variables are fixed. Published tables (Harvey, 1945, p. 59, 62, 64–66) make it possible to work out, for any given salinity and temperature, the amounts of dissolved carbon dioxide, carbonic acid, bicarbonate and carbonate ions in equilibrium with different amounts of carbon dioxide in the atmosphere, and also to find the resulting hydrogen-ion concentration in the sea water. In the left-hand diagram of Figure 1, the amounts of carbonate range from high values, which would cause precipitation of calcium carbonate, at low pressures of CO_2, to low concentrations of carbonate, which would cause solution of calcium carbonate, at high pressures of CO_2. The vertical line through the middle of both graphs represents the amount of CO_2 in the present atmosphere.

These values at low and high pressures of CO_2 would not represent stable conditions in nature. If the solubility product of $CaCO_3$ was greatly exceeded, carbonate would be precipitated from the sea water, either directly or by organic agencies, and it would continue to precipitate until equilibrium was re-established. If, on the other hand, much of the water was significantly unsaturated with $CaCO_3$, carbonate sediments lying on the sea floor would be dissolved, or Ca would simply accumulate from incoming stream water until the point of saturation was reached. From Smith's values of the solubility product for 20° and 30° C and 36‰ salinity (Smith, 1940, p. 182) and from Wattenberg's temperature and salinity coefficients (Wattenberg, 1936, p. 176), the solubility product for 8° C and 35‰ salinity may be estimated at 2.1×10^{-6}.

The right-hand diagram of Figure 1 shows, for comparison, the conditions when sea water is saturated with calcium carbonate. Note that the total carbon dioxide dissolved in sea water increases much more with increase in the partial pressure than on the other diagram.

TABLE 4.—COMPOSITION OF NORMAL SEA WATER
(Salinity 35‰, 8°C, $\rho = 1.025$, pH 8.17)

	Dissolved matter (g/kg)	Milli-mols / liter	Milli-equivalents / liter
Cl^-	19.360	560.70	560.70
$SO_4^=$	2.701	28.88	57.76
Br^-	0.066	0.85	0.85
F^-	0.001	0.07	0.07
CO_2 }H_2CO_3	0.001	0.01	—
HCO_3^-	0.116	1.90	1.90 } "Carbonate
$CO_3^=$	0.012	0.20	0.40 } alkalinity"
H_3BO_3	0.022	0.35	—
$H_2BO_3^-$	0.005	0.08	0.08
			619.38 2.38
Na^+	10.770	480.80	480.80
Mg^{++}	1.298	54.78	109.56
Ca^{++}	0.408	10.46	20.92
K^+	0.387	10.18	10.18
Sr^{++}	0.014	0.15	0.30
	35.161		621.76 − 619.38 = 2.38 = "Excess base"

Recalculated for Cl = 19.37‰, $\rho = 1.025$, and pH = 8.17 from Sverdrup, Johnson, and Fleming (1946, p. 173).

Also the proportion of bicarbonate is much greater, and the pH varies much less.

These quantities of dissolved carbon dioxide in sea water and of partial pressure in the atmosphere may, if we wish, be converted into exposed to the atmosphere in a period of from 2000 to 5000 years (Callendar, 1938, p. 224). The solid line in Figure 2 gives the pH values that would follow, perhaps more slowly, as calcium carbonate was precipitated or dis-

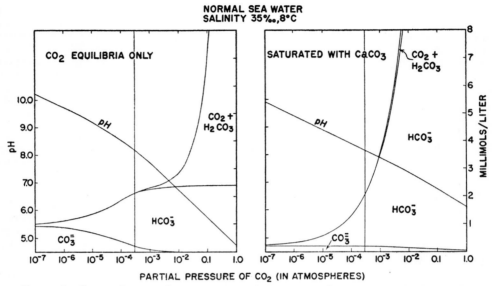

FIGURE 1.—CARBON DIOXIDE COMPONENTS AND HYDROGEN-ION CONCENTRATION OF SEA WATER AT DIFFERENT PARTIAL PRESSURES OF CARBON DIOXIDE

total carbon dioxide of the atmosphere and ocean combined. This can be done simply by multiplying through by the quantities of sea water and of atmosphere on the earth today and then adding these together. We then may see more readily the effects of changes in the total amount of carbon dioxide in the atmosphere-ocean system.

Figure 2 shows the effects that changes in the total CO_2 in the atmosphere-ocean system would have upon the hydrogen-ion concentration. Note that the dashed line gives the pH values based merely on the carbon-dioxide equilibria and that these are the values that would follow relatively soon, geologically speaking, after sudden changes of total carbon dioxide. The rate at which sea water could re-establish equilibrium after a change in atmospheric CO_2 depends mainly on the rate at which the deeper water is brought into contact with the air. From what is known about oceanic circulation, it has been estimated that the entire volume of sea water is solved. Figure 3 shows the effects that changes in the total CO_2 would have upon the partial pressure of CO_2 in the atmosphere. The values of total CO_2 shown in Figures 2 and 3 are the original ones before the adjustment that would be caused by solution or precipitation of $CaCO_3$.

The relationships shown in Figures 2 and 3 are for the present-day volume of sea water, a salinity of $35^0/_{00}$, and a temperature of 8° C. The effects of reducing the volume of sea water, of decreasing the salinity, or of raising the temperature are to shift the lines for pH in Figure 2 and for P_{CO_2} in Figure 3 somewhat to the left; but the essential relationships are unchanged.

Effects on Organisms

What would be the effects of such changes in total carbon dioxide on organisms living in the sea? We have already considered the narrow tolerance ranges of some animals for total salinity and for composition of the waters in which they live. For many forms of life, the concentration of bicarbonate and hydrogen ions and the carbon-dioxide tension are among the most critical factors in their chemical environment. A number of higher marine animals (the herring, for example) are extremely sensitive to small changes in the pH of their environment. A large proportion of the eggs of some marine animals remains unfertilized if the acidity of sea water departs more than about 0.5 pH from normal. Lower organisms are commonly less sensitive; but many species of mollusks, sea urchins, Medusa, diatoms, bacteria, algae, and others seem unable to tolerate a range of more than about 1 unit of pH. Recent workers attribute a larger part of these observed biologic effects to the carbon-dioxide tension or to the concentration of bicarbonate ions than to hydrogen-ion concentration directly (McClendon, 1916, p. 148; Shelford, 1918, p. 101–102; Gail, 1919, p. 288, 295, 297; Powers, 1920, p. 381–382; 1939, p. 73; Atkins, 1922, p. 734–735; Legendre, 1925, p. 213; Singh Pruthi, 1927, p. 743; Valley, 1928, p. 215–216, 218–220; Davidson, 1933; Rogers, 1938, p. 97, 285, 286, 294, 430, 653–656, 679, 680; Edmondson, 1944, p. 43–45, 63, 64; Allee and others, 1949, p. 175, 197).

Having thus reviewed in outline the part played by carbon dioxide in the chemistry of sea water and in the environment of marine organisms, let us now return to the inventory of carbon dioxide and to the significance of the fact that more than 600 times as much of it is buried in ancient sedimentary rocks as there is now in circulation in all the atmosphere, hydrosphere, and biosphere.

If only one-one hundredth of all this buried carbon dioxide were suddenly added to today's atmosphere and ocean (that is, if the amount in the present atmosphere and ocean were suddenly increased $\left(\frac{600}{100} + 1 =\right)$ sevenfold, from 1.3×10^{20} to 9.1×10^{20} g), it would have profound effects on the chemistry of sea water and on the organisms living in the sea. The first effect would be to change the average pH of sea water from about 8.2 to 5.9 (Fig. 2, a). This acid water would be much less than saturated with calcium carbonate, and thus further changes would follow. Eventually, when equilibrium was re-established, the partial pressure of carbon dioxide in the atmosphere would be about 110 times its present value, and the pH of the sea water would end up at an average of about 7.0 (Fig. 2, b).

The two values of pH, 5.9 and 7.0, are those for present volume and salinity of sea water and a mean water temperature of 8° C. For three-fourths of the present volume of sea water but with present salinity and temperature, these values are 5.8 and 6.9, respectively. For a salinity of 27°/oo instead of 35°/oo but with present volume and temperature of water, the corresponding values are 5.8 and 7.1. For a mean water temperature of 30° C but with present volume and salinity of sea water, these values are 5.9 and 6.8.

The effects of these changes on living organisms would be drastic. If the supposed increase of carbon dioxide happened suddenly, it would probably mean wholesale extinction of many of the marine species of today. If, however, the increase were gradual, so that organisms could adapt themselves by generations of evolutionary changes, the effects would be much less disastrous—but perhaps no less clearly recorded in the physiological adaptations of the surviving forms. From the paleontologic record it appears improbable that any change so drastic as an abrupt sevenfold increase of carbon dioxide has happened, at least since the beginning of the Cambrian.

Effects on Composition of Sediments

The conclusion that the amount of carbon dioxide in the atmosphere and ocean cannot have varied widely through much of the geologic past does not rest solely on the narrow bicarbonate and pH tolerances of many organisms. The mineralogical and chemical compositions of sedimentary rocks tell much the same story.

Carbon dioxide is constantly being added to the present atmosphere and ocean by the weathering of limestones and other rocks, by artificial combustion (Callendar, 1940, p. 399), and from volcanoes, geysers, and hot springs. At the same time, it is also being subtracted by the deposition and burial of calcium carbonate and of organic carbon in sediments.

Let us, for the moment, ignore the contributions of carbon dioxide from artificial combustion and from volcanoes and hot springs, and compare only the rates of addition by rock weathering and of subtraction by sedimentation. These rates are not known accurately, of course, but from several lines of evidence it appears that the loss by sedimentation must exceed the gain from weathering by something like 10^{14} grams of carbon dioxide each year. At this rate of net loss, the total carbon dioxide in the atmosphere and ocean would be reduced to about one-fourth its present value in one million years, and the concentration of hydroxyl ions in sea water would be so high that magnesium hydroxide—the mineral brucite—would be precipitated and almost completely take the place of calcite as a common marine sediment. Even if liberal allowance is made for

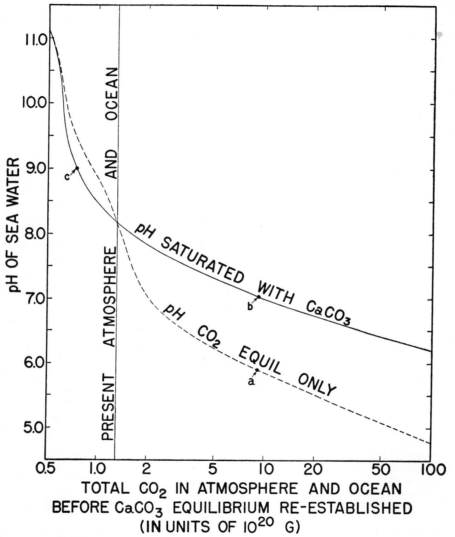

FIGURE 2.—EFFECTS OF CHANGES IN THE TOTAL CARBON DIOXIDE IN THE ATMOSPHERE-OCEAN SYSTEM UPON THE HYDROGEN-ION CONCENTRATION OF SEA WATER (AT 35°/₀₀ SALINITY AND 8° C)

a probable slowing down of weathering and of the synthesis of organic matter as carbon dioxide is subtracted from the atmosphere-ocean system, it would still be only about 2 million years until brucite would largely take the place of calcite deposition. No such occurrences of sedimentary brucite are known to me in rocks of any age, and it appears unlikely that this condition was ever reached in the geologic past. It seems much more likely that the net losses by sedimentation have been roughly balanced by a source neglected in the foregoing calculation—that is, by continuous additions of carbon dioxide from volcanoes and hot springs.

It seems clear that the loss of CO_2 from the atmosphere-ocean system by sedimentation must, in the long run, somewhat exceed the gain from rock weathering, but the amount of the difference is uncertain. Sedimentary rocks contain, on the average, roughly equivalent amounts of CaO and CO_2 (Clarke, 1907a, p. 169; 1907b, p. 269; 1924, p. 30). Their decomposition by weathering or precipitation by sedimentation does not disturb this equivalent ratio significantly. But igneous rocks contain many times more CaO than CO_2 (Clarke, 1924, p. 29). Thus the proportion of igneous rocks to all rocks undergoing erosion at any time is a rough measure of the amount of CO_2 that must be supplied from some other source in order to maintain the observed ratio in newly formed sediments. Several different methods of estimation (based on the probable quantities of CO_2, organic C, and Ca in rocks undergoing erosion, in material being transported by streams, and in sediments being deposited, both today and in the geologic past) yield values for the net loss of CO_2 from the atmosphere-ocean system of from 0.3×10^{14} to 4.0×10^{14} g/yr. The estimates of 1×10^{14} to 2×10^{14} g appear somewhat the more reliable, and, for purposes of the following calculations, a value of 1×10^{14} g CO_2/yr has been adopted.

At this rate of net loss, the total CO_2 would be reduced from 131×10^{18} to 31×10^{18} g in $\frac{(131-31) \times 10^{18} \text{ g}}{1 \times 10^{14} \text{ g/yr}} = 1,000,000$ years, and the pH of sea water would then be about 9.0. Wattenberg and Timmermann (1938, p. 87–88) found the solubility product of $Mg(OH)_2$ to be 5×10^{-11} in sea water; and Harvey remarks that "when the hydrogen-ion concentration of sea water falls and the pH rises above circa pH 9, magnesium hydroxide separates as a precipitate with calcium carbonate" (1945, p. 25). If CO_2 subtraction continues after the precipitation of $Mg(OH)_2$ has begun, the relative amounts of Mg and Ca in sea water and the various CO_2 equilibria cause $Mg(OH)_2$ to accumulate in far greater amounts than $CaCO_3$.

This estimate of the time required to reduce the total CO_2 in the atmosphere and ocean to the point when brucite would be deposited neglects several factors—one that would operate to decrease the time and two that would operate to increase it. The several chemical relations involved require that, as the pH of sea water is raised from 8.2 to 9.0, 70 mg/l of $CaCO_3$ must, on the average, be subtracted by precipitation to maintain the various equilibria. With the present volume of sea water, this would mean the deposition as carbonate of about 43×10^{18} g of CO_2 in addition to the average net subtraction of 10^{14} g/yr based on present conditions. The total subtraction of $(131-31=)$ 100 units of CO_2 is thus made up in part of 57 units lost because of normal sedimentation. If this is not replaced by CO_2 from other sources, this primary loss causes the precipitation, as carbonate, of 43 additional units of CO_2 in order to maintain the various equilibria. Figure 2 shows the effects of primary rather than secondary changes of total CO_2 in the system; and point c (pH 9.0) is therefore plotted at $(131-57=)$ 74 units instead of at 31 units. For purposes of the present calculation, we are interested in the time required for the primary loss of 57 units of CO_2. That is, if there were no other corrections to be considered, the time required to reduce the pH to 9.0 would be only

$$\frac{(131 - 31 - 43) \times 10^{18} \text{ g}}{1 \times 10^{14} \text{ g/yr}} = 570,000 \text{ years.}$$

The total rate of loss of CO_2 by sedimentation is the rate of deposition of carbonates and that of organic carbon. Over the years, the mean rate of carbonate deposition depends upon the rate at which streams bring new Ca to the sea, and this in turn depends upon the rate of rock weathering on the lands. It seems reasonable to assume that the rate of chemical weathering is controlled in large measure by the partial pressure of CO_2 in the atmosphere (Van Hise, 1904, p. 465, 476; Clarke, 1924, p. 110–111) and by the quantity of decomposing organic matter (Jensen, 1917, p. 255–258, 267–268) which likewise depends upon atmospheric CO_2, as is discussed somewhat more fully below. As the atmospheric CO_2 would decrease under the assumed conditions of this calculation, the rate of rock weathering and hence the rate of carbonate deposition would likewise decrease.

Similarly the rate of accumulation and burial of organic carbon is probably roughly proportional to the rate at which new organic matter is produced, both on the lands and in the sea. This in turn depends largely if not entirely on the rate of carbon fixation by photosynthesis which, at optimum light intensities, varies almost directly with the partial pressure of CO_2, up to a P_{CO_2} of about 3, 4, or 5 times that of the present (Brown and Escombe, 1905, p. 40–41; Hoover, Johnston, and Brackett, 1933, p. 10–17; Rabinowitch, 1945, p. 330–331). Thus as the partial pressure of CO_2 decreased under the assumed conditions, the rates of deposition of both carbonate sediments and organic carbon would probably decrease in roughly the same proportion.

From the relationship between CO_2 and P_{CO_2}, it may be found that, when the total CO_2 is reduced to 31×10^{18} g (Fig. 3, d, plotted at 74×10^{18} g to show the effects of primary loss), the P_{CO_2} would be reduced to only one-forty-fourth its present value. Numerical integration indicates that, over this range, the P_{CO_2} would average about 28% of its present value. Thus the mean rate of CO_2 deposition would be only about 0.28×10^{14} g/yr. Combining the effects of these several corrections, a revised estimate of the time required to reduce the total CO_2 to the point where $Mg(OH)_2$ would begin to precipitate would be $\frac{(131 - 31 - 43) \times 10^{18} \text{ g}}{0.28 \times 10^{14} \text{ g/yr}} =$ 2,000,000 years. This is a relatively brief interval in geologic time, and it seems necessary to conclude that CO_2 must have been supplied to the atmosphere-ocean system more or less continuously from some extraneous source in order to account for the absence of brucite as a common marine sediment. The foregoing calculations are based on the constants for sea water of normal salinity at a mean temperature of 8°C. The calculated values are somewhat different if other salinities and temperatures are assumed but not enough so to modify this conclusion significantly.

Possible variation in the amount of CO_2 in the atmospheres of the past might be estimated, if sufficiently reliable data were available on the rate at which organic carbon has been deposited during different geologic periods. As stated above, the amount of organic carbon that becomes buried along with sediments and thus removed from circulation at any time is probably roughly proportional to the amount of it then in existence as organic matter. This in turn depends, through a narrow but significant range, on the partial pressure of atmospheric CO_2. The rate of deposition of organic carbon is thus a rough measure of the partial pressure of CO_2 in the atmosphere. For this reason, it is tempting to suppose that the CO_2 content of the atmosphere may have been much greater than at present during those parts of the Carboniferous, Cretaceous, and Tertiary periods when great quantities of coal and other organic deposits accumulated.

Available data do not support this possibility, however. To begin with, the total amount of carbon in all coal, petroleum, oil shale, etc., is negligible in comparison with the carbon that is present (to the extent of only a few tenths of 1 per cent) in the far more abundant ordinary sedimentary rocks (Table 3). Furthermore, local deposits of highly organic sediments do not necessarily mean unusually large accumulations at that same time over the entire earth. Finally, even widespread deposits of highly organic sediments are not in themselves sufficient evidence; the time elapsed during accumulation of the sediments is an essential factor in determining whether or not the rate of carbon deposition was abnormally rapid.

Merely to list the data needed for this calculation is sufficient to demonstrate how far we now are from being able to appraise at all accurately, from this type of evidence, the rates of burial of organic carbon through the geologic past. One needs (a) reliable information on the organic carbon content of different types of sediment of different ages, (b) reasonably good estimates of the relative amounts of different sediment types deposited during a given interval of time, and (c) data on the number of years required for deposition. Of these several categories of essential data, the third is probably now known better than the other two.

Nevertheless, something of interest may still be obtained, even from such data as now exist. The average composition of the principal types of sedimentary rocks (Clarke, 1924, p. 30, 547, 552) affords one of the bases for estimation; the relative abundance of sandstone, shale, and limestone in rocks of different ages may be taken from such compilations as Schuchert's (1931, p. 49); and the estimated durations, in millions of years, of the different periods are available from several sources (for example, Holmes, 1947b, p. 144). When these various estimates are combined, along with estimates of the amounts of contemporaneous deep-sea sediments, one finds no indication whatever that the rate of carbon burial has ever exceeded that of today. In fact this evidence, taken at face value, appears to show that the rate of carbon accumulation has been increasing gradually and rather uniformly ever since the beginning of geologic time.

This apparent evidence of an increasing rate of carbon deposition needs to be viewed with considerable skepticism, however. It may readily be recognized as simply one other aspect of the widely observed relationship that the maximum thicknesses of sedimentary rocks deposited during a unit of time appear to increase progressively as one ascends through the geologic column. This general relationship has been noted and variously interpreted by a number of geologists. Gilluly (1949, p. 574–582) has clearly pointed out that it does not necessarily mean any real increase in the rate of sedimentation with time but may be explained equally well as the result of the accidents of preservation and exposure of ancient sediments. Whatever may be the correct explanation of this commonly observed relationship, one finds no indication from this type of evidence that the average rate of deposition of organic carbon was greater at any time in the past than it is today.

Possible Limits of Variation in the Past

Even if the rate of carbon deposition, averaged over fairly long intervals of time, has remained approximately constant through much of the past, it still seems likely that the rate would have varied somewhat from this average, as the result of irregularities of volcanic eruptions and other possible sources of supply. The observed fact that many plants are capable of a higher rate of carbon-dioxide assimilation and photosynthesis, if the carbon-dioxide content of the atmosphere is several times greater than normal, is perhaps most simply explained as the result of the adaptation of ancestral forms to an atmosphere that was at times somewhat richer in CO_2 than at present. If we adopt this interpretation and take a P_{CO_2} of about 0.0015 (or five times that of the present) as an optimum condition for photosynthesis (Brown and Escombe, 1905, p. 40–41; Hoover, Johnston, and Brackett, 1933, p. 15–16), we have a basis for estimating what may have been the total amount of CO_2 in the atmosphere-ocean system at such times.

With the present volume and salinity of sea water, a mean water temperature of about 8° C, and a prevailing P_{CO_2} of 0.00030, the quantity of CO_2 in the atmosphere and ocean is calculated to be 2.33×10^{18} and 129×10^{18} g, respectively. If, with the other conditions remaining constant, the total quantity of CO_2 in the atmosphere and ocean were increased from 131×10^{18} to 157×10^{18} g, the various CO_2 equilibria would distribute the CO_2 between atmosphere and water so that the P_{CO_2} would rise to the assigned value of 0.00150 (Fig. 3, e) and the pH would fall to 7.5. This sea water would then be much less than saturated with $CaCO_3$, and, as a result, $CaCO_3$ would, on the average, be dissolved from the ocean floor until equilibrium was re-established. At that stage, the total quantity of CO_2 in the atmosphere and ocean would be about 177×10^{18} g; the pH would rise to 8.0 (nearly its original value), and the P_{CO_2} would fall back to about 0.00059 (Fig. 3, f).

It would probably give a more reasonable estimate of the limiting conditions to assume that the arbitrary P_{CO_2} of 0.00150 prevailed *after* rather than before adjustmsnt to $CaCO_3$ equilibrium, which would mean that the P_{CO_2} was at times well above that limit. On this alternative assumption, the total quantity of CO_2 would have increased from 131×10^{18} to 212×10^{18} g, the addition having come from some unspecified source. The CO_2 equilibria alone would distribute this total between the atmosphere (49×10^{18} g) and sea water (163×10^{18} g) so that the P_{CO_2} would rise to 0.00630 (Fig. 3, g) and the pH would fall to 6.9. Sea water of this acidity would dissolve $CaCO_3$ from the ocean floor until about 62×10^{18} g of CO_2 had been added to the system, thereby bringing the total to $(212 + 62 =) 274 \times 10^{18}$ g. At that stage, when equilibrium was re-established, the quantity of CO_2 in the atmosphere and ocean would be about 11.6×10^{18} and 262×10^{18} g, respectively; the P_{CO_2} would fall back to the assigned value of 0.00150 (Fig. 3, h); and the pH would end up at 7.8. It is worth nothing that, although the quantity of CO_2 in the atmosphere would have increased by a factor of $\frac{11.6}{2.33} = 5$, the quantity in the ocean would have increased by only $\frac{262}{129} = 2$.

The volume and salinity of sea water and the temperature of the water have been held constant in the foregoing calculation. The probable effects of their variation may be estimated by similar calculations. If the volume of the ocean were only three-fourths as great as at present, the quantity of CO_2 dissolved in sea water would be only 194×10^{18} instead of 262×10^{18} g when the P_{CO_2} was 0.00150. For salinity ranges of as much as 20 per cent above and below those of present average sea water, the net effects on the distribution of CO_2 between atmosphere and sea water are almost negligible. If the mean water temperature were 30°C instead of 8°C, the quantity of CO_2 dissolved in sea water would be about 138×10^{18} instead of 262×10^{18} g when the P_{CO_2} was 0.00150. Thus the effects of each of these other controlling variables is such that, if the P_{CO_2} were 5 times that of today, the total CO_2 in the atmosphere-ocean system would probably not have been much more than about twice the present total of 131×10^{18} g.

photosynthesis, which in turn is proportional to the P_{CO_2} through this range. This would mean that, as the P_{CO_2} increased by a factor of 5, the quantity of CO_2 equivalent to organic carbon in the biosphere

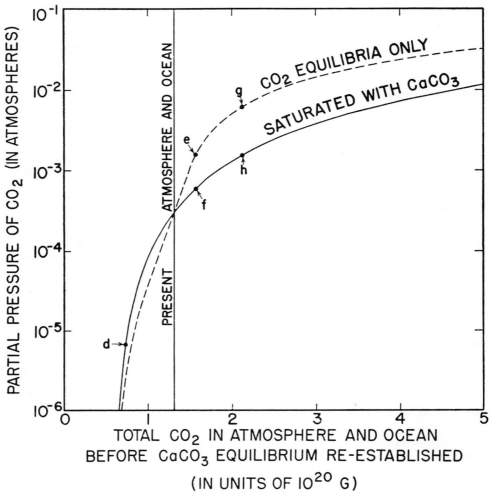

FIGURE 3.—EFFECTS OF CHANGES IN THE TOTAL CARBON DIOXIDE IN THE ATMOSPHERE-OCEAN SYSTEM UPON THE PARTIAL PRESSURE OF CARBON DIOXIDE IN THE ATMOSPHERE (AT 35⁰/₀₀ SALINITY AND 8° C)

All these calculations have considered only the amounts of CO_2 in the atmosphere and ocean and have ignored any changes in the total quantity of living organisms and undecayed organic matter in the biosphere, as the P_{CO_2} increased from 0.00030 to 0.00150. By an extension of the reasoning followed above, in the section entitled Effects on Composition of Sediments, it might be estimated that the quantity of organic matter in the biosphere at any time would be roughly proportional to the rate of

would also have increased from the estimated 14.5×10^{18} g of today (Table 3) to about 72.5×10^{18} g. This would bring the total quantity of C in circulation in the atmosphere, hydrosphere, and biosphere under these limiting conditions to about $(274 + 72 =) 346 \times 10^{18}$ g CO_2. A similar "correction" might be made for organic C in the biosphere in each of the other calculations. None of the conclusions would be modified, however, and the basis for the "correction" itself seems too uncertain to warrant this refinement.

The possible effect of variation in the CO_2 content of the atmosphere upon world climate has been discussed by meteorologists and others for many years. The hypothesis, developed in some detail by Arrhenius (1896) and T. C. Chamberlin (1897, p. 680; 1899), was that an increase of atmospheric CO_2 would act as a trigger mechanism, retarding radiation of heat from the earth, and thereby starting a train of events that would cause general warming. Physical measurements subsequently indicated that the heat absorption by water vapor far outweighs that by CO_2, and the hypothesis fell into general disfavor (reviews in Clarke, 1924, p. 52–53, 147–149; Humphreys, 1940, p. 584–586, 621–622). Recent work, however, seems to have reopened the question to some extent (Callendar, 1938, p. 231–232; 1941; Dobson, 1942; Elsasser, 1942, p. 64). Callendar's later estimates (1938, p. 231; 1941, p. 32) call for an increase of 20 to 30 fold over the present atmospheric CO_2 in order to account for the warmest climates indicated by Tertiary floras, and they call for a decrease to about one-tenth the present content (corresponding to a decrease to about 0.4 of the present total CO_2 in atmosphere and ocean) to account for the coldest epochs of continental glaciation. Other lines of evidence, discussed elsewhere in this paper, make it seem unlikely that the P_{CO_2} has been 20 to 30 times greater than at present, at least in later parts of geologic time. But it appears possible, if CO_2 has the climatic effects some have claimed for it, that decreases in the amount of atmospheric CO_2 may have been a contributory cause of some of the epochs of glaciation.

From several lines of evidence, it seems difficult to escape the conclusion that, for a large part of geologic time, carbon dioxide has been supplied to the atmosphere and ocean gradually and at about the same rate that it has been subtracted by sedimentation.

Sources of Supply

If the bulk of the carbon dioxide in the atmosphere, hydrosphere, biosphere, and sedimentary rocks cannot have been residual from a primitive atmosphere, one must look for other sources of this gradual and continuous supply. At first glance, the possibility seems tempting that this unidentified source may have been the great reserve of carbonates and organic carbon in the ancient sediments. Crustal movements bring these older sedimentary rocks to the surface, where they undergo weathering and gradually release their contained carbon to the streams and atmosphere. If sedimentary deposits accumulated only on the continents and continental shelves and if no igneous rocks were eroded, this problem of the supply of carbon would not concern us, for the quantities of older sedimentary rocks eroded from and new sediments deposited on the continents have probably maintained a reasonably close balance through much of geologic time. But igneous rocks have also been eroded, and sediments have accumulated not only on the continents but also in large quantity in the deep-sea basins (Twenhofel, 1929, p. 394, 400; Kuenen, 1937; 1941; 1946). It is evident, therefore, that the weathering of sedimentary rocks alone could not suffice to maintain the supply of carbon dioxide to the atmosphere-ocean system.

If weathering of sedimentary rocks is not sufficient, what about the carbon released by metamorphism or by actual melting of the older sediments? Many of the successive steps in metamorphism of siliceous carbonate rocks involve progressive decarbonation of the mineral species (Bowen, 1940, p. 245, 256–257, 266); and actual melting of sediments would drive off all carbon dioxide or carbon monoxide that could not be dissolved in the resulting melt. In many areas, volcanic rocks and associated schists of Precambrian age have largely been converted into carbonates (Van Hise, 1904, p. 972; Collins and Quirke, 1926, p. 31–32; Macgregor, 1927, p. 159–162), possibly as the result of high concentrations of carbon dioxide driven off from near-by sources.

The extent to which ancient sediments have undergone decarbonation by metamorphism or melting is not known. It seems very improbable, however, that such processes are adequate to account for the continuous supply of carbon dioxide to the atmosphere and ocean. As an outside estimate, we may assume that *all* Precambrian and *half* of all Paleozoic sedimentary rocks now remaining uneroded on the continents have undergone complete (not merely partial) decarbonation and that *all* the resulting carbon dioxide escaped into the atmosphere-ocean system. This would reduce the estimated amount of CO_2 now buried in sedimentary rocks from 920×10^{20} (Table 1) to about 780×10^{20} g and would mean that the amount now buried is about 530 (instead of 600) times as great as that in circulation in the present atmosphere, hydrosphere, and biosphere. In other words, even if decarbonation of sedimentary rocks has been far more extensive than seems likely, the essential problem of accounting for this huge discrepancy still remains.

It may be worth noting also that, judged by the estimated relative abundances shown in Table 1, the volatile constituents that could be released on complete decarbonation, dehydration, etc., of average sedimentary rock would be very different in over-all composition from the "excess" volatiles.

One other possible source of carbon dioxide may be worth mentioning. Stony and iron meteorites contain an average of 0.04 per cent and 0.11 per cent carbon, respectively (Brown, 1949b, p. 626). Conceivably the supply of carbon dioxide to the earth's surface might have been replenished continuously by showers of meteoritic dust from interplanetary or interstellar space (Van Hise, 1904, p. 970, 973, 974). A brief consideration, however, of the exceedingly slow rate at which meteoritic debris is reaching the earth today (less than 1 mm/billion years—Jeffreys, 1933—to about 10 cm/billion years—Nininger, 1940, p. 461) and the apparent absence of geologic evidence that the rate has been significantly higher any time since sedimentary rocks began to accumulate makes it seem highly improbable that extra-terrestrial sources have made important contributions to the earth's supply of carbon dioxide since very early in the history of the planet.

The conclusion, arrived at so laboriously here, that carbon dioxide has been supplied to the atmosphere and ocean gradually and at about the same rate that it has been subtracted by sedimentation, is not particularly new. The Swedish geologist, A. G. Högbom, clearly recognized in 1894 (summarized in Arrhenius, 1896, p. 269–273) that carbon dioxide has been supplied continuously to the earth's surface since early in geologic history and that the chief source of this supply must have been volcanic or juvenile. Essentially this same conclusion has been restated many times (T. C. Chamberlin, 1897, p. 654–656; Fairchild, 1904, p. 97, 109–110; Clarke, 1924, p. 58–59; Macgregor, 1927, p. 156–157; Jeffreys, 1929, p. 312; Goldschmidt, 1934, p. 415; 1938, p. 101; Eskola, 1939; Conway, 1943, p. 170–174; Cotton, 1944; Hutchinson, 1944, p. 180, 192; 1947, p. 300–301); but it is a significant fact about the history of the earth that has sometimes been ignored and, as such, it deserves frequent repetition.

Similarity of "Excess" Volatiles and Magmatic Gases

The evidence bearing on the probable source and rate of supply of carbon dioxide has been emphasized in the preceding discussion because the chemical and physiological effects of carbon dioxide are relatively well known. Perhaps for this reason, carbon dioxide appears to be of particular significance in our understanding of the history of the atmosphere and ocean. Comparable data on the effects of other volatiles are apparently not available; but, from scattered information, approximately the same conclusions are suggested for the source and rate of supply of chlorine, oxygen, and several other constituents of the atmosphere and hydrosphere.

In summarizing the evidence on the source of carbon dioxide, Hutchinson put it succinctly when he stated that "it seems unreasonable to accept juvenile addition in the case of one constituent, and, in the absence of very strong reasons, to deny it in the case of the others" (1944, p. 180). A number of other writers have reached much the same conclusion (Fairchild 1904, p. 98, 103; Evans, 1919; Jeffreys, 1929, p. 147, 312; Eskola, 1932b, p. 66–67, 69–70; Lane, 1932, p. 318; Gilluly, 1937, p. 440–441; Goldschmidt, 1938, p. 17, 24, 101; Wildt, 1939, p. 143; R. T. Chamberlin, 1949, p. 253, 255–256). (See also references regarding the earth's loss of inert gases in section on Dense Primitive Atmosphere-Solution in the Melt.)

Let us now return to the "excess" volatiles we started with and see how they compare in composition with gases that are escaping from volcanoes, fumaroles, and hot springs and with gases that are occluded in igneous rocks. Considering that these gases have come in contact and presumably reacted with diverse rock

TABLE 5.—VOLUME PERCENTAGES OF GASES FROM VOLCANOES, ROCKS, AND HOT SPRINGS

Volcano gases from Kilauea and Mauna Loa* (26 samples)			Gases from rocks						Gases from fumaroles of the Katmai region† and from steam wells and geysers of California and Wyoming.** (23 samples)			
			Basaltic lava and diabase* (13 samples)			Obsidian, andesitic lava, and granite* (17 samples)						
	Minimum	Maximum	Median	Minimum	Maximum	Median	Minimum	Maximum	Median		Minimum	Maximum	Median
CO₂	0.87	47.68	11.8	0.89	15.30	8.1	0.08	20.26	2.0	CO₂	0.03	1.24	0.02
CO	0.00	3.92	0.5	0.02	8.28	0.2	0.01	2.22	0.5	CO	—	0.01	tr
H₂	0.00	4.22	0.4	0.38	6.18	1.2	0.08	11.60	0.4	O₂	0.00	0.08	tr
N₂	0.68	37.84	4.7	0.27	7.21	2.0	0.03	3.90	1.2	CH₄	0.00	0.30	0.11
A	0.00	0.66	0.2	0.00	0.04	tr	0.00	0.02	tr	H₂	0.00	0.29	0.15
SO₂	0.00	29.83	6.4	—	—	—	—	—	—	N₂ + A	0.00	0.31	0.02
S₂	0.00	8.61	0.2	0.08	1.96	1.1	0.00	2.89	0.2	NH₃	—	0.02	0.01
SO₃	0.00	8.12	2.3	—	—	—	—	—	—	H₂S	0.00	0.10	0.02
Cl₂	0.00	4.08	0.05	0.06	1.33	0.5	0.01	10.59	0.5	HCl	0.01	0.57	0.06
F₂	—	—	—	0.00	14.12	3.8	0.25	7.80	2.3	HF	0.00	0.10	0.03
H₂O	17.97	97.09	73.5	71.32	92.40	83.1	69.44	98.55	92.9	H₂O	98.04	99.99	99.58
			100.0			100.0			100.0				100.00

* Analyses by E. S. Shepherd (1938, p. 321, 326).
† Allen and Zies (1923, p. 126, 142).
** Allen and Day (1927, p. 76; 1935, p. 87).

COMPOSITION OF GASES AND OF "EXCESS" VOLATILES

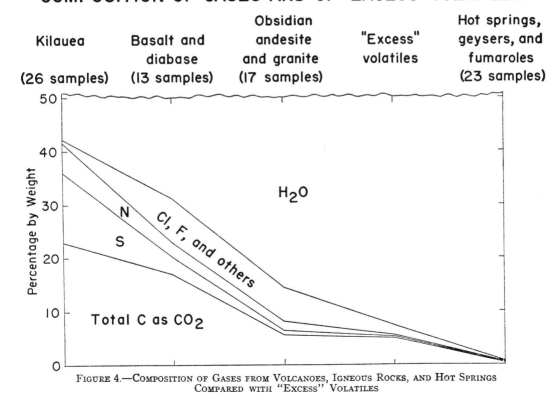

FIGURE 4.—COMPOSITION OF GASES FROM VOLCANOES, IGNEOUS ROCKS, AND HOT SPRINGS COMPARED WITH "EXCESS" VOLATILES

types and ground waters and that they probably contain at least some recirculated or *resurgent* volatiles that have been absorbed from earlier sedimentary rocks (Shepherd, 1938, p. 312–313), it is to be expected that they will be

TABLE 6.—COMPOSITION OF GASES FROM VOLCANOES, IGNEOUS ROCKS, AND HOT SPRINGS, AND OF "EXCESS" VOLATILES IN ATMOSPHERE, HYDROSPHERE, AND SEDIMENTARY ROCKS

(Median analyses, recalculated from volume to weight percentages)

	Kilauea and Mauna Loa	Basalt and diabase	Obsidian, andesite, and granite	Fumaroles, steam wells, and geysers	"Excess" volatiles
H_2O	57.8	69.1	85.6	99.4	92.8
Total C as CO_2	23.5	16.8	5.7	0.33	5.1
S_2	12.6	3.3	0.7	0.03	0.13
N_2	5.7	2.6	1.7	0.05	0.24
A	0.3	tr	tr	tr	tr
Cl_2	0.1	1.5	1.9	0.12	1.7
F_2	—	6.6	4.4	0.03	tr
H_2	0.04	0.1	0.04	0.05	0.07
Others	—	—	—	—	tr
	100.04	100.0	100.04	100.01	100.04

variable in composition. The extreme ranges of composition of these gases are in fact wide; but median compositions by groups indicate consistent differences in the gases in rocks and lavas of different types (Tables 5 and 6). If we compare these median gas compositions with the relative abundance of the constituents of the "excess" volatiles, we see that an average of the gases in granitic rocks and those from hot springs and fumaroles approximates closely the over-all composition of the "excess" volatiles (Fig. 4). Judged solely by compositions then, it is conceivable that all the hydrosphere and atmosphere may have come from such gases (Allen, 1922, p. 39–42, 52). Several questions must be answered, however, before we may consider such a conclusion.

ESCAPE OF VOLATILES DURING CRYSTALLIZATION OF MAGMAS

Is there some reasonable mechanism by which these gases from the interior could be the source of the "excess" volatiles? In the cooling and gradual crystallization of complex silicate melts, certain minerals crystallize out first and thus enrich the remaining melt in other constituents, such as silica, alumina, and the alkalies (Bowen, 1928, p. 293, 297–298) and water, carbon dioxide, and other volatiles (Allen, 1922, p. 57). As crystallization proceeds, the concentration of the remaining volatiles rises and therefore also their vapor pressures. As Morey (1922; 1924) and others (Goranson, 1931, p. 499; Ingerson, 1940, p. 784) have pointed out, when these vapor pressures become greater than the weight of overlying rocks, escape to the surface may be violent, as in volcanic eruptions; or where the crystallizing melt lies at greater depths, the volatiles may escape more gradually along fractures in the overlying rocks.

How adequate is such a mechanism to explain the quantity of volatiles that must have risen to the earth's surface in the course of geologic time? Probably the total quantity of extrusive rocks that has risen to the earth's surface in all of geologic time, although large, is still insufficient to furnish more than a few per cent of the water of the ocean, unless extrusive rocks bulk larger than the total of all sedimentary rocks and unless they originally contained more than 10 per cent of water. (*See also* Verhoogen, 1946, p. 746.) Intrusive rocks, on the other hand, appear to be a more adequate source. Goranson's experimental work has shown that a melt of granitic composition can dissolve as much as 9.3 per cent of water at 900° C and at a pressure equivalent to a depth of 15 kilometers (Goranson, 1931, p. 493). The amount of water in natural magmas, however, is a subject about which petrologists are not agreed. Gilluly's estimates, which are fairly representative of others[5], are that the water content of many magmas has been "fully 4 per cent for basalts and perhaps as much as 8 per cent for deep granites" (1937, p. 441).

Let us estimate 4 per cent water in an average magma. Then, as igneous rocks now contain an average of about 1 per cent water, this would mean that magmas have commonly given off about 3 per cent of water during crystallization.

---

[5] In a paper that appeared after this was written, Ingerson (1950, p. 813–814) has reviewed the estimates made by a number of geologists. Gilluly's estimates lie near the middle of the entire range.

On these assumptions, the crystallization of a shell of igneous rock 40 kilometers thick would account for the entire volume of water in the ocean. In a similar calculation, Gilluly (1937, p. 440–441) estimated that the ocean could have been derived from the escape of 5.8 per cent of water from the sialic rocks of the crust.

Hot Springs as Possible Channels of Escape

How might this quantity of volatiles have escaped to the earth's surface? Hot springs are much less spectacular than volcanoes, but they are more numerous and more widely distributed, and they may be the principal channels by which volatiles escaping from crystallizing magmas have reached the surface. For example, Allen and Day (1935, p. 40, 87; Allen, 1935, p. 6, 12; Day, 1939, p. 328–329) have estimated, from the temperature of escaping steam in Yellowstone Park, that the hot springs of that area are discharging ground water that has been heated by mixing with about 10 or 15 per cent of magmatic steam[6].

Hot springs are much more abundant in the area of the Idaho batholith than in that of the much younger lava flows of the adjacent Snake River Plains (Meinzer, 1924, p. 296–297; Stearns, Stearns, and Waring, 1937, Fig. 11, p. 82). More than 100 hot springs and groups of hot springs are known within the limits of the Idaho batholith, and the discharge and temperature of a number of them have been measured (Stearns, Stearns, and Waring, 1937, p. 136–151). If these springs are heated by the same process of mixing of ground water and magmatic steam that is believed to operate in the Yellowstone, then one may estimate roughly that the average percentage of magmatic water is about 2½ per cent. The Idaho batholith is thought to be of Cretaceous age (Ross, 1936, p. 382–383), yet it is still discharging hot waters. Can it be that batholiths cool much more slowly than is generally supposed? If so, this may be one explanation for the long-continued and relatively uniform supply of volatiles that seems to be indicated by the geologic record.

The average temperature of the hot springs of Idaho that have been measured is about 39°C, and the mean annual temperature there is near 8°C. Making the same assumptions about original temperature and mixing as those used in the similar calculation that follows, we find that 1 g of magmatic steam would be sufficient to heat 35.5 g of ground water to 39°C. This would mean 2.7 per cent magmatic water in the final hot-spring mixture.

A crude estimate of the total quantity of water being discharged today from all the hot springs of the earth may be made from available data on the springs of a smaller area. Stearns, Stearns, and Waring (1937, p. 115–191) list more than 1000 thermal springs or groups of springs within the limits of the continental United States. Undoubtedly some of these are heated by deep artesian circulation, but probably only a small proportion, for many are located in areas of igneous rock. To allow roughly for such artesian heating, the hot springs of the eastern two-thirds of the United States are here omitted from calculation. If the total area of the States, as thus "corrected", is a fair sample of the continents and if the oceans contribute about half as much per unit area, then the annual discharge of all hot springs, multiplied by 3 billion years, gives a total volume somewhat more than 100 times that of the present ocean. That is to say, if hot springs are delivering to the surface an average of only 0.8 per cent of juvenile water, they could, in the course of geologic time, account for the entire volume of the ocean. This is not counting any contribution of water directly from volcanic eruptions.

Stearns, Stearns, and Waring consider only springs with water more than about 8°C above local mean annual temperature. Of the 1059 hot springs they list for the entire United States, 721 show estimated discharges, and 616 show water temperatures. The total discharge of the 721 measured or estimated springs is 511,600 gal/min. The 338 springs for which no discharge is recorded include the 96 hot springs and geysers of Yellowstone

---

[6] Rubey and Murata (1941) have investigated a small but apparently typical group of hot springs farther south along the Wyoming-Idaho border. From relations between the compositions of water and of contained gases and the temperatures at different outlets, it appears that approximately 5 per cent of the water of these springs is of magmatic origin. For estimates of the amount of magmatic water in other areas, see Zies (1929, p. 73–74) and White and Brannock (1950, p. 573).

National Park, for which an estimate of 49,400 gal/min is available from another source (Allen, 1935, p. 5). Estimating separately by States for the unmeasured springs, it appears that a total of approximately 800,000 gal/min is a fair approximation for the discharge of all 1059 springs. Comparing the water temperatures with estimated mean annual temperatures and weighting the differences in proportion to the quantities discharged, one arrives at the estimate that the springs are heated, on the average, about 20°C above mean annual temperature.

By the same methods, the hot springs of the western one-third of the country (Ariz., Calif., Colo., Idaho, Mont., Nev., N. Mex., Ore., Utah, Wash., and Wyo.) yield a total of approximately 700,000 gal/min or 1.4×10^{15} g/yr of water. Taking the area of all continental U. S. as one-thirtieth that of all the continents and using the discharge from only these 11 western States would give a total from all continents of 42×10^{15} g H_2O/yr. The rate of hot-spring discharge on the ocean floor is unknown. Presumably it is lower than that on the continents but, to judge by the evidence of extensive lava flows there (Pettersson, 1949, p. 186), it is probably considerable. If the mean rate per unit area from the oceans is taken as half that from the continents, then the hot-spring discharge from the continents and oceans combined is about 66×10^{15} g H_2O/yr. This rate, continued over a period of 3 billion years, the assumed age of the earth (Holmes, 1947a; Bullard and Stanley, 1949; Ahrens, 1949, p. 254, 257–258; H. N. Russell, 1949, p. 11, 12, 15, 20, 24), would yield a total of 2.0×10^{26} g H_2O. The $16,600 \times 10^{20}$ g of "excess" water (Table 1) would be 0.0083 of this amount.

It is of some interest to compare this rough estimate of 0.8 per cent juvenile water with another estimate arrived at in quite a different way. Judged by those measured, the hot springs in the 11 western States have an average temperature of about 29° C, which is roughly 20° above the local mean annual temperatures. For purposes of calculation, let us assume that these springs are heated by deep-seated magmatic steam at an original mean temperature of, let us say, 600° C, and that this steam is cooled to 29° by mixing with ground water at the mean annual temperature of 9° C. In cooling from 600° to 29°, steam and water would release about 1100 gram-calories for each gram of magmatic water (estimated from data in Keenan and Keyes, 1936, p. 79–81). This would be sufficient to heat $\left(\frac{1100}{20} = \right)$ 55 grams of ground water from 9° to 29°. On these assumptions, the hot springs of the United States are being heated by mixing with an average of $\left(\frac{1}{56} = \right)$ 1.8 per cent of magmatic water.

It would, of course, be rash to attach any great significance to the numerical results of these calculations, for the difference between the two estimates is subject to several possible explanations. Nevertheless, it may be worth reminding ourselves that the estimate of 0.8 per cent is for new *juvenile* water and the 1.8 per cent estimate is for total *magmatic* water of whatever source. It is conceivable that the difference of 1.0 per cent between the two estimates may be a rough measure of the *resurgent* water, not newly risen from the earth's interior for the first time but recirculated water that magmas have resorbed from earlier sedimentary rocks—"second-hand" water that is now being returned to the surface.

Generation of Local Magmas

Even if we should grant all that has gone before about the probable juvenile origin of the "excess" volatiles, there still remains the question of why the magmas that released the volatiles have not long since cooled off and crystallized completely. In spite of seismic evidence that indicates the possibility, at least in some areas, of a layer of molten rock at a depth of about 80 km (Gutenberg and Richter, 1939b; Gutenberg, 1945, p. 302–307), serious difficulties restrain us from assuming a world-encircling layer (Daly, 1946, p. 712, 721–722) of slowly crystallizing magma from which volatiles have been escaping continuously ever since the world began. However, the abundant field evidence of local intrusive masses, some of them very large, that have been generated repeatedly throughout the past suggests a possible answer.

If we are willing to adopt the eclectic method of borrowing from various theories those parts of each that we like best, then a possible mechanism suggests itself—a mechanism by which local masses of hydrous magma might be generated more or less continuously throughout geologic time. The geothermal gradient varies from place to place (Van Orstrand, 1939, p. 132–141; Landsberg, 1946); but probably below a depth of 50 to 100 kilometers rocks are heated to significantly higher temperatures (Holmes, 1915, p. 111; Adams, 1924, p. 468–472; Jeffreys, 1929, p. 154; Gutenberg, 1939, p. 162; Urry, 1949, p. 176; or one may calculate his own thermal gradient from revised data on the radioactive generation of heat in different rock types—Keevil, 1943, p. 299; Birch, 1950a, p. 612–613, 616, 619), temperatures at which they would melt, if the confining pressures were not so great (Holmes, 1916, p. 269; Adams, 1924, p. 462, 468–472; Anderson, 1938, p. 50–51, 56–57, 72–73; Benfield, 1940, p. 157–158; Buddington, 1943, p. 139).

Under these conditions, several processes may operate to cause local melting. As sediments accumulate in a region, their low thermal conductivity makes them act as a blanket to impede the upward flow of heat (Jeffreys, 1930, p. 328). And in geosynclinal troughs, where sediments accumulate to great thicknesses, the entire column of underlying rocks is depressed into zones where the temperature is significantly higher (Eskola, 1932a, p. 456–457, 468–469). Temperatures may thus be reached at which some of the minerals in the deeper rocks are melted. Moreover, as strains accumulate in the earth's crust and the rocks eventually fail by fracture, sudden localized relief of pressure and shearing stresses may, without increasing the temperature, bring some of the minerals above their melting points (Johnston and Adams, 1913, p. 210–223, 248–253; Johnston and Niggli, 1913, p. 602–603, 613–615; Bowen, 1928, p. 314; Holmes, 1932, p. 556; Wahl, 1949, p. 148). As has been shown by Bowen (1928, p. 311–320), Eskola (1932a, p. 474), and others (Holmes, 1926, p. 315–316; 1932, p. 545, 556; Kennedy, 1938, p. 38–39; Buddington, 1943, p. 132–133, 137–139; Wahl, 1949, p. 161–163), such selective fusion of the lower-melting minerals would form magmas more granitic and at the same time richer in volatiles than the original rock.

Fusion would cause increase of volume, and the resulting hydrous, granitic magmas would tend to rise, partially recrystallizing as they moved into higher, cooler zones. This process would "pump" nearly all the volatile materials from the original rock; in recrystallizing these magmas could form suites of intermediate and granitic rocks; and the volatiles would largely escape to the earth's surface. By this mechanism, the volume of granitic rocks would increase near areas of sedimentation and where the crust was for any other reason unstable. As these unstable areas migrated with the continental margins and with the sites of mountain making, the granitic rocks would thus increase progressively in volume (Eskola, 1932a, p. 468–469, 480; Lawson, 1932, p. 358–361; Wilson, 1950, p. 101, 108, 111; Bucher, 1950, p. 500–504; Hurley, 1950).

Corollaries and Tests of the Suggested Mechanism

Isostatic Considerations

This leads to an important corollary of the suggested mechanism. If granitic rocks have grown progressively in volume through geologic time, then isostatic equilibrium would require more or less continuous readjustments, by which the thickening blocks of lighter, granitic or continental types of rocks would rise higher and higher above adjacent blocks of undifferentiated material. In another terminology, this would mean that the ocean basins must sink deeper as the continental blocks thicken. If the ocean basins have sunk at approximately the same rate as the volumes of granitic rocks and of sea water have grown, then it is no coincidence, but the effects of one single process, that the surface of the sea has oscillated back and forth near the same level on the continental shelves throughout the geologic past.

Isostatic balance between the oceanic and continental blocks affords a basis for estimating the relative quantities of water and granitic rocks and of the original subcrustal matter from which they may have been derived by selective fusion. The area of the earth's surface that rises above sea level is roughly 1.49×10^{18} cm^2; but the actual margins of the continental blocks lie much farther oceanward, beyond the continental shelves and near the base of the steep continental slopes. If 3000 meters below sea level is taken as approximately the base of the continental slopes, then the total area of the continental blocks is about 2.32×10^{18} cm^2 (Sverdrup, Johnson, and Fleming, 1946, p. 15, 19, 21).

The mean elevation of the continental blocks, as thus defined, is about 0.12 km above sea level, and the mean depth of the ocean basins (below the -3000 m contour) is about 4.6 km below sea level. The average density of sea water *in situ* is approximately 1.04 (Sverdrup, Johnson, and Fleming, 1946, p. 219). From independent estimates, mentioned briefly elsewhere in this paper, it appears that the mean thickness of deep-sea sediments may be about 1.3 km and their mean density, when filled with interstitial water, about 2.3. For purpose of this calculation, the average density of rocks to several tens of kilometers below the continents (the sedimentary, "granitic", and "intermediate" shells) may be taken as 2.9, and that of undifferentiated ultrabasic rocks to similar depths below the deep-sea sediments as 3.2.

From these several estimates, the average depth of compensation below the continents may be found to be

$$\frac{0.12 + (0.12 \times 2.9) + 4.6 (3.2 - 1.04) + 1.3 (3.2 - 2.3)}{3.2 - 2.9}$$

$$= 38.3 \text{ km}.$$

This value accords reasonably well with the average depth to the Mohorovičić discontinuity, which is found widely distributed beneath the continents but is not recognized with certainty over extensive areas beneath the oceans (Macelwane, 1939, p. 237–239; Gutenberg and Richter, 1939a, p. 321–322). This major discontinuity is often interpreted as representing the approximate base of the crustal materials that make up the continents.

It seems permissible to assume that these densities of continental and oceanic materials have remained approximately the same throughout the past. It follows then that, if there was a time when the mean depth of the ocean basins and the mean thickness of deep-sea sediments were half their present values, isostatic equilibrium would require that the mean elevation of the continents and the mean depth of compensation below the surface of the continental blocks would, at that time, also be just half their present values. If the total areas of the continental blocks and of the oceanic basins have remained approximately the same (peripheral growth of the continents having approximately balanced subtraction by continental foundering), this strictly proportionate deepening of the oceanic basins and thickening of the continental crust, which is required by isostatic balance, would accord with the suggested mechanism of selective fusion which likewise would require that the volumes of sea water and of continental rocks have grown in strict proportionality.

The total mass of the continental blocks today may be estimated from the results of the previous calculation at $(2.32 \times 10^{18} \text{ cm}^2) (38.3 \times 10^5 \text{ cm}) \times 2.9 \text{ g/cm}^3 = 258 \times 10^{23}$ g. Allowing for the probable quantity of crystalline rocks that has been eroded from the continents in the past and is now represented as deep-sea sediment on the ocean floor, one obtains an estimate of 2.65×10^{25} g for the total mass of crustal rocks that have been formed on and in the continental blocks. It is to be noted that this estimate makes no allowance for any similarly differentiated crustal rocks that may now be lying foundered beneath the sea.

According the the eclectic hypothesis here suggested, this 2.65×10^{25} g of crustal rock has been formed, largely by selective fusion and subsequent recrystallization of the resulting magma, from an unknown quantity of subcrustal rock materials. The total mass of the earth's silicate mantle that lies directly below the continents, to a depth of 2900 km and above the earth's "iron core," may be estimated at about 185×10^{25} g. Much of this subcrustal matter may still be in very nearly its original condition (Birch, 1950b). But that part of it lying above the maximum depth (of about 700 km) to which deep-focus earthquakes are recorded (Gutenberg and Richter, 1941, p. 4–11; Benioff, 1949, p. 1844–1854) may have been worked over and subjected, at one time or another during the course of geologic history, to at least some selective fusion. The total mass of this subcrustal matter below the continental areas and between depths of 38 and 700 km is about 50×10^{25} g.

On the assumption that this is roughly the quantity of subcrustal matter that may have yielded both the "granitic" magmas that became the continental rocks and the volatile substances that have formed the atmosphere and ocean, it is of interest to compare the relative amounts of each. With an average water content of 1.15 per cent (Clarke, 1924, p. 29), the 2.65×10^{25} g of present crustal rocks would contain 3050×10^{20} g H_2O. This, added to the $16,600 \times 10^{20}$ g of "excess" H_2O in the present atmosphere, ocean, and sedimentary rocks, gives a total of $19,650 \times 10^{20}$ g H_2O that has escaped, along with the 2.65×10^{25} g of "granitic" magmas, from the subcrustal matter. Thus the original $(50 + 2.65) \times 10^{25}$ g of subcrustal matter would have yielded, by selective fusion, $\frac{2.65}{52.65} =$ 5.0 per cent, or 1 part in 20, of magmas that have risen to form the crustal rocks and $\frac{0.1965}{52.65} = 0.37$ per cent H_2O. If the "granitic" magmas originally held all this water in solution, they would have contained at that time an average of $\frac{0.1965}{2.65} = 7.4$ per cent H$_2$O, of which $\frac{0.166}{2.65} = 6.3$ per cent subsequently escaped to the surface during recrystallization of the magmas.

These estimates are, of course, only crude ones based upon very uncertain foundations. They make no allowance for possible "granitic" rocks now foundered beneath the oceanic depths; to whatever extent any continental rocks now lie beneath the ocean floor, the calculated percentages of water in the "granitic" magmas and of water lost from the subcrustal matter must be reduced proportionately.

A rough but independent check on the reasonableness—or otherwise—of this estimate of the amount of subcrustal matter that has been selectively fused can be made by considering certain substances that may have been concentrated in the crustal rocks or at the surface as the result of the supposed differentiation. In the absence of better information, the average composition of the original undifferentiated peridotitic material may be taken as similar to that of stony meteorites (Bowen, 1928, p. 315–316; 1950. *See also* references in section on Dense Primitive Atmosphere—Solution in the Melt). The silicate phase of stony meteorites contains an average of 0.063% H [equivalent to 0.56% H$_2$O] and 0.199% K [equivalent to 0.24% K$_2$O] (Brown, 1949b, p. 626). Thus the calculation that 0.37% H$_2$O has escaped from the subcrustal matter to a depth of 700 km would mean that, on the average, $\frac{0.37}{0.56} = 66\%$ of this mass of material has undergone selective fusion.

If the residual unmelted fraction remaining after selective fusion has an average composition similar to that of dunite, it would contain about 0.04% K$_2$O (Daly, 1933, p. 20). The average igneous rocks now exposed on the continents contain 3.13% K$_2$O (Clarke, 1924, p. 29). From these average compositions, we may calculate that the original subcrustal matter has yielded about $\frac{(0.24 - 0.04)}{(3.13 - 0.04)} = 6.5\%$, or 1 part in 15, of magmas that have crystallized to form the present crustal rocks. This ratio of 1 part of magma to 15 parts of subcrustal matter, when compared with the previous estimate of 1 part in 20 (based on total material to 700 km), would mean that on the average about 75% of the mass of subcrustal matter to a depth of 700 km has undergone selective fusion.

It is only fair to add that ratios of much more and much less than 1 part in 15 may be obtained by using for comparison other elements than K and other rock types (peridotite instead of dunite and diorite instead of the average igneous rock now exposed at the upper surface of the continental blocks). Yet when one considers that the process of selective fusion would act unequally on different constituents of the subcrustal matter and would probably remove much larger proportions of the readily volatile substances, such as water, than of the more refractory rock-forming minerals, it is surprising to find even such rough agreements as those just mentioned. About all that may safely be concluded is that the general order of magnitude of the ratio of "granitic" magmas to subcrustal matter appears not unreasonable.

Other Geologic Tests

The least one can ask of any hypothesis is that it be consistent with the known facts. But it is not always as easy as it sounds to apply this test, because much information and hard work are sometimes required to formulate corollaries of a hypothesis so that they can be tested by the known facts. I have thought of a few geologic tests of this suggested mechanism, but I am sure that many others have not occurred to me.

At first thought, the widespread epicontinental seas of the past seem to contradict the suggestion flatly. But areas are not to be confused with volumes. If the ocean basins have been sinking relative to the continental blocks, then one must look largely to the ocean floor, rather than to the continents, for evidence of a growing volume of sea water. Unfortunately, our information about the ocean floor is still very sketchy. The flat-topped sea mounts, or guyots, discovered by Hess (1946) indicate a considerable sinking of the ocean floor, some part of which may mean a general rise of sea level and not merely local deformation. Other lines of evidence, such as submerged terraces and great thicknesses of shallow-water limestones on some oceanic islands (Carsey, 1950, p. 376), suggest, although likewise they do not prove, the same conclusion. It has long been recognized that very few areas that once were part of the deep sea have later risen to become part of the continents (Walther, 1911, p. 60–61). Furthermore, the truncation of structural trends at some continental margins and the geographic distribution of certain animals and plants seem to require that parts of some land masses have broken off and foundered to oceanic depths

(Barrell, 1927, p. 303–305; Moore, 1936, p. 1786). These are all rather obvious, general tests that yield results consistent with, but certainly not proving, an increasing volume of ocean basins through geologic time.

Undoubtedly other and more rigorous tests should be looked for in the stratigraphic and structural record on the continents, where many more facts are available than we yet know about the ocean basins. A gradually increasing volume of sea water and of marine sediments would mean a gradual rise of sea level and relatively slow marine transgressions, modified of course by the effects of local deformation, and interrupted by relatively rapid withdrawals of the sea, when the ocean floor sank to new positions of equilibrium. Such sudden withdrawals of the sea would affect parts of all continents at the same time (Moore, 1936, p. 1803–1805, 1808). On the other hand, if hydrous granitic magmas have formed by the mechanism indicated, then mountain-making movements, as contrasted with emergences of the continental shelves, need not have been even approximately contemporaneous from one continent to another (Gilluly, 1949).

How well or how poorly do the geologic facts bear out such corollaries as these? Or if these are not valid tests of the suggested mechanism, I trust that others will think up much better ones.

As I warned at the beginning, this subject ramifies almost endlessly into many problems of earth history. A satisfactory solution of the problem of the source and history of the earth's air and water depends on the solution of a great many other questions. Because it is so closely related to many others, this problem is likely long to remain one of those hardy perennials that need a new look and reappraisal every few years, as new observations accumulate. Perhaps what I have presented here is simply another case of putting 2 and 2 together and getting 22, instead of 4. But this review and these tentative conclusions will have served their purpose if they help stimulate—or if they provoke—critical observations and critical thinking about the history and significance of sea water.

REFERENCES CITED

Adams, L. H. (1924) *Temperatures at moderate depths within the earth*, Washington Acad. Sci., Jour., vol. 14, p. 459–472.

Ahrens, Louis (1949) *Measuring geologic time by the strontium method*, Geol. Soc. Am., Bull., vol. 60, p. 217–266.

Allee, W. C.; Emerson, A. E.; Park, Orlando; Park, Thomas; and Schmidt, K. P. (1949) *Principles of animal ecology*, 837 pp.

Allen, E. T. (1922) *Chemical aspects of volcanism, with a collection of analyses of volcanic gases*, Franklin Inst., Jour., vol. 193, p. 29–80.

—— (1935) *Geyser basins and igneous emanations*, Econ. Geol., vol. 30, p. 1–13.

Allen, E. T.; and Day, A. L. (1927) *Steam wells and other thermal activity at "The Geysers", California*, Carnegie Inst. Washington, Pub. no. 378, 106 pp.

—— —— —— (1935) *Hot springs of the Yellowstone National Park* (Microscopic examinations by H. E. Merwin), Carnegie Inst. Washington, Pub. no. 466, 525 pp.

—— and Zies, E. G. (1923) *A chemical study of the fumaroles of the Katmai region*, Nat. Geog. Soc., Contrib. Tech. Papers, Katmai ser., no. 2, p. 75–155.

Anderson, E. M. (1938) *Crustal layers and the origin of magmas. Part II, Geophysical data applied to the magma problem*, Bull. Volcanologique, ser. 2, tome 3, p. 42–82.

Arrhenius, Svante (1896) *On the influence of carbonic acid in the air upon the temperature of the ground*, Philos. Mag., 5th ser., vol. 41, p. 237–276.

Association for Planning and Regional Reconstruction (1942) *Coal in relation to world energy supplies*, Broadsheet No. 10, London, W. C. I., 32 Gordon Square, 4 pp.

Aston, F. W. (1924) *The rarity of the inert gases on the earth*, Nature, vol. 114, p. 786.

Atkins, W. R. G. (1922) *The hydrogen ion concentration of sea water in its biological relations, Part I*, Marine Biol. Assoc., Jour., vol. 12, p. 717–771.

Barrell, Joseph (1927) *On continental fragmentation and the geologic bearing of the moon's surficial features*, Am. Jour. Sci., 5th ser., vol. 13, p. 283–314.

Benfield, A. E. (1940) *Thermal measurements and their bearing on crustal problems*, Am. Geophys. Union Tr., pt. II, p. 155–159.

Benioff, Hugo (1949) *Seismic evidence for the fault origin of oceanic deeps*, Geol. Soc. Am., Bull., vol. 60, p. 1837–1856.

Birch, Francis (1950a) *Flow of heat in the Front Range*, Geol. Soc. Am., Bull., vol. 61, p. 567–630.

——(1950b) *Elasticity and composition of the earth's interior* (Abstract), Science, vol. 112, p. 453.

——; Schairer, J. F.; and Spicer, H. C., editors (1942) *Handbook of physical constants*, Geol. Soc. Am., Spec. Papers, no. 36, 325 pp.

Bowen, N. L. (1928) *The evolution of the igneous rocks*, Princeton, 334 pp.

——(1940) *Progressive metamorphism of siliceous limestone and dolomite*, Jour. Geol., vol. 48, p. 225–274.

——(1947) *Magmas*, Geol. Soc. Am., Bull., vol. 58, p. 263–280.

——(1950) *Petrologic-cosmogonic dilemma* (Abstract) Science, vol. 112, p. 453–454.

Brown, Harrison (1949a) *Rare gases and the formation of the earth's atmosphere*, In *The atmosphere of the earth and planets*. Univ. Chicago Press, p. 260–268.
——(1949b) *A table of relative abundances of nuclear species*, Rev. Mod. Phys., vol. 21, p. 625–634.
Brown, Horace T., and Escombe, F. (1905) *Researches on some of the physiological processes of green leaves, with special reference to the interchange of energy between the leaf and its surroundings*, Royal Soc. London, Pr., ser. B, vol. 76, p. 29–111.
Bruce, E. L. (1945) *Pre-Cambrian iron formations*, Geol. Soc. Am., Bull., vol. 56, p. 589–602.
Bucher, W. H. (1950) *Megatectonics and geophysics*, Am. Geophys. Union, Tr., vol. 31, p. 495–507.
Buddington, A. F. (1943) *Some petrological concepts and the interior of the earth*, Am. Mineral., vol. 28, p. 119–140.
Bullard, E. C., and Stanley, J. P. (1949) *The age of the earth*, Helsinki, Veröff. Finn. Geod. Inst., no. 36, p. 33–40.
Bullen, K. E. (1940) *The problem of the earth's density variation*, Seismol. Soc. Am., Bull., vol. 30, p. 235–250.
—— (1942) *The density variation of the earth's central core*, Seismol. Soc. Am., Bull., vol. 32, p. 19–29.
Callendar, G. S. (1938) *The artificial production of carbon dioxide and its influence on temperature*, Roy. Meteor. Soc., Quart. Jour., vol. 64, p. 223–240.
—— (1940) *Variations of the amount of carbon dioxide in different air currents*, Roy. Meteor. Soc., Quart. Jour., vol. 66, p. 395–400.
—— (1941) *Atmospheric radiation*, Roy. Meteor. Soc., Quart. Jour., vol. 67, p. 31–32.
Carsey, J. B. (1950) *Geology of Gulf Coastal area and continental shelf*, Am. Assoc. Petrol. Geol., Bull., vol. 34, p. 361–385.
Chamberlin, R. T. (1908) *The gases in rocks*, Carnegie Inst. Washington, Pub. no. 106, 80 pp.
—— (1949) *Geological evidence on the evolution of the earth's atmosphere*, In *The atmospheres of the earth and planets*, Chicago, p. 250–259.
Chamberlin, T. C. (1897) *A group of hypotheses bearing on climatic changes*, Jour. Geol., vol. 5, p. 653–683.
—— (1899) *An attempt to frame a working hypothesis of the cause of glacial periods on an atmospheric basis*, Jour. Geol., vol. 7, p. 545–584, 667–685, 751–787.
Clarke, F. W. (1907a) *The composition of red clay*, Roy. Soc. Edinburgh, Pr., vol. 27, p. 167–171.
—— (1907b) *The composition of terrigenous deposits*, Roy. Soc. Edinburgh, Pr., vol. 27, p. 269–270.
—— (1924) *The data of geochemistry*, U. S. Geol. Survey, Bull. 770, 841 pp.
Cloud, P. E., and Barnes, V. E. (1948) *Paleo-ecology of early Ordovician sea in central Texas*, Nat. Research Council, Rept. of Comm. on Treatise on Marine Ecology and Paleo-ecology, 1947–1948, no. 8, p. 29–83.
Collins, W. H., and Quirke, T. T. (1926) *Michipicoten iron ranges*, Geol. Survey Canada, Mem. 147, 141 pp.
Conway, E. J. (1942) *Mean geochemical data in relation to oceanic evolution*, Royal Irish Acad., Pr., vol. XLVIII, Sect. B., no. 8, p. 119–159.
—— (1943) *The chemical evolution of the ocean*, Royal Irish Acad., Pr., vol. XLVIII, Sect. B, no. 9, p. 161–212.
—— (1945) *Mean losses of Na, Ca, etc. in one weathering cycle and potassium removal from the ocean*, Am. Jour. Sci., vol. 243, p. 583–605.
Cooper, L. H. N. (1937) *Some conditions governing the solubility of iron*, Roy. Soc. London, Pr., ser. B., vol. 124, p. 299–307.
Cotton, C. A. (1944) *Volcanic contributions to the atmosphere and ocean*, Nature, vol. 154, p. 399–400.
Crawford, J. G. (1940) *Oil-field waters of Wyoming and their relation to geologic formations*, Am. Assoc. Petrol. Geol., Bull., vol. 24, p. 1214–1329.
Currier, L. W.; Gwinn, G. R.; and Waggaman, W. H. (1947) *Graphite*, p. 247–249, In *Mineral position of the United States*, Hearings before a Subcommittee of the Comm. on Public Lands, U. S. Senate, 80th Cong., 1st session.
Dakin, W. J. (1935) *The aquatic animal and its environment*, Linnean Soc. New South Wales Pr., vol. 60, p. 7–32.
Daly, R. A. (1909) *First calcareous fossils and the evolution of the limestones*, Geol. Soc. Am., Bull., vol. 20, p. 153–170.
—— (1933) *Igneous rocks and the depths of the earth*, McGraw-Hill, 598 pp.
—— (1946) *Nature of the asthenosphere*, Geol. Soc. Am., Bull., vol. 57, p. 707–726.
Davidson, F. A. (1933) *Temporary high carbon dioxide content in an Alaskan stream at sunset*, Ecology, vol. 14, p. 238–240.
Day, A. L. (1939) *The hot spring problem*, Geol. Soc. Am., Bull., vol. 50, p. 317–336.
Dobson, G. M. B. (1942) *Atmospheric radiation and the temperature of the lower stratosphere*, Roy. Meteor. Soc., Quart. Jour., vol. 68, p. 202–204.
Edmondson, W. T. (1944) *Ecological studies of sessile Rotatori, Part I, Factors affecting distribution*, Ecol. Monog., vol. 14, p. 31–66.
Elsasser, W. M. (1942) *Heat transfer by infra-red radiation in the atmosphere*, Harvard Meteor. Studies, Blue Hill Meteor. Observ., no. 6, 107 pp.
Eskola, Pentti (1932a) *On the origin of granitic magmas*, Mineralog. petrog. Mitt., vol. 42, p. 455–481.
—— (1932b) *Conditions during the earliest geological times, as indicated by the Archaean rocks*, Suomalaisen Tiedeakatemian Toimituksia, Sarja A, Nid. 36, no. 4, p. 1–70.
——(1939) *Maapallon hiili, happi ja elämä*, Suomalaisen Tiedeakat. Esitelmät ja Pöytäkirjat, p. 70. (Cited by Rankama and Sahama, 1950, p. 394, 401, 407, 535, 539, 540, 614).
Evans, J. W. (1919) *A modern theory of the earth*, Observatory, vol. 42, p. 165–167.
Ewart, A. J. (1896) *On assimilatory inhibition in plants*, Linnean Soc. Botany, Jour., vol. 31, p. 364–461.
Fairchild, H. L. (1904) *Geology under the new hypothesis of earth-origin*, Am. Geol., vol. 33, p. 94–116.
Farrington, O. C. (1915) *Meteorites*, Chicago, 233 pp.
Fenner, C. N. (1926) *The Katmai magmatic province*, Jour. Geol., vol. 34, p. 673–772.
Foster, M. D. (1942) *Base-exchange and sulphate reduction in salty ground waters along Atlantic and Gulf Coasts*, Am. Assoc. Petrol. Geol., Bull., vol. 26, p. 838–851.

REFERENCES CITED

Gail, F. W. (1919) *Hydrogen ion concentration and other factors affecting the distribution of Fucus*, Puget Sound Biol. Sta., Pub., vol. 2, p. 287-306.

Gibson, D. T. (1949) *The terrestrial distribution of the elements*, Chem. Soc. London, Quart. Rev., vol. 3, p. 263-291.

Giese, A. C. (1945) *Ultraviolet radiations and life*, Physiol. Zoöl., vol. 18, p. 223-250.

Gilluly, James (1937) *The water content of magmas*, Am. Jour. Sci., 5th ser., vol. 33. p. 430-441.

—— (1949) *Distribution of mountain building in geologic time*, Geol. Soc. Am. Bull., vol. 60, p. 561-590.

Goldschmidt, V. M. (1933) *Grundlagen der quantitativen Geochemie*, Fortsch. Mineral. Krist. Petrog., vol. 17, p. 112-156.

—— (1934) *Drei Vorträge über Geochemie*, Geol. Fören. Förhandl., Bd. 56, p. 385-427.

—— (1938) *Geochemische Verteilungsgesetze der Elemente. IX. Die Mengenverhältnisse der Elemente und der Atom-Arten*, Skrifter Norske Videnskaps-Akad. Oslo, I. Mat. Natur. Klasse 1937, no. 4, p. 1-148.

Goranson, R. W. (1931) *The solubility of water in granitic magmas*, Am. Jour. Sci., 5th ser., vol. 22, p. 481-502.

—— (1932) *Some notes on the melting of granite*, Am. Jour. Sci., 5th ser., vol. 23, p. 227-236.

Gruner, J. W. (1922) *The origin of sedimentary iron ore formations, The Biwabik formation of the Mesabi Range*, Econ. Geol., vol. 17, p. 407-460.

Gunter, Gordon (1947) *Paleo-ecological import of certain relationships of marine animals to salinity*, Jour. Paleon., vol. 21, p. 77-79.

Gutenberg, Beno (1939) *The cooling of the earth and the temperature of its interior*, p. 153-164, In *Internal constitution of the earth*, Physics of the Earth, VII.

—— (1945) *Variations in the physical properties within the earth's crustal layers*, Am. Jour. Sci., vol. 243-A (Daly vol.), p. 285-312.

—— and Richter, C. F. (1939a) *Structure of the crust. Continents and oceans*, p. 301-327, In *Internal constitution of the earth*, Physics of the Earth, VII.

—— —— (1939b) *New evidence for a change in physical conditions at depths near 100 kilometers*, Seismol. Soc. Am., Bull., vol. 29, p. 531-537.

—— —— (1941) *Seismicity of the earth*, Geol. Soc. Am., Spec. Papers, no. 34, 131 pp.

Harvey, H. W. (1945) *Recent advances in the chemistry and biology of sea water*, Cambridge Univ. Press, 164 pp.

Heck, E. T.; Hare, C. E.; and Hoskins, H. A. (1940) *Origin and geochemistry of connate waters in West Virginia* (Abstract), Geol. Soc. Am., Bull., vol. 51, p. 1995.

Henderson, L. J. (1927) *The fitness of the environment; an inquiry into the biological significance of the properties of matter*, Macmillan, 317 pp.

Hess, H. H. (1946) *Drowned ancient islands of the Pacific Basin*, Am. Jour. Sci., vol. 244, p. 772-791.

Högbom, A. G. (1894) *Om sannolikheton för sekulära förandringar i atmosfärens kolsyrehalt*, Svensk. Kem. Tid., vol. 6, p. 169 (Summarized in Arrhenius, 1896, p. 269-273).

Holmes, Arthur (1915) *Radioactivity and the earth's thermal history. II. Radioactivity and the earth as a cooling body*, Geol. Mag., n. ser., Decade 6, vol. 2, p. 102-112.

—— (1916) *Radioactivity and the earth's thermal history. III. Radioactivity and isostasy*, Geol. Mag., n. ser., Decade 6, vol. 3, p. 265-274.

—— (1926) *Contributions to the theory of magmatic cycles*, Geol. Mag., vol. 63, p. 306-329.

—— (1932) *The origin of igneous rocks*, Geol. Mag., vol. 69, p. 543-558.

—— (1947a) *A revised estimate of the age of the earth*, Nature, vol. 159, p. 127-128.

—— (1947b) *The construction of a geological time-scale*, Geol. Soc. Glasgow, Tr., vol. 21, pt. 1. p. 117-152.

Hoover, W. H.; Johnston, E. S.; and Brackett, F. S. (1933) *Carbon dioxide assimilation in a higher plant*, Smithson. Misc. Coll., vol. 87, no. 16, 19 pp.

Hubbert, M. K. (1950) *Energy from fossil fuels*, Am. Assoc. Adv. Sci., Centennial vol., p. 171-177.

Humphreys, W. J. (1940) *Physics of the air*, 3d ed., McGraw-Hill, 676 pp.

Hurley, P. M. (1950) *Progress report on geologic time measurement* (Abstract), Science, vol. 112, p. 453.

Hutchinson, G. E. (1944) *Nitrogen in the biogeochemistry of the atmosphere*, Am. Scientist, vol. 32, p. 178-195.

—— (1947) *The problems of oceanic geochemistry*, Ecol. Mon., vol. 17, p. 299-307.

Ingerson, Earl (1940) *Nature of the ore-forming fluid; a discussion*, Econ. Geol., vol. 35, p. 772-785.

—— (1950) *The water content of primitive granitic magma*, Am. Mineral., vol. 35 (Larsen vol.), p. 806-815.

Jeffreys, Harold (1924) *The rare gases of the atmosphere*, Nature, vol. 114, p. 934.

—— (1929) *The earth; its origin, history and physical constitution*, 2d ed., 346 pp.

—— (1930) *The thermal effects of blanketing by sediments*, Roy. Astron. Soc., Monthly Notices, Geophys. Suppl., vol. 2, p. 323-329.

—— (1933) *Quantity of meteoric accretion*, Nature, vol. 132, p. 934.

Jensen, C. A. (1917) *Effect of decomposing organic matter on the solubility of certain inorganic constituents of the soil*, Agric. Research, Jour., vol. 9, p. 253-268.

Johnston, John, and Adams, L. H. (1913) *On the effect of high pressures on the physical and chemical behavior of solids*, Am. Jour. Sci., 4th ser., vol. 35, p. 205-253.

—— —— Niggli, Paul (1913) *The general principles underlying metamorphic processes*, Jour. Geol., vol. 21, p. 481-516, 588-624.

Jones, Sir H. S. (1950) *The evolution of the earth's atmosphere*, Sci. Prog., vol. 38, p. 417-429.

Keenan, J. H., and Keyes, F. G. (1936) *Thermodynamic properties of steam*, John Wiley and Sons, 89 pp.

Keevil, N. B. (1943) *Radiogenic heat in rocks*, Jour. Geol., vol. 51, p. 287-300.

Kennedy, W. Q. (1938) *Crustal layers and the origin of magmas; Part I, Petrological aspects of the problem*, Bull. Volcanologique, ser. 2, tome 3, p. 24-41.

Krogh, August (1934a) *Conditions of life in the ocean*, Ecol. Mon., vol. 4, p. 421-429.

Krogh, August (1934b) *Conditions of life at great depths in the ocean*, Ecol. Mon., vol. 4, p. 430–439.

Kuenen, P. H. (1937) *On the total amount of sedimentation in the deep sea*, Am. Jour. Sci., vol. 34, p. 457–468.

—— (1941) *Geochemical calculations concerning the total mass of sediments in the earth*, Am. Jour. Sci., vol. 239, p. 161–190.

—— (1946) *Rate and mass of deep-sea sedimentation*, Am. Jour. Sci., vol. 244, p. 563–572.

Kuiper, G. P. (1949) *Survey of planetary atmospheres*, In *The atmospheres of the earth and planets*, Univ. Chicago Press, p. 304–345.

Ladd, H. S. (1944) *Reefs and other bioherms*, Nat. Research Council, Rept. of Comm. on Marine Ecology as related to Paleontology, 1943–1944, no. 4, p. 26–29.

Landergren, Sture (1945) *Contribution to the geochemistry of boron. II. The distribution of boron in some Swedish sediments, rocks and iron ores. The boron cycle in the upper lithosphere*, Arkiv. Kemi. Mineral. Geol., vol. 19A, no. 26, p. 1–31.

Landsberg, H. E. (1946) *Note on the frequency distribution of geothermal gradients*, Am. Geophys. Union, Tr., vol. 27, p. 549–551.

Lane, A. C. (1908) *Mine waters*, Lake Superior Min. Inst., Pr., vol. 13, 63–152.

—— (1932) *Eutopotropism*, Geol. Soc. Am., Bull., vol. 43, p. 313–330.

—— (1945) *The evolution of the hydrosphere*, Am. Jour. Sci., vol. 243–A (Daly vol.), p. 393–398.

Latimer, W. M. (1938) *The oxidation states of the elements and their potentials in aqueous solutions*, Prentice-Hall, New York, 352 pp.

—— (1950) *Astrochemical problems in the formation of the earth*, Science, vol. 112, p. 101–104.

Lawson, A. C. (1932) *Insular arcs, foredeeps, and geosynclinal seas of the Asiatic coast*, Geol. Soc. Am., Bull., vol. 43, p. 353–381.

Legendre, René (1925) *La concentration en ions hydrogène de l'eau de mer*, Paris, 291 pp.

Leith, C. K. (1934) *The pre-Cambrian*, Geol. Soc. Am., Pr. 1933, p. 151–180.

—— and Mead, W. J. (1915) *Metamorphic geology*, Henry Holt and Co., 337 pp.

Lowenstam, H. A. (1948) *Biostratigraphic study of the Niagaran inter-reef formations in northeastern Illinois*, Ill. State Mus. Sci. Papers, vol. 4, p. 1–146.

Macallum, A. B. (1904) *The paleochemistry of the ocean in relation to animal and vegetable protoplasm*, Royal Canad. Inst., Tr., vol. 7, p. 535–562.

—— (1926) *Paleochemistry of body fluids and tissues*, Physiol. Rev., vol. 6, p. 316–357.

Macelwane, J. B. (1939) *Evidence on the interior of the earth derived from seismic sources*, p. 219–290, In *Internal constitution of the earth*, Physics of the Earth, VII.

Macgregor, A. M. (1927) *The problem of the pre-Cambrian atmosphere*, S. African Jour. Sci., vol. 24, p. 155–172.

McClendon, J. F. (1916) *The composition, especially the hydrogen-ion concentration, of sea-water in relation to marine organisms*, Jour. Biol. Chem., vol. 28, p. 135–152.

Meinzer, O. E. (1924) *Origin of the thermal springs of Nevada, Utah, and southern Idaho*, Jour. Geol., vol. 32, p. 295–303.

Merrill, G. P. (1930) *Composition and structure of meteorites*, Smithson. Inst., Bull. 149, 62 pp.

Miholić, Stanko (1947) *Ore deposits and geologic age*, Econ. Geol., vol. 42, p. 713–720.

Mills, R. van A., and Wells, R. C. (1919) *The evaporation and concentration of waters associated with petroleum and natural gas*, U. S. Geol. Survey, Bull. 693, 104 pp.

Moore, E. S., and Maynard, J. E. (1929) *Solution, transportation and precipitation of iron and silica*, Econ. Geol., vol. 24, p. 272–303, 365–402, 506–527.

Moore, R. C. (1936) *Stratigraphic evidence bearing on problems of continental tectonics*, Geol. Soc. Am., Bull., vol. 47, p. 1785–1808.

Morey, G. W. (1922) *Development of pressure in magmas as a result of crystallization*, Washington Acad. Sci., Jour., vol. 12, p. 219–230.

—— (1924) *Relation of crystallization to the water content and vapor pressure of water in a cooling magma*, Jour. Geol., vol. 32, p. 291–295.

—— and Fleischer, Michael (1940) *Equilibrium between vapor and liquid phases in the system CO_2–H_2O–K_2O–SiO_2*, Geol. Soc. Am., Bull., vol. 51, p. 1035–1058.

Nash, L. K., and Baxter, G. P. (1947) *The determination of the gases in meteoritic and terrestrial irons and steels*, Am. Chem. Soc., Jour., vol. 69, p. 2534–2544.

Newcombe, R. B. (1933) *Oil and gas fields of Michigan*, Mich. Geol. Survey, Pub. 38, 293 pp.

Nininger, H. H. (1940) *Collecting meteoritic dust*, Sci. Mo., vol 50, p. 460–461.

Osborne, F. F. (1931) *Non-metallic mineral resources of Hastings County*, Ontario Dept. Mines, 39th Ann. Rept., vol. 39, pt. 6, p. 22–59.

Pettersson, Hans (1949) *The floor of the ocean*, Endeavor, vol. 8, p. 182–187.

Pettijohn, F. J. (1943) *Archean sedimentation*, Geol. Soc. Am., Bull., vol. 54, p. 925–972.

Piper, A. M. (1933) *Ground water in southwestern Pennsylvania*, Penna. Geol. Survey, Bull. W1, 4th ser., 406 pp.

Poole, J. H. J. (1941) *The evolution of the atmosphere*, Royal Dublin Soc., Sci. Pr., vol. 22 (n.s.), p. 345–365.

Powers, E. B. (1920) *The variation of the condition of sea-water, especially the hydrogen ion concentration, and its relation to marine organisms*, Wash. State Univ., Puget Sound Biol. Sta., Pub., vol. 2, p. 369–385.

—— (1939) *Chemical factors affecting the migratory movements of the Pacific Salmon*, In *The migration and conservation of salmon*, Am. Assoc. Adv. Sci., Pub., vol. 8, p. 72–85.

Rabinowitch, E. I. (1945) *Photosynthesis and related processes. Vol. I. Chemistry of photosynthesis, chemosynthesis and related processes in vitro and in vivo*, New York, 599 pp.

Rankama, Kalervo (1948) *New evidence of the origin of pre-Cambrian carbon*, Geol. Soc. Am., Bull., vol. 59, p. 389–416.

—— and Sahama, T. G. (1950) *Geochemistry*, Univ. Chicago Press, 912 pp.

Rayleigh, Lord (R. J. Strutt) (1939) *Nitrogen, argon, and neon in the earth's crust with applications to cosmology*, Roy. Soc. London, Pr., ser. A, vol. 170, p. 451–464.

Revelle, Roger, and Fleming, R. H. (1934) *The solubility product constant of calcium carbonate in sea water*, 5th Pacific Sci. Cong., Canada 1933, Pr., vol. 3, p. 2089–2092.

Riley, G. A. (1944) *The carbon metabolism and

photosynthetic efficiency of the earth as a whole, Am. Scientist, vol. 32, p. 132–134.

Rogers, C. G. (1938) *Textbook of comparative physiology*, 2d ed., McGraw-Hill, 715 pp.

Ross, C. P. (1936) *Some features of the Idaho batholith*, Inter. Geol. Cong., Rept. XVI Session, United States, vol. 1, p. 369–385.

Rubey, W. W., and Murata, K. J. (1941) *Chemical evidence bearing on origin of a group of hot springs* (Abstract), Washington Acad. Sci., Jour., vol. 31, no. 4, p. 169–170.

Russell, H. N. (1949) *The time scale of the universe*, In *Time and its mysteries*, series III, New York Univ. Press, p. 3–30.

—— and Menzel, D. H. (1933) *The terrestrial abundance of the permanent gases*, Nat. Acad. Sci., Pr., vol. 19, p. 997–1001.

Russell, W. L. (1933) *Subsurface concentration of chloride brines*, Am. Assoc. Petrol. Geol., Bull., vol. 17, p. 1213–1228.

Rutherford, R. L. (1936) *Geologic age of potash deposits*, Geol. Soc. Am., Bull., vol. 47, pp. 1207–1216.

Schuchert, Charles (1931) *Geochronology or the age of the earth on the basis of sediments and life*, p. 10–64, In The age of the earth, Nat. Research Council, Bull. 80, Physics of the Earth, IV.

Sederholm, J. J. (1925) *The average composition of the earth's crust in Finland*, Commiss. Géol. Finlande, Bull., tome 12, no. 70, p. 3–20.

Shelford, V. E. (1918) *The relation of marine fishes to acids with particular reference to the Miles acid process of sewage treatment*, Puget Sound Biol. Sta., Pub., vol. 2, p. 97–111.

Shepherd, E. S. (1938) *The gases in rocks and some related problems*, Am. Jour. Sci., 5th ser., vol. 35A, p. 311–351.

Singh Pruthi, H. (1927) *The ability of fishes to extract oxygen at different hydrogen ion concentrations of the medium*, Marine Biol. Assoc., Jour., vol. 14, p. 741–747.

Slichter, L. B. (1941) *Cooling of the earth*, Geol. Soc. Am., Bull., vol. 52, p. 561–600.

—— (1950) *The Rancho Santa Fe conference concerning the evolution of the earth*, Nat. Acad. Sci., Pr., vol. 36, p. 511–514.

Smith, C. L. (1940) *The Great Bahama Bank. II. Calcium carbonate precipitation*, Jour. Marine Research, vol. 3, p. 171–189.

Stearns, N. D., Stearns, H. T., and Waring, G. A. (1937) *Thermal springs in the United States*, U. S. Geol. Survey, W. S. Paper 679-B, p. 59–206.

Stetson, H. T. (1942) *The sun and the atmosphere*, Sigma Xi Quart., vol. 30, p. 16–35.

Suess, H. E. (1949) *Die Häufigkeit der Edelgase auf der Erde und im Kosmos*, Jour. Geol., vol. 57, p. 600–607.

Sverdrup, H. U.; Johnson, M. W.; and Fleming, R. H. (1946) *The oceans, their physics, chemistry, and general biology*, Prentice-Hall, 1087 pp.

Twenhofel, W. H. (1926) *Treatise on sedimentation*, 1st ed., Baltimore, 661 pp.

Twenhofel, W. H. (1929) *Magnitude of the sediments beneath the deep sea*, Geol. Soc. Am., Bull., vol. 40, p. 385–402.

—— (1936) *Organisms and their environment*, Nat. Research Council, Rept. of Committee on Paleo-ecology, 1935-36, p. 1–9.

Urry, W. D. (1949) *Significance of radioactivity in geophysics—Thermal history of the earth*, Am. Geophys. Union, Tr., vol. 30, p. 171–180.

Valley, George (1928) *The effect of carbon dioxide on bacteria*, Quart. Rev. Biol., vol. 3, p. 209–224.

Van Hise, C. R. (1904) *A treatise on metamorphism*, U. S. Geol. Survey, Mon. 47, 1286 pp.

Van Orstrand, C. E. (1939) *Observed temperatures in the earth's crust*, p. 125–151, In *Internal constitution of the earth*, Physics of the Earth, VII.

Vaughan, T. W. (1940) *Ecology of modern marine organisms with reference to paleogeography*, Geol. Soc. Am., Bull., vol. 51, p. 433–468.

Verhoogen, Jean (1946) *Volcanic heat*, Am. Jour. Sci., vol. 244, p. 745–771.

Wahl, Walter (1949) *Isostasy and the origin of sial and sima and of parental rock magmas*, Am. Jour. Sci., vol. 247, p. 145–167.

Walther, Johannes (1911) *Origin and peopling of the deep sea* (Translation by C. M. Le Vene), Am. Jour. Sci., vol. 31, p. 55–64.

Wattenberg, Hermann (1936) *Kohlensäure und Kalziumkarbonat im Meere*, Fortschritte d. Mineral., Kristal. u. Petrographie, vol. 20, p. 168–195.

—— and Timmermann, E. (1938) *Die Löslichkeit von Magnesiumkarbonat und Strontiumkarbonat im Seewasser*, Kieler Meeresforchungen, Bd. 2, p. 81–94.

Weeks, L. G. (1950) *Concerning estimates of potential oil reserves*, Am. Assoc. Petrol. Geol., Bull., vol. 34, p. 1947–1953.

White, D. E., and Brannock, W. W. (1950) *The sources of heat and water supply of thermal springs, with particular reference to Steamboat Springs, Nevada*, Am. Geophys. Union, Tr. vol. 31, no. 4, p. 566–574.

Wildt, Rupert (1939) *Constitution of the planets*, Am. Philos. Soc., Pr., vol. 81, p. 135–152.

—— (1942) *The geochemistry of the atmosphere and the constitution of the terrestrial planets*, Rev. Modern Physics, vol. 14, p. 151–159.

Wilson, J. Tuzo (1950) *Recent applications of geophysical methods to the study of the Canadian Shield*, Am. Geophys. Union, Tr., vol. 31, p. 101–114.

Zies, E. G. (1929) *The Valley of Ten Thousand Smokes. II. The acid gases contributed to the sea during volcanic activity*, Nat. Geog. Soc., Contrib. Tech. Papers, Katmai series, vol. 1, no. 4, p. 61–79.

U. S. GEOLOGICAL SURVEY, WASHINGTON 25, D. C.
MANUSCRIPT RECEIVED BY THE SECRETARY OF THE SOCIETY, MAY 10, 1951.
PUBLISHED WITH THE PERMISSION OF THE DIRECTOR, U. S. GEOLOGICAL SURVEY.

The geologic history of sea water—an attempt to solve the problem*

HEINRICH D. HOLLAND

Department of Geological and Geophysical Sciences, Princeton University,
Princeton, New Jersey, U.S.A.

Abstract—Thermodynamics and kinetics are both important in the operation of the ocean–atmosphere system. The carbonate chemistry of the oceans can be treated in terms of perturbations of a well-defined equilibrium state. The equilibrium state for silicates in sea water is less well defined, and thermodynamic arguments are much less persuasive. Kinetics are of overwhelming importance in determining—among other parameters—the nitrate concentration of sea water and the pressure of oxygen and nitrogen in the atmosphere. It is likely that a similar mix of attainment and non-attainment of equilibrium prevailed during the entire history of the oceans, and that the sedimentary record must be interpreted accordingly.

Marine evaporites are among the most helpful sediments in reconstructing the geologic history of sea water. The relative abundance of minerals in marine evaporites is a complex function of the history of the enclosing basins, but their sequence, and the absence of many minerals typical of non-marine evaporites set rather interesting limits on possible excursions of the composition of sea water since late Precambrian time. It is likely that the concentration of the major constituents of sea water has rarely, if ever, been more than twice or less than half their present concentration during the past 700 million years.

The similarity of the mineralogy of sedimentary rocks during the past 3500 million years suggests that the chemistry of sea water has been highly conservative. However, the extensive bedded cherts of the Precambrian, and the frequently associated banded iron formations do indicate a real evolution of the ocean–atmosphere system. They are explained most readily in terms of a higher silica content in Precambrian sea water and a lower oxygen pressure in the atmosphere.

Fossil fuel burning will probably introduce some ten times the present content of CO_2 into the atmosphere during the next two centuries. The effects of the additional CO_2 on climate and on the shallow marine biota are presently very difficult to predict. Other man-made chemicals are produced in much smaller quantities than CO_2. However a number of these are so toxic, that their integrated effects are potentially much more damaging than those of fossil fuel burning.

THE PRESIDENTIAL address delivered to the Geological Society of America at its Washington meetings in November 1950 was unusual. Unlike most presidential addresses W. W. Rubey's address and its published version (RUBEY 1951) have been remembered, reprinted and served as points of departure for later studies. The title of my paper this morning is almost identical to Rubey's. Both of us address ourselves to the geologic history of sea water. However, Rubey modestly subtitled his paper: "An attempt to state the problem". I will leave it up to you to decide whether my subtitle: "An attempt to solve the problem" is premature, or whether the change in verbs is an appropriate measure of progress in our understanding of the evolution of ocean water.

Most of this progress has been made somewhat indirectly. KRAMER's (1965) direct approach to the history of sea water via the analysis of inclusion fluids in sedimentary rocks failed to be convincing because it is so difficult to establish the

---

* Presidential Address delivered to the Geochemical Society, November 1, 1971 at Washington, D.C., U.S.A.

pedigree of these inclusion fluids. CHAVE's (1960) approach via the chemistry of connate brines was hampered by the difficulty of seeing through the complex chemical history of these brines. The most convincing pieces of evidence have come through an improved understanding of the chemical dynamics of the present oceans and from new ways of looking at the sedimentary record.

In this talk I would like to deal first with the improved understanding of the solid Earth–ocean–atmosphere system as it is operating presently. I will then speak about the implications of these insights for the interpretation of the historical record, and will finally venture a few comments on the oceans during the coming century.

Not long after Rubey's address to the Geological Society of America T. F. W. BARTH (1952) first used the concept of residence time to analyze the behavior of elements in the oceans. All of these residence times have turned out to be short compared to the age of the oceans. The longest, those of sodium, chlorine and bromine are on the order of 10^8 yr; the shortest, those of beryllium and aluminum, are only on the order of 100 yr. The dissolved salts are obviously passing through the oceans, and the concentration of an element at any given time is determined largely by the relationship of input to output during a prior time interval approximately equal to the residence time of the element. The residence time of calcium in the oceans today is close to one million years. The calcium concentration of sea water today is therefore determined largely by the inflow of calcium with river water and by the output of calcium from the oceans as a constituent of a variety of minerals and of interstitial water during the past million years.

The river inflow of calcium has depended on a host of variables, both climatic and geographic—in the broadest sense. The composition of river waters is a sensitive function of the annual runoff in their drainage basins. It turns out, however, that the product of the annual runoff times the concentration of most dissolved elements is much more constant than either the runoff or the concentration alone, and that the annual river input of dissolved salts to the oceans has probably varied a good deal less than a factor of ten during much of geologic time.

The response of the oceans to these inputs is complicated, and has been the source of one of the livelier controversies in geochemistry during the past decade. In his classic paper of 1961 (SILLÈN, 1961) and in several subsequent papers SILLÈN (1963, 1967a, b) proposed that the composition of sea water is controlled largely by chemical equilibria between sea water and oceanic sediments. Much the same concept is at the heart of one of my papers (HOLLAND 1965) and those of MACKENZIE and GARRELS (1966a, b) and SIEVER (1968). On the other hand WEYL (1966) and BROECKER (1971) among others have been much more impressed with the influence of kinetics than with the influence of thermodynamics on the chemistry of sea water. In part the controversy seems to have been of the doughnut variety, in which one side emphasized the doughnut, the other the hole. The first slide, taken from the 1969 paper of LI, TAKAHASHI and BROECKER (1969) shows the degree of saturation of water in the Atlantic and Pacific Oceans as a function of depth below the surface. Surface ocean water is apparently everywhere supersaturated with respect to calcite, whereas deep waters are usually undersaturated with respect to calcite. This state of affairs has obvious and important consequences for the distribution of carbonate sediments within ocean basins, and on the entire dynamics of the marine cycle of

The geologic history of sea water—an attempt to solve the problem 639

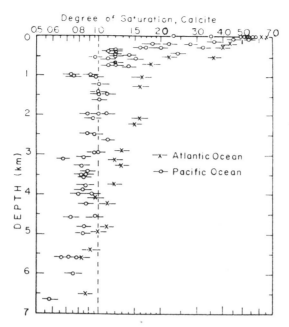

Fig. 1. The degree of saturation of Atlantic and Pacific sea water with respect to calcite as a function of depth below the surface (LI, TAKAHASHI, and BROECKER, 1969).

calcium and carbon. Thermodynamic equilibrium does not prevail, and chemical kinetics aided and hindered by physical and biological processes are clearly important. So much for the hole. The doughnut view, to which I am somewhat partial, much to the detriment of my figure, if not my science, maintains that the carbonate chemistry of sea water is clearly constrained by thermodynamics, and that the data in this slide are interpreted most reasonably in terms of perturbations of chemical equilibrium by the tossings and turnings of a restless Earth. Both views have a good deal to recommend them. In the quest for the chemical history of sea water holders of the doughnut view tend to emphasize that the presence of limestones in sedimentary rocks of all ages implies near-saturation of ocean water with respect to calcite during the past 3 billion years. Holders of the hole view tend to emphasize that the oceans were probably never precisely at saturation with respect to calcite, and that the deviations from equilibrium have always been critical in determining the deposition and hence the subsequent history of carbonate sediments. It seems to me that these views are not only compatible but complementary, and that they are both useful in reconstructing the geologic history of sea water from the sedimentary record.

Such complementarity cannot be claimed for many of the other components of sea water. The search for the oceanic sinks of potassium, sodium, and magnesium has been unexpectedly frustrating (see for instance RUSSELL, 1970). Since we have not yet identified these sinks with certainty, discussions of their chemical kinetics is premature. However, neither laboratory experience with silicate reactions at higher temperatures, nor the evidence from measurements of the isotopic composition of clays, or of the chemical composition of interstitial water encourage a belief in the rapid equilibration of sea water with silicate sediments. The silicate mineralogy of sedimentary rocks is therefore still a rather uncertain starting point for arguments concerning the geologic history of sea water. This is even more true for several other components of sea water. The nitrate concentration of sea water is some 11 orders of magnitude smaller than its value at equilibrium with atmospheric oxygen and nitrogen at a pH of 8·0. Even the staunchest Gibbsian tends to blanch at the prospect of trying to explain such a difference in terms of perturbations from equilibrium. Kinetics are clearly of overwhelming importance; thermodynamic calculations serve simply to underscore this fact and to define some extremely distant boundary conditions.

It could be argued that nitrate is a very minor component of sea water, and that its kinetics are of no particular concern to Earth history. But this is simply not true. Nitrate is a limiting nutrient today, perhaps the most limiting nutrient. As such it probably plays a critical note in determining the total rate of photosynthesis in the oceans and thence influences the oxygen content of the atmosphere.

BROECKER (1970) has pointed out that atmospheric oxygen is used up in weathering reactions on a time scale of a few million years. On the other hand the continued existence of land animals during the past several hundred million years implies that the oxygen content of the atmosphere was never so low as to produce death by anoxia nor so high as to produce death by oxidative tissue damage. Oxygen used by weathering reactions has therefore been replaced rather exactly during the last several hundred million years by the net supply of oxygen produced via photosynthesis. The exact nature of the control mechanisms are still somewhat obscure. BROECKER (1971) has suggested that marine phosphate is a critical link in the feedback system which balances the rate of oxygen production with the rate of oxygen use. In a somewhat more detailed analysis (HOLLAND, 1972) I have proposed that both marine nitrate and marine phosphate are important components of the feedback control mechanism. Whichever view turns out to be correct, thermodynamics exercises virtually no control over the redox balance between the atmosphere and oceans, and has probably failed to do so during much of geologic time.

These insights into the present workings of the ocean–atmosphere–solid Earth system have a number of important consequences for attempts to reconstruct the geologic history of sea water. In a system near steady state changes in the concentration of a particular component are apt to be small in periods of time smaller than its residence time. Thus, major changes in the calcium concentration of sea water are unlikely in less than a million years, and major changes in the sodium concentration are unlikely in less than one hundred million years if there are no drastic changes in the rate of ocean input and output of these elements.

The "if" is somewhat haunting. The sulfate locked up in sediments of the Zechstein basin is approximately equal to 10 per cent of the present sulfate content of the oceans (SCHAEFFER, 1971). It follows that several Zechstein Basins filling simultaneously could seriously deplete the sulfate content of the oceans in less than 20 million years.

Reasonable constancy of the concentration of the components of ocean water during a period equal to their present residence time is surely likely but is hardly guaranteed. Reasonable constancy for periods exceeding the residence time can sometimes be inferred from the linking of the geochemical cycle of an element with a short residence time to one of longer residence time; generally, however, independent data are required to define the concentration of components in ocean water more than one residence time ago.

The sedimentary record is the most fruitful source of such data. The two gentlemen in the next slide are expressing the fervent desire of all seekers after the history

Fig. 2. The fervent desire of all seekers after the history of sea water.

of sea water; "Boulder, boulder on the ground, tell us something real profound". Recently, of course, the boulder has been studied not only outdoors but indoors as well, and has been subjected to detailed mineralogical, chemical and isotopic scrutiny. The results have been encouraging.

Evaporites should be the perfect sediments for reconstructing the history of sea water (HOLLAND, 1972). In the ideal case they represent the residue from the evaporation of a single batch of sea water. In reality periods of net evaporation alternate with periods of net dilution in evaporite basins, and the relative proportion of evaporite minerals rarely, if ever, equals their relative proportion in contemporary sea water. The early evaporite minerals are typically overrepresented, and the later minerals are typically underrepresented. To make matters worse late brines frequently modify the earlier evaporites, so that it is impossible to determine the composition of sea water precisely from the mineralogy and chemistry of marine-evaporites.

Nonetheless marine evaporites set some rather interesting limits on excursions in the composition of sea water since late Precambrian time. During this entire period marine evaporites typically began with a carbonate section; this passes upward into a gypsum–anhydrite section, which in turn is followed by a halite section. Differences between marine evaporites tend to be pronounced only during the crystallization of chlorides and sulfates from the final bitterns.

The evaporation of present-day sea water yields just this sequence of minerals.

Understandably, students of evaporites (see for instance BRAITSCH, 1962) have tended to show little interest in possible variations in the composition of sea water, since no changes in composition are needed to explain the mineralogy of marine evaporites from late Precambrian to the present time.

As historians of sea water we must invert the problem, and must define the limits on possible excursions of sea water which are set by the conditions of an invariant sequence of evaporite minerals. We can describe these limits in geometric terms. Let the concentration of the seven major components of sea water: sodium, magnesium, calcium, potassium, chloride, sulfate, and bicarbonate be axes in seven-dimensional space. We then have to circumscribe a volume within this space such that all points lying within the volume yield typical marine evaporite sequences. The composition of present day sea water obviously lies within this volume. In fact, it probably lies near the center of the volume.

To simplify matters we will first consider some two dimensional sections through the seven-dimensional volume. Figure 3 is such a section. The logarithm of the calcium concentration has been plotted against the logarithm of the bicarbonate

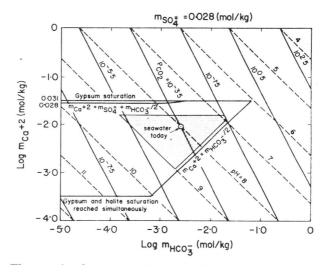

Fig. 3. The permitted concentration range of calcium and bicarbonate in sea water, consistent with the mineralogy of marine evaporites when $m_{SO_4^{2-}} = 0.028$ mole/kg and $m_{NaCl} = 0.55$ mole/kg.

concentration, and the point near the center of the diagram represents the present day position of sea water in the diagram. The sulfate and sodium chloride concentrations have been set equal to those in present day sea water. I would like to propose that the composition of sea water has not strayed outside the stippled area in this section since late Precambrian time.

A strong upper limit for the calcium concentration of sea water is set by the solubility of gypsum. Figure 4 shows the concentration product $m_{Ca^{2+}} \cdot m_{SO_4^{2-}}$ for solutions saturated with respect to gypsum at 25°C as a function of the sodium chloride concentration of the solution. A three-fold increase in the calcium concentration of present day sea water would produce an ocean saturated with respect to gypsum. Gypsum would then be a constant companion of limestones in slightly evaporative settings. It would also become a common component of non-evaporative sediments in cold, shallow areas, because the solubility of gypsum decreases with decreasing temperature. Neither of these events has been observed in the sedimentary record. Gypsum deposition has apparently always been preceded by a fair

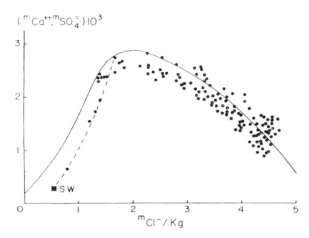

Fig. 4. The concentration product $m_{Ca^{2+}} \cdot m_{SO_4^{2-}}$ = in NaCl solutions saturated with respect to gypsum and in sea water undergoing evaporation; courtesy of D. J. J. Kinsman.

amount of evaporative concentration of sea water. Within the confines of this diagram an increase of a factor of three is therefore a strong upper limit to upward excursion in the calcium concentration of sea water.

A strong lower limit to the calcium concentration in sea water is set by the condition that gypsum always precedes halite in marine evaporites. If the calcium concentration in sea water were reduced by a factor of thirty, sea water would become saturated simultaneously with gypsum and halite during evaporative concentration. Any further reduction in the calcium concentration would reverse the order of appearance of the two minerals; this would be at complete variance with the geologic record.

Other lines of evidence suggest strongly that sea water has never been close to this southern boundary. Toward the southeast sea water is limited by the line $m_{Ca^{2+}} = m_{HCO_3^-}/2$. In present day sea water the calcium concentration is some four times greater than the bicarbonate concentration. During calcium carbonate deposition bicarbonate is therefore exhausted, and most of the calcium is available for gypsum deposition. If the concentration of calcium ever dropped below half that of bicarbonate, calcium would be exhausted during calcium carbonate deposition, and bicarbonates of other cations would become common products of the later stages in the evaporative concentration of sea water. The deposition of dolomite would help in removing excess bicarbonate, but the calcium required to form gypsum would largely offset this gain. The boundary as drawn is therefore probably quite generous. It is certain that the oceans have never reached the composition of soda lakes since late Precambrian times.

The southwestern boundary of the sea water composition field is set by pH limitations. The continued presence of dolomite as a primary or nearly primary mineral in marine carbonate sections demands that the ratio of the magnesium concentration to the calcium concentration in sea water has never been less than unity. The oceans would therefore always have become saturated with respect to brucite at a pH between 9·0 and 9·5. Authigenic brucite has not been reported in marine sediments, and a pH of 9·5 has surely never been exceeded. Even a pH of 9·0 seems extreme, since it is likely that sepiolite would become a common authigenic marine mineral if sea water pH were to rise above 8·6. The contours of constant pH and P_{CO_2} in the diagram have been drawn for sea water saturated with respect to calcite. I have suggested earlier that sea water must have been close to or at saturation during the past 3 billion years. In our diagram excursions toward the southwest are therefore quite severely limited by the requirement that values near 9·0 probably represent the upper extreme for the pH of sea water since late Precambrian time.

A final boundary is worth mentioning. If the calcium concentration of sea water ever exceeded the sum of the sulfate concentration and half the bicarbonate concentration, late bitterns would be enriched in calcium and strongly depleted in sulfate. This has apparently never happened. Typically, calcium becomes strongly depleted, and both sulfate and magnesium are concentrated in the later brines. However, the boundary

$$m_{Ca^{2+}} = m_{SO_4^{2-}} + \frac{m_{HCO_3^-}}{2}$$

is so close to the gypsum saturation line that its presence adds only slightly to the constraints on the calcium and bicarbonate concentration at a sulfate concentration of 0·028 mole/kg.

Toward lower sulfate concentrations this boundary becomes progressively more important. Figure 5 shows the permitted range of sea water compositions at a sulfate concentration of 0·010 mole/kg., about one third of the present day value. The maximum value of the calcium concentration permitted by sea water saturation with gypsum has increased by a factor of three. The minimum value of the calcium concentration demanded by the requirement that gypsum precipitate before halite has similarly increased by a factor of three. The position of the eastern and western boundaries has remained unchanged, but the boundary

$$m_{Ca^{2+}} = m_{SO_4^{2-}} + \frac{m_{HCO_3^-}}{2}$$

has become a strong upper limit for $m_{Ca^{2+}}$. The point representing the present day calcium and bicarbonate content of sea water is now at the edge of the permissible area.

At a sulfate concentration one tenth that of present day sea water the permissible area for the calcium and bicarbonate concentration of sea water is virtually eliminated

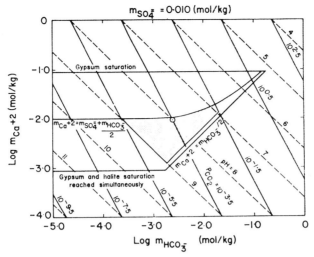

Fig. 5. The permitted concentration range of calcium and bicarbonate in sea water, consistent with the mineralogy of marine evaporites when $m_{SO_4^{2-}} = 0.010$ mole/kg. and $m_{NaCl} = 0.55$ mole/kg.

between the northward movement of the lower boundary for the calcium concentration, and the southward movement of the upper boundary for the calcium concentration. At sulfate concentrations only slightly below 0·003 mole/kg the permissible area pinches out altogether, and no combination of calcium and bicarbonate concentrations can yield the mineral sequence of marine evaporites.

The permitted compositional area also decreases toward higher sulfate concentrations. This is shown in Fig. 7, drawn for a sulfate concentration of

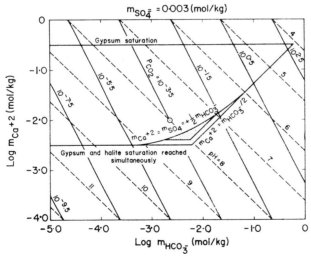

Fig. 6. The permitted concentration range of calcium and bicarbonate in sea water, consistent with the mineralogy of marine evaporites when $m_{SO_4^{2-}} = 0.003$ mole/kg. and $m_{NaCl} = 0.55$ mole/kg.

0·100 mole/kg. The upper limit placed on the calcium concentration by gypsum saturation has cut deeply into the permissible composition triangle. Only a minor additional increase in $m_{SO_4^{2-}}$ would wipe out the permitted area altogether.

We could now combine all of these sections into a single diagram with axes $\log m_{Ca^{2+}} - \log m_{HCO_3^-} - \log m_{SO_4^{2-}}$. Within this diagram there would be a volume containing all sea water compositions compatible with the geologic record of marine

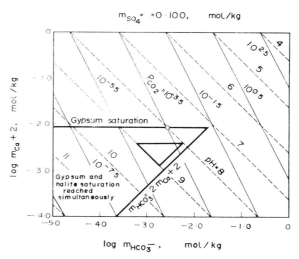

Fig. 7. The permitted concentration range of calcium and bicarbonate in sea water, consistent with the mineralogy of marine evaporites when $m_{SO_4^{2-}} = 0.100$ mole/kg and $m_{NaCl} = 0.55$ mole/kg.

evaporites. The volume has roughly the shape of a trigonal bipyramid, because horizontal sections are roughly triangular and decrease in area both toward higher and toward lower sulfate concentrations. Rather than projecting the three dimensional diagram in all its glory, I would like to show you the superposition of the permitted areas at various sulfate concentrations. The largest area, corresponding to the center of the bipyramid is near the present sulfate concentration of sea water. Toward higher concentrations the area shifts and shrinks. This also happens toward lower sulfate concentrations.

The permitted range in the concentration of calcium, bicarbonate, and sulfate is quite large. The upper limit of the sulfate concentration is near 0·30 mole/kg, the lower limit near 0·002 mole/kg. The upper limit of the calcium concentration is 0·030 mole/kg, the lower limit 0·0001 mole/kg. The upper limit of the bicarbonate concentration is 0·020 mole/kg, the lower limit 0·0001 mole/kg. The extremes are, however, quite unlikely compositions for sea water, because they demand precisely fixed values for the concentration of all three constituents. It seems more reasonable to ask how far the concentration of the three constituents could vary independently without leaving the permitted composition volume. Geometrically this is equivalent to constructing within the bipyramid on ellipsoid that is tangent or nearly tangent to all six faces. It turns out that the composition of sea water presently lies near the center of this ellipsoid, and that the sphere contains all compositions in which $m_{Ca^{2+}}$, $m_{HCO_3^-}$, and $m_{SO_4^{2-}}$ differ by less than a factor of 2 from their present day value.

The diagrams which you have just seen were constructed for the present day concentration of NaCl in sea water. Doubling and halving the NaCl concentration has very little affect on the shape, size, and position of the permitted volumes. Larger variations in the salinity of sea water since late Precambrian time are unlikely. Upward excursions are limited by the availability of NaCl in the Earth's

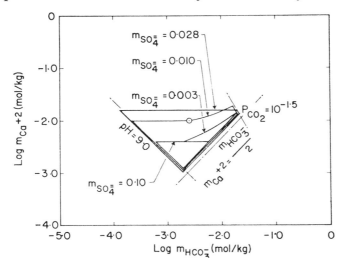

Fig. 8. The permitted concentration range of calcium and bicarbonate in sea water, consistent with the mineralogy of marine evaporites when $0.100 \geqslant m_{SO_4^{2-}} \geqslant 0.003$ mole/kg and $m_{NaCl} = 0.55$ mole/kg.

crust. The quantity of halite in sediments and sedimentary rocks is probably roughly equal to the quantity of NaCl dissolved in ocean water. A doubling of the salinity of sea water would therefore require the complete absence of evaporites at some period since late Precambrian time. This is ruled out by the presence of evaporites in sediments from all periods since the late Precambrian. Upward excursions in salinity were therefore surely minor during the past 700 million years.

The extent of downward excursions in sea water salinity is more difficult to gauge. One could imagine that during a period such as the Permian, in which salt withdrawal appears to have been unusually extensive, the salt content of the oceans dropped significantly below its present level. Perhaps one half the present concentration is a reasonable estimate for the minimum salinity of ocean water during the Phanerozoic, but some independent criteria are obviously needed to lend substance to this estimate. HOLSER (1966) has explored the use of the bromine content of basal halites in evaporite basins for this purpose. Unfortunately the data and their interpretation are not yet clear-cut; the scatter in the observed bromine concentrations neither demands nor precludes a change in the salinity of sea water during the Phanerozoic era.

The evaporite record extends no further back in time than the late Precambrian, and our knowledge of the earlier history of sea water is poorer for their absence.

Fortunately, the earliest well dated marine evaporites predate the faunal revolution of the early Phanerozoic. The mineralogy of the late Precambrian evaporites of the Amadeus Basin in central Australia (see for instance Schopf, 1968) is quite similar to that of later marine evaporites, and suggests that the major changes in the marine biota some 600 million years ago did not have a profound effect on the chemistry of sea water.

It is most likely that evaporites were formed during early and mid-Precambrian time, and that they have been recycled—to use a well worn phase—as suggested by Garrels, on a time scale of a few hundred million years. Casts of gypsum and halite crystals and suggestions of solution collapse structures in early and mid-Precambrian sediments from evaporitic settings are all the remains that have been discovered to date. The casts are reassuring tid-bits of evidence in favor of the compositional constancy of sea water; but they can hardly be classed as convincing evidence.

The constancy or near-constancy of the silicate mineralogy of sedimentary rocks from the Paleozoic at least to the base of the Proterozoic, some 2·5 billion years ago also supports the notion that the composition of sea water has been very conservative. However, uncertainties still cloud the connection between the silicate mineralogy of sediments and the chemistry of sea water, and caution against assigning too much weight to the constancy of silicate mineralogy in pronouncements concerning the early history of sea water.

Even the earliest known sediments, those of the Barberton Mountain Land in South Africa are similar to later sediments in similar geologic settings. The mineralogy and chemistry of the Fig Tree shales are virtually indistinguishable from those of, say, the Lower Paleozoic Martinsburg formation in eastern Pennsylvania, and indicate that the intensity of chemical weathering some 3·4 billion years ago was already similar to that of the present day.

There are, however, some real differences between Precambrian and Phanerozoic sediments. The extensive bedded cherts of the Precambrian, and the frequently associated banded iron formations, seem to have no Phanerozoic analogs. The bedded cherts can be explained most readily in terms of the response of the silica content of sea water to the absence of organisms with siliceous tests. The concentration of dissolved silica in sea water today is almost certainly controlled by the silica demand of siliceous organisms. In their absence silica was probably removed as amorphous silica from sea water containing roughly 100 ppm. dissolved silica. Cold, shallow water areas and evaporitic settings must have been the preferred removal sites. The trapping and the superb preservation of microorganisms in silica gels deposited from sea water has earned for these cherts the well deserved title "The Amber of the Precambrian".

It is hard to gauge precisely the effect of a higher SiO_2 concentration on the concentration of the other components of sea water. It is likely that the pH was somewhat lower, perhaps a unit or so, but that otherwise the oceans were able to digest the river input of dissolved salts without major differences in composition. This is certainly suggested by the similarity between the Precambrian Bitter Springs evaporites and their Phanerozoic counterparts.

For some time now the Precambrian banded iron formations have served as delightful sources of controversy. Proponents of a volcanic origin have been baffled by the lack of equivalent iron formations in association with Phanerozoic volcanism. A rival hypothesis proposes that the oxygen content of the atmosphere was sufficiently low to permit the persistence of ferrous iron during weathering, and river transport, followed by the oxidation and precipitation of iron in nearshore marine settings. However this seems to require huge river deltas with a minimal detrital input for which there are no later counterparts.

A third hypothesis (BORCHERT, 1960), which I find rather attractive, suggests that during the earlier part of the Precambrian much of the ocean floor was somewhat reducing. Under slightly reducing conditions siderite is stable, rather than goethite on the one hand or the iron sulfides on the other. The concentration of iron in such bottom waters could well have been several parts per million. If upwelling currents brought these waters in contact with an atmosphere containing small amounts of free oxygen, the ferrous iron would probably have been oxidized and precipitated as hydrated ferric oxides. If precipitation took place in areas receiving little or no land detritus, such sediments would be extremely iron rich. In areas of amorphous silica precipitation all the components for producing siliceous iron formation would be on hand. If sufficient organic matter were buried with the hydrous iron oxides, reduction of ferric iron could have produced the siderite and sulfide facies of Precambrian iron formations. The isotopic data of PERRY and TAN (1971) for the Biwabik iron formation lend strong support to this point of view.

A smaller oxygen pressure in the atmosphere and a lower mean oxidation state of bottom waters and sediments would probably have had only minor effects on the major components of sea water. A rather highly reducing state almost certainly has existed at some time in the past, but the presence of carbonates and the evidence for extensive chemical weathering in sediments more than 3 billion years old show that a highly reduced ocean–atmosphere system must have been confined to the first billion years of Earth history.

In the absence of any remnants of sedimentary rocks from that distant shore of time our views of the chemistry of sea water must rely on a less than perfect understanding of the atmosphere–ocean–solid Earth system in the reducing mode, and on the still rather ill-defined biochemical requirements for the origin of life. A sizable oil slick, or at least a layer of insoluble organic compounds probably covered the surface of the ocean (LASAGA et al., 1971). The thickness of the slick is hard to estimate, but it could well have been on the order of 10 meters, and probably exerted a profound effect on the chemistry of the near surface marine environment.

In a curious way this excursion into the distant past turns out to be highly contemporary. I would hardly dare to suggest that true conservationism implies a return to the early state of the Earth and hence the accumulation of a thick marine oil slick. But the prebiotic state of the Earth does bring to mind what man hath wrought and is apt to wreak during the next few centuries. The burning of fossil fuels has been described as man's greatest geochemical experiment. It is likely that during the next 200 years some ten times the present content of carbon dioxide will be added to the atmosphere. Most of this CO_2 will find its way to the ocean, but it now looks as if the CO_2 content of the atmosphere will climb to something like five times its present pressure before the exhaustion of the supply of fossil fuels puts an end to the great experiment. The climatic impact of the added CO_2 is still difficult to evaluate, because the intensity of the greenhouse effect depends critically on the nature and extent of the cloud cover in an atmosphere containing several times the present quantity of CO_2. We obviously need a better understanding of cloud physics.

The addition of CO_2 to the oceans will reduce the supersaturation of the surface layers with respect to calcite. Upwelling water masses are most supersaturated with respect to calcite. During their passage across the surface of the ocean they can be expected to take up CO_2 from the atmosphere, to become progressively less supersaturated, and ultimately to become undersaturated with respect to calcite. The physiological effects of this process on the marine biota will probably be slight, but data supporting this conjecture would be extremely comforting.

Man's other geochemical experiments are on a smaller scale, but their effects on living things may well be more spectacular. Rivers draining heavily populated and industrialized areas have gradually turned into grand trunk sewers; the integrated damage per pound of a number of man made chemicals discharged into the oceans is extremely large. During the next decades a new level of care and of watchfulness is absolutely essential. The old slogan "Better living through Chemistry" is currently regarded with deep suspicion. We must never let it become an epitaph for a dead ocean.

REFERENCES

BORCHERT H. (1960) Genesis of marine sedimentary iron ores. *Trans. Inst. Min. Metall.* **69**, 261–277.

BRAITSCH O. (1962) *Entstehung und Stoffbestand der Salzlagerstätten*, vol. 3 of *Mineralogie und Petrographie in Einzeldarstellung*. Springer Verlag.

BROECKER W. S. (1970) Man's oxygen reserves. *Science* **168**, 1537–1538.

BROECKER W. S. (1971) A kinetic model for the chemical composition of sea water. *Quaternary Research* **1**, 188–207.

CHAVE K. E. (1960) Evidence on history of sea water from chemistry of deeper subsurface waters of ancient basins. *Bull. Amer. Assoc. Petr. Geol.* **44**, 357–370.

HOLLAND H. D. (1965) The history of ocean water and its effect on the chemistry of the atmosphere. *Proc. Nat. Acad. Sci.* **53**, 1173–1182.

HOLLAND H. D. (1972) Marine evaporites and the composition of sea water during the Phanerozoic. S.E.P.M. Symposium, Houston, 1971.

HOLLAND H. D. (1972) Ocean water, nutrients, and atmospheric oxygen; Proceedings of the International Association of Geochemistry and Cosmochemistry, Tokyo, 1970. In Press.

HOLSER W. T. (1966) Bromide geochemistry of salt rocks. *Second Symposium on Salt* (editor J. L. Rau), Vol 1, pp. 248–275, Northern Ohio Geological Society.

LASAGA A. C., HOLLAND H. D., and DWYER M. J. (1971) Primordial oil slick. *Science* **174**, 53–55.

LI T. H., TAKAHASHI T. and BROECKER W. S. (1969) The degree of saturation of $CaCO_3$ in the oceans. *J. Geophys. Res.* **74**, 5507–5525.

KRAMER J. R. (1965) History of sea water: Constant temperature pressure equilibrium models compared to fluid inclusion analyses. *Geochim. Cosmochim. Acta* **29**, 921–945.

MACKENZIE F. T. and GARRELS R. M. (1966a) Silica–bicarbonate balance in the ocean and early diagenesis. *J. Sed. Pet.* **36**, 1075–1084.

MACKENZIE F. T. and GARRELS R. M. (1966b) Chemical mass balance between rivers and oceans. *Amer. J. Sci.* **264**, 507–525.

PERRY E. C., JR. and TAN F. C. (1971) Significance of carbon isotope variations in carbonates from the Biwabik Iron Formation, Minnesota. Kiev Conference on iron formations.

RUBEY W. W. (1951) The geologic history of sea water—An attempt to state the problem. *Bull. Geol. Soc. Amer.* **62**, 1111–1148.

RUSSELL K. L. (1970) Geochemistry and halmyrolysis of clay minerals, Rio Ameca, Mexico. *Geochim. Cosmochim. Acta* **34**, 893–907.
SCHOPF J. W. (1968) Microflora of the Bitter Springs Formation, Late Precambrian, Central Australia. *J. Paleontol.* **42**, 651–688.
SIEVER R. (1968) Sedimentological consequences of a steady-state ocean-atmosphere. *Sedimentology* **11**, 5–29.
SILLÉN L. G. (1961) The physical chemistry of sea water. In *Oceanography* (editor M. Sears) pp. 549–581. American Association for the Advancement of Science, Publ. 67.
SILLÉN L. G. (1963) How has sea water got its present composition. *Svensk. Kem. Tidskr.* **75**, 161–177.
SILLÈN L. G. (1967a) The ocean as a chemical system. *Science* **156**, 1189–1197.
SILLÈN L. G. (1967b) Gibbs phase rule and marine sediments. In *Equilibrium Concepts in Natural Water Systems* (editor W. Stumm) Advanced Chemistry Series 67, Amer. Chem. Soc.
WEYL P. K. (1966) Environmental stability of the Earth's surface; chemical considerations. *Geochim. Cosmochim. Acta* **30**, 663–679.

Copyright © 1962 by the Geological Society of America

Reprinted from *Petrologic Studies — A Volume to Honor A. F. Buddington*, A. E. J. Engel, H. L. James, and B. F. Leonard, eds., Geological Society of America, Boulder, Colorado, 1962, pp. 447–477

Model for the Evolution of the Earth's Atmosphere

HEINRICH D. HOLLAND

Princeton University, Princeton, New Jersey

ABSTRACT

The history of the earth's atmosphere has been divided into three stages. The first covers the earliest part of earth history, prior to the formation of the core; during this time native iron was probably present in the upper part of the mantle. Volcanic gases ejected during the first stage probably contained a large amount of hydrogen, and the atmosphere was highly reduced. Carbon was probably present in the form of methane, nitrogen may have been present as ammonia, and free oxygen was absent in all but nonequilibrium quantities.

The second stage of atmospheric history began when iron was removed from the upper mantle. Volcanic gases became much less reducing and probably approached rather rapidly the present oxidation state of Hawaiian volcanic gases. The atmosphere responded to this change largely in terms of the chemistry of atmospheric carbon. Carbon dioxide took the place of methane and began to participate in the processes of weathering and chemical precipitation. Free oxygen was probably absent during the second stage except in trace, nonequilibrium quantities.

The third stage began when the rate of production of oxygen by photosynthesis exceeded the rate needed to oxidize injected volcanic gases completely. During this, the present, stage oxygen has probably accumulated at a roughly constant rate.

The earth's core was probably formed quite early, and the first stage of the atmosphere probably lasted less than 0.5 b.y. (billion years). The second stage may have continued until about 1.0 b.y. ago. The presence of detrital uraninite in mid-Precambrian sediments suggests that the second stage lasted at least until 2.0 b.y. ago, and the oxygen requirements of late Paleozoic insects suggest that a fairly high oxygen pressure obtained 0.3 b.y. ago.

INTRODUCTION

During the last decade our understanding of the evolution of the earth's atmosphere has been advanced very materially by the contributions of Urey (1952a; 1952b; 1959) and Rubey (1951; 1955). The hypothesis of a strongly reducing atmosphere during at least the earliest part of earth history has been established by Urey. The hypothesis of the gradual accumulation of the ocean and of excess volatiles in the hydrosphere and atmosphere has been virtually established by Rubey. Nevertheless, a good deal of uncertainty remains concerning the composition of the atmosphere between the earliest period and the late Paleozoic. Urey has pointed out that a highly reducing state may have prevailed during much of geologic time. The evidence for this suggestion came from the work of Thode *et al.* (1953) on the fractionation of sulfur isotopes in the Precambrian. Since that time Ault and Kulp (1959) have shown that isotopic fractionation of sulfur extends very far into the Precambrian, so that the basis for Urey's proposal has been removed.

Rubey (1955) on the other hand maintains that the excess volatiles now present at the earth's surface do not contain enough hydrogen to permit a methane-hydrogen atmosphere during geologic time. This evidence is not at all conclusive. If a methane-hydrogen atmosphere existed in the past, the free hydrogen and that bound to carbon would since have escaped. Rubey's excess volatiles must therefore be regarded as residual excess volatiles.

We are therefore left with only scant evidence for the chemistry of the atmosphere during much of geologic time. This paper is an attempt to present a unified model for the evolution of the atmosphere. The model must be regarded as a first approximation. However, it brings together a number of mutually supporting lines of evidence and explains some otherwise puzzling data.

FIRST STAGE OF THE ATMOSPHERE

There seems little question now that the constituents of the earth's atmosphere have been largely, if not wholly, evolved from the interior of the earth. The strong-

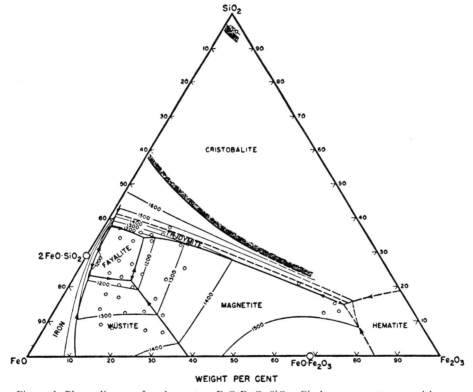

Figure 1. Phase diagram for the system FeO-Fe_2O_3-SiO_2. Circles represent compositions, as shown by analysis, of mixtures at liquidus temperatures. Light lines are liquidus isotherms at 100° C intervals, and heavy lines are boundary curves, with arrows pointing in the directions of falling temperatures. The tridymite-cristobalite boundary is dashed. Medium lines with stippling on one side indicate limit of two-liquid region (Muan, 1955).

est evidence for this is the very much smaller abundance of the rare gases in the earth than in average cosmic matter (Brown, 1952), silicon being the standard for comparison. The chemistry of the earliest atmosphere was therefore dependent largely on the nature of volcanic gases ejected at that time, and on the rate of escape of hydrogen from the upper atmosphere. There is still considerable controversy concerning the temperature during and soon after accretion of the earth (Urey, 1959). It seems likely, however, that iron was present in the outer parts of the earth. Sufficient melting must then have taken place to permit the accumulation of iron at the center of the earth. The oxidation state of magmas generated while iron was still present in the outer part of the mantle was buffered by the pair Fe^0-Fe^{+2}. We can obtain a first approximation of the partial pressure of oxygen in a vapor phase in equilibrium with magmas produced by partial melting in the upper mantle from Muan's (1955) data on the system FeO-Fe_2O_3-SiO_2 and from Muan and Osborn's (1956) data on the system MgO-FeO-Fe_2O_3-SiO_2.

Two of Muan's (1955) diagrams have been reproduced in Figures 1 and 2. Melts in equilibrium with solid iron, fayalite, and either tridymite or wüstite near

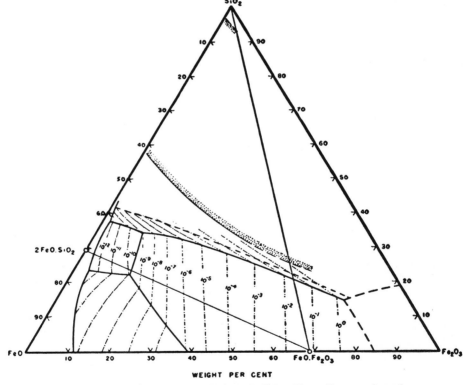

Figure 2. Phase diagram for the system FeO-Fe_2O_3-SiO_2. Heavy lines are boundary curves, and dash-dot lines are lines of equal O_2 pressures for points on liquidus surface. Lines with stippling on one side indicate limit of two-liquid region (Muan, 1955).

1200° C are in equilibrium with a vapor phase having an oxygen pressure of about $10^{-12.5}$ atm. The oxygen pressure defines the ratio of p_{H_2O} to p_{H_2} and of p_{CO_2} to p_{CO} since

$$(H_2)_g + \tfrac{1}{2}(O_2)_g \rightleftharpoons (H_2O)_g \qquad (1)$$

$$K_1 = \frac{p_{H_2O}}{p_{H_2} \cdot p_{O_2}^{1/2}}$$

and

$$(CO)_g + \tfrac{1}{2}(O_2)_g \rightleftharpoons (CO_2)_g \qquad (2)$$

$$K_2 = \frac{p_{CO_2}}{p_{CO} \cdot p_{O_2}^{1/2}}$$

Values of these equilibrium constants are listed in the Appendix. At 1200° C

$$K_1 = 10^{5.89} = \frac{p_{H_2O}}{p_{H_2} \cdot p_{O_2}^{1/2}}.$$

Thus, when $p_{O_2} = 10^{-12.5}$ atm

$$\frac{p_{H_2O}}{p_{H_2}} = 0.44.$$

Also

$$K_2 = 10^{5.47} = \frac{p_{CO_2}}{p_{CO} \cdot p_{O_2}^{1/2}}$$

and thus

$$\frac{p_{CO_2}}{p_{CO}} = 0.17.$$

In gases in equilibrium with such a melt, hydrogen is therefore roughly twice as abundant as water, and CO more than five times as abundant as CO_2. Methane would normally be a minor constituent. Its pressure would be determined by the equilibrium

$$(CO_2)_g + 4(H_2)_g \rightleftharpoons (CH_4)_g + 2(H_2O)_g, \qquad (3)$$

for which the equilibrium constant, K_3, is

$$K_3 = \frac{p_{CH_4} \cdot p_{H_2O}^2}{p_{CO_2} \cdot p_{H_2}^4}.$$

At 1200° C

$$K_3 = 10^{-3.15} = \left(\frac{p_{CH_4}}{p_{CO_2}}\right)\left(\frac{p_{H_2O}}{p_{H_2}}\right)^4 \cdot \frac{1}{p_{H_2O}^2}$$

$$\frac{p_{CH_4}}{p_{CO_2}} = 10^{-3.15} \cdot \frac{1}{(0.44)^4} \cdot p_{H_2O}^2$$

$$= 1.9 \times 10^{-2} p_{H_2O}^2.$$

TABLE 1. COMPARISON OF RUBEY'S (1951) "EXCESS VOLATILES" AND EATON AND
MURATA'S (1960) AVERAGE COMPOSITION OF HAWAIIAN VOLCANIC GASES

	Rubey (1951) (gm per 1000 gm H_2O)	Eaton and Murata (1960) (gm per 1000 gm H_2O)
Total C as CO_2	55	370
Cl	18	2.5
N	2.5	25
S	1.3	155
H	0.6	0.8
B, Br, Ar, F, etc.	0.2	1.1

The total pressure on ejection was probably greater than 1 and less than 10 atmospheres. Methane was therefore probably much less abundant than CO_2.

Rubey (1951) has estimated the total amount of water and the total amount of carbon as CO_2 which have been ejected from volcanoes and hot springs. If the early volcanoes were "average" in this respect

$$R = \frac{p_{H_2O}}{p_{CO} + p_{CO_2} + p_{CH_4}} \approx 43.$$

It is of interest to note that in Hawaiian gases today (Eaton and Murata, 1960)

$$R = \frac{p_{H_2O}}{p_{CO} + p_{CO_2} + p_{CH_4}} \approx 6.6.$$

Sulfur and its compounds were undoubtedly present in these early volcanic gases. In Hawaiian volcanic gases today SO_2 is the dominant molecule containing sulfur. Just as CO_2 and CO are more abundant relative to water in these gases than in the total "excess volatiles," so the rate of sulfur ejection is abnormally high relative to water. Rubey's (1951) results and the data of Eaton and Murata (1960) have been recast in Table 1 to point out these differences. It is difficult to assess the extent to which the Hawaiian samples are representative of volcanic gases through time. If they are representative, a large fraction of water now in the hydrosphere must have reached its present position by a route other than volcanic ejection. It may be that volatiles released at depth from cooling intrusives lose much of their sulfur as sulfides and much of their carbon as carbonates. On the other hand, there is no evidence that the very large quantities of sulfide and carbonate demanded by this alternative are present in the earth's crust. At present it therefore seems likely that the samples of Hawaiian gases studied by Eaton and Murata (1960) were abnormally enriched in compounds of carbon and sulfur.

Regardless of the ratio of sulfur and sulfur compounds to water in the earliest gases, H_2S should have been the dominant molecular species. The equilibrium constant K_4 for the reaction

$$(H_2S)_g + 2(H_2O)_g \rightleftharpoons (SO_2)_g + 3(H_2)_g \tag{4}$$

at 1200° C is

$$K_4^{1200°C} = \frac{p_{SO_2} \cdot p_{H_2}^3}{p_{H_2S} \cdot p_{H_2O}^2} = 10^{-3.29};$$

thus

$$\frac{p_{SO_2}}{p_{H_2S}} \approx 10^{-3.29} \cdot (0.44)^3 \cdot p_{H_2O}$$

$$\approx 10^{-4.36} \cdot p_{H_2O}.$$

At all reasonable values of the water vapor pressure, H_2S would therefore have far exceeded SO_2 in abundance. The equilibrium constant K_5 for the reaction

$$(H_2)_g + \tfrac{1}{2}(S_2)_g \rightleftharpoons (H_2S)_g \tag{5}$$

at 1200° C is

$$K_5^{1200°C} = \frac{p_{H_2S}}{p_{H_2} \cdot p_{S_2}^{1/2}} = 10^{+0.61}.$$

Since hydrogen was probably the most abundant constituent of these gases, p_{H_2} would have been in the neighborhood of 1 to 10 atm. Therefore

$$p_{S_2} \approx 10^{-2} \cdot p_{H_2S}^2.$$

The H_2S pressure was probably well below 1 atm if the sulfur content of Hawaiian gases can be taken as a guide, so that normally p_{S_2} would have been smaller than p_{H_2S}.

We can conclude then, that volcanic gases in equilibrium with iron, fayalite, and either tridymite or wüstite at an oxygen pressure of $10^{-12.5}$ atm at 1200° C would consist largely of H_2, H_2O, CO, and H_2S. Carbon dioxide and sulfur molecules would be present in minor amounts, and the partial pressure of CH_4 and SO_2 would be very small.

Equivalent data for the system MgO-FeO-Fe_2O_3-SiO_2 are not available, but the effect of MgO on the oxygen pressure of a gas in equilibrium with a melt and solid iron would probably be well below one order of magnitude. As an indication that this is the case, consider the oxygen pressure at the ternary invariant point magnetite-fayalite-tridymite in the system FeO-Fe_2O_3-SiO_2 and at the corresponding quaternary invariant point magnesioferrite-olivine-pyroxene-tridymite in the system MgO-FeO-Fe_2O_3-SiO_2. The ternary point lies at 1140° C and at an oxygen pressure of $10^{-9.0}$ atm (Muan, 1955). At this temperature and p_{O_2}

$$\frac{p_{H_2O}}{p_{H_2}} = K_1^{1140°C} \cdot p_{O_2}^{1/2} = 10^{6.30} \times 10^{-4.50} = 10^{+1.80}.$$

The quaternary invariant point lies at 1255° C and at an oxygen pressure of $10^{-7.1}$ atm (Muan and Osborn, 1956). At this temperature and p_{O_2}

$$\frac{p_{H_2O}}{p_{H_2}} = K_1^{1255°C} \cdot p_{O_2}^{1/2} = 10^{5.57} \times 10^{-3.55} = 10^{+2.02}.$$

Thus volcanic gases ejected under these two sets of conditions have nearly the same ratio of water pressure to hydrogen pressure.

At temperatures above 1300° C the presence of magnesium affects the ratio

p_{H_2O}/p_{H_2} considerably. This effect is probably due to the presence of large amounts of MgO in the spinel phase.

Undoubtedly, the presence of other components will modify gas compositions. We cannot determine the magnitude of these effects at present. They are probably small, because the concentration of metal oxides other than FeO, Fe_2O_3, MgO, and SiO_2 in a homogeneous earth based on a chondrite model is only about 8 per cent (Urey and Craig, 1953). The present oxidation state of volcanic gases can be explained reasonably well on the basis of data on the system MgO-FeO-Fe_2O_3-SiO_2 alone. It seems probable, then, that the composition of gases in equilibrium with a melt in the system FeO-Fe_2O_3-SiO_2 and with solid iron, fayalite, and either tridymite or wüstite, is not very different from the composition of volcanic gases in the pre-core stage of the earth.

If the temperature structure of the pre-core (first stage) atmosphere was similar to the present-day structure, much of the water contained in these gases would condense on cooling. The fate of H_2, CO, CO_2, H_2S, and CH_4 would depend in large part on the escape rate of hydrogen from the upper atmosphere, on the rate of reaction of CO and CO_2 with H_2, and on the rate of reaction of H_2S with iron silicates. Reaction rates in the upper atmosphere are fast today, and it seems very unlikely that it would have taken more than 1 year to approach chemical equilibrium in the pre-core atmosphere. Such reaction rates would have been much faster than H_2 escape from the exosphere. Thus equilibrium in the reactions

$$(CO)_g + 3(H_2)_g \rightleftharpoons (CH_4)_g + (H_2O)_g \tag{6}$$

and

$$(CO_2)_g + 4(H_2)_g \rightleftharpoons (CH_4)_g + 2(H_2O)_g \tag{7}$$

would have determined the final distribution of carbon among the three compounds CO, CO_2 and CH_4. The value of K_6 at 25° C is

$$K_6^{25°C} = 10^{24.90} = \frac{p_{CH_4} \cdot p_{H_2O}}{p_{CO} \cdot p_{H_2}^3}.$$

Thus

$$\frac{p_{CH_4}}{p_{CO}} = 10^{24.90} \cdot \frac{p_{H_2}^3}{p_{H_2O}}.$$

The water pressure at 25° C will have approximately the value in equilibrium with an ocean. Thus at 25° C for all values of p_{H_2} greater than ca. 10^{-6} atm, the ratio p_{CH_4}/p_{CO} would be greater than 1. The same will be true of the ratio p_{CH_4}/p_{CO_2}.

The hydrogen pressure in this primeval atmosphere is difficult to calculate. However, broad limits can be set on its value. The total amount of water evolved during geologic time (Rubey, 1951) is about 1.6×10^{24} gm. The mean rate of evolution is therefore about 3.6×10^{14} gm/year. Since most of the water today is in the ocean, recycling through the sedimentary-metamorphic-igneous sequence has presumably not been of great importance. If evolution of volcanic gases took place at an average rate, the hydrogen would have been evolved approximately at the rate of

$$\frac{3.6 \times 10^{14}}{0.44} \times \frac{2}{18} = 9 \times 10^{13} \text{ gm/year.}$$

The actual rate was probably no smaller than this. It was probably no greater than this by more than a factor of 10; otherwise all the water in the oceans would have appeared in less than 5×10^8 years. Thus hydrogen evolution took place at a rate of about $10^{14.5 \pm 0.5}$ gm/year. In these gases

$$p_{H_2} \gg 3 p_{CO} + 4 p_{CO_2}.$$

Therefore the production of methane would not decrease markedly the hydrogen available for escape. The amount of hydrogen in the atmosphere at any time after equilibrium had been established would have been defined by the equation

$$N_{H_2} = \frac{1}{\lambda_{H_2}} \cdot \frac{dN_{H_2}}{dt},$$

where dN_{H_2}/dt equals the rate of introduction, and

$$\lambda_{H_2} = \frac{0.693}{(T_{1/2})_{H_2}}$$

where $(T_{1/2})_{H_2}$ is the half-life for hydrogen escape from the atmosphere. Today $(T_{1/2})_{H_2}$ must be less than 5×10^5, the escape half-life of He^3 (Damon and Kulp, 1958). It is probably greater than 10^3 years (Spitzer, 1952). It is very difficult to estimate the temperature structure in the early atmosphere of the earth. Zabriskie (1960, Ph.D. thesis, Princeton Univ., Princeton, N. J.) has discussed the similar problem of the present atmosphere of Jupiter. It seems unlikely, however, that $(T_{1/2})_{H_2}$ was less than 10^3 or more than 10^6 years in the early atmosphere. Thus the minimum weight of H$_2$ in the atmosphere would have been

$$(N_{H_2})_{\min} \approx \frac{10^3}{0.693} \times 10^{14} \approx 1.4 \times 10^{17} \text{ gm,}$$

and the maximum weight

$$(N_{H_2})_{\max} \approx \frac{10^6}{0.693} \times 10^{15} \approx 1.4 \times 10^{21} \text{ gm.}$$

These values correspond to a minimum value for p_{H_2} of 2.7×10^{-5} and a maximum of 0.27 atmosphere; it seems unlikely that the hydrogen pressure ever exceeded one atm.

If the hydrogen pressure was close to the lower limit during this period of earth history, then CO$_2$ was already an important part of the atmosphere. This can be seen from the relationships in Figure 3. Line I represents the relationship between p_{CH_4} and p_{H_2} in equilibrium with graphite at 25° C. Its position is determined by the equilibrium

$$\langle C \rangle_c + 2(H_2)_g \rightleftharpoons (CH_4)_g$$

$$K_8^{25°C} = \frac{p_{CH_4}}{p_{H_2}^2} = 10^{8.90}. \tag{8}$$

Line II represents the equilibrium at 25° C between p_{CO_2} and p_{H_2} in equilibrium with graphite and water-saturated air.

$$\langle C \rangle_c + 2(H_2O)_g \rightleftharpoons (CO_2)_g + 2(H_2)_g$$

$$K_9^{25°C} = \frac{p_{H_2}^2 \cdot p_{CO_2}}{p_{H_2O}^2} = 10^{-11.01}. \tag{9}$$

It can be shown that p_{CO} is always much less than 10^{-6} atm.

If now p_{H_2} is 3×10^{-6} atm, p_{CH_4} is nearly equal to p_{CO_2}, and the atmosphere is in equilibrium with graphite. If the prevailing hydrogen pressure had an intermediate value of, say, 10^{-3} atm, methane would be the dominant gas. Under these conditions, the methane pressure would be determined by the total amount of $CO + CO_2 + CH_4$ evolved up to that time. The maximum pressure might be obtained by converting all C and CO_2 now in the crust to methane. If we again take Rubey's (1951) figure for the total amount of C as CO_2,

$$(p_{CH_4})_{max} \approx \frac{920 \times 10^{20} \times 16}{44 \times 5.1 \times 10^{21}} \approx 6.6 \text{ atm.}$$

This is obviously a very strong maximum.

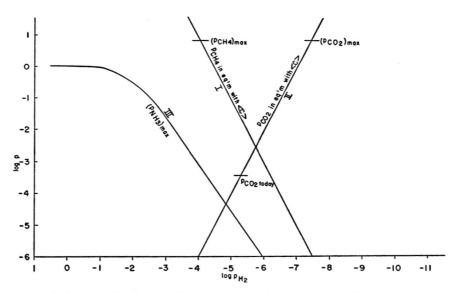

Figure 3. Relationship between H_2 pressure and the pressure of CH_4, Co_2, and NH_3 during the first stage of the earth's atmosphere

Nitrogen would be evolved from volcanoes, together with the compounds already discussed. The reaction with hydrogen to ammonia

$$(N_2)_g + 3(H_2)_g \rightleftharpoons 2(NH_3)_g$$

$$K_{10}^{25°C} = \frac{p_{NH_3}^2}{p_{N_2} \cdot p_{H_2}^3} = 10^{5.83} \tag{10}$$

depends on the hydrogen pressure and on the total amount of nitrogen evolved. The maximum ammonia pressure at a given p_{H_2} increases with the total amount of nitrogen in the atmosphere. Since the atmosphere has served nearly as a simple accumulator for nitrogen during geologic time, the maximum amount of ammonia at any time is defined by p_{H_2} and $(p_{N_2} + p_{NH_3}) = 0.8$ atm. Curve III has been constructed under this assumption. It follows that ammonia may have been, but was probably not, a major constituent of the earliest atmosphere.

The H_2S pressure was probably determined largely by the balance between the rate of ejection from volcanoes and the rate of reaction of atmospheric H_2S with iron oxides and silicates to form pyrite or pyrrhotite. In this sense the situation of H_2S is similar to that of CO_2 in the present atmosphere.

The maximum H_2S pressure can be calculated like that of methane, on the basis of Rubey's estimate of "excess volatiles":

$$(p_{H_2S})_{max} \approx \frac{22 \times 10^{20} \times 34}{5.1 \times 10^{21} \times 32} \approx 0.46 \text{ atm.}$$

As in the case of methane, this is a very strong maximum. It seems most unlikely that p_{H_2S} was ever greater than 0.05 atm.

In discussing the pressure of methane, of nitrogen plus ammonia, and of hydrogen sulfide we have to set some value on the amount of degassing during this earliest stage of earth history. The time span was probably less than 0.5×10^9 years, because in the parent body (or bodies) of meteorites fractionation into silicate and metal phase apparently took place faster than this. The very large release of gravitational energy during settling of the metallic phase makes such a time span appear reasonable. If the rate of degassing was nearly average, less than 10 per cent of the total volume of volatiles would have evolved during this time. However, the rate of evolution of volatiles may have been much faster than the mean rate, so that little can be said on this basis concerning the total quantity of volatiles evolved during the pre-core period.

One line of evidence seems to offer hope of setting some limits to speculation concerning this matter. Kokubu, Mayeda, and Urey (1961) have recently confirmed the result that the deuterium-hydrogen ratio in the hydrosphere is about 5 per cent greater than that in volcanic waters. The explanations offered by them seem inadequate. To account for the observed effect by assuming a large amount of water dissociation followed by protium loss involves the production of embarrassingly large quantities of oxygen and disagrees with Urey's (1959) calculations on the rate of photodissociation of water and the subsequent escape of hydrogen. To account for the effect by assuming the presence of surface water "at the beginning" seems out of harmony with the data on rare gas abundances. On the other hand, the observed effect would follow reasonably logically from the model of volcanic gases and atmospheric chemistry proposed here for the pre-core stage of the atmosphere.

Hydrogen evolving from volcanoes would escape from the atmosphere. Unless the temperature in the atmosphere was very high, the deuterium escape rate would have been smaller than that of protium by at least a factor of 10 and possibly by a factor of 100. There would therefore have been a tendency to build up in the atmosphere a deuterium-protium ratio well in excess of that in volcanic gases. At equilibrium, and neglecting all other factors, the ratio of deuterium to protium in the atmosphere would have been equal to the ratio of deuterium to protium in volcanic gases, multiplied by their respective escape half-lives. However, such a ratio could not be maintained at present surface temperatures since water of isotopic composition equal to that of volcanic water would be cycled through the atmosphere. The amount of water passing annually through the atmosphere would have been determined by the surface area of the oceans at the time.

We can then write down the equations controlling the amount of deuterium, D, and protium, P, in the atmosphere and in the oceans at this time. If x_1 is the number of moles of protium in the atmosphere, x_2 the number of moles of deuterium in the atmosphere, y_1 the number of moles of protium in the oceans, and y_2 the number of moles of deuterium in the oceans, then

$$\frac{dx_1}{dt} = k_1 - \lambda_1 x_1 + \beta M \left(\frac{x_2}{x_1} - \frac{y_2}{y_1}\right)$$

$$\frac{dx_2}{dt} = k_2 - \lambda_2 x_2 - \beta M \left(\frac{x_2}{x_1} - \frac{y_2}{y_1}\right)$$

$$\frac{dy_1}{dt} = K_1 - \beta M \left(\frac{x_2}{x_1} - \frac{y_2}{y_1}\right)$$

$$\frac{dy_2}{dt} = K_2 + \beta M \left(\frac{x_2}{x_1} - \frac{y_2}{y_1}\right)$$

where

k_1 = number of moles of protium injected into the atmosphere per year
k_2 = number of moles of deuterium injected into the atmosphere per year
K_1 = number of moles of protium combined as water injected into the atmosphere per year
K_2 = number of moles of deuterium combined as water injected into the atmosphere per year
λ_1, λ_2 = escape rate of protium and deuterium from the atmosphere per year
M = number of moles of hydrogen circulated annually as water through the atmosphere
β = probability that a water molecule circulating through the atmosphere attains isotopic equilibrium with the atmosphere.

Since the D/P ratio in volcanic gases is very small, the effect of the exchange term on the protium content of the atmosphere and oceans can be neglected. Thus at equilibrium

$$x_1 \approx \frac{k_1}{\lambda_1}$$

and
$$y_1 \approx K_1 t;$$

the value of x_2 depends on the relative magnitude of the term $\lambda_2 x_2$ and of the exchange term

$$\beta M \left(\frac{x_2}{x_1} - \frac{y_2}{y_1} \right).$$

An approximate relationship between these terms can be obtained from a consideration of the maximum permissible difference between the isotopic composition of the atmosphere and ocean, since

$$\left(\frac{x_2}{x_1} - \frac{y_2}{y_1} \right) \leq \frac{k_2}{\beta M}.$$

Now

$$k_2 \approx k_1/6700 \approx \frac{10^{14.5 \pm 0.5}}{2 \times 6700} \approx 0.7 \times 10^{10.5 \pm 0.5}.$$

The total mass of water circulating annually through the atmosphere today is 4×10^{20} gm, corresponding to 2.2×10^{19} moles of hydrogen. The mass was almost certainly smaller in the pre-core stage of the earth. Perhaps a reasonable lower limit would be 10^{16} moles/year. Today the exchange in the upper atmosphere would probably be essentially complete. Mixing between the upper and lower atmosphere today is reasonably fast, so that β might well be between 0.01 and 1. In the absence of an ozone layer, dissociation due to ultraviolet rays would have been more important in the troposphere than today, so that the value of β may well have been greater during the first stage of the atmosphere. If we use the present-day value of M and an estimated value of 0.1 for β, we obtain

$$\frac{k_2}{\beta M} \approx \frac{10^{10}}{2 \times 10^{19} \times 10^{-1}}$$

$$\approx 0.5 \times 10^{-8}.$$

But

$$\frac{y_2}{y_1} \approx 1.4 \times 10^{-4}.$$

Therefore, under present-day conditions the difference

$$\left(\frac{x_2}{x_1} - \frac{y_2}{y_1} \right)$$

would be a very small fraction of y_2/y_1; i.e., the isotopic composition of the atmosphere would be essentially identical with that of the oceans. Only if the product βM was less than 1/1000 of the present value would any appreciable difference in the isotopic composition of these reservoirs have been possible.

Today isotopic exchange would therefore probably overshadow deuterium escape as a means of removing deuterium from the atmosphere. It follows that, as protium escapes from the atmosphere, deuterium enters the oceans, so that the D/P ratio in the atmosphere remains nearly the same as in the ocean reservoir. When the ratio λ_1/λ_2 is large, nearly all the deuterium enters the oceans. A steady state will thus be established in which the D/P ratio in the oceans is determined by the D/P ratio and by the ratio p_{H_2O}/p_{H_2} in volcanic gases. If, under these conditions

$$\frac{p_{H_2O}}{p_{H_2}} = 0.44,$$

then

$$\left(\frac{D}{P}\right)_{\text{oceans}} \approx \left(\frac{D}{P}\right)_{\text{volcanic gases}} \times \left(1 + \frac{1}{0.44}\right)$$

$$\approx 3.3 \left(\frac{D}{P}\right)_{\text{volcanic gases}}$$

At present, only a very small amount of hydrogen is ejected from volcanoes. This situation has probably prevailed for a long time. Any "heavy" water from the pre-core stage has, therefore, been diluted with water of volcanic composition. The amount of dilution required to reduce the difference between ocean water and volcanic water is shown in graphical form in Figure 4. If the ratio p_{H_2O}/p_{H_2} in

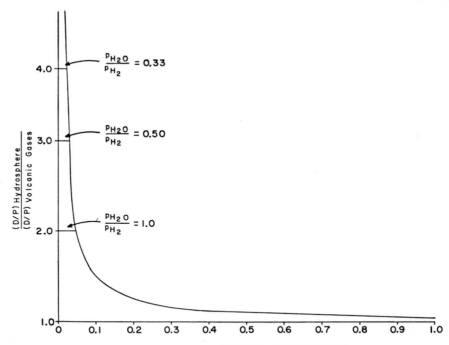

Figure 4. Ratio of the deuterium-protium ratio in the hydrosphere to that in volcanic gases as a function of the size of the hydrosphere

pre-core volcanic gases was 1.0, the fraction of water ejected at that time was about 5 per cent of the total amount now present in the hydrosphere. If the ratio p_{H_2O}/p_{H_2} was 0.50, the total amount ejected would have been only 2.5 per cent of the total amount now present. If the ratio had been 0.33, only 1.6 per cent of the present water would have been ejected in the pre-core stage.

There are far too many uncertainties in the propositions concerning the early atmosphere to permit us to take these results too seriously. If exchange between ocean water and atmospheric hydrogen were incomplete, then the fraction of water ejected in the pre-core stage would have been larger than the calculated value. Until more data become available on the probable temperature structure of the pre-core atmosphere, the matter must remain in an unsatisfactory state. What we can conclude is that the difference between the deuterium content of the hydrosphere and of volcanic gases could be explained satisfactorily in terms of the proposed mechanism.

If the proposed mechanism and atmospheric conditions are correct, approximately 1–5 per cent of the total water now present in the hydrosphere was ejected during this high-hydrogen stage. If the remainder of the water has been ejected at a nearly uniform rate, an age effect on the D/H ratio of ancient waters would probably be observed only in rocks more than 2 b.y. old. On the other hand, a strong age effect should be observed if a large fraction of the present hydrosphere has accumulated since the beginning of the Paleozoic Era.

SECOND STAGE

During the pre-core first stage, the chemistry of the atmosphere was probably determined in large part by the high percentage of hydrogen in volcanic gases. After the formation of the core, essentially all metallic iron was probably removed from at least the upper part of the mantle.

If we again take the system $MgO\text{-}FeO\text{-}Fe_2O_3\text{-}SiO_2$ as our model, the crystalline phases in the upper mantle would probably have become olivine, pyroxene, and magnetite. Other phases were probably present as well, and at depths of more than a few tens of kilometers polymorphic transformations would probably have altered the prevailing mineralogy. Nevertheless, it seems worth while to assess the consequences of a simplified model based on Muan and Osborn's (1956) data on the system $MgO\text{-}FeO\text{-}Fe_2O_3\text{-}SiO_2$. Magmas formed in the upper mantle by partial fusion would have an oxidation state defined by equilibrium between olivine, pyroxene, and magnetite:

$$olivine + O_2 \rightleftharpoons pyroxene + magnetite.$$

None of the phases would be pure. If we take Urey and Craig's (1953) average for the composition of chondritic meteorites as the best estimate of the composition of the mantle, the olivine would have had a composition near $Fo_{75}Fa_{25}$ and the pyroxene the approximate composition of $En_{75}Fs_{25}$. The magnetite would have contained at least some MgO, Cr_2O_3, and TiO_2 in solid solution. A precise statement concerning the oxidation state of gases in equilibrium with these mineral phases is not yet possible. It does seem unlikely, however, that p_{O_2} would differ markedly from $10^{-7.1}$ atm, its value at the quaternary invariant point in the system $MgO\text{-}FeO\text{-}Fe_2O_3\text{-}SiO_2$, if melting started near 1255° C.

At this temperature and oxygen pressure

$$\frac{p_{H_2O}}{p_{H_2}} \approx 105$$

and

$$\frac{p_{CO_2}}{p_{CO}} \approx 37.$$

In these gases water would therefore be much more abundant than hydrogen, and CO_2 much more abundant than CO. Methane should be virtually absent. SO_2 would be the dominant sulfur species. If the total amount of melting in the mantle has been small, the chemistry of volcanic gases might be expected to remain nearly constant at this composition with time. Table 2 compares the composition of volcanic gases predicted on the basis of this model with the composition of average volcanic gases from Kilauea (Eaton and Murata, 1960). The temperature of the associated lavas was between ca. 1150° C and 1200° C. The agreement is exceedingly good and supports the validity of the simplified model.

The model also demands that the oxidation state of volcanic gases has not changed appreciably with time. Perhaps the best indication of the oxidation state of the gases can be obtained from the Fe^{+3}/Fe^{+2} ratio in magmas (Kennedy, 1948). The safest indication of this ratio is probably the Fe^{+3}/Fe^{+2} ratio in the chilled borders of intrusive rocks. Table 3 contains analyses of chill-facies samples from Triassic diabases of New Jersey, U. S. A., and of Nipissing diabase from Ontario, Canada, 1.8 ± 0.3 b.y. old. There is no significant difference between the oxidation state of these samples, and of Hawaiian tholeiitic basalt (Eaton and Murata, 1960) and therefore none between the oxidation state of the gases that would have been evolved from extrusive phases of these units. Green and Poldervaart (1955) have pointed out the lack of a time effect on the Fe_2O_3 content of basaltic lavas, so that the hypothesis of reasonably constant oxidation state of volcanic gases

TABLE 2. ACTUAL AND PREDICTED OXIDATION STATE OF AVERAGE HAWAIIAN VOLCANIC GASES

Average Hawaiian volcanic gases (volume per cent)	Ratios found	Ratios predicted
H_2O .... 79.31 H_2 .... 0.58	$\frac{p_{H_2O}}{p_{H_2}} = 137$	105
CO_2 .... 11.61 CO 0.37	$\frac{p_{CO_2}}{p_{CO}} = 31$	37
SO_2 .... 6.48 S_2 .... 0.24 H_2S .... —		

TABLE 3. COMPOSITION OF THE CHILLED BORDERS OF SOME DIABASIC INTRUSIVE ROCKS

	Palisades diabase (H. H. Hess, unpub.)					Rocky Hill diabase (H. H. Hess, unpub.)	Nipissing diabase (Hriskevich, unpub.)	
	Pl 1	Pl 2	Pl 3	Pl 6	Pl 9	RH 2	312	166
SiO_2	52.11	51.98	52.41	51.74	52.14	52.21	51.89	51.98
Al_2O_3	14.41	14.26	13.91	14.31	14.51	14.59	14.79	14.69
Fe_2O_3	1.25	1.28	1.02	1.35	1.70	1.46	1.60	0.85
FeO	8.91	8.96	9.10	8.90	8.51	8.75	7.95	8.70
MgO	7.68	7.94	8.48	7.74	7.35	7.35	7.54	7.50
CaO	10.58	10.56	11.10	10.11	10.26	10.79	10.93	11.46
Na_2O	2.04	1.93	1.93	2.07	2.07	2.02	1.74	1.83
K_2O	0.73	0.80	0.54	1.11	0.86	0.63	0.89	0.51
H_2O^+	0.77	0.86	0.26	1.08	1.01	0.60	1.29	1.35
H_2O^-	0.10	0.11	0.13	0.22	0.25	0.16	0.05	0.08
P_2O_5	0.13	0.14	0.13	0.13	0.15	0.14	0.06	0.06
TiO_2	1.14	1.14	1.08	1.15	1.16	1.17	0.62	0.62
Cr_2O_3	—	—	—	—	—	—	—	—
MnO	0.18	0.18	0.18	0.18	0.20	0.17	0.18	0.18
NiO	—	—	—	—	—	—	—	—
CO_2							0.22	0.04
	100.03	100.14	100.27	100.09	100.17	100.04	99.75	99.85
	G. Kahan, analyst						H. Baadsgaard, analyst	

emitted during all but the pre-core stage of the earth's history appears to be tenable.

Thus, the chemistry of the atmosphere must have been controlled in large part by the composition of such volcanic gases after the formation of the earth's core. This should have been true at least until the appearance of organisms capable of photosynthesis. The time when life processes became important for atmospheric chemistry is not known with any degree of certainty, but mineralogical evidence cited below suggests that free oxygen appeared more recently than 2.0 b.y. ago. If this is correct, there was a long period of time between the formation of the core and the appearance of free oxygen during which the chemistry of the atmosphere was controlled by the injection of Hawaiian-type volcanic gases. This period of atmospheric history is here called the "second stage".

The composition of the atmosphere during the second stage can be defined reasonably well. There is ample evidence that between 2 and 3 b.y. ago water was present at the earth's surface as standing bodies, and the assumption of temperatures similar to those of the present seems justified. Water from volcanoes would then join the hydrosphere, and the water content of the atmosphere would depend in large measure on the surface temperature. Since the amount of hydrogen was very much smaller than that required to reduce CO and CO_2 to methane, carbon dioxide would have remained to participate in the weathering process much as it is doing today. The concentration of CO_2 in the atmosphere would have been controlled by the kinetics of weathering and would probably not have risen to values much above its present value. Carbonates of calcium and magnesium would have precipitated from the oceans.

CO is unstable in such an atmosphere in all but trace amounts. Numerous reactions probably took place in its destruction, but two deserve special mention:

$$(CO)_g + (H_2)_g \rightleftharpoons (C)_c + (H_2O)_g \tag{11}$$

and

$$(CO)_g + (H_2O)_g \rightleftharpoons (CO_2)_g + (H_2)_g . \tag{12}$$

Equation (11) is important at higher hydrogen pressures than equation (12). The crossover point is defined by the two equilibrium constants:

$$K_{11} = \frac{p_{H_2O}}{p_{H_2} \cdot p_{CO}}$$

$$K_{12} = \frac{p_{CO_2} \cdot p_{H_2}}{p_{CO} \cdot p_{H_2O}} .$$

When equilibrium is established in both reactions at 25° C,

$$\frac{K_{11}^{25°C}}{K_{12}^{25°C}} = \frac{10^{16.00}}{10^{5.00}} = \frac{p_{H_2O}^2}{p_{CO_2} \cdot p_{H_2}^2}$$

$$p_{H_2} = \frac{p_{H_2O}}{p_{CO_2}^{1/2}} \cdot 10^{-5.50}$$

If we use

$$p_{H_2O} = 3 \times 10^{-2} \text{ atm}$$
$$p_{CO_2} = 3 \times 10^{-4} \text{ atm}$$

then

$$p_{H_2} = 10^{-5.25} \text{ atm.}$$

At hydrogen pressures greater than $10^{-5.25}$ atm, CO and H_2 will react to form graphite and water. A slow shower of graphite may therefore have been associated with volcanic eruptions during a part of the second stage of atmospheric evolution.

One of the principal uncertainties in predictions concerning the chemistry of the atmosphere during the second state is introduced by the assumption concerning the SO_2 content of volcanic gases. If these gases contained as much SO_2 as Hawaiian gases, then p_{SO_2} would have been more than ten times greater than p_{H_2}. Hydrogen would have been used up by reaction (13):

$$(SO_2)_g + 3(H_2)_g \rightleftharpoons (H_2S)_g + 2(H_2O)_l ; \tag{13}$$

at 25° C

$$K_{13}^{25°C} = 10^{+36.4} = \frac{p_{H_2S}}{p_{SO_2} \cdot p_{H_2}^3}$$

and

$$p_{H_2} = 10^{-12.1} \left(\frac{p_{H_2S}}{p_{SO_2}}\right)^{1/3}.$$

Since p_{H_2S} would have been smaller than p_{SO_2}, the hydrogen pressure would have been less than $10^{-12.1}$ atm. (Fig. 5). At this H_2 pressure CO would react with water to form CO_2 and H_2, and the small amount of hydrogen thus generated would be used up in reacting with SO_2. Therefore if the SO_2 content of volcanic gases was similar to that of the Hawaiian gases, hydrogen was essentially absent from the atmosphere during the second stage.

On the other hand, it seems much more likely that the ratio of p_{H_2O} to p_{SO_2} in volcanic gases during this stage was equal to the value demanded by Rubey's (1951) "excess volatiles". Then p_{SO_2} would have been less than one tenth of p_{H_2}. Under such circumstances the rate of escape of H_2 from the atmosphere probably was the most important single factor in determining p_{H_2} in the atmosphere.

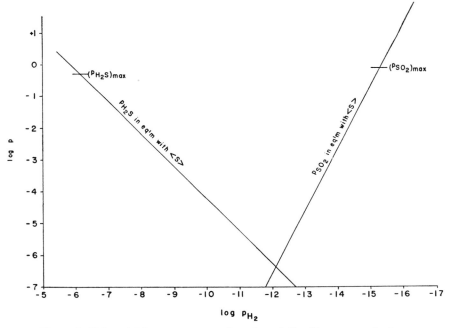

Figure 5. H_2S and SO_2 pressure as a function of the H_2 pressure during the second stage of the atmosphere

We can calculate the hydrogen pressure in the atmosphere on this basis. As before we have

$$N_{H_2} = \frac{1}{\lambda_{H_2}} \cdot \left(\frac{dN_{H_2}}{dt}\right).$$

If the escape half-life was 10^4 years, the rate of introduction of water vapor $10^{14.5}$ gm/year, and the ratio p_{H_2O}/p_{H_2} in volcanic gases equal to 100, then

$$p_{H_2} \approx \frac{10^4}{0.693} \cdot \frac{10^{14.5}}{9 \times 100} \cdot \frac{1}{5.1 \times 10^{21}} \text{ atm}$$

$$\approx 1 \times 10^{-6} \text{ atm}.$$

A variety of reactions involving CO and SO_2 probably modified the effect of escape on the hydrogen pressure. But it seems very likely that hydrogen was an extremely minor constituent of the atmosphere. Its actual partial pressure may therefore have been influenced strongly—as it is today—by the persistence of nonequilibrium conditions.

Photodissociation of water vapor in the upper atmosphere would have affected the hydrogen content of the atmosphere. Kuiper (1952, p. 312) finds that the present rate of oxygen production due to dissociation followed by hydrogen escape is about 2×10^{12} gm per year, an amount nearly sufficient to oxidize the calculated annual amount of hydrogen ejected from volcanoes. Recently Urey (1959) has again looked at this problem and concludes that Kuiper's figure is much too high. The photodissociation of water vapor followed by hydrogen escape has apparently affected the oxidation state of the atmosphere only very slightly.[1]

Oxidation of sediments was already taking place rather early in the Precambrian. At 25° C in the presence of hydrogen pressures less than $10^{-7.5}$ atm, the reaction

$$2\langle Fe_3O_4\rangle_c + (H_2O)_l \rightleftharpoons 3\langle Fe_2O_3\rangle_c + (H_2)_g \tag{14}$$

proceeds to the right. It can be doubted that this reaction would proceed at a geologically significant rate in the absence of free oxygen. Possibly, trace quantities of H_2O_2 produced by the effect of ultraviolet radiation on surface waters catalyzed the reaction.

The SO_2 would undoubtedly have participated in oxidizing reactions. In a very schematic way we could summarize these reactions by the equations:

$$11\langle Fe_2SiO_4\rangle_c + 4(SO_2)_g \rightleftharpoons 10\langle Fe_2O_3\rangle_c + 2\langle FeS_2\rangle_c + 11\langle SiO_2\rangle_c \tag{15}$$

and

$$\langle CaSiO_3\rangle_c + (SO_2)_g + (H_2O)_l \rightleftharpoons (Ca^{+2})_{aq} + (SO_4^{-2})_{aq} + \langle SiO_2\rangle_c + (H_2)_g. \tag{16}$$

The equilibrium pressure of SO_2 demanded by these equilibria is less than 10^{-10} atm. Undoubtedly, however, p_{SO_2} would have been greater than this, just as the

---

[1] This conclusion might require some modification if the increased intensity of ultraviolet radiation due to the absence of an ozone layer during the first and second stage resulted in a very much larger amount of H_2O dissociation in the lower parts of the atmosphere.

value of p_{CO_2} is greater today than the pressure in equilibrium with calcium silicates and calcium carbonate.

The most abundant constituent of the atmosphere during the second stage would probably have been nitrogen, which has been accumulating in the atmosphere somewhat like argon. Its abundance must have increased with time but could not have exceeded its present value. A probable composition for the atmosphere during the second stage is presented in Table 4. The total pressure was almost certainly less than 1 atmosphere.

Available geologic evidence suggests that the atmosphere was sufficiently oxidized to permit formation of calcite in the earliest known sediments (Armstrong, 1960). The oxidation state was such that hematite was the stable oxide of iron in equilibrium with the atmosphere at least as early as 2.0 ± 0.2 b.y. ago in Huronian sediments (James, 1951; 1954) and probably as early as Keewatin time (Klinger, 1956), more than 2.5 b.y. ago (Goldich et al., 1958). In order to appear, neither hematite nor calcite demands the presence of free oxygen, but the stability of hematite demands that the hydrogen pressure must have fallen very low. Both minerals are, of course, stable under present atmospheric conditions as well, so that their presence merely defines a lower limit for the degree of oxidation of the atmosphere during mid-Precambrian time.

The upper limit for the oxidation state of the atmosphere 2.0 to 2.5 b.y. ago is not so well defined. An indication that free oxygen was not present in the atmosphere comes from the apparent persistence of uraninite in detrital sediments formed during this period. Today uraninite, UO_2, normally weathers rapidly to a variety of hydrated oxides of U^{+6}. Oxidation is often followed by solution in surface waters and transport as a complex ion of U^{+6}. There is evidence that this sequence of events was not characteristic of the weathering of uraninite in the early Precambrian. Ramdohr (1958) has shown that the uranium-gold deposits of the Dominion Reef and Witwatersrand systems in South Africa, at Blind River in Canada, and at Serra de Jacobina in Brazil had a placer origin, and that they were later partially recrystallized during a period of metamorphism. Ramdohr's

TABLE 4. SOME PROBABLE LIMITS FOR THE COMPOSITION OF THE ATMOSPHERE DURING THE SECOND STAGE

Component	Pressure range (atm)
N_2	>0.01; <0.8
H_2O	3×10^{-3} to 3×10^{-2}
CO_2	ca. 3×10^{-4}
Ar	$>10^{-4}$; $<10^{-2}$
Ne	$>10^{-6}$; $<10^{-4}$
He	$>10^{-6}$; $<10^{-4}$
CH_4	$>10^{-6}$; $<10^{-4}$
NH_3	trace
SO_2	trace
H_2S	trace
O_2	nonequilibrium trace

conclusions, based mainly on evidence from ore microscopy, reinforce the conclusion reached by Liebenberg (1955; 1958) on similar grounds for the South African gold-uranium deposits. Liebenberg (1958) concludes that "there is no doubt as to the origin of the uraninite grains, the available evidence showing that the uraninite is one of the detrital minerals of a heavy-mineral suite which was deposited simultaneously with the other components of the conglomerates". Nel (1960) has recently summarized the pertinent field evidence and rules out a hydrothermal origin for the South African gold-uranium deposits.

Recently, Mair, Maynes, Patchett, and Russell (1960) have presented isotopic data on the age of monazite, zircon, uraninite, pyrite, pyrrhotite, sericite, and feldspar of the Blind River uranium deposits. They report that the apparent age of the uraninite (*ca.* 1.7 b.y.) is much less than that of the associated monazite and zircon (*ca.* 2.5 b.y.) and could be identical with the time of formation of the sediments. It is possible, however, that the later metamorphism of the sediments recorded by Mair *et al.* (1960) has reduced the apparent age of the uraninite from an original value in excess of 2.0 b.y. Goldich *et al.* (1958) suggest that an event 1.7 ± 0.1 b.y. ago marked the end of the Huronian period. This is supported by a 1.62 b.y. K/Ar age for a fresh biotite from sheared basal Huronian quartzite (Porter Township, between Blind River and Sudbury; R. K. Wanless, personal communication, 1960). Aldrich and Wetherill (1960) report a Rb/Sr age of 2.23 b.y. and a K/Ar age of 1.99 b.y. for pre-Cobalt biotite from Coleman Township, Ontario. These results indicate that Bruce Series sedimentation took place between 2.2 and 1.7 b.y. ago, and that the apparent age of the uraninite has been affected by post-depositional metamorphism.

A similar instance is recorded in the careful study by de Villiers, Burger, and Nicolaysen (1958) of galenas from the Witwatersrand System. They suggest very strongly that the 2.1 b.y. age found for the uraninites from this system was reduced from an initial value equal to or greater than 2.5 b.y. during a "chemical rejuvenation" of the ores 2.1 b.y. ago. This event did not produce a gross change in the morphology of the uraninite grains. Despite the arguments of Davidson (1957), it seems to me that the evidence for a sedimentary origin of these deposits is overwhelming. The only reasonable explanation for the observed difference in the behavior of uraninite seems to be that the oxygen content of the atmosphere during the formation of these deposits was much lower than at present.

Uraninite is an extremely rare mineral in sediments today. Only two occurrences have been described (Steacy, 1953; Zeschke, 1960); neither is similar to those of the South African uraninite deposits, but Zeschke's (1960) observations show that rivers can transport uraninite in the form of minute grains.

If we could assume that the uraninite of the Witwatersrand type was in equilibrium with the atmosphere during weathering and transportation, we could set an upper limit to the prevailing oxygen pressure in the atmosphere. The system uranium-oxygen has been studied most recently by Grønvold (1955) and by Blackburn (1958) and Blackburn *et al.* (1958). At high temperatures UO_2, U_3O_8, and UO_3 are the stable oxides of uranium. Below 300° C several phases in the range $UO_{2.20}$ and $UO_{2.40}$ can be produced, but these may be metastable and are of no particular concern here. The boundary of importance is that between the stability field of U_3O_8 and UO_3 as defined by the reaction

$$2 \langle U_3O_8 \rangle_c + (O_2)_g \rightleftharpoons 6 \langle UO_3 \rangle_c \qquad (17)$$

and by the equilibrium constant

$$K_{17} = \frac{1}{p_{O_2}}.$$

The value of this equilibrium constant can be calculated from the data of Blackburn *et al.* (1958). The older data of Coughlin (1954) are included in Table 5. The agreement between the two sets of data for UO_3 is satisfactory; the discrepancy for U_3O_8 is quite large. The values of p_{O_2} calculated from the two sets of data are shown in Table 6.

Both sets of data show that U_3O_8 can be in equilibrium with the atmosphere only when essentially no free oxygen is present. However, the rate of oxidation of U_3O_8 to UO_3 at oxygen pressures somewhat above the equilibrium pressure may be slow enough at 25° C to permit the accumulation of uraninite in placers. The values of p_{O_2} in Table 6 are therefore not definite maximum values for p_{O_2} during the formation of detrital uraninite deposits. They show, however, that U_3O_8 can oxidize to UO_3 even in the presence of minute traces of oxygen, and this suggests the virtual absence of atmospheric oxygen during the deposition of these sediments.

Davidson (1957) has argued against the possibility of a low oxygen pressure during the accumulation of the Witwatersrand System. He has pointed out that hematite occurs in the same assemblage of rocks and that this implies the availability of free atmospheric oxygen during sedimentation. The argument is not

TABLE 5. THERMOCHEMICAL DATA FOR THE FORMATION OF U_3O_8 AND UO_3 FROM U AND O_2 AT 25° C

	Blackburn *et al.* (1958)		Coughlin (1954)	
	$-\Delta H°$	$-\Delta S°$	$-\Delta H°$	$-\Delta S°$
U_3O_8	846.6	150.9	851.0	159.7
UO_3	290.1	59.3	290.5	60.0

TABLE 6. VALUES OF THE OXYGEN PRESSURE AT THE STABILITY BOUNDARY U_3O_8-UO_3 CALCULATED FROM THE DATA OF BLACKBURN *et al.* (1958) AND COUGHLIN (1954)

T (° C)	p_{O_2}	
	Blackburn	Coughlin
0	$10^{-25.8}$	$10^{-23.9}$
10	$10^{-24.5}$	$10^{-22.8}$
20	$10^{-23.3}$	$10^{-21.7}$
30	$10^{-22.1}$	$10^{-20.7}$

valid, because the equilibrium constants of reactions (14) and (17) show that there is a range of oxygen pressure (10^{-72} to 10^{-21}) in which U_3O_8 and Fe_2O_3 are stable together.

The conclusion that atmospheric oxygen was virtually absent during mid-Precambrian time could be checked by studying the oxidation state of other metals. Manganese is a suitable example since three oxidation states are represented in manganese minerals. Data for oxygen pressures at the stability boundary of lower oxides and silicates of manganese have been published by Muan (1959) and by Hahn and Muan (1960). Klingsberg and Roy (1959) have studied some of the reactions in the Mn-O-H system and have reported recently (Klingsberg and Roy, 1960) on the univariant equilibrium curves between Mn_3O_4 and Mn_2O_3 and between Mn_2O_3 and β-MnO_2 (pyrolusite). Goldsmith and Graf (1957) have studied the decomposition of rhodochrosite.

If Klingsberg and Roy's (1960) data are extrapolated to 25° C, the boundary between the stability field of bixbyite and pyrolusite lies at an oxygen pressure of about 10^{-3} atm. This figure may be uncertain by one or two orders of magnitude. Nevertheless, the bixbyite-pyrolusite boundary lies at the highest oxygen pressure of any of the mineral pairs so far considered and should prove a sensitive indicator of lower oxygen pressures in the Precambrian. Unfortunately Precambrian manganese deposits have normally been recrystallized so thoroughly by subsequent metamorphism that the original mineralogy has been obscured. The relationships in the system Mn-O-C-H are extremely pertinent to the problem of the history of the earth's atmosphere, and a further search for suitable, unmetamorphosed deposits seems amply warranted.

Several possible oxygen barometers can also be built around the oxidation of S^{-2} to SO_4^{-2}. For example, the equilibrium between galena and anglesite

$$\langle PbS \rangle_c + 2(O_2)_g \rightleftharpoons \langle PbSO_4 \rangle_c \tag{18}$$

depends on the prevailing oxygen pressure. The equilibria between sphalerite and zinkosite ($ZnSO_4$) or goslarite ($ZnSO_4 \cdot 7H_2O$) and the equilibrium between pyrrhotite and melanterite ($FeSO_4 \cdot 7H_2O$) are similarly a function of the oxygen pressure, although the partial pressure of water also influences the stability of the hydrates.

Kellogg and Basu (1960) have recently re-examined the thermodynamic properties of the system Pb-S-O. The free energy of reaction for the oxidation of one mole of galena to anglesite at 25° C according to these authors is -172.1 kcal, and the equilibrium p_{O_2} at this temperature is therefore 10^{-63} atm. In the presence of liquid water p_{H_2} would be $10^{-10.1}$ atm. For calculating the equilibrium between sphalerite and zinkosite we can take the data of Kubaschewski and Evans (1955), since the redetermination of the free energy of formation of ZnS by Curlook and Pidgeon (1958) is in reasonable agreement with that of the earlier compilation. A calculation similar to that for the galena-anglesite equilibrium shows that sphalerite and zinkosite are in equilibrium at 25° C at an oxygen pressure of 10^{-59} atm, and in equilibrium with water at a p_{H_2} of $10^{-12.1}$ atm.

These results show that magnetite oxidizes to hematite at a lower oxygen pressure than galena to anglesite or sphalerite to zinkosite, but that galena is oxidized to anglesite and sphalerite to zinkosite before U_3O_8 is oxidized to UO_3. Perhaps this explains the presence of a normal "hydrothermal" assemblage of minerals in the uranium-gold ores of Witwatersrand type. Galena and sphalerite may well have been present as detrital minerals in these sediments. Rounded pebbles of pyrite are certainly common, and there may well have been other sulfides which are today oxidized rapidly at the outcrop of ore deposits. During metamorphism these sulfides have been partially or wholly recrystallized. I believe that the hypothesis of the virtual absence of oxygen during much of the Precambrian would solve the mineralogic problem of these deposits that has been posed by the proponents of a hydrothermal origin. The search for detrital sulfides in less metamorphosed sedimentary rocks more than 1.5 b.y. old should prove fruitful in this connection.

The evidence thus suggests that between about 1.8 and 2.5 b.y. ago the atmosphere had a composition within the range predicted in Table 4 for the second stage of the evolution of the earth's atmosphere. There are no geologic data to check the proposal that the second stage started much before 2.5 b.y. ago. The fact that detrital uraninite persisted during mid-Precambrian time suggests that oxidation due to the photodissociation of water vapor followed by hydrogen escape was not sufficiently rapid to permit the accumulation of free oxygen in the atmosphere. This observation also suggests that the combined oxidative effect of the photodissociation of water vapor and of photosynthesis was not sufficiently pronounced to yield appreciable amounts of free atmospheric oxygen. This does not, of course, demand that life originated more recently than 1.8 b.y. ago but only that the rate of oxygen production was insufficient to overcome the reducing effect of the introduction of volcanic gases and of the loss of oxygen due to the oxidation of surface materials. The end of the second stage of atmospheric evolution can be set as the time when the combined effect of photodissociation of water vapor and photosynthesis first produced free oxygen in equilibrium amounts. If the evidence from detrital uraninites is valid, this took place more recently than about 1.8 b.y. ago.

THIRD STAGE

The beginning of the third stage of the history of the atmosphere has been defined as the time when the combined rate of oxygen production due to photosynthesis and photodissociation of water vapor exceeded the rate of consumption of oxygen by the oxidation of volcanic gases, organic remains, and sediments and rocks exposed to weathering. The accumulation of free oxygen has been contingent on production of a continuing surplus of oxygen, and the present amount of free oxygen must have accumulated very gradually. At the moment, we can set only rather broad limits on the variation of p_{O_2} with time since the beginning of the third stage.

A balance sheet for the production of atmospheric oxygen and for its present distribution is a convenient starting point. The most recent compilation of this kind (Hutchinson, 1954) must now be modified and can be expanded (Table 7). Oxygen production by photosynthesis followed by incomplete decay of organic

TABLE 7. PRODUCTION AND USE OF OXYGEN

Total estimated production	
By photosynthesis in excess of decay	181×10^{20} gm
By photodissociation of water vapor followed by hydrogen escape	1
Total	182×10^{20} gm
Total estimated use	
Oxidation of ferrous iron to ferric iron during weathering	14
Oxidation of S^{-2} to SO_4^{-2} during weathering	12
Oxidation of volcanic gases	
CO to CO_2	10
SO_2 to SO_3	11
H_2 to H_2O	<150
Total	$<197 \times 10^{20}$ gm
Free in atmosphere	$\sim 12 \times 10^{20}$ gm

matter appears to be much more important than oxygen production by photodissociation of water vapor followed by hydrogen escape. The estimate for the former source comes from Rubey's (1951) estimate of the total amount of organic carbon in sediments. The estimate for photolysis of water vapor is taken from Urey's (1959) analysis.

The estimate of the use of oxygen in the oxidation of Fe^{+2} to Fe^{+3} is taken from Hutchinson's (1954) table. The oxygen consumption in the oxidation of S^{-2} to SO_4^{-2}, CO to CO_2, SO_2 to SO_3, and H_2 to H_2O is based on Rubey's (1951) estimate of total excess C, excess S, and H_2O and on the oxidation state of volcanic gases predicted for all but the first stage of the earth's atmosphere. The consumption of oxygen in the oxidation of H_2 has been computed assuming no loss of H_2 from the upper atmosphere, and therefore must approximate an upper limit.

Despite its many shortcomings, Table 7 points out important relationships, particularly the relatively small amount of oxygen now in the atmosphere. Further, it seems that the oxidation of H_2 to H_2O has been a major user of oxygen and that the oxidation of Fe^{+2}, S^{-2}, CO, and SO_2 has played a rather subordinate role.

Production and use can be balanced in Table 7 by permitting some hydrogen escape from the upper atmosphere, but no reliable figure for the mean percentage of hydrogen escape can be obtained because the estimate both of the total amount of organic carbon in sediments and of the H_2O/H_2 ratio in volcanic gases is uncertain by a factor of 2. The data do however rule out the idea that the oceans have been produced by the oxidation of hydrogen captured by the earth from interplanetary space. About $12{,}000 \times 10^{20}$ gm of oxygen would have been needed for this purpose.

The time dependence of the oxygen content of the atmosphere during the third stage is difficult to estimate accurately, chiefly because the rate of accumulation of oxygen was determined by the small difference between two much faster rates.

Nevertheless, there is evidence that the oxygen content of the atmosphere increased essentially linearly with time. The rate of oxygen production by photosynthesis in excess of plant decay during a particular period is reflected in the carbon content of sediments deposited during that period. There seems to be no well-defined trend of the carbon content in sediments of similar origin (Pettijohn, 1943). This suggests that the rate of oxygen production by this mechanism has been roughly constant during the third stage. Since there is no good reason to suspect variations in the rate of photodissociation of water vapor in the upper atmosphere during the third stage, the total rate of oxygen production during this time was probably reasonably constant.

The rate of consumption of oxygen in the past was determined largely by the intensity of volcanic activity. This in turn was related to the intensity of orogenic activity. No definitive data on the time dependence of the intensity of orogenic activity on a world-wide basis are available. Certainly the intensity has fluctuated with time, but it seems unlikely that there has been a continuing trend in this rate for a large part of the third stage. Perhaps the best indication of this is the reasonably steady rate of precipitation of calcium and magnesium carbonates, at least since the beginning of the Paleozoic Era. The rate of precipitation of these carbonates is directly related to the rate of volcanic activity. The time lag between changes in volcanic activity and carbonate sedimentation is geologically insignificant, so that a precise determination of past volcanic activity is possible, at least in principle, if the contribution of CO_2 from weathering is known.

The hypothesis of a rate of oxygen production essentially constant when taken over periods on the order of 100 million years seems attractive. The duration of the third stage is not known and is difficult to assess. Certainly the oxygen pressure in the atmosphere must have been an appreciable fraction of its present value when land animals evolved during late Paleozoic time. The oxygen demands of the large Carboniferous insects cannot be determined accurately, but it seems most unlikely that p_{O_2} could have been less than one tenth of its present value during their lifetime. Unfortunately this restriction on the possible range of p_{O_2} is not particularly helpful in limiting our thinking on the time of origin of the third stage. If stage three started 1.5 b.y. ago, p_{O_2} would have been about 80 per cent of its present value during Carboniferous time; if the start of stage three was only 0.5 b.y. ago, p_{O_2} would have been about 50 per cent of its present value during Carboniferous time, provided the oxygen pressure increased at a linear rate. It seems to me that the beginning of stage three cannot at present be bracketed more closely than between 1.8 and 0.5 b.y. ago.

During the accumulation of free oxygen, the nitrogen, neon, argon, krypton, and xenon content of the atmosphere must have approached its present value gradually. The other constituents should have been present in about their present concentration.

SUMMARY, A CRITIQUE AND SOME IMPLICATIONS

The history of the atmosphere has been divided into three stages. During the first, the input into the atmosphere consisted largely of volcanic gases in equilibrium with magmas generated in a mantle which still contained metallic iron. The partial pressure of oxygen in such gases was probably about $10^{-12.5}$ atm, and the temperature on ejection from volcanoes on the order of 1200° C. These gases consisted largely of H_2, H_2O, and CO, with minor amounts of N_2, CO_2, and H_2S. Hydrogen was probably the most important constituent. On cooling, the CO and CO_2 reacted with H_2 to form methane. Nitrogen may have reacted to form ammonia, provided the rate of escape of hydrogen from the atmosphere was sufficiently slow to permit the build-up of an appreciable hydrogen pressure. If surface temperatures were similar to those at present, nearly all the water condensed, and the main constituents of the atmosphere were CH_4 and H_2 (Table 8).

The duration of the first stage was determined by the time that elapsed between the accretion of the earth and the removal of metallic iron from the upper mantle. It seems unlikely that this process took more than 0.5 b.y. After the removal of the iron phase, crystallization in the upper mantle probably proceeded until olivine, pyroxene, and magnetite were among the mineral phases. Subsequent melting and volcanism was accompanied by the ejection of gases much more highly oxidized that those of the first stage. If the available data on the system MgO-FeO-Fe_2O_3-SiO_2 apply to this situation, the oxidation state of these gases was very similar to that of Hawaiian volcanic gases today. It seems likely, therefore, that the oxidation state of volcanic gases has not changed a great deal since the beginning of the second stage.

Water was the dominant component in these volcanic gases; CO_2, CO, H_2, SO_2, and N_2 were minor constituents. The atmosphere during the second stage contained largely N_2 with minor amounts of CO_2 and H_2O (Table 8). Free oxygen was not present in equilibrium amounts, since the rate of oxygen production by

TABLE 8. SUMMARY OF DATA ON THE PROBABLE CHEMICAL COMPOSITION OF THE ATMOSPHERE DURING STAGES 1, 2, AND 3

	Stage 1	Stage 2	Stage 3
Major components $P > 10^{-2}$ atm	CH_4 H_2 (?)	N_2	N_2 O_2
Minor components $10^{-2} < P < 10^{-4}$ atm	H_2 (?) H_2O N_2 H_2S NH_3 Ar	H_2O CO_2 Ar	Ar H_2O CO_2
Trace components $10^{-4} < P < 10^{-6}$ atm	He	Ne He CH_4 NH_3 (?) SO_2 (?) H_2S (?)	Ne He CH_4 Kr

photodissociation was less than that required to oxidize volcanic gases introduced into the atmosphere.

The second stage ended when oxygen input exceeded oxygen use; that is, when the combined rate of oxygen production by photosynthesis, followed by burial of organic carbon with sediments, and by photodissociation of water vapor, exceeded the rate of use in oxidizing surface rocks and the products of volcanoes and hot springs. The time of transition from the second to the third stage is not yet well determined. Rather scant evidence from the mineralogy of Precambrian sedimentary uranium deposits suggests that free oxygen in appreciable amounts was not present in the atmosphere *ca.* 1.8 b.y. ago. The oxygen requirements of Paleozoic land animals suggest that the oxygen content of the atmosphere at the end of the Paleozoic Era was a large fraction of its present value. Thus the second stage probably ended between 0.5 and 1.8 b.y. ago.

During the third stage oxygen gradually accumulated in the atmosphere. It is difficult to be sure of the time function describing the oxygen pressure. The increase may well have been roughly linear, since there is little evidence that the rate of volcanic activity and the rate of burial of carbon with sediments differed markedly in successive time intervals of 100 m.y.

The validity of the proposed model for the development of the atmosphere depends on the validity of the assumptions used in its construction. Chief among these are the assumptions that the data on the system MgO-FeO-Fe_2O_3-SiO_2 can be applied directly to melting phenomena in the upper mantle, and that the temperature structure of the atmosphere has always been similar to that of the present day. Phase studies at high pressures and in systems approaching more closely the composition of the upper mantle are needed to assess the validity of the first assumption.

It should soon be possible to check the second assumption, at least in semi-quantitative fashion, by applying to this problem the results of recent data on the chemistry of the upper atmosphere of the earth, and possibly data yet to come on the temperature structure of the atmosphere of Mars, Venus, and the major planets.

If the proposed model is nearly correct, a highly reducing atmosphere existed for only a small fraction of earth history. If the processes leading to the first forms of life required such a highly reducing environment, then life on earth must be a very ancient phenomenon. There seems to be no evidence at present to suggest that life did not originate more than 3 b.y. ago.

The proposed model suggests that the chemistry of the oceans has not changed a great deal since the beginning of the second stage in early Precambrian time. This model suggests a history very different from that proposed by Krynine (1960) and is in essential agreement with that proposed by Rubey (1951).

ACKNOWLEDGMENTS

The growth of the ideas in this paper has been stimulated by discussions with members of the staff at Princeton. In particular, I thank Professor A. G. Fischer for guidance regarding biologic and paleontologic matters, Professor H. H. Hess for suggestions regarding the oxidation state of magmas, and Professor E. Sampson for help with the section on uranium deposits. Professor A. F. Buddington's comments were always helpful, and this paper is dedicated to him with deep affection.

REFERENCES CITED

Aldrich, L. T., and Wetherill, G. W., 1960, Rb-Sr and K-A ages of rocks in Ontario and northern Minnesota: Jour. Geophys. Research, v. 65, p. 337–340
Armstrong, H. S., 1960, Marbles in the "Archean" of the southern Canadian shield: 21st Internat. Geol. Cong. (Norden) Rept., pt. 9, p. 7–20
Ault, W. U., and Kulp, J. L., 1959, Isotopic geochemistry of sulphur: Geochim. et Cosmochim. Acta, v. 16, p. 201–235
Blackburn, P. E., 1958, Oxygen dissociation pressures over uranium oxides: Jour. Phys. Chemistry, v. 62, p. 897–902
——— Weissbart, J., and Gulbransen, E. A., 1958, Oxidation of uranium dioxide: Jour. Phys. Chemistry, v. 62, p. 902–908
Brown, H., 1952, Rare gases and the formation of the earth's atmosphere, p. 258–266 *in* Kuiper, G. P., *Editor*, The atmospheres of the earth and planets, 2d ed.: Chicago, Univ. Chicago Press, 434 p.
Coughlin, J. P., 1954, Contributions to the data on theoretical metallurgy. XII. Heats and free energies of formation of inorganic oxides: U. S. Bur. Mines Bull. 542, 80 p.
Curloook, W., and Pidgeon, L. M., 1958, Determination of the standard free energies of formation of zinc sulfide and magnesium sulfide: Am. Inst. Min. Met. Eng. Trans., v. 212, p. 671–676
Damon, P. E., and Kulp, J. L., 1958, Inert gases and the evolution of the atmosphere: Geochim. et Cosmochim. Acta, v. 13, p. 280–292
Davidson, C. F., 1957, On the occurrence of uranium in ancient conglomerates: Econ. Geology, v. 52, p. 668–693
de Villiers, J. W. L., Burger, A. J., and Nicolaysen, L. O., 1958, The interpretation of age measurements on the Witwatersrand uraninite: 2d United Nations Internat. Conf. on the Peaceful Uses of Atomic Energy Proc., v. 2, p. 237–238
Eaton, J. P., and Murata, K. J., 1960, How volcanoes grow: Science, v. 132, p. 925–938
Goldich, S. S., Nier, A. O., Krueger, H. W., and Hoffman, J. H., 1958, K-A dating of Precambrian iron formations (Abstract): Am. Geophys. Union Trans., v. 39, p. 516
Goldsmith, J. R., and Graf., D. L., 1957, The system $CaO\text{-}MnO\text{-}CO_2$: Solid-solution and decomposition relations: Geochim. et Cosmochim. Acta, v. 11, p. 310–334
Green, J., and Poldervaart, A., 1955, Some basaltic provinces: Geochim. et Cosmochim. Acta, v. 7, p. 177–188
Grønvold, F., 1955, High-temperature x-ray study of uranium oxides in the $UO_2\text{-}U_3O_8$ region: Jour. Inorg. and Nuclear Chem., v. 1, p. 357–370
Hahn, W. C., Jr., and Muan, A., 1960, Studies in the system Mn-O: the $Mn_2O_3\text{-}Mn_3O_4$ and $Mn_3O_4\text{-}MnO$ equilibria: Am. Jour. Sci., v. 258, p. 66–78
Hutchinson, G. E., 1954, The biochemistry of the terrestrial atmosphere, p. 371–433 *in* Kuiper, G. P., *Editor*, The earth as a planet: Chicago, Univ. Chicago Press, 751 p.
James, H. L., 1951, Iron formation and associated rocks in the Iron River district, Michigan: Geol. Soc. America Bull., v. 62, p. 251–266
——— 1954, Sedimentary facies of iron-formation: Econ. Geology, v. 49, p. 235–293
Kellogg, H. H., and Basu, S. K., 1960, Thermodynamic properties of the system Pb-S-O to 1100° K: Am. Inst. Min. Met. Eng. Trans., v. 218, p. 70–81
Kennedy, G. C., 1948, Equilibrium between volatiles and iron oxides in igneous rocks: Am. Jour. Sci., v. 246, p. 529–549
Klinger, F. L., 1956, Geology of the Soudan mine and vicinity, p. 120–134 *in* Schwartz, G. M., *Editor*, Precambrian of northeastern Minnesota: Geol. Soc. America Guidebook, 235 p.
Klingsberg, C., and Roy, R., 1959, Stability and interconvertibility of phases in the system Mn-O-OH: Am. Mineralogist, v. 44, p. 819–838

———— 1960, Solid-solid and solid-vapor reactions and a new phase in the system Mn-O: Am. Ceram. Soc. Jour., v. 43, p. 620–626

Kokubu, Nobuhide, Mayeda, T., and Urey, H. C., 1961, Deuterium content of minerals, rocks and liquid inclusions from rocks: Geochim. et Cosmochim. Acta, v. 21, p. 247–256

Krynine, P. D., 1960, Primeval ocean (Abstract): Geol. Soc. America Bull., v. 71, p. 1911

Kubaschewski, O., and Evans, E. L., 1955, Metallurgical thermochemistry, 2d ed., London, Pergamon Press, 410 p.

Kuiper, G. P., 1952, Planetary atmospheres and their origin, p. 306–405 *in* Kuiper, G. P., *Editor*, The atmospheres of the earth and planets: Chicago, Univ. Chicago Press, 434 p.

Liebenberg, W. R., 1955, The occurrence and origin of gold and radioactive minerals in the Witwatersrand system, the Dominion reef, the Ventersdorp contact reef and the Black reef: Geol. Soc. South Africa Trans., v. 58, p. 101–227

———— 1958, The mode of occurrence and theory of origin of the uranium minerals and gold in the Witwatersrand ores: 2d United Nations Internat. Conf. on the Peaceful Uses of Atomic Energy Proc., v. 2, p. 379–387

Mair, J. A., Maynes, A. D., Patchett, J. E., and Russell, R. D., 1960, Isotopic evidence on the origin and age of the Blind River uranium deposits: Jour. Geophys. Research, v. 65, p. 341–348

Muan, A., 1955, Phase equilibria in the system $FeO\text{-}Fe_2O_3\text{-}SiO_2$: Jour. Metals, v. 7, p. 965–976

———— 1959, Stability relations among some manganese minerals: Am. Mineralogist, v. 44, p. 946–960

Muan, A., and Osborn, E. F., 1956, Phase equilibria at liquidus temperatures in the system $MgO\text{-}FeO\text{-}Fe_2O_3\text{-}SiO_2$: Am. Ceram. Soc. Jour., v. 39, p. 121–140

Nel, L. T., 1960, The genetic problem of uraninite in the South African gold-bearing conglomerates: 21st Internat. Geol. Cong. (Norden) Rept., pt. 15, p. 15–25

Pettijohn, F. J., 1943, Archean sedimentation: Geol. Soc. America Bull., v. 54, p. 925–972

Ramdohr, P., 1958, Die Uran- und Goldlagerstätten Witwatersrand-Blind River District-Dominion Reef-Serra de Jacobina: Erzmikroskopische Untersuchungen und ein geologischer Vergleich: Deutsche Akad. Wiss. Berlin Abh., Klasse für Chemie, Geologie und Biologie, No. 3, 35 p.

Rubey, W. W., 1951, Geologic history of sea water. An attempt to state the problem: Geol. Soc. America Bull., v. 62, p. 1111–1148

———— 1955, Development of the hydrosphere and atmosphere, with special reference to probable composition of the early atmosphere, p. 631–650 *in* Poldervaart, Arie, *Editor*, Crust of the earth: Geol. Soc. America Special Paper 62, 762 p.

Spitzer, L., Jr., 1952, The terrestrial atmosphere above 300 km, p. 211–247 *in* Kuiper, G. P., *Editor*, The atmospheres of the earth and planets, 2d ed.: Chicago, Univ. Chicago Press, 434 p.

Steacy, H. R., 1953, An occurrence of uraninite in a black sand: Am. Mineralogist, v. 38, p. 549–550

Thode, H. G., Macnamara, J., and Fleming, W. H., 1953, Sulphur isotope fractionation in nature and geological and biological time scales: Geochim. et Cosmochim. Acta, v. 3, p. 235–243

Urey, H. C., 1952a, On the early chemical history of the earth and the origin of life: Nat. Acad. Sci. Proc., v. 38, p. 351–363

———— 1952b, The planets, their origin and development: New Haven, Yale Univ. Press, 245 p.

———— 1959, The atmosphere of the planets: Handbuch der Physik, v. 52, p. 363–418

Urey, H. C., and Craig, H., 1953, The composition of the stone meteorites and the origin of the meteorites: Geochim. et Cosmochim. Acta, v. 4, p. 36–82

Zeschke, G., 1960, Transportation of uraninite in the Indus River, Pakistan: South Africa Geol. Soc. Trans. 63 (preprint)

APPENDIX

Values for Some Equilibrium Constants

$T(°C)$	$\log K_1$	$\log K_2$	$\log K_3$	$\log K_4$	$\log K_5$
25	40.07	45.08	19.93	−33.91	12.92
900	8.16	8.02	−1.65	−5.32	1.45
1000	7.28	7.03	−2.23	−4.54	1.13
1100	6.54	6.19	−2.72	−3.87	0.85
1200	5.89	5.47	−3.15	−3.29	0.61
1300	5.32	4.83	−3.54	−2.78	0.40
1400	4.83	4.28	−3.85	−2.34	0.21

Copyright © 1963 by the Swedish Society of Chemists and the Swedish Association of Chemical Engineers

Reprinted from *Svensk Kemisk Tidskrift*, **75**(4), 161–177 (1963)

HOW HAS SEA WATER GOT ITS PRESENT COMPOSITION?*

LARS GUNNAR SILLÉN

Royal Institute of Technology, Stockholm, Sweden

The system (sea water + sediments) is thought to have resulted from the interaction of primary igneous rock with volatile substances, mainly H_2O and HCl, from the interior of the earth. The real system is compared with an imaginary model in which these reactions have come to equilibrium. It is concluded that the pH of sea water, and the concentrations of the major cations, are determined by equilibria with the aluminosilicates in the sediments, and that the role of the carbonate system is best described as that of an indicator. The solid phases on the sea floor, and the order of magnitude of the concentrations of various metal ions in sea water, are usually what one could expect at equilibrium. Exceptions—such as the unexpectedly high concentrations of Fe and Mn—indicate gaps in our knowledge of solution chemistry.

Some future economic implications of marine chemistry are indicated.

Some facts

The total area of the Earth's surface is 510×10^6 km²; Fig. 1 gives the distribution of heights and depths. The sea occupies 361×10^6 km², that is 71 per cent. A great part of the ocean floor lies around the average depth of the ocean, almost 4 km, but there are a few pits as deep as 10—11 km.

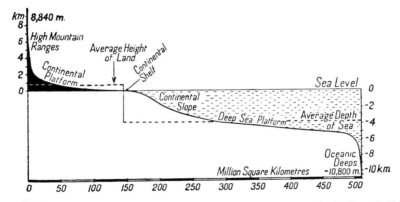

Fig. 1. The distribution of the Earth's surface over various heights and depths (from A. Holmes, Principles of physical geology, 1944, p. 10. Reproduced by permission of the editors, Thomas Nelson and Sons, who wish to point out that the present depth record is 11,516 meters (the Cook deep).

The total volume of the ocean has been calculated as 1.37×10^9 km³ = = 1.37×10^{21} liters. A laboratory chemist will find this volume somewhat

---

* This paper is based on a lecture given at Elfte Nordiska Kemistmötet (11th meeting of Scandinavian chemists) at Åbo (Turku), Finland, in August 1962 and on a Berzelius lecture in Stockholm in Januari 1963. The general ideas, and the detailed calculations were given in a plenary lecture[1] at the first international conference on oceanography in New York, Sept. 1959. The English has been revised by Dr Leonard Newman.

unwieldy, and in the following, we shall divide all quantities related to the sea—water, air, sediments—by the factor 1.37×10^{21}. We shall thus consider one liter of sea water with the corresponding amounts of air and sediments.

TABLE 1. Main components of sea water, average concentrations (c) in mole/liter (M).

	c	log c		c	log c
H_2O	54.90	1.74	Cl^-	0.54830	— 0.26
Na^+	0.47015	— 0.33	SO_4^{2-}	0.02824	— 1.55
Mg^{2+}	0.05357	— 1.27	HCO_3^-	0.00234	— 2.63
Ca^{2+}	0.01024	— 1.99	Br^-	0.00083	— 3.08
K^+	0.00996	— 2.00	F^-	0.00007	— 4.15
Sr^{2+}	0.00015	— 3.82	(H_3BO_3)	0.00043	— 3.37

Table 1 gives the average composition of sea water, expressed in the unit M = mole/liter. It can easily be checked that—not unexpectedly—the total charge of the positive ions in one liter of sea water is equal to the total charge of the negative ions. The pH value may vary by 0.1 or 0.2; we shall use the average value pH = 8.1 ± 0.2. The oxidizing power can be expressed by the quantity pE = 12.5 ± 0.2 [defined in eq. (11)].

It can easily be shown that to one liter of sea water there correspond around 3 liters of air at 1 atm. There is around 1 kg of air above each cm^2 of the earth's surface and the total weight of the atmosphere above the 510×10^6 km^2 has been calculated as 5.13×10^{21} g. If we divide this number by the average molecular weight of the air (28.97) and by 1.37×10^{21} (the volume of the ocean in liters), we find that for each liter of sea water there is 0.1293 mole air (0.1010 mole N_2 + 0.0271 mole O_2 + 0.0012 mole Ar + 0.000039 mole CO_2). From the general gas law we find that this amount of air would take up, at 1 atm. pressure, 2.90 l at 0° and 3.16 l at 25°.

The balance of materials according to Goldschmidt

How did sea water come to have the composition it now has?

It seems a natural suggestion that the metal ions, Na^+, Mg^{2+} etc., in some way or another come from primary igneous rock. There is a never-ceasing interaction between the continents and the sea water. By the action of wind and running water new materials—solid particles, dissolved ions—are continually carried down to the sea, where new sediments are formed.

The sediments are, however, not always allowed to remain on the ocean floor. During the cretaceous period, for instance, the borderlines between sea and continents (and in all likelihood also the north pole and the south pole) were not at all where they are now. Where the Alps are now there was then ocean floor, on which thick sediments were formed. A little wrinkle on the earth and these sediments were lifted up from the ocean as a mountain chain. By the action of rain and wind the material in the Alps is now slowly on its way back to the sea.

The sea water also attacks directly, both primary rock and old marine sediments. Victor Moritz Goldschmidt, the famous Norwegian geochemist, tried to make a balance-sheet for the formation of the sea.[2] He used the following figures, given in the order of increasing uncertainty: the average composition of sea water, the volume of the sea, the average composition of the primary rock in the earth crust, the average composition and the total amount of each type of sediment. The figures for the sediments were partly adjusted so as to make the balance come out even.

Goldschmidt found that for each liter of sea water, about 600 g of primary rock must have been decomposed chemically. In this process a practically equal amount of sediments was formed. The amounts of Ca, Mg and Na dissolved were balanced by the CO_2 that was taken up by the sediments.

TABLE 2. Balance of materials for the formation of one liter of sea water, calculated from the estimates of V. M. Goldschmidt.
Unit = mole/liter sea water.

The sediments are assumed to contain 0.46 mole $CaCO_3$ and 0.09 mole $MgCO_3$. Only carbonate carbon has been counted, not coal and other forms with lower oxidation state.

	From 0.6 kg primary rock	Volatile	In 1 liter sea water (c)	In 0.6 kg sediments (n)
H_2O	—	54.90	54.90	—
Si	6.06	—	—	6.06
Al	1.85	—	—	1.85
Cl	0.01	0.54	0.55	—
Na	0.76	—	0.47	0.29
Ca	0.56	—	0.01	0.55
Mg	0.53	—	0.05	0.48
K	0.41	—	0.01	0.40
C	0.02	0.53	—	0.55
O_2	—	0.03	(0.03, g)	—
Fe	0.55	—	—	0.55
Ti	0.06	—	—	0.06
S	0.01	0.06	0.03	0.04
F	0.03	?	—	0.03 + ?
P	0.02	—	—	0.02
Mn	0.01	—	—	0.01
N_2	—	0.10	(0.10, g)	—

The balance according to Goldschmidt is given in Table 2, where it has been recalculated to mole per 1 liter of sea water. For each one of the 17 most important components Table 2 gives the number of moles in 600 g primary rock (column 2), the number of moles in one liter of sea water (column 4) and the number of moles in 600 g sediments (column 5). In order to make the balance come out even, Goldschmidt had to assume that certain volatile substances (column 3)—especially H_2O, but also HCl, CO_2, SO_3, HF etc.—had been added from some other source than the 600 g of primary rock.

So, we can sum up Goldschmidt's balance as follows:

$$0.60 \text{ kg primary rock} + \approx 1 \text{ kg volatile substances} \rightarrow 1 \text{ liter sea water} + \\ + 0.60 \text{ kg sediments} + 3 \text{ liters air} \quad (1)$$

From where did the volatile substances come?

Theories on the origin of the earth

When I went to school we were told that the earth had once been a ball of incandescent gases, probably a sample of solar matter which had been detached from the sun in one way or another. When the primary earth had cooled down somewhat it consisted of a red-hot liquid sphere surrounded by an atmosphere which contained, among other gases, all the water vapor that was to become the ocean. On further cooling a crust was formed on the melt and at last the crust was so cool that water could condense. Then came the first rain of the earth and the ocean was born. This birth of the ocean has been depicted with great dramatic power by writers of popular science.

However, on closer inspection one finds many facts which do not fit in with this picture. The earth does not contain the volatile substances that one would have expected, if a sample of solar matter had been allowed to cool. Of course, a great part of the volatile substances would have been lost by evaporation to outer space, but it is very hard to explain for instance why we have got so little of the noble gases, in relation to other volatile substances.

All scientists who have made accurate calculations during the last decade on the composition of the earth have concluded that the earth was originally built up from small, cool particles, and that all the volatile substances that we now have were brought here bound in solid substances. Water and chloride ions, for instance, can have been bound in solid hydroxides and chlorides. Even argon has arrived here in a solid form, namely as radioactive ^{40}K. Among those scientists who have made such calculations I must especially mention Harold Urey in Chicago and Aleksandr Pavlovič Vinogradov in Moscow.

Let us thus make a quick review of the evolution of the earth, according to them and others. We do not know in detail what happened once upon a time when the Lord decided to start the cosmos that we are now living in. At any rate, a few thousand million years ago the sun and perhaps 10^{11} other stars condensed as small drops of matter in our galaxy. In the interior of each star the temperature was kept high by nuclear reactions (around 10^7 degrees). The outside space was considerably cooler. We can imagine that the sun was originally surrounded by a rotating, flat cloud of condensed dust. Urey[3] has made a serious attempt to calculate the temperature and pressure fields and the chemical composition of the dust and gas at various distances from the original sun.

Through vortex formation in the dust cloud, particles gradually agglomerated to larger and larger units and at last planets were formed. According to Weizsäcker[4] this is a quick process by astronomical measures. The transition from a dust cloud to a system of planets need not have taken more than around 20 million years.

The surface of the primary earth, perhaps 3×10^9 years ago, may have looked like the present surface of the moon. According to Urey[3] the craters on the moon have the shape one could expect if the moon was hit by big lumps of matter in the last period of its formation. The "seas" (maria) on the moon were formed when hitting lumps of high kinetic energy covered a large area of the surface with molten material. Good pictures show splashes, which sometimes make a full turn around the moon.

In the interior of the primary planets there were considerable amounts of radioactive atoms — Th, U, and in addition much more ^{40}K than nowadays. When they gradually disintegrated, energy was set free and accumulated so that the temperature increased. The processes that then occurred with the solids have been summarized by Vinogradov[5] under the headings zone melting and degassing. Volatile substances, especially H_2O and HCl, were driven to the surface. The gravitational field of the moon is so weak that it has not been able to hold these gases. On the earth they have however remained to a great part.

These processes are still active. Volcanoes send out large amounts of the volatile substances that were needed in Goldschmidt's balance. A great part of the water molecules which nowadays emerge through the volcanoes have certainly circulated so that they have recently, in some way or another, penetrated the earths' surface. However, it seems likely that a fraction of the volcanic water now passes the earth's surface for the first time. The magnitude of this fraction of primary water is still uncertain. It is also uncertain whether the water added from the interior of the earth and the hydrogen that may be added from the sun, outweighs the hydrogen that earth looses to outer space. For instance, it has been claimed that the amount of water in the sea is continually increasing and that the continents will have disappeared in two hundred million years or so.

What determines the pH of sea water?

Hence, the sea would have been formed from primary rock and volatile substances from the interior of the earth. This process can be compared to an immense acid-base titration: the water and the acids (especially HCl) come from the interior of the earth, the "bases" NaOH, $Ca(OH)_2$, $Mg(OH)_2$ etc. come from the primary rock.

We can now ask: Why has sea water got the composition it has by now table 1)? Why is it so close to neutral (pH $\approx$ 8.1)? First let us see what determines the pH of sea water.

The usual explanation, which is found in every book that gives any attention to this problem, is that the pH of sea water is determined by a buffer system consisting of the ions HCO_3^- and CO_3^{2-} in sea water and CO_2 in the air.

However, if one considers how small a fraction the dissolved carbonate is of all the components in our "acid-base titration", one may begin to doubt that this can be the determining factor. To be sure, one liter of sea water contains 2.34 mmole HCO_3^-, but what really determines the buffer capacity of the carbonate system is the amount of the other forms, 0.2 mmole CO_3^{2-} and (0.04 mmole CO_2 in the air + 0.02 mmole H_2CO_3 in the water). This can be compared with 470 mmole Na^+, 54 mmole Mg^{2+}, 548 mmole Cl^- etc. in one liter of sea water.

Such considerations offer two alternatives. One is that the titration has happened to stop less than 0.5 percent from the equivalence point. (This, by the way, is better than most students of chemistry can achieve in their first attempts.) If it were so the pH of sea water would be labile and could easily be changed for instance by addition of HCl during periods of high volcanic activity. The other alternative is that the pH of sea water is determined by something else than the carbonate system. In the following I hope to be able to show the probable determining factor.

An imaginary experiment

In order to understand better the processes in the making of the sea water we make an imaginary experiment. Fig. 2 shows to the right the "real" system: 1 liter of sea water and the corresponding 3 liters of air + 0.6 kg sediments.

We assume that the "real" system contains as many moles of each component as are given by the balance sheet in table 2. (Minor changes that have been proposed after Goldschmidt will not affect the following argument).

Now let us assume that we use the same amounts of these components to mix a "model" system (to the left in fig. 2). We start with the components that are most abundant, and we assume that true equilibrium is achieved after each addition. Finally we compare this imaginary model with the real system.

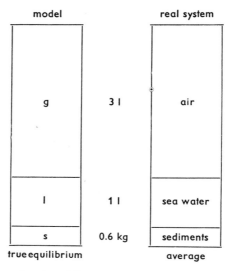

Fig. 2. To the right the real system, 1 liter of sea water with the corresponding 0.6 kg sediments and 3 liter air. To the left equilibrium model with the same amounts of the various components at true equilibrium (with the exception of N_2).

Unfortunately a complete comparison cannot be made since we do not have all the equilibrium data needed for the model[6] and not all analytical data for the real system.[7] Even so it seems that many conclusions of interest can be drawn.

Let us start with $\approx$ 54 mole H_2O and 6.06 mole SiO_2. At equilibrium the system would consist of quartz and a saturated aqueous solution, containing around $10^{-4.0}$ M $Si(OH)_4$. This is approximately the Si concentration in deep-sea water (at the ocean surface it is often considerably lower since some Si has been eaten by diatoms (silica algae)).

Then we add 1.85 mole $Al(OH)_3$. At equilibrium we would, in addition to quartz, have another solid phase, probably kaolinite $Al_2Si_2O_5(OH)_4$(s). The solution, besides $Si(OH)_4$, would also contain a very low concentration of Al; I wish I could tell you which species of Al would be present.

Now we shall make the addition which, in my opinion, is the most important one to determine the character of the "model sea water". We add the chloride ions, and the main cations, in the form of 0.55 mole HCl + 0.76 mole NaOH + 0.10 mole $Ca(OH)_2$ + 0.44 mole $Mg(OH)_2$ + 0.41 mole KOH. (We have set aside so much of Ca and Mg as is needed for the carbonate sediments.)

We stir vigorously and wait until equilibrium has been achieved (which is easier in imagination than in the laboratory). We can be sure that quartz is left as a stable phase—we had a large excess of SiO_2—and that a number of new phases, aluminosilicates, have been formed. In the aluminosilicates, such as micas or zeolites, Al, Si and O form sheets or a three-dimensional network. In the interstices are the new cations: Na^+, K^+, Mg^{2+}, Ca^{2+}. The equilibrium reactions in our mixture can be divided into two groups:

Phase equilibria involving several solid phases such as:

$$3Al_2Si_2O_5(OH)_4(s) + 4SiO_2(s) + 2K^+ + 2Ca^{2+} + 9H_2O \rightleftharpoons$$
$$\rightleftharpoons 2KCaAl_3Si_5O_{16}(H_2O)_6(s) + 6H^+ \quad (2)$$

As a rule, $[H^+]$ will enter the equilibrium condition. For the schematic equilibrium (2) we have for instance:

$$6 \log [H^+] = 2 \log [K^+] + 2 \log [Ca^{2+}] + \text{constant} \quad (3)$$

Ion exchange equilibria between the solution and aluminosilicates, for instance (R represents the aluminosilicate lattice):

$$Mg(R) + Ca^{2+} \rightleftharpoons Ca(R) + Mg^{2+} \quad (4)$$

$$Na(R) + K^+ \rightleftharpoons K(R) + Na^+ \quad (5)$$

Laboratory experience on aluminosilicates indicates that equilibria of the types (4) and (5) are both shifted to the right.

In our model, not only pH but also the concentrations of the main cations, $[Na^+]$, $[K^+]$, $[Ca^{2+}]$ and $[Mg^{2+}]$ are fixed by such phase equilibria and ion exchange equilibria. I have allowed myself to advance the hypothesis[1] that the same is true also for sea water, although we do not as yet have sufficient equilibrium data to make a strict proof.

If we begin with the ion exchange equilibria, we can easily see that sea water is enriched in Na and Mg in comparison with K and Ca. The ratio Na/K is $10^{1.67}$ in sea water but $10^{-0.14}$ in the sediments. The ratio Mg/Ca is $10^{0.72}$ in sea water and $10^{-0.06}$ in the sediments.

A simple pH-stat model

We now consider a simpler system in order to show how the silicate equilibria can determine the pH and composition of sea water. This system is $KOH + HCl + H_2O + SiO_2 + Al(OH)_3$. It has the additional advantage that those parts of this system which are especially related to the sea water system have recently been studied by Garrels.[8] In a large region of the system, an aqueous solution is at equilibrium with the three solid phases quartz (SiO_2), kaolinite [$Al_2Si_2O_5(OH)_4$], and K-mica [$KAl_3Si_3O_{10}(OH)_2$]. The condition for equilibrium is:

$$1.5\, Al_2Si_2O_5(OH)_4(s) + K^+ \rightleftharpoons KAl_3Si_3O_{10}(OH)_2(s) + 1.5H_2O + H^+ \quad (6)$$

$$[H^+]/[K^+] \approx 10^{-6.4}\ (25°) \quad (7)$$

Now let us assume that an aqueous solution is at equilibrium with these three phases. The composition of the solution is determined by the equilibrium condition (7), and by the charge condition:

$$[K^+] + [H^+] = [Cl^-] + K_w/[H^+] \quad (8)$$

[Cl⁻] is determined by the ratio HCl/H₂O that was used in mixing the solution (for instance was added with the volcano gases). Then [H⁺] and [K⁺] in the solution are fixed by the conditions (7) and (8). To such a solution one can add considerable amounts of KOH or (HCl + H₂O) without any shift in the equilibrium pH. The only change is that some more K-mica or some more kaolinite is formed by reaction (6).

In this case we have not a buffer but a pH-stat. If we want to have a different pH we have to get rid of one of the three phases. This means that the numbers, n, of moles of the components must violate at least one of the very liberal conditions: $n_{Al} < n_{Si}$ and $n_{Cl} < n_K < n_{Cl} + \frac{1}{3} n_{Al}$. To be strict we should add that a change in the ratio Cl⁻/H₂O brings with it a (small) change in pH.

The "model ocean" that we have considered contains, besides the K⁺ of the simple system, also the ions Na⁺, Mg²⁺ and Ca²⁺, which means that up to six solid phases can be formed at equilibrium; in all likelihood most of them are solid solutions. As yet we do not have sufficient information to calculate the final pH and the final cation concentrations in our model. In the following calculations we shall, however, make the bold assumption that at equilibrium our model will have the same pH (8.1) and the same concentrations of Na⁺, K⁺, Mg²⁺ and Ca²⁺ as sea water.

The ratio $[H^+]/[K^+] = 10^{-6.1 \pm 0.2}$ in sea water is not too far from the equilibrium condition (7). This indicates that our assumption may not be completely unreasonable. We could not expect a complete agreement since in the sea the temperature is different, and there is not pure K-mica but a solid solution.

Carbonate

Only now we add the carbonates, in the form 0.46 mole CaCO₃ + 0.09 mole MgCO₃. We shall refrain from discussing the controversial equilibria of dolomite and we shall consider only CaCO₃, which obviously exists in large amounts in marine sediments.

If we add solid CaCO₃ to our model system, the changes in solution and gas phase will not be very dramatic. In our model [H⁺] and [Ca²⁺] are already given by the silicate equilibria. In addition, we get a few more equilibria:

$$\text{CaCO}_3(s) + 2H^+ \rightleftharpoons Ca^{2+} + H_2O + CO_2(g); \; p(CO_2) = [H^+]^2 K/[Ca^{2+}] \quad (9)$$

$$\text{CaCO}_3(s) + H^+ \rightleftharpoons Ca^{2+} + HCO_3^-; \; [HCO_3^-] = K'[H^+]/[Ca^{2+}] \quad (10)$$

Kramer[9,10] has measured these equilibria for solutions with the same concentrations of the main ions as sea water; for instance he found that the first reaction (9) has log K = 10.29(5°) and 10.93(25°). Since, according to our assumption, [H⁺] = 10⁻⁸·¹ and [Ca²⁺] = 10⁻²·⁰ are already determined through the silicate equilibria, one can calculate that at equilibrium in the model system we would have in the gas phase log $p(CO_2)$ = −3.3(5°) or −3.9(25°). In the real system the average value in air is log $p(CO_2)$ = −3.52.

Furthermore we calculate log [HCO₃] = −2.6(5°) or −3.3(25°) whereas the average value in the sea is −2.63. The agreement is, here also, better than one might expect considering that the figures are rather crude. One could say that [HCO₃⁻] in sea water and the partial pressure of CO₂ in the air are not far from those corresponding to equilibrium with sea water and CaCO₃ at some temperature near 5° C.

It is worth noting that, in our model, the carbonate system does not determine the pH; pH is already determined by the silicate equilibria. On the other hand one could say that $CO_2(g)$ and HCO_3^- are an *indicator* of the values of $[H^+]$ and $[Ca^{2+}]$ that have been fixed by the silicate equilibria and which are not changed by reasonable additions of acid or base. We shall once more remember the difference between the quantities involved. For each liter of sea water there is 2.34 mmole HCO_3^-, 0.06 mmole $(CO_2 + H_2CO_3)$ and 0.2 mmole CO_3^{2-}. This can be compared with 6060 mmole SiO_2 and 1850 mmole Al. The aluminosilicates have an overwhelming capacity for taking up acids or bases.

Climate in danger?

Many people have been concerned about the "large" amounts of CO_2 set free by burning fossil fuels. It has been stated that the concentration of CO_2 in the air has increased since 1900 by 10 per cent (thus by 0.004 mmole/liter sea water) and that the surface of the Earth must become increasingly warmer because of the insulating effect of the CO_2. In due time, it has been said, the polar ice caps will melt, and the sea will rise by more than 100 meters, after which many harbor cities will find it difficult to use their present installations.

However, if the real sea behaves like our equilibrium model, the danger is not very great. To be sure, mankind may achieve a somewhat higher concentration of carbon dioxide in the air, but this will be in a struggle against those processes that tend to re-establish equilibrium: CO_2 dissolves in the sea, and there is a little shift in the solids so that some Ca leaves the aluminosilicates and is transformed to $CaCO_3$. This involves only a very small fraction of the total amounts of material present, and when equilibrium is again established, the pressure of carbon dioxide must be practically the same as earlier. — If this reasoning is correct, one must try to find some explanation other than a change in $p(CO_2)$ to account for the variations in the climate: such as for instance variations in solar radiation.

Oxygen and pE

Next addition is three liters of gas with 0.027 mole O_2. At equilibrium a part will dissolve so that the concentration in the aqueous solution will be around 0.0002 M. At present there seems to be a general opinion that the earth originally had a reducing atmosphere but that the atmosphere became oxidizing at some stage, either by photochemical processes in the atmosphere or through biological processes. This early history does not influence the following discussion.

The addition of oxygen will fix the redox potential e, of the solution; we shall prefer to express it by means of the quantity pE.

$$pE = e/(RTF^{-1} \ln 10) = -\log \{e^-\} \qquad (11)$$

This quantity is fully analogous to

$$pH = -\log \{H^+\} \qquad (12)$$

In calculations on redox equilibria it is very practical to treat the electron e^- in the same way as H^+ and other reagents. By using the electron activity (or pE) one can apply directly the law of mass action to redox equilibria.

There is a simple relationship between the equilibrium constants and the standard potentials.

In our model system we find for instance from the equilibrium

$0.5\ O_2(g) + 2H^+ + 2e^- \rightleftharpoons H_2O$;
$\log K = 41.35 = \log \{H_2O\} - 0.5 \log p(O_2) - 2 \log \{H^+\} - 2 \log \{e^-\} =$
$= -0.01 - 0.5 \log 0.21 + 2pH + 2 pE$.
Introducing pH $= 8.1 \pm 0.2$ we find pE $= 12.5 \pm 0.2$ (25°).

At this point someone may ask whether it is permissible to treat sea water as homogeneous. Fig. 3 shows three maps of the same vertical section through the Atlantic from 70° S to 60° N. (Note the high mountains on the sea floor.)

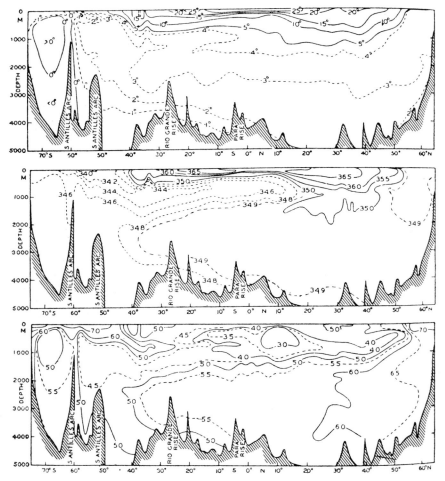

Fig. 3. The distribution of temperature, salinity and oxygen concentration in a north-south section through the Atlantic (according to Wüst). From H. U. Sverdrup, N. W. Johnson and R. H. Fleming, The oceans, their physics, chemistry and general biology, Prentice-Hall Inc. 1942, p. 748. Reproduced by permission.

The uppermost map shows isotherms: curves with the same temperature. In a small region around the equator, the surface water has a comfortable temperature above 25° C. However, one needs only go to a depth around 600 m to find a temperature around 5°. Closer to the sea floor the water is still colder; the average temperature seems to be below 5°.

The middle map shows curves with the same salinity; the salinity varies between 3.4 per cent by weight (in most of the ocean) and 3.6 per cent in the surface layer at the equator, where the evaporation is strong.—The last map finally shows lines with same concentration of oxygen, in ml/liter sea water. The figures do not vary more than what corresponds to a shift in pE of ± 0.1 or 0.2.

For our purpose we can, in first approximation, consider sea water as homogeneous. This need not be surprising since the stirring is good; with an average circulation time of 500—1,000 years we have had at least 1/2 million circulations in the last 500 million years.

Iron

Now we add 0.55 mole FeOOH(s), which is the stable form of iron at equilibrium under these conditions, and which occurs in large amounts in marine sediments. Sea water contains some iron, not only in suspended particles, and even after careful filtration there remains a concentration of around $10^{-7.2}$M Fe.

In which form does this iron exist? An obvious suggestion is hydroxo complexes, but on applying the equilibrium constants that have been determined one finds, at pH = 8.1, the equilibrium concentration $[Fe(OH)_2^+] = 10^{-10.5}$, and still much lower concentrations for Fe^{3+}, $FeOH^{2+}$ and $Fe_2(OH)_2^{4+}$.

It might be suggested that iron exists as Fe^{2+}. The equilibrium concentration of iron (II) can be easily calculated from the equilibrium conditions:

$FeOOH(s) + 3 H^+ \rightleftharpoons Fe^{3+} + 2 H_2O$	$\log K = 4.0$
$Fe^{3+} + e^- \rightleftharpoons Fe^{2+}$	$\log K = 13.0$
$FeOOH(s) + 3 H^+ + e^- \rightleftharpoons Fe^{2+} + 2 H_2O$	$\log K = 17.0$

Introducing pH = 8.1 and pE = 12.5 we find $\log [Fe^{2+}] = 17 - 3pH - pE = -20$.

If iron is to exist at an equilibrium concentration of $10^{-7.2}$ M, then it must exist in some form that cannot be found in present text-books. One might suggest a complex with phosphate ions, or with unknown organic substances in sea water, but the most plausible guess is the uncharged trihydroxo complex, $Fe(OH)_3$. This would be an interesting research problem for inorganic chemists.

Titanium, fluorine, sulfur, phosphorus

Let us pass quickly through the last components in table 2. We add 0.06 mole $TiO_2(s)$, 0.03 mole HF, 0.07 mole H_2SO_4, 0.02 mole H_3PO_4, 0.01 mole $MnO_2(s)$ and 0.10 mole $N_2(g)$. In our equilibrium model, *titanium* probably would exist as $TiO_2(s)$ and in solid solution in some silicate phases. *Fluorine* in the solution would exist partly as MgF^+ partly as F^-. The solubility product for $CaF_2(s)$ is not exceeded, and the fluorine in the sediments probably exists largely in the silicate phases, where it takes up the place of oxygen atoms. At equilibrium, *sulfur* in solution exists as sulfate ions, partly as weak sulfato complexes with for instance Ca^{2+} and Mg^{2+}. The stable *phosphate* phase will be some solid solution of hydroxo apatite, $Ca_5(PO_4)_3OH(s)$ and fluoro apatite, $Ca_5(PO_4)_3F(s)$ (the equilibrium conditions do not seem to be too well known). It is hard to predict the total concentration of phosphate in the solution at equilibrium since the formation constants for phosphate complexes are little known.

For *manganese* $MnO_2(s)$ would be stable. Among known dissolved species one calculates the highest equilibrium concentrations for Mn^{2+} and MnO_4^-, as 10^{-16} M for each, whereas in sea water Mn concentrations from $10^{-6.7}$ to $10^{-7.9}$ M have been measured. Some scientists claim that Mn exists as Mn^{2+} in sea water, and that its concentration is kept 10^9 times larger than the equilibrium concentration by some continuously acting process. I find it hard to subscribe to this belief, considering that sea water contains so many silicate and oxide particles and other nuclei for crystal growth. In my opinion, Mn in sea water exists in some form that cannot be found in the text-books; probably it is a hydroxo complex, although one can neither exclude an organic complex nor a phosphato complex.

Nitrogen at equilibrium would exist practically completely as nitrate ions in the sea water. It does not, which shows that equilibrium arguments have a limitation. In our model we shall make an exception and let the nitrogen remain in the non-equilibrium form N_2.

Minor components

Now we have treated the 17 most abundant components of our system. In addition there are many minor components. Table 3 gives data for altogether 54 elements. For some elements an interval is given for the concentration; I would not dare to state whether the difference depends on real variations between different parts of the sea, or on variations in methods and analytical skill. The fourth column gives the form of the element that is likely to predomiate in the solution according to available equilibrium data. The last column gives a solid phase in which the element probably exists at equilibrium.

TABLE 3. Survey of the number of moles of various elements in one liter sea water (c) and the corresponding 0.6 kg sediments (n), calculated from V M Goldschmidts estimates.

The tentative formulas given for "main species" and solids are not always certain. Many elements probably exist mainly in solid solutions (see text).

Element	-log n (solid)	-log c (solution)	main species in solution	solid phase
H_2O	—	−1.74	H_2O	
Si	−0.78	4.0—4.5	$Si(OH)_4$	SiO_2
Al	−0.27	7.0—7.7	$Al(OH)_3$?	$Al_2Si_2O_5(OH)_4$*
Cl		0.26	Cl^-	—
Na	0.54	0.33	Na^+	
Ca	0.26	1.99	Ca^{2+}	Aluminosilicates
Mg	0.32	1.27	Mg^{2+}	
K	0.40	2.00	K^+	

* The kaolinite may disappear when the following elements are added.

Element	-log n solid	-log c solution	main species in solution	solid phase
C	0.26	2.63	HCO_3^-	$CaCO_3$, $CaMg(CO_3)_2$
O_2	—	(1.57 g)	(O_2, g)	—
Fe	0.26	6.0—7.9; 7.2	$Fe(OH)_3$?	FeOOH
Ti	1.25	6.7—8	$Ti(OH)_4$?	TiO_2
F	1.54	4.15	F^-, MgF^+	—
S	1.4	1.55	SO_4^{2-}	—
P	1.63	5.5—6.0	HPO_4^{2-}	$Ca_5(PO_4)_3(OH, F)$
Mn	1.95	6.7—7.9	$Mn(OH)_{3(4)}$?	MnO_2
N_2	—	(1.00, g)	(N_2, g)	—
Li	2.24	4.3—5.0	Li^+	—
B	1.5	3.37	$B(OH)_3$	—
V	2.7	6.9—8.4	$VO_2(OH)_3^{2-}$	—
Cr	2.6	7.3—9.1	(OH)?	—
Zn	2.9	6.5—7.7	Zn^{2+}	—
Rb	2.7	5.3—6.4	Rb^+	—
Sr	2.7	3.82	Sr^{2+}	$SrCO_3$
Zr	2.8	+	(OH)?	—
Ba	3.0	6.2—7.4	Ba^{2+}	$BaSO_4$
Be	3.4	?	(OH)?	—
B	1.5 ?	3.37	$B(OH)_3$	—
Co	3.6	7.9—8.8	Co^{2+}	CoOOH
Ni	3.1	7.0—8.9	Ni^{2+}	—
Cu	3.2	5.8—8.4	Cu^{2+}, $CuOH^+$	$Cu(OH)_{1.5}Cl_{0.5}$?
Ga	3.9	8.1	(OH)?	—
Br	?	3.08	Br^-	—
Y	3.7	8.5	(OH)?	YPO_4?
Sn	3.7	7.6	(OH)?	SnO_2?
Ce	3.7	8.5	Ce^{3+}?	CeO_2?
Sc	4.2	9	(OH)?	$ScPO_4$?
Ge	4.2	> 8.9	$Ge(OH)_4$	—
As	4.4	6.4—7.5	$HAsO_4^{2-}$	—
Mo	4.0	6.8—8.5	MoO_4^{2-}	—
Cs	4.5	7.1—8.5	Cs^+	—
La	4.1	8.7	La^{3+}	$LaPO_4$?
Pb	4.3	7.6—8.0	Pb^{2+}, $PbOH^+$, $PbCl^+$	PbO_2
Th	4.5	> 8	(OH)?	—
Sb	5.3?	> 8.4	$Sb(OH)_6^-$?	—
I	6?	6.3—6.7	IO_3^-	—
W	5?	9.3	WO_4^{2-}	—
Hg	5.8	9.8	$HgCl_4^{2-}$	—
Tl	5—6	> 10.3	Tl^+	—
U	5.0	7.9—9.2	$UO_2(CO_3)_3^{4-}$?	—
Se	6.2	7.1—7.4	SeO_4^{2-}	—
Ag	6.2	7.6—8.9	$AgCl_3^{2-}$	—
Cd	6.1	9.3—9.5	$CdCl_2$?	—
Bi	6.2	9	?	—
Au	7.8	8—10.7	$AuCl_2^-$	—

Let us add to our model the elements from Li on, as a whole, for instance as chlorides or hydroxides. We can then try to calculate the final state of real equilibrium. A few indications will be given here. Somewhat more detailed calculations have been given elsewhere.[1]

Solid solutions

If an element exists only in small amounts, it will rarely form a separate solid phase. In general it will hide in solid solution in a compound of some more abundant element with a similar ionic radius and charge. For instance one can expect Sr, Pb and Co to hide in solid solutions (Ca, Sr)CO$_3$, (Mn, Pb)O$_2$ and (Fe, Co)OOH. A "hidden" element may have an apparent solubility product which is considerably lower than if it had existed in a pure compound; this is important for the calculations on equilibrium.

Let us for instance assume that we have a solid solution (A, B)L, where the molar fraction is x (small) for BL and $(1 - x)$ for AL. The solubility products for the pure substances are K_{AL} and K_{BL}. The equilibrium conditions for the solid solution are then $[A][L] = (1 - x)K_{AL} \approx K_{AL}$ and $[B][L] = xK_{BL} \ll K_{BL}$. (In our first approximation we have neglected the activity factors.) The apparent solubility product for BL will be very much influenced by the fact that a solid solution has formed.

Complex formation

To what extent do the various elements exist as complexes in the aqueous solution? The best numerical data are available for chloro and hydroxo complexes. One may easily set up the conditions for the chloro complexes to predominate over the hydroxo complexes at the [Cl$^-$] and [H$^+$] in sea water; one will then find that these conditions are fulfilled only for a small number of elements in the "covalent triangle" in the center of the long periodic table: AgCl$_2^-$, AgCl$_3^{2-}$, AuCl$_2^-$, CdCl$^+$, CdCl$_2^-$, HgCl$_3^-$, HgCl$_4^{2-}$, PbCl$^+$. In sea water, however, these elements exist only in very low concentrations. For other metals, the hydroxo complexes are more important than the chloro complexes.

The incomplete equilibrium data for fluoro complexes do not allow us to state that any complex other than MgF$^+$ will exist in considerable amounts. Of sulfato complexes MgSO$_4$ and CaSO$_4$ exist but are rather weak. At least UO$_2^{2+}$ and perhaps Th^{4+} exist as carbonato complexes. Unfortunately too little is known about the phosphato complex equilibria, but they may be important for many multivalent ions.

Sea water in addition contains a certain amount of organic substances with a total concentration of reactive groups between 10^{-4} and 10^{-5} M. It is not known exactly what these substances are, and thus one cannot say whether or not they form strong metal complexes. This again gives some scope for work for solution chemists.

Redox equilibria

From the pH and pE in sea water ($=$ the model) and from known equilibrium constants we can calculate which form of each element can be expected to predominate at equilibrium. For instance, for the reaction ClO$_3^-$ + 6H$^+$ + + 6e$^-$ $\rightleftharpoons$ Cl$^-$ + 3H$_2$O one can calculate that at equilibrium [ClO$_3^-$]/Cl$^-$] $\approx$ $\approx 10^{-23}$. Practically the same ratio is found for [BrO$_3^-$]/[Br$^-$] $\approx 10^{-23}$. Similar low ratios are found for other Cl and Br species, so that these elements, not unexpectedly, should exist predominantly in the form of Cl$^-$ and Br$^-$. On the other hand, for iodine one finds for instance, [IO$_3^-$]/[I$^-$] $\approx 10^{13.5}$. Under the conditions of sea water it seems that the iodate ion, IO$_3^-$ is the most stable form of iodine. In a similar way one finds that HAsO$_4^{2-}$ is the predominating form of As, SeO$_4^{2-}$ of Se etc.

Solution equilibria

For *barium* one can calculate that BaSO$_4$(s) should be the stable solid phase and not BaCO$_3$(s). Indeed, barium sulfate has been found as a separate phase in marine sediments. The equilibrium concentration of Ba in the model at 25° can be calculated as around $10^{-7.0}$ M, the analytical values for sea water are between $10^{-6.2}$ and $10^{-7.4}$.

For *cobalt* one finds that, at the pH and pE of sea water (= the model) CoOOH(s) is more stable than CoO_2(s) or $Co(OH)_2$(s). Of the known dissolved species, Co^{2+} has the highest calculated equilibrium concentration. From the equilibrium constants for

$$CoOOH(s) + 3H^+ + e^- \rightleftarrows Co^{2+} + 2H_2O, \log K = 29.3 \;(25°)$$

one finds $\log [Co^{2+}] = -6.7$, whereas analyses of sea water give -7.9 to -8.8. This is plausible since CoOOH probably exists in solid solution in FeOOH(s). Hence the equilibrium concentration of Co^{2+} must be multiplied by the activity for CoOOH which is considerably smaller than unity.

For *lead* one finds that PbO_2(s) should be stable as compared with $PbSO_4$. Furthermore PbO_2 should be stabilized by solid solution in MnO_2 which has the same structure. With regard to this, and to the formation of complexes, $PbOH^+$ and $PbCl^+$, one calculates an equilibrium concentration for Pb of the same order of magnitude as that found in sea water.

The concentration of *gold* is considerably lower than that Fritz Haber had hoped for when he planned to pay the German war debt after World War I with gold extracted from sea water. A conservative figure is 10^{-10} M Au in sea water in general, even if 10^{-8} has been claimed in certain places. As usual one can ask whether this reflects real variations in gold concentration or analytical difficulties.

It has often been assumed that gold in sea water would exist as a colloid, thus as small particles of solid gold. Let us assume that this is true. At equilibrium with solid gold at least some gold should exist as complexes in aqueous solution, and through the work of Niels Bjerrum we have the necessary equilibrium constants available (18—20°):

$AuCl_2^- + e^- \rightleftarrows Au(s) + Cl^-$ $\qquad \log K = 19.2$
$AuCl_4^- + 3e^- \rightleftarrows Au(s) + 4Cl^-$ $\qquad \log K = 51.3$
$Au(OH)_4^- + 4H^+ + 4Cl^- \rightleftarrows AuCl_4^- + 4H_2O$ $\quad \log K = 29.64$

If we insert the values for sea water (model), pE = 12.5, pH = 8.1, $\log [Cl^-] = -0.26$, and neglect the activity factors, we find that $AuCl_2^-$ gives the highest equilibrium concentration of all gold complexes, namely $10^{-7.2}$ M. The equilibrium concentrations would be lower for $Au(OH)_4^-$ ($10^{-11.0}$ M) and $AuCl_4^-$ ($10^{-14.8}$ M).

Thus, if sea water were at equilibrium with solid gold, we could expect a concentration of $AuCl_2^-$ around $10^{-7.2}$ M. The real concentration of gold in sea water is, however, much lower, around 10^{-10} M, whence sea water ought to have a capacity to dissolve considerable amounts of gold. Perhaps the ocean is not the best place for long-time storage of gold treasures.

Comparison between model and real system

We have now discussed in some detail a model in which sea water has reached true equilibrium with its surroundings, and we have tried to compare this model with the real system. It is not claimed that true equilibrium exists between the sea and its surroundings. On the contrary, practically everything of interest that occurs in and around the sea is connected with deviations from equilibrium: the shifting weather, the currents in the ocean, the various forms of life. The present state in the sea is not a true equilibrium but a steady state which exists through the interaction of disturbances and restoring reactions. We must especially consider two sorts of disturbances that tend to remove the real system (sea + air + sediments) from equilibrium.

Sea water flows along or against gradients of pressure and temperature. Around the poles it sinks to the ocean floor where the pressure is higher, around the equator it ascends toward lower pressure and higher temperature. Since the equilibrium constants vary with pressure and temperature, the result may be the precipitation or dissolution of solids, such as calcium carbonate, or a transport of gases such as CO_2 across the ocean surface. Such processes may lead to a transport of considerable amounts of material, but the concentration for each substance can be expected to be somewhere in between the equilibrium values at the extreme values for pressure and temperature.

More serious for the equilibrium picture are the non-equilibrium processes: biological processes, photochemical processes, etc. For instance, the surface water is depleted of Si and P through various organisms; the restoring processes are dissolving suspended solids and sediments. Those concentrations which result from such opposing processes need not be related to the equilibrium concentration at any set of values for pressure and temperature.

At any rate, it can do no harm to try to find out what the equilibrium state would look like, toward which the ocean system is striving against the transport and non-equilibrium processes. Perhaps our discussions have shown that this equilibrium state (the model) is sufficiently similar to the real system to be of interest.

Future ocean chemistry

We are living in a time of Mamon worship and hence it is proper to end this article with a pious contemplation of the economic profit that mankind can derive in the future from the chemistry of the ocean.

The ocean contains very large deposits of minerals. The metal oxides which have been mentioned above—$FeOOH$, MnO_2, $CoOOH$, PbO_2—and many others are found in the deep-sea clay which covers a large part of the ocean floor, especially in the Pacific but also in the Indian Ocean and in the Atlantic. The oxides are finely dispersed, and thus easy to transport, and to treat chemically. The only disadvantage that someone might wish to point out is that they are stored at a depth of 4 kilometers or more. However, once the technical difficulties have been solved that are involved in dredging or pumping this sludge, the sea floor contains mineral deposits which are so rich that there seems to be no reason for the great powers to quarrel about them.

We can also expect that in the future the ocean will be still more important than now as a source for food. At present we use the sea mainly as hunting ground; we hunt and kill whales, fish and other sea animals, and eat them. Now, for an area on dry land, the most economic way of using it is not considered to be for hunting: usually one prefers agriculture or cattle-breeding. Could not the sea be used as farmland or as pasture land?

There has been a good deal of speculation around the possibility of using the ocean for agriculture, for instance growing algae of high nutrition value at the ocean surface in suitable regions, perhaps protected from below and from the sides by sheets of plastics. In due time one could then harvest the algae; the most practical procedure would probably be to collect them together with the attached shrimps and other small fish. Provided the spicing and advertising is well done, the soup could certainly be sold and eaten. However, there are chemical problems in this project: which are the critical minor elements that must be added in a fertilizer to get the best possible crops from the farming sea?

The same type of problems appear if one wants to use the sea for pasture. We do catch fish in the Baltic, but one may ask why we cannot get for instance a hundred times as many salmons and cod as now. If one looks for limiting factors (excuse the fumbling of a non-biologist) one will perhaps find that the first limit is set by the supply of certain small animals or algae, but if the causal chain is followed, one will probably end up with a chemical cause. (It seems unlikely that even in the summer of 1962, the supply of sunlight would be the limiting factor.)

If the limiting factor would come out to be a substance which otherwise exists in a comparatively high concentration, such as phosphate or silicate, it may not be economically feasible to add it to the sea water. On the other hand one may try to improve the circulation, so that water from the depth, with sufficient concentration of the substances needed, can reach the surface more quickly. Either one can blast away hindering mountain chains on the seafloor, or introduce suitable pumping stations.

On the other hand, it may turn out—in analogy to what one has found for certain deficiency diseases on shore—that the limits are set by some minor element, the concentration of which has the order of 10^{-8} M, for instance Zn, Co or Mn. Then it may prove economical to fertilize a restricted region of the sea, such as the Baltic or some part of it. This, however, raises certain practical questions, the two fundamental of which are: Who is going to pay for the fertilizer? and Who is going to eat the fish? By the way, where are the boundaries of individual states in such a pasture sea?

When the chemistry of ocean life has become so well known that these questions have a practical meaning, then let us hope that the leaders of the nations have learned to live in friendship better than they seem to do right now.

References

1. Sillén, L. G., Oceanography (Invited lectures presented at the international oceanographic congress in New York, 31 Aug.—12 Sept. 1959), p. 549, 1959.
2. Goldschmidt, V. M., *Fortschr. Mineral. Krist. Petrol.* **17**, 112 (1933).
3. Urey, H. C., The planets, their origin and development. Yale University Press, New Haven, 1952.
4. von Weizsäcker, C. F., *Z. Astrophys.* **22**, 319 (1944); *Naturwissenschaften* **33**, 8 (1946).
5. Vinogradov, A. P., Chimičeskaja evoljucija zemli, Izdatel'stvo Akad. Nauk, Moskva, 1959.
6. Bjerrum, J., Schwarzenbach, G. and Sillén L. G., Stability constants, II, inorganic ligands, Chem. Soc. Spec. Publication **7** (1958), and manuscript for new edition (main source for equilibrium constants).
7. Harvey, H. W., The chemistry and fertility of sea water, 2 ed. Cambridge University Press (1957). (Important source for analyses of sea water, unfortunately several printing errors in the tables).
8. Garrels, R. M., *Am. Mineralogist* **42**, 780 (1957) and later private communications.
9. Kramer, J. R., thesis (1958), University Microfilms 58-7747.
10. Kramer, J. R., *J. Sediment. Petrol.* **29**, 465 (1959).

Abundance of the elements, areal averages and geochemical cycles*

Tom. F. W. Barth

We have no quantitative knowledge of the composition of the earth as a whole, or the composition of the individual parts. The concept of geochemical cycles explains some of the migration of the chemical species in the earth; but the quantitative relations of the cyclic processes are not known. Our ignorance is really shocking; the underlying data are inadequate and no final solution would seem possible today. So little is known that, in order to get a start, a series of hypotheses have to be introduced. Such hypotheses must be critically examined—this is the main object of the present paper. If the hypotheses can be shown to be plausible one may at least go ahead; if not, there is no use to proceed with subsequent calculations.

Our interpretation of the geochemical cycles is highly influenced by inherited notions, and our ideas of rock genesis have been through many and often sinuous changes during the 200 years growth of the earth sciences, from Werner's neptunism and Hutton's plutonism to present day. Whereas it is possible for us to see Werner and Hutton in a historical perspective, this is not yet possible in the case of the Rosenbusch system: In modern geochemistry the Rosenbusch ideas have prevailed; igneous rocks are regarded as primary, they come from the depths of the earth, from "die ewige Teufe", i.e., they are "juvenile", whereas sedimentary (and metamorphic) rocks are secondary and ultimately derived from igneous material.

Clarke and Washington adhered to this theory: The average composition of the crust is in effect that of igneous rocks; crustal abundances of the elements can be computed, therefore, by averaging the chemical analyses of the igneous rocks. The overall average of the 5159 analyses of igneous rocks compiled by Washington is still the accepted basis for nearly all geochemical calculations. V. M. Goldschmidt, a scientist of unusual capacity and founder of modern geochemistry was likewise deeply rooted in Rosenbusch's school; so was N. L. Bowen, an equally great man and the founder of modern petrology.

Thus the dead hand of the past is still directing our thinking: The average composition of the sedimentary rocks must correspond to that of the igneous rocks; for ultimately all sediments derive from igneous material. However, if this theory is applied to actual rocks, discrepancies are found, and sundry hypotheses have appeared to explain away the facts. A particularly disturbing difference is that of sodium: according to current calculations, the igneous rocks contain about three times as much sodium as do the sedimentary rocks.

In a paper of great consequence given before the German Mineralogische Gesellschaft in 1933 and repeated in his Hugo Müller lecture delivered before the

---

* Presidential Address presented before the Geochemical Society November 1, 1960, during its annual meeting in Denver, Colorado.

Chemical Society in London in 1937, and again repeated in his posthumous book of 1954, GOLDSCHMIDT maintained in principle*:

$$\text{Age of the Sea} = \frac{\text{Total amount of Na in the sea}}{\text{Na supplied per year}} \quad (1)$$

As a consequence of this GOLDSCHMIDT assumed:

$$\text{Total Weight of weathered rocks } W = \frac{\text{Total amount of Na in the sea}}{\text{Na\% of igneous rocks} - \text{Na\% of Sediments}} \quad (2)$$

His reasoning is well known: He thought that the sodium missing in the sediments is to be found in the sea.† The concentration of sodium in igneous rocks and in sediments are about 3 per cent and 1 per cent respectively. The total amount of sodium in the ocean corresponds to a mass of 3 kg per cm^2 of the earth's surface. From (2) we thus obtain, W total $= \frac{3}{3\% - 1\%} = 150$ kg per cm^2. By introducing the small corrections mentioned above GOLDSCHMIDT actually arrived at 160 kg per cm^2.

I have several objections to this kind of calculation, the most important are:
  I. I do not believe that only igneous rocks or their derivatives make up the crust.
  II. I do not believe that the weathered sodium stays in the ocean.

Ad. I

The source of the sediments is obviously the surface of the earth. In addition to older sediments which will be reworked, there are, at the surface, much more that 50 per cent of rocks of metamorphic origin (see Table 1). Metamorphites, in particular crystalline schists and gneisses, have been changed chemically by "juvenile" matter which derive from the same source as that of the igneous rocks. Consequently, there should *not* be a direct chemical correspondence between igneous and sedimentary rocks: all previous quantitative computations of crustal abundance are erroneous (see Table 2).

The composition of the surface of the earth at any one time, in particular that of the surface of the dry continents, is of the utmost importance. Not until such data are at hand, can we hope for a more quantitative knowledge of the geochemical migrations.

Unfortunately, we know very little about areal averages. This was again and most recently emphasized by FLEISCHER and CHAO before the International Geological Congress last summer (1960): "The need for better information was clearly stated by Daly in 1914, who pointed out that such data were required for many petrological problems. Essentially nothing has been attempted since. We can only repeat his hope that some international group should encourage the

---

* He made a modification by assuming that 33% of the sodium was locked up in evaporites and other sediments. It should also be added that he did not think that the present supply of Na (as observed in river water, etc.) should be used.

† He made a small correction in accordance with the fact that the total amount of sediments (shales) is slightly less than that of the igneous rocks (v.i.).

Abundance of the elements, areal averages and geochemical cycles

Table 1. Norway south of 65° N. 231500 km²

Relative Areas of Rock Species

		I			
Caledonian Zone		144100 km²	62·3%		
1. T.hjemite-Opdalite	14·4%	2600 km²			
2. Gabbro-Amphibolite	16·0%	2900 km²			
3. Ultrabasics	3·9%	700 km²			
4. Thrust masses, acid	18·8%	3400 km²			
5. Thrust masses basic	47·0%	8500 km²			
Intrusives				18100 km²	12·5%
Devonian				2300 km²	1·6%
6. Valdres sparagmite		1100 km²			
7. Schists-Jotun-W. Coast		19500 km²			
8. Schists-T.hjem area		27000 km²			
9. Schists-Grong area		4300 km²			
Cambro-Silurian Supracrustals				51900 km²	36·0%
10. Sparagmite (arkoses)		18600 km²			
11. Meta-sparagmite in N.W.		1100 km²			
Eocambrian				19700 km²	13·7%
12. Fosen area		11000 km²			
13. Møre-Romsdal		40000 km²			
14. Haugesund		1100 km²			
Migmatite				52100 km²	36·2%
		II			
Oslo Graben		9900 km²	4·3%		
1. Subvolcanics	77·2%	6100 km²			
2. Basic lavas	3·8%	300 km²			
3. Romb Porphyries	17·7%	1400 km²			
4. Tuffs, etc.	1·3%	100 km²			
Igneous rocks				7900 km²	80·0%
Sedimentary rocks				2000 km²	20·0%
		III			
Precambrian		77500 km²	33·4%		
1. Mixed gneisses	48·0%	26000 km²			
2. Gneissic granites	33·2%	18000 km²			
3. Quartzites	12·0%	6500 km²			
4. Post-Granites	4·2%	2300 km²			
5. Anorthosite, etc.	2·2%	1200 km²			
6. Amphibolite	0·4%	200 km²			
W. of Oslo Graben				54200 km²	70·0%
Östfold-Trysil				19000 km²	24·5%
"Windows"				4300 km²	5·5%

Table 2. Chemical composition of the crust

Some Areal Averages
(cation per cent)

Averages for Norway

	Caledonian Zone	Precambrian	Oslo Igneous
Si	60·9	65·2	55·2
Ti	0·5	0·5	0·8
Al	17·1	15·5	18·3
Fe	1·6	1·0	2·1
Fe	2·8	2·0	1·6
Mn	0·1	—	0·1
Mg	4·0	2·3	1·7
Ca	3·7	2·8	3·1
Na	5·2	6·0	9·6
K	3·7	4·3	5·3
P	0·1	0·2	0·2
	100·0	100·0	100·0

Other Averages

	S. Norway Grand Total	Finland Sederholm 1925	Norw. Loams Goldschmidt 1933	Ign. Rocks Clark and Washington
Si	62·3	63·8	58·3	55·8
Ti	0·5	0·3	0·6	0·7
Al	16·6	16·8	18·4	17·0
Fe	1·5	0·9	2·5	2·2
Fe	2·6	2·3	3·0	3·0
Mn	0·1	0·1	0·1	0·1
Mg	3·4	2·4	4·9	4·9
Ca	3·4	3·5	3·2	5·2
Na	5·6	5·6	3·9	7·1
K	3·9	4·3	4·9	3·8
P	0·1	0·1	0·2	0·2
	100·0	100·0	100·0	100·0

compilation, country by country, of estimates of the relative areal extent of the various types of igneous rocks. Until this is done, estimates of abundance of elements must be of the nature of guesses".

Indeed, not only the igneous rocks, but also the areal extent of the various types of the sedimentary and metamorphic rocks must be estimated. I hope soon to finish the work of such a survey of Southern Norway.

Ad II

The accumulation of sodium in the ocean would mean a one way traffic in geochemistry, or, rather worse: The ocean becomes a kind of digestive organ of geochemistry: fed by rivers, wind and ice, it receives the weathered material, digesting certain elements for ever, whereas others are rejected and returned to the endless cycle of geochemistry.

This is not so! The ocean is just a channel through which the migrating elements are cycled. Some of the elements run fast through this channel, others are slow, and according to the speed the concentration will vary. There is no one-way traffic, but a dynamic equilibrium between supply and removal. Consequently it becomes in principle impossible to determine the age of the sea from its salt content, and all previous quantitative computations of the total mass of sediments are erroneous.

I prefer to look at the problem in a different way: If we assume perfect balance between input and output, it is possible to calculate how long each element remains in the sea before it is removed.* The period of passage, τ_i may be defined as the average time expressed in years necessary for an element i to pass through the sea in the geochemical cycle. From our assumptions we derive

$$\tau_i = \frac{\text{Total amount of an element in the sea}}{\text{Amount supplied per year}} \tag{I}$$

Thus we find that a sodium ion spends an average of 208 million years in the sea but potassium only 10 million years (see data in Table 3). The attainment of such balances must have taken some time. So the query is: Is the span from the earliest to modern time long enough? Computations show that for all practical purposes equilibrium for an element will be attained after three times the period of passage. Thus for sodium within about 600 million years, for potassium within only 30 million years.

Equation I gives a totally different meaning to the Goldschmidt equation here quoted as (1). Obviously Goldschmidt's second equation will also be influenced. To simplify the following calculations I shall in the first approximation, limit myself to the typical geosynclinal sediments, shales and schists.

The problem may be formulated as follows:

Let W be the total weight of weathered rock in unit time (mainly derived from

---

* Many questions could be asked as to the justification of this assumption. But time does not permit to discuss it fully here. The main question is, perhaps, how sodium, which is conspicuously low in the sediments, can find its way from the sea and back into continental rocks. My answer to this is: By formation of evaporites, and by occlusion in the pore spaces of practically all sedimentary rocks (the efficiency of both processes will be proportional to the concentration in the sea). Part of the occluded sea water is leached out of the sediments and sodium carried in solution back to the sea, part of it manifests in sodium-bearing minerals of the metamorphosed sediments.

5

Table 3. Data for calculation of the amount of weathered rocks

	Ign. Rocks	Canada	Finland	Norway	Caledonian	Shale × 0.85
Mg	2.09	1.07	1.02	1.45	1.70	1.19
Ca	3.63	2.91	2.42	2.41	2.60	0.34
Na	2.84	2.70	2.27	2.28	2.15	0.85
K	2.59	2.55	2.95	2.72	2.60	2.16
	\multicolumn{5}{c}{$c_i - 0.85 \cdot s_i$}	D_i
Mg	0.90	—	—	0.26	0.51	0.016
Ca	3.29	2.57	2.08	2.07	2.26	0.108
Na	1.99	1.85	1.42	1.43	1.30	0.014
K	0.43	0.39	0.79	0.56	0.44	0.011
	\multicolumn{5}{c}{$D_i/c_i - 0.85 \cdot s_i$}	Average
Mg	1.8	neg.	neg.	6.2	3.1	—
Ca	3.3	4.2	5.2	5.2	4.8	4.5
Na	0.7	0.8	1.0	1.0	1.1	0.9
K	2.6	2.8	1.4	2.0	2.5	2.3
Average	2.1	—	—	3.6	2.9	2.7

The content of Mg, Ca, Na, and K expressed in weight per cent is given for Igneous rocks (CLARKE and WASHINGTON), Canadian Shield (GROUT, 1938), Findland (SEDERHOLM, 1925), Norway S of 65°N, and Caledonian rocks in Scandinavia S of 65°N. D_i is the amount delivered by rivers to the sea expressed as kg per cm² in a millon years, according to data by CONWAY (1942).

the continents). It will split into two parts: $W = \Sigma D_i + S$. ΣD_i is the sum of all dissolved ions that are carried into the sea, S is mainly the hydrolyzates that form the shales and schists. Let c_i and s_i be the average weight percentages of an element, i, in the continents and in the shales respectively. The following equation will be satisfied for each element.*

$$c_i W = D_i + s_i S$$

By making certain assumptions WICKMAN (1954) has found that $S = 0.845W$, this factor does not affect the principles of the calculations, for convenience we shall use 0.85, and we obtain:

$$W = \frac{D_i}{c_i - 0.85 s_i} \quad (II)$$

If the total amount of an element in the sea is designated by M_i, and we use the other symbols as introduced above, we obtain in principle two sets of equations of similar form but of different meaning.

Goldschmidt: Age of the Sea $= \dfrac{M_{Na}}{D_{Na}}$ (1) $\qquad W_{Total} = \dfrac{M_{Na}}{c_{Na} - s_{Na}}$ (2)

Barth: Period of passage, $\tau_i = \dfrac{M_i}{D_i}$ (1) $\qquad W_{Year} = \dfrac{D_i}{c_i - s_i}$ (II)

---
* Provided there are no gains or losses in the geochemical cycle.

Combining (I) and (II) shows that $\tau_i W_{\text{Year}} = \dfrac{M_i}{c_i - s_i}$, or that GOLDSCHMIDT'S equation (2) does not give the total amount of weathered rock but, for each element, gives the amount that weathered during the "period of passage" of that particular element. Thus Goldschmidt's result, computed from the sodium balance, in the amount of 160 kg "total" weathered rock per cm² of the earth (and extensively quoted in the geochemical literature) actually applied to the amount of weathered rock during 208 million years (because $\tau_{\text{Na}} = 208$ million years is the period of passage of sodium through the sea).

My equations as distinct from those of Goldschmidt's apply in principle to all elements. But using shales (and schists) as the standard sediment they are, as previously explained, limited to the elements that go into ionic solutions virtually: magnesium, calcium, sodium, and potassium. Any of these elements used in equation II should give the same figure for W.

The amounts of these elements delivered by the rivers of the world into the sea were first estimated by Clarke. The latest and most accurate estimates are those of CONWAY (1943), they are used here. The composition of the average shale is taken from WICKMAN (1954). Based on these figures one may calculate from equation II the weight of weathered rocks per unit time, for example per a million years (see table 3). The figures refer to kg per cm² of the surface of the earth. The average comes out as 2·7 kg/cm² corresponding to an average erosion of 10 meters per million years. Putting the geological age of the earth at $3 \cdot 10^9$ years, the "total" erosion becomes 30 kilometers. This is a large figure, but it harmonizes quite well with other estimates of the total erosion based on geological evidence.

However, this figure is obviously inaccurate. It will have to be adjusted as more knowledge becomes available. But the principles involved in the present calculations come closer to the true conditions than did those of GOLDSCHMIDT. They expose to us a vista of the possibilities of some day arriving at an entirely new petrology.

If we assume that weathering and sedimentation have been restricted to the continents and their shelves, while the ocean troughs were passive, it means that a total erosion corresponding to a depth of more than 60 km must have taken place in the continents. If the order of magnitude of this figure is correct, we have to conclude that practically all continental matter has been reworked, i.e., there is no "juvenile" matter in the continents; all rocks have been sediments but modified and changed by metamorphism, metasomatism, and remelting under plutonic conditions, then again raised to the surface to become sediments. Said Hutton "In the economy of the world, I can find no traces of a beginning, no prospect of an end".

REFERENCES

CLARKE F. W. and WASHINGTON H. S. (1924) The composition of the Earth's crust. *U.S. Geol. Survey, Prof. Paper*, **127**.

CONWAY E. J. (1943) Mean geochemical data in relation to oceanic evolution. *Proc. Irish Royal Acad.* **48**, sect. B, No. 8.

FLEISCHER M. and CHAO E. C. T. (1960) Some problems in the estimation of the abundances of elements in the Earth's crust. XXI Int. Geol. Congress, Pt. 1, p. 141. (Copenhagen 1960).

GOLDSCHMIDT V. M. (1933) Grundlagen der quantitativen Geochemie. *Fortschr. der Mineral. Krist. und Petr.* **17,** p. 112.
GOLDSCHMIDT V. M. (1937) Principles of distribution of chemical elements in minerals and rocks. *Jour. Chem. Soc. London*, April, 1937, p. 655.
GOLDSCHMIDT V. M. (1954) *Geochemistry*. Clarendon Press, Oxford.
GROUT F. F. (1938) Petrographic and Chemical data on the Canadian Shield. *J. Geol.* **46,** p. 486.
SEDERHOLM J. J. (1925) The average composition of the Earth's crust in Finland. *Bull. comm. geol.* Finlande No. **70**.
WASHINGTON H. S. (1917) Chemical analyses of igneous rocks. *U.S. Geol. Survey, Prof. Paper* **99**.
WICKMAN F. E. (1954) The total amount of sediments and the composition of the average igneous rock. *Geochim. et Cosmochim. Acta* **5,** p. 97.

CHEMICAL MASS BALANCE BETWEEN RIVERS AND OCEANS

FRED T. MACKENZIE* and ROBERT M. GARRELS**

ABSTRACT. The assumption of constancy of the chemical composition of ocean water requires that the excesses of dissolved constituents carried by streams to the ocean be removed. The chemical mass balance between streams and oceans is an attempt to evaluate this removal process, to see what reactions are necessary to accomplish this removal, and to maintain present ocean water composition.

Postulation of a steady-state model for the ocean leads to the removal of stream-derived dissolved solids as minerals in marine sediments, as dissolved constituents in sediment pore waters, and as materials cycled through the atmosphere. Disposal of the dissolved solids requires synthesis of typical clay minerals from degraded aluminosilicates before burial in sediments, by reactions of the type:

X-ray amorphous Al silicate $+ SiO_2 + HCO_3^- +$ cations $=$ cation Al silicate $+ CO_2 + H_2O$.

Such chemical, if not structural, synthesis implies control of the major ion ratios in sea water and the CO_2 pressure of the atmosphere by equilibria involving aluminosilicates. These "reverse weathering" reactions, by removing HCO_3^- and alkali metals from the oceanic system, prevent the ocean from attaining the composition of a soda lake. Also, chemical synthesis of typical clay minerals in the oceanic system implies that these minerals are in equilibrium, or nearly so, with an aqueous solution of the composition of average ocean water and should not be significantly altered upon entering the ocean as allogenic particles.

INTRODUCTION

Although the ocean receives materials from several sources, the rates of flow of water and dissolved constituents into the ocean from streams far exceeds that from any other source. Postulation of a steady-state oceanic system (Rubey, 1951; Sillén, 1961, 1963; Holland, 1965; Kramer, 1965) requires that these stream-derived constituents be removed as rapidly as they enter the ocean. Garrels and Mackenzie (in press) showed that simple isothermal evaporation, at the P_{CO_2} of the Earth's atmosphere, of waters derived from felsic igneous rocks and isolated from solid weathering products leads to the formation of a soda lake water, high in alkali cations, low in alkaline-earth cations, and with a pH of about 10. The ocean certainly is not a giant soda lake, but the Sierra waters are similar to river waters, which definitely give rise to water of the composition of the ocean. Therefore, the question arises as to why the ocean is not a soda lake.

The purpose of this paper is to present a simple chemical mass balance between streams and oceans. This mass balance is an attempt to evaluate the processes that lead to the removal of stream-derived dissolved constituents from the ocean system and to see what reactions are necessary to accomplish this removal and to maintain constant present ocean water composition.

* Bermuda Biological Station for Research, St. George's West, Bermuda.
** Department of Geology, Northwestern University, Evanston, Illinois.

SOME PRELIMINARY CONSIDERATIONS

As background for the mass balance between river water and ocean water, we will consider some of the relations between streams and oceans.

Dissolved constituents.—More than 99 percent of the dissolved material in the ocean is accounted for by the constituents Na, Mg, Ca, K, Cl, SO_4, HCO_3, and SiO_2; the same constituents make up more than 99 percent of the total dissolved solids brought in by rivers (table 1). Ocean water is several to many times more concentrated with respect to the major constituents than river water, except for silica, which is much lower in the ocean.

The rates of flow of stream-derived dissolved constituents into the ocean are rapid. The major constituents, except for Na, have residence times in the ocean of less than 10^8 years (Barth, 1952; Goldberg and Arrhenius, 1958), a geologically short time. Livingstone (1963a) has made the most recent study of discharges and dissolved loads of rivers of the world and estimated that 3.9×10^{12} kilograms of dissolved solids are carried to the ocean annually by rivers. This amount of dissolved materials, if precipitated as solids, would approximately equal the total volume of the ocean (1.37×10^{21} liters) in 10^9 years, assuming no net removal of materials from the oceanic system. Also, if the ocean basins were not filled with water and the rivers at their present discharge began to fill them, it would require only 40,000 years for the rivers to fill the ocean basins to their present level. Assuming that the composition of ocean water has remained essentially constant for the past 10^8 years

TABLE 1

Major constituents of river water and sea water

Constituent	River water[1]		Sea water[2]	
	ppm	millimoles/liter	ppm	millimoles/liter
Cl^-	7.8	0.220	19,000	535.2
Na^+	6.3	0.274	10,500	456.5
Mg^{++}	4.1	0.171	1,300	54.2
$SO_4^=$	11.2	0.117	2,650	27.6
K^+	2.3	0.059	380	10.0
Ca^{++}	15	0.375	400	9.7
HCO_3^-	58.4	0.958	140	2.3
SiO_2	13.1	0.218	6	0.1
NO_3^-	1	0.016	----	----
Fe	0.67	0.012	----	----
Br^-	----	----	65	0.8
$CO_3^=$	----	----	18	0.3
Sr^{++}	----	----	8	0.1

[1] Livingstone (1963a)
[2] Goldberg (1957)

(Lowenstam, 1961, presents some evidence for constancy of ocean water composition for at least the past 2×10^8 years) and that present stream discharge and content are characteristic of the geologic past, the amount of water brought to the ocean by rivers in 10^8 years would be equal to 2400 times the amount now in the ocean. By similar reasoning, the amount of Cl carried to the ocean in 10^8 years would just about equal the amount currently in the ocean, Na would have been "renewed" 1.4 times, Mg 7 times, SO_4 10 times, K 15 times, Ca 81 times, HCO_3 1000 times, and SiO_2 5300 times (table 2). The number, 10^8 years, is not a "magic" number but is used to illustrate that the rates of delivery of dissolved constituents to the ocean by rivers are rapid and that if the ocean suddenly became fresh, it would not take long relative to 10^8 years for it to exceed its present salinity, if all the dissolved solids carried to the ocean by rivers remained in solution.

Suspended material.—Great tonnages of suspended material are carried by rivers to the ocean. If we assume a present average annual world-wide river runoff to the ocean of 3.3×10^{16} liters/year (Livingstone, 1963a) and an average suspended-sediment concentration equal to that of the Mississippi River, 250 milligrams/liter (Edwards, Kister, and Scarcia, 1956), we calculate 8.3×10^{12} kilograms as the amount of suspended sediment brought to the ocean annually. Kuenen (1950) estimates 32.5×10^{12} kilograms. Thus, the world-wide ratio of river suspended load to dissolved load probably is somewhere between 2/1 (Mississippi data) and 8/1 (Kuenen's estimate). Assuming a density of 2.7 grams/cubic centimeter for the suspended particles of rivers and therefore a minimum volume of compacted sediment, the quantity of

TABLE 2

Number of times river constituents have "passed through" the ocean in 10^8 years assuming present annual world-wide river discharge, mean dissolved constituent concentration of rivers and ocean, and ocean volume of 1.37×10^{21} liters

Constituent	Amount delivered by rivers to the ocean annually	Amount in ocean	Amount delivered by rivers to ocean in 10^8 years	Number of times constituents have been "renewed" in 10^8 years
SiO_2	42.6×10^{10} kg	0.008×10^{18} kg	42.6×10^{18} kg	5300
HCO_3^-	190.2×10^{10} kg	0.19×10^{18} kg	190.2×10^{18} kg	1000
Ca^{++}	48.8×10^{10} kg	0.6×10^{18} kg	48.8×10^{18} kg	81
K^+	7.4×10^{10} kg	0.5×10^{18} kg	7.4×10^{18} kg	15
$SO_4^=$	36.7×10^{10} kg	3.7×10^{18} kg	36.7×10^{18} kg	10
Mg^{++}	13.3×10^{10} kg	1.9×10^{18} kg	13.3×10^{18} kg	7
Na^+	20.7×10^{10} kg	14.4×10^{18} kg	20.7×10^{18} kg	1.4
Cl^-	25.4×10^{10} kg	26.1×10^{18} kg	25.4×10^{18} kg	1
H_2O	$3,333,000 \times 10^{10}$ kg	1370×10^{18} kg	$3,333,000 \times 10^{18}$ kg	2400

suspended material delivered to the ocean in 10^9 years would amount to 2 to 9 times the volume of the ocean, if all this material remained in the oceanic system.

Type and reactivity of suspended material.—Unfortunately, complete chemical and mineralogical analyses of the suspended load of modern streams are lacking, but some general conclusions can be made concerning the composition and reactivity of the suspended sediment. This material is dominantly composed of the aluminosilicate minerals: montmorillonite, kaolinite, illite, chlorite, mixed-layer clay, and dioctahedral vermiculite; and of X-ray amorphous aluminosilicates, gibbsite, fine-grained fragments of quartz, feldspar, carbonate, Fe-Ti minerals, rock fragments, and organic matter (compare Powers, 1954; Grim and Johns, 1954; Nelson, 1963; Taggart and Kaiser, 1960; Griffin, 1962; Kennedy, 1963, 1965). It is interesting to note that these same materials, primarily the aluminosilicates, compose the bulk of modern, fine-grained, non-biogenic sediments in the ocean (compare Correns, 1937; Revelle, 1944; Griffin, 1962; Griffin and Goldberg, 1963; Arrhenius, 1963; Biscaye, 1965) and are the major constituents found in ancient shales. The *gross* mineralogical similarity between ancient shales and modern marine argillaceous sediments has led some investigators to conclude that the bulk of the suspended material carried by rivers to the ocean is unreactive and consequently has little effect on oceanic chemistry.

Studies of the distribution patterns of clay minerals in modern stream and nearshore marine sediments (Grim, Dietz, and Bradley, 1949; Favajee, 1951; Grim and Johns, 1954; Powers, 1957, 1959; Nelson, 1963) show progressive changes in the clay mineral types from fresh to saline waters. For example, Grim and Johns (1954) found that illite and chlorite increased at the expense of montmorillonite down the Guadalupe River into the Gulf of Mexico. They interpreted this distribution to be the result of the conversion of montmorillonite to chlorite and illite in the higher salinity environments. The experiments of Whitehouse and McCarter (1958), although not conclusive, lend a certain amount of support to this conclusion. They immersed montmorillonitic material in artificial sea water for periods of up to five years. Analyses of the material showed that montmorillonite was transformed to illitic and chloritic clays. Similar experiments with kaolinite and illite showed that these minerals were unaltered after immersion.

In rebuttal of the above evidence for the reactivity of land-derived clay minerals in the marine environment, Weaver (1959) suggested that the distribution patterns of clay minerals in modern sediments may be due to such physical factors as preferential flocculation, current sorting, effect of various source areas, floods, and periodic variations in the concentration and composition of river detritus. Griffin (1962) presented convincing evidence that the clay mineral distributions in the northeastern Gulf of Mexico are a function of parent rock, stream drainage patterns, and current patterns in the Gulf. Griffin and Goldberg (1963) studied clay mineral distributions in the Pacific Ocean. Their data show that land-derived clay minerals are not markedly altered on entering the Pacific Basin, although K^+- and Mg^{++}—uptake by degraded clays may occur. Biscaye (1965) reported that most of the recent deep-sea clay from the Atlantic Ocean is detritus from the continents, and in situ formation of clay minerals is unimportant in the Atlantic Basin. Hurley and others (1963) determined K-Ar ages for K-bearing minerals, principally illite, in pelagic sediments from the North Atlantic. Their age values range from 200 to 400 million years and indicate that the bulk of oceanic illites is detrital.

Recent experimental work (Mackenzie and Garrels, 1965), in which aluminosilicates were placed in low-silica sea water and the pH and silica content of the systems monitored with time, showed that silicate minerals rapidly release silica to sea water. The silica concentrations reached in these experiments were in the order of the average silica content of the ocean (approximately 6 ppm). We concluded that the rates observed indicate that the ocean must be considered as a chemical system with a rapid compositional response to added detrital silicates.

PROCEDURES AND CALCULATIONS

In the steady-state oceanic model, the dissolved constituents carried by streams to the ocean are removed as rapidly as they are added. To simulate this model, we will imagine a hugh batch process in which we allow the rivers to flow for 10^8 years into an ocean of present-day constant composition, maintaining a constant ocean volume by removing

TABLE 3

Amount of dissolved constituents transported to the ocean by rivers in 10^8 years, assuming present mean annual total dissolved load and mean composition of world's rivers

Constituent	Mean composition of river waters of the world		Amount of dissolved constituents brought to the ocean annually, assuming the annual total dissolved load is 390.5×10^7 metric tons		Amount of dissolved constituents carried to the ocean in 10^8 years
	ppm	millimoles liter	$(\times 10^{-7}$ metric tons)	$(\times 10^{-13}$ moles)	$(\times 10^{-21}$ mmoles)
HCO_3^-	58.4	0.958	190.2	3.118	3118
Ca^{++}	15	0.375	48.8	1.220	1220
Na^+	6.3	0.274	20.7	0.900	900
Cl^-	7.8	0.220	25.4	0.715	715
SiO_2	13.1	0.218	42.6	0.710	710
Mg^{++}	4.1	0.171	13.3	0.554	554
$SO_4^=$	11.2	0.117	36.7	0.382	382
K^+	2.3	0.059	7.4	0.189	189

only H_2O. If we assume that present stream discharge and content remain constant for 10^8 years, we may calculate the total discharge into the ocean in this period of time of each major stream-derived dissolved constituent (table 3). The river data used in this calculation were taken from Livingstone (1963a), who has made the most recent and comprehensive study of the chemical composition of rivers and lakes. He discusses thoroughly the problems involved in making estimates of the mean dissolved constituent composition of rivers and of the total amount of dissolved material carried to the ocean annually, and we will not pursue these problems further in this paper. His estimates are probably the best to date. Furthermore, our balance is not significantly affected by changes in estimates of the mean world-wide salinity of rivers but only by relatively large changes in the estimates of the mean ratios of dissolved constituents. Table 4 shows that the mean percentage estimates of dissolved constituents in river water as given by several investigators are similar.

We will now dispose of these excesses of dissolved constituents, recognizing that we have simulated a steady-state rate process by integration over 10^8 years of the rates of flow of dissolved constituents into the oceanic system. From the excesses, we have chosen to make the most common minerals observed to precipitate from the ocean, or those reported as most common in marine sediments. Some of the excess material is removed in the pore waters of sediments or through the atmosphere. The balance sheet is shown in table 5. Details of the removal process are given in further sections.

Reactions involving sulfate.—River waters contain 11.2 parts per million $SO_4^=$ derived from (1) oxidation of pyrite and solution of gypsum or anhydrite during rock weathering, (2) sea-spray particles generated at the ocean-atmosphere interface and removed from the atmosphere by rain or as dry fallout, (3) oxidation by photochemical processes in the atmosphere of H_2S formed by the decomposition of organic materials, and (4) industrial activities. About 2.5 parts per million $SO_4^=$ is added to the rivers by atmospheric processes (data in Gorham, 1961; Junge and Werby, 1958); thus 8 to 9 parts per million $SO_4^=$ in rivers is derived from the weathering of rocks, and a negligible amount from human activities in which sulfur-bearing wastes are added directly to the rivers.

Most of the sulfur in sediments is found as gypsum or anhydrite in evaporites and as pyrite in shales and other sedimentary rocks. Unfortunately, the amount of sulfur in the geologic column tied up in pyrite versus that in gypsum or anhydrite is difficult to determine. For the purposes of our balance, we have removed 50 percent of the $SO_4^=$ in making pyrite and 50 percent as $CaSO_4$. The relative proportions used have little effect on the balance; the excess HCO_3^-, generated by forming pyrite, is used up to form $CaCO_3$ by reacting the HCO_3^- with the Ca^{++} left over after making $CaSO_4$.

TABLE 4

Comparison of estimates of the mean composition of river waters of the world

Constituent	Salinity, 181 ppm (based on Murray's data, 1887)		Salinity, 187 ppm (based on Clarke's river content data 1924; and Murray's H₂O runoff estimate, 1887)		Salinity, 199 ppm (Conway's, 1942, estimate of Clarke's data)		Salinity, 120 ppm based on Livingstone's data, 1963)	
	ppm	Percent of total dissolved solids	ppm	Percent of total dissolved solids	ppm	Percent of total dissolved solids	ppm	Percent of total dissolved solids
HCO_3^-	152	58.7	71.5	52.3	104	52.3	58.4	48.7
Ca^{++}	37.1	14.3	20.4	14.9	29.8	14.9	15	12.5
Na^+	6.3	2.4	5.8	4.2	8.4	4.2	6.3	5.3
Cl^-	3.3	1.3	5.7	4.2	8.3	4.2	7.8	6.5
SiO_2	19.5	7.6	11.7	8.6	17.1	8.6	13.1	10.9
Mg^{++}	8.4	3.2	3.4	2.5	5	2.5	4.1	3.4
$SO_4^=$	14.9	5.8	12.1	8.9	17.7	8.9	11.2	9.4
K^+	2.4	0.9	2.1	1.6	3.1	1.6	2.3	1.9
NO_3^-	5.1	2	0.9	0.7	1.3	0.7	1	0.8
Fe	8.6 (Fe + Al)	3.5	2.8 (Fe + Al)	2.0	4 (Fe + Al)	2.0	0.67	0.6
Others	0.7	0.3	0.2	0.1	0.3	0.1	—	—

TABLE 5

Mass balance calculation for removal of river-derived constituents from the ocean

Reaction (balanced in terms of mmoles of constituents used)	Constituent balance ($\times 10^{-21}$ mmoles)								HCO_3^- Consumed (−) / Evolved (+)	CO_2 Consumed (−) / Evolved (+)	Products ($\times 10^{-21}$ mmoles)	Percentage of total products formed (mole basis)
	$SO_4^=$	Ca^{++}	Cl^-	Na^+	Mg^{++}	K^+	SiO_2	HCO_3^-				
Amount of material to be removed from ocean in 10^8 years ($\times 10^{-21}$ mmoles)	382	1220	715	900	554	189	710	3118				
95.5 $FeAl_6Si_6O_{20}(OH)_4$ + 191 $SO_4^=$ + 47.8 CO_2 + 55.7 $C_6H_{12}O_6$ + 238.8 H_2O = 286.5 $Al_2Si_2O_5(OH)_4$ + 95.5 FeS_2 + 382 HCO_3^-	191	1220	715	900	554	189	710	3500	+382	−48	96 Pyrite / 287 Kaolinite	3% / 8%
191 Ca^{++} + 191 $SO_4^=$ = 191 $CaSO_4$	0	1029	715	900	554	189	710	3500			191 "$CaSO_4$"	5%
52 Mg^{++} + 104 HCO_3^- = 52 $MgCO_3$ + 52 CO_2 + 52 H_2O	0	1029	715	900	502	189	710	3396	−104	+52	52 $MgCO_3$ in Magnesian Calcite	2%
1029 Ca^{++} + 2058 HCO_3^- = 1029 $CaCO_3$ + 1029 CO_2 + 1029 H_2O	0	0	715	900	502	189	710	1338	−2058	+1029	1029 Calcite and/or Aragonite	29%
715 Na^+ + 715 Cl^- = 715 $NaCl$	0	0	0	185	502	189	710	1338			715 "$NaCl$"	20%
71 H_4SiO_4 = 71 $SiO_{2(s)}$ + 142 H_2O	0	0	0	185	502	189	639	1338			71 "Free" Silica	2%
138 $Ca_{0.17}Al_{2.33}Si_{3.67}O_{10}(OH)_2$ + 46 Na^+ = 138 $Na_{0.33}Al_{2.33}Si_{3.67}O_{10}(OH)_2$ + 23.5 Ca^{++}	0	24	0	139	502	189	639	1338			138 Sodic Montmorillonite	4%
24 Ca^{++} + 48 HCO_3^- = 24 $CaCO_3$ + 24 CO_2 + 24 H_2O	0	0	0	139	502	189	639	1290	−48	+24	24 Calcite and/or Aragonite	1%
486.5 $Al_2Si_2O_5(OH)_4$ + 139 Na^+ + 361.4 SiO_2 + 139 HCO_3^- = 417 $Na_{0.33}Al_{2.33}Si_{3.67}O_{10}(OH)_2$ + 139 CO_2 + 625.5 H_2O	0	0	0	0	502	189	278	1151	−139	+139	417 Sodic Montmorillonite	12%
100.4 $Al_2Si_2O_5(OH)_4$ + 502 Mg^{++} + 60.2 SiO_2 + 1004 HCO_3^- = 100.4 $Mg_5Al_2Si_3O_{10}(OH)_8$ + 1004 CO_2 + 301.2 H_2O	0	0	0	0	0	189	218	147	−1004	+1004	100 Chlorite	3%
472.5 $Al_2Si_2O_5(OH)_4$ + 189 K^+ + 189 SiO_2 + 189 HCO_3^- = 378 $K_{0.5}Al_{2.5}Si_{3.5}O_{10}(OH)_2$ + 189 CO_2 + 661.5 H_2O	0	0	0	0	0	0	29	−42	−189	+189	378 Illite	11%

Pyrite in marine sediments results from the reduction of $SO_4^=$ and subsequent reactions with mineral phases containing iron (Berner, 1964). Concomitant oxidation of organic materials accompanies the $SO_4^=$ reduction. The source of iron is detrital sediments, primarily clays (Kaplan, Emery, and Rittenberg, 1963). There are several ways to write the $SO_4^=$ reduction reaction:

$$3\ SO_4^= + C_6H_{12}O_6 = 3\ H_2S + 6\ HCO_3^- \qquad (1)$$

where $C_6H_{12}O_6$ represents organic compounds, or

$$24\ Fe(OH)_3 + 48\ SO_4^= + 15\ C_6H_{12}O_6 + 6\ CO_2 =$$
$$24\ FeS_2 + 78\ H_2O + 96\ HCO_3^- \qquad (2)$$

where $Fe(OH)_3$ represents iron obtained from detrital iron-bearing materials and mobilized as ferric hydroxide and

$$12\ FeAl_6Si_6O_{20}(OH)_4 + 24\ SO_4^= + 6\ CO_2 + 7\ C_6H_{12}O_6 + 30\ H_2O =$$
$$36\ Al_2Si_2O_5(OH)_4 + 12\ FeS_2 + 48\ HCO_3^- \qquad (3)$$

where $FeAl_6Si_6O_{20}(OH)_4$ represents clay minerals as a source of iron. The important aspect of all three reactions is that two times as much HCO_3^- is released as $SO_4^=$ consumed. This is a crude estimate of the amount of HCO_3^- released during oxidation of organic materials and sulfate reduction, because the actual amount depends on the carbon to hydrogen ratio of the organic compounds involved. For the purpose of the balance, we will use reaction (3) which, using 50 percent of the available $SO_4^=$ in our balance to make pyrite, is written

$$95.5\ FeAl_6Si_6O_{20}(OH)_4 + 191\ SO_4^= + 47.8\ CO_2 + 55.7\ C_6H_{12}O_6 +$$
$$238.8\ H_2O = 286.5\ Al_2Si_2O_5(OH)_4 + 95.5\ FeS_2 + 382\ HCO_3^-.$$

Thus 191×10^{21} millimoles of $SO_4^=$ are consumed in making 95.5×10^{21} millimoles of pyrite and releasing 382×10^{21} millimoles HCO_3^-.

The rest of the $SO_4^=$ is removed as $CaSO_4$. The reaction is

$$Ca^{++} + SO_4^= = CaSO_4$$

and similarly in millimoles of material available in the balance

$$191\ Ca^{++} + 191\ SO_4^= = 191\ CaSO_4.$$

The "$CaSO_4$" formed represents that precipitated in evaporite basins as gypsum or anhydrite, $SO_4^=$ contained in the pore waters of sediments and balanced by Ca^{++}, and $SO_4^=$ and Ca^{++} deposited on the continents by atmospheric processes. At this point in our balance, all the $SO_4^=$ is used up.

Reactions involving carbonate minerals.—Calcium and magnesium are found in modern marine sediments and ancient rocks in several phases. We have removed some of the Ca^{++} in making "$CaSO_4$". The great mass of limestone and dolomite in the geologic column and the large carbonate provinces of the modern seas suggest that most of the rest of the Ca^{++} is removed from the oceanic system in minerals that form carbonate sediments—aragonite, calcite, magnesian calcite, and dolomite. Dolomite is not a major mineral component of modern carbonate sediments. Textural, structural, and geochemical relations suggest that most dolomite in modern and ancient sediments is not precipitated directly from solution. Therefore, for the purposes of the mass balance, we will remove the remaining Ca^{++} as aragonite or calcite and magnesian calcite.

Chave (1954) plotted the $MgCO_3$ content of modern calcareous shelf sediments against number of samples and concluded that the Mg^{++} in shelf sediments is derived mainly from skeletal material. His plot reveals that modern calcareous shelf sediments contain about 5 percent $MgCO_3$ as magnesian calcite, and therefore, an amount of Mg^{++} equal to 5 percent of the Ca^{++} is removed from our balance as magnesian calcite. This percentage represents a maximum for the percentage of $MgCO_3$ tied up in modern calcareous sediments since Turekian (1964) showed that most of the calcium carbonate is now being deposited in the deep sea, and the magnesium concentration of this calcium carbonate is 0.05 to 0.1 percent. The reaction is

$$Mg^{++} + 2\ HCO_3^- = MgCO_3 + CO_2 + H_2O$$

and in millimoles

$$52\ Mg^{++} + 104\ HCO_3^- = 52\ MgCO_3 + 52\ CO_2 + 52\ H_2O.$$

The remaining Ca^{++} is removed as calcite and/or aragonite according to the reaction

$$1029\ Ca^{++} + 2058\ HCO_3^- = 1029\ CaCO_3 + 1029\ CO_2 + 1029\ H_2O.$$

At this stage in the removal of materials, $SO_4^=$ and Ca^{++} have been completely used up, 9 percent of the Mg^{++} has been reacted, 43 percent of the original HCO_3^- is unreacted, and the products are pyrite, "$CaSO_4$", aragonite and/or calcite, and magnesian calcite. Almost half the HCO_3^- remains to be accounted for; even by assuming that dolomite is directly precipitated in the ratio to calcite found in ancient rocks (⅕), we could not reduce the HCO_3^- to less than 30 percent of the original amount.

The important point is that the amount of HCO_3^- carried to the ocean by streams cannot be disposed of by deposition of carbonate sediments. Assuming that all the Ca^{++} is removed with an equivalent amount of HCO_3^- and no Ca^{++} leaves the oceanic system balanced by $SO_4^=$, the most conservative estimate of the bicarbonate excess that can be calculated from the estimates of the mean composition of river waters (table 4) is 15 percent of the original amount transported to the ocean by streams. The following sections are concerned with a mechanism for removal of this HCO_3^- excess and with the possible processes and reactions that lead to the disposal of this excess HCO_3^- and remaining dissolved constituents, in accordance with the volumes of sedimentary rocks in the geologic column and the types of minerals found in marine sediments.

Reactions involving chloride and sodium.—All the Cl⁻ is removed from the system with an equal amount of Na⁺. Part of this "NaCl" leaves the system as halite, part becomes dissolved in the pore waters of sediments, and part goes into the atmosphere. Some Cl⁻ may be removed in the exchange positions of clay minerals, but we do not have a reasonable estimate of the amount of Cl⁻ that can be disposed of by this process. Livingstone (1963b) recently recalculated the sodium balance in nature and the sodium age of the ocean. His data show that 65 to 75 percent of the Na⁺ carried by rivers to the ocean since post-Alkongian time can be accounted for by deposition of halite in evaporite deposits, as a dissolved constituent in the pore waters of sediments or as material cycled through the atmosphere. He suggests that the rest of the Na⁺ is removed in the exchange positions of clay minerals. Our value of 80 percent, obtained by removing equal amounts of Cl⁻ and Na⁺, is in reasonable agreement with Livingstone's.

The NaCl reaction is

$$Na^+ + Cl^- = NaCl$$

and in millimoles

$$715\ Na^+ + 715\ Cl^- = 715\ NaCl.$$

The removal of the remaining Na⁺ will be discussed in the next section.

Reactions involving silica and aluminosilicate minerals.—At this stage in the balance, we must remove Mg^{++}, K^+, Na^+, and a large amount of dissolved SiO_2 and HCO_3^- from the system. Bicarbonate and SiO_2 are the clues to this removal. As shown in a previous section, HCO_3^- cannot be disposed of simply by deposition of carbonate minerals. Accumulation of this HCO_3^- would lead to an increase in the alkalinity of the oceans, and ocean water composition would evolve toward that of a soda lake. Also, at the present rate of flow of dissolved SiO_2 into the ocean by streams, the volume of silica, primarily as chert, would be much greater than observed in the geologic column if all the SiO_2 were precipitated chemically or biochemically as "free" silica (silica not bound with alumina and essentially devoid of alkaline-earth or alkali cations). This relationship is developed in detail by Mackenzie and Garrels (in press); figure 1 is adapted from that work.

To resolve the HCO_3^- excess and to be consistent with the volume of "free" silica in the geologic column, we suggest that HCO_3^-, SiO_2, and cations react with degraded aluminosilicates of the suspended load of rivers to form new minerals, according to reactions of the type

X-ray amorphous Al silicate + HCO_3^- + SiO_2 + cations =
  cation Al silicate + CO_2 + H_2O.

These reactions may be considered "reverse weathering" reactions in that new minerals are formed from degraded clays and dissolved constituents and CO_2 released; whereas in weathering reactions, degraded

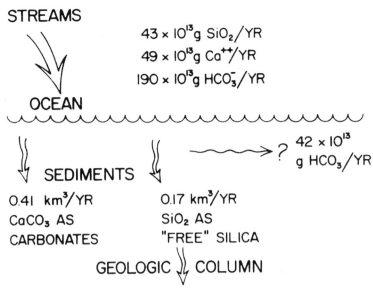

Fig. 1. Silica-bicarbonate balance assuming present stream discharge and content (after Livingstone 1963a), and all Ca^{++} deposited as carbonate minerals. About 1 to 5 percent of the volume of sedimentary rocks is chert. The present rate of delivery of dissolved SiO_2 to the ocean would produce a much greater volume of "free" silica, mainly as chert, than is observed in the geologic column.

products are formed, CO_2 is consumed, and dissolved constituents released. Sillén (1961) and Deffeyes (1963) suggested that reactions involving clays in the oceanic system could be important in determining the pH of the ocean and the CO_2 pressure of the atmosphere. Garrels (1965) emphasized the importance to oceanic chemistry of the above general reaction.

Ion exchange and reconstitution reactions could also account for removal of some of the excess material. Except for the removal of some of the Na^+ by exchange for other cations on montmorillonite, these reactions could not account for the large excesses of constituents that must be removed from a steady-state oceanic system. Hydrogen ion in the exchange sites of river-derived clays could help to change HCO_3^- to CO_2 but would not help to resolve the SiO_2 balance.

The bulk composition of the degraded aluminosilicates that best fits our model is similar to kaolinite ($Al_2Si_2O_5(OH)_4$) but contains more silica ($Al_2Si_{2.4}O_{5.8}(OH)_4$). This material is compositionally similar to the X-ray amorphous aluminosilicates carried by rivers to the ocean

(Moberly, 1963; Kennedy, 1963, 1965). These "amorphous aluminosilicates" constitute from about 1 to 10 percent of the suspended-sediment load of rivers and consist of structurally disordered alumina, silica, and aluminosilicates.

Some of the SiO_2 initially is removed from the system as "free" silica. Most of this silica is removed from the natural environment by organisms, principally diatoms, and reorganized during diagenesis. The amount of SiO_2 that leaves the system as "free" silica is difficult to determine. Bien, Contois, and Thomas' (1958) data for the Mississippi River and the immediate Gulf of Mexico suggest that during the initial mixing of river water with sea water, biological removal of soluble SiO_2 is not an important process. At chlorosities greater than 8 grams Cl^-/liter, uptake of SiO_2 by organisms increases rapidly (Bien, Contois, and Thomas, 1958, fig. 11). They also point out that removal of soluble SiO_2 from river water as it enters the ocean could be entirely accounted for by inorganic processes. In contrast to this conclusion, Stefánnson and Richard's (1963) data from the Columbia River plume show that inorganic removal of SiO_2 is not an important process in this region of the ocean.

On the basis of the above discussion and because of insufficient knowledge of the amount of stream-derived SiO_2 presently being removed from the ocean as "free" silica, we will dispose of an amount of SiO_2 in our balance that is reasonably consistent with the volume of "free" silica in the geologic column (chert constitutes approximately 1 to 5 percent of the volume of sedimentary rocks). Removal of 10 percent of the SiO_2 from our balance as "free" silica would be consistent with a volume of chert about equal to 2 percent of the volume of sedimentary rocks. The reaction is

$$H_4SiO_4 = SiO_{2(s)} + 2\ H_2O$$

and in millimoles of SiO_2 in our balance

$$71\ H_4SiO_4 = 71\ SiO_{2(s)} + 142\ H_2O.$$

The only reasonable solution to the removal of the remaining Na^+ is by exchange with Ca^{++} on montmorillonites and by synthesis of new sodic montmorillonites from degraded aluminosilicates. Although we recognize that river-borne montmorillonites are not pure Ca-montmorillonites and oceanic montmorillonites are not pure Na-montmorillonites, the exchange positions in the two environments are occupied primarily by Ca^{++} and Na^+ respectively. Therefore, we will use the compositions $Na_{0.33}Al_{2.33}Si_{3.67}O_{10}(OH)_2$ and $Ca_{0.17}Al_{2.33}Si_{3.67}O_{10}(OH)_2$ to represent oceanic and river-borne montmorillonites respectively.

If the suspended load of streams is $32,500 \times 10^6$ tons/year, and montmorillonite with an exchange capacity of 80 milliequivalents per 100 grams is 10 percent of the load, then about 25 percent of the Na^+

carried by rivers to the ocean can be removed by exchange with Ca^{++}. This is a crude but reasonable estimate. The exchange reaction is

$$Ca_{0.17}Al_{2.33}Si_{3.67}O_{10}(OH)_2 + 0.33\ Na^+ =$$
$$Na_{0.33}Al_{2.33}Si_{3.67}O_{10}(OH)_2 + 0.17\ Ca^{++}$$

and, in millimoles, using 25 percent of the available Na^+

$$138\ Ca_{0.17}Al_{2.33}Si_{3.67}O_{10}(OH)_2 + 46\ Na^+ =$$
$$138\ Na_{0.33}Al_{2.33}Si_{3.67}O_{10}(OH)_2 + 23.5\ Ca^{++}.$$

The Ca^{++} released is combined with HCO_3^- to form calcite and/or aragonite.

The remaining Na^+ is removed by reacting with degraded aluminosilicates, SiO_2, and HCO_3^- to form sodic montmorillonite, according to the reaction

$$3.5\ Al_2Si_{2.4}O_{5.8}(OH)_4 + Na^+ + 2.6\ SiO_2 + HCO_3^- =$$
$$3\ Na_{0.33}Al_{2.33}Si_{3.67}O_{10}(OH)_2 + CO_2 + 4.5\ H_2O$$

and in millimoles of material available

$$486.5\ Al_2Si_{2.4}O_{5.8}(OH)_4 + 139\ Na^+ + 361.4\ SiO_2 + 139\ HCO_3^- =$$
$$417\ Na_{0.33}Al_{2.33}Si_{3.67}O_{10}(OH)_2 + 139\ CO_2 + 625.5\ H_2O.$$

The remaining Mg^{++} and all the K^+ are disposed of by reaction with degraded aluminosilicates, SiO_2, and HCO_3^- to form chlorite and illite respectively. Both of these minerals are found in abundance in modern and ancient marine sediments. Once more, the phases chosen are considered to be pure, but in the natural environment they are compositionally more complex. The reactions for Mg^{++} and K^+ are

$$Al_2Si_{2.4}O_{5.8}(OH)_4 + 5\ Mg^{++} + 0.6\ SiO_2 + 10\ HCO_3^- =$$
$$Mg_5Al_2Si_3O_{10}(OH)_8 + 10\ CO_2 + 3\ H_2O,$$

$$2.5\ Al_2Si_{2.4}O_{5.8}(OH)_4 + K^+ + SiO_2 + HCO_3^- =$$
$$2\ K_{0.5}Al_{2.5}Si_{3.5}O_{10}(OH)_2 + CO_2 + 3.5\ H_2O;$$

and in millimoles of material available

$$100.4\ Al_2Si_{2.4}O_{5.8}(OH)_4 + 502\ Mg^{++} + 60.2\ SiO_2 + 1004\ HCO_3^- =$$
$$100.4\ Mg_5Al_2Si_3O_{10}(OH)_8 + 1004\ CO_2 + 301.2\ H_2O,$$

$$472.5\ Al_2Si_{2.4}O_{5.8}(OH)_4 + 189\ K^+ + 189\ SiO_2 + 189\ HCO_3^- =$$
$$378\ K_{0.5}Al_{2.5}Si_{3.5}O_{10}(OH)_2 + 189\ CO_2 + 661.5\ H_2O.$$

In the reactions involving the synthesis of new minerals from degraded aluminosilicates, all the K^+, 90 percent of the Mg^{++}, 90 percent of the SiO_2, 15 percent of the Na^+, and 43 percent of the original HCO_3^- are removed from the system.

It could be argued that we have synthesized pure clay minerals in our balance rather than ones with the actual compositions of those in oceanic sediments. The proportions of K^+, Mg^{++}, Na^+, and SiO_2 in the clay minerals formed in our balance are very similar to the proportions of these constituents found in average shale and average deep-sea argillaceous sediment (fig. 2). The resemblance between our bulk "shale" composition and that of argillaceous sediments suggests that our balance would not be greatly affected by the use of more exact aluminosilicate mineral compositions involving extensive cation substitution.

Assuming (1) degraded aluminosilicates constitute 10 percent of the suspended load of rivers, (2) the suspension to solution load ratio in streams is 5/1, and (3) the relative amounts of shale, sandstone, and carbonates in the geologic column are 0.6, 0.2, 0.2 respectively, the amount of newly formed illite, chlorite, and montmorillonite would compose only about 7 percent of the total mass of sediments at any one time. Such a small amount of authigenic illite would not affect

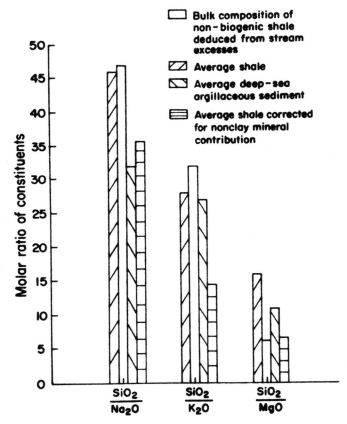

Fig. 2. Bulk composition, expressed as molar ratios, of average shale (Clarke, 1924, p. 34), average deep-sea argillaceous sediment (carbonate, water, and organic C-free basis; Wakeel and Riley, 1961, p. 123), and average shale less nonclay mineral contribution (Shaw and Weaver, 1965, p. 222) compared with the bulk composition of the aluminosilicate minerals ("shale") formed in our balance from stream-derived dissolved constituents and degraded aluminosilicates.

the K-Ar ages of 200 to 400 million years obtained by Hurley and others (1963) from K-bearing minerals in modern marine sediments.

The balance works out well considering the assumption that the average river water analysis represents adequately the dissolved material carried by streams to the ocean. The apparent deficiency of HCO_3^- to form montmorillonite represents the electrical imbalance in the initial excesses of dissolved constituents in which cation equivalents exceed anion equivalents. The final excess in SiO_2 is negligible and could be disposed of by increasing the amount of "free" silica removed from the system or decreasing slightly the silica content of the degraded aluminosilicates.

SUMMARY AND CONCLUSIONS

In recent years several writers (Sillén, 1961, 1963, 1965; Deffeyes, 1963; Garrels, 1965; Holland, 1965; Kramer, 1965; Mackenzie and Garrels, 1965) have discussed the role of silicate reactions in considerations of oceanic chemistry. The mass balance calculation between stream-derived dissolved constituents and a steady-state oceanic system clearly demonstrates the importance of aluminosilicate reactions in the oceanic system and leads to some interesting suggestions and conclusions.

The ocean, when considered over millions of years, is a nearly steady-state system, in which the major constituents carried in by streams are lost at approximately the same rate as they are added. Furthermore, the materials lost by chemical or biochemical precipitation or reaction can be removed as minerals that correspond to those found in sedimentary rocks. However, this removal process requires that typical clay minerals be synthesized compositionally, if not structurally, from degraded aluminosilicates before burial in marine sediments. It is possible that, instead of aluminosilicate clays being synthesized, simple hyroxylated silicates are formed, according to reactions of the type

$$\text{cations} + SiO_2 + HCO_3^- = \text{cation OH silicate} + CO_2.$$

However, we consider this unlikely because simple silicates of this type are not found in abundance in modern or ancient marine sediments.

These "reverse weathering" reactions prevent the alkali metals and bicarbonate from accumulating in the ocean. Consequently, ocean water composition is not simply the result of evaporation of river water, as this would lead to a water of the composition of a soda lake (Garrels and Mackenzie, in press), but is controlled to a large extent by interactions between stream-derived dissolved constituents and suspended sediments. Also, because these reactions involve the release of CO_2 to the atmosphere, it is likely that the long-term carbon dioxide pressure of the atmosphere is maintained by such reactions.

The compositional synthesis of the minerals illite, chlorite, and montmorillonite in the oceanic system implies that these phases are in equilibrium, or nearly so, with an aqueous solution of the composition of average ocean water. Consequently, the reactions involving these minerals probably determine to a large extent the major ionic ratios of ocean water. Also, the minerals illite, chlorite, and montmorillonite carried to the ocean in the suspended load of streams should not be markedly altered chemically or structurally upon entering the ocean, although, a small part of the suspended sediment of rivers, the "amorphous aluminosilicates", does react to form new minerals.

We are well aware that innumerable chemical balance models could be devised and that there are many aspects of our model open to question. However, the fact that the proportions of Mg^{++}, K^+, Na^+, and SiO_2 in our synthesized clay minerals are in the approximate proportions found in shales, and that the reaction to form typical shale minerals from these excesses simultaneously solves the problems of the HCO_3^- excess and the silica balance cannot be attributed to chance.

ACKNOWLEDGMENTS

We have benefited greatly from discussions with many colleagues in many places; we especially want to thank Drs. Roland Wollast of the University of Brussels, Harold Helgeson of Northwestern University, Charles Christ of the United States Geological Survey, and Owen Bricker of The Johns Hopkins University for reading the manuscript and for offering many useful suggestions and criticisms.

This work was supported by grants from the National Science Foundation (GP-4140) and the Petroleum Research Fund of the American Chemical Society. Contribution No. 389, Bermuda Biological Station for Research.

REFERENCES

Arrhenius, G., 1963, Pelagic sediments, *in* Hill, M. N., ed., The Sea, v. 3: New York, Intersci. Pub., p. 655-727.
Barth, T. F. W., 1952, Theoretical petrology: New York, John Wiley and Sons, Inc., 387 p.
Berner, R. A., 1964, Distribution and diagenesis of sulfur in some sediments from the Gulf of California: Marine Geology, v. 1, p. 117-140.
Bien, G. S., Contois, D. E., and Thomas, W. H., 1958, The removal of soluble silica from fresh water entering the sea: Geochim. et Cosmochim. Acta, v. 14, p. 35-54.
Biscaye, P. E., 1965, Mineralogy and sedimentation of recent deep-sea clay in the Atlantic Ocean and adjacent seas and ocean: Geol. Soc. America Bull., v. 76, p. 803-832.
Chave, K. E., 1954, Aspects of the biogeochemistry of magnesium. Pt. 2, Calcareous sediments and rocks: Jour. Geology, v. 62, p. 587-599.
Clarke, F. W., 1924, The data of geochemistry, 5th ed.: U.S. Geol. Survey Bull. 770, 841 p.
Conway, E. J., 1942, Mean geochemical data in relation to oceanic evolution: Roy. Irish Acad. Proc., v. 48, sec. B, p. 119-159.
Correns, C. W., 1937, Die Sedimente des äquatorialen Atlantischen Ozeans: Deutsche Atlantische Exped. auf dem Forschungsschiff- und Vermessungsschiff "Meteor" (1925-1927), Wissenschaftliche Ergebnisse, v. 3, pt. 3, p. 1-42, 135-298.
Deffeyes, K. S., 1963, Role of oceanic carbonate and silicate sedimentation in determining atmospheric carbon dioxide pressure [abs.]: Geol. Soc. America Spec. Paper 76, p. 39-40.
Edwards, M. L., Kister, L. R., and Scarcia, G., 1956, Water resources of the New Orleans area, Louisiana: U.S. Geol. Survey Circ. 374, 41 p.

Favajee, J. C. L., 1951, The origin of the Wadden mud: Wageningen Landbouwhoogeschool Mededel., v. 51, p. 113-141.

Garrels, R. M., 1965, The role of silica in the buffering of natural waters: Science, v. 148, p. 69.

Garrels, R. M., and Mackenzie, F. T., in press, Origin of the composition of some springs and lakes: Am. Chem. Soc., Advances in Chemistry Ser., in press.

Goldberg, E. D., 1957, Biogeochemistry of trace metals, in Hedgpeth, J. W., ed., Ecology, v. 1 of Treatise on Marine Ecology and Paleocology: Geol. Soc. America Mem. 67, p. 345-358.

Goldberg, E. D., and Arrhenius, G. O. S., 1958, Chemistry of Pacific pelagic sediments: Geochim. et Cosmochim. Acta, v. 13, p. 153-212.

Gorham, E., 1961, Factors influencing supply of major ions to inland waters with special reference to the atmosphere: Geol. Soc. America Bull., v. 72, p. 795-840.

Griffin, G. M., 1962, Regional clay mineral facies—products of weathering intensity and current distribution in the northeastern Gulf of Mexico: Geol. Soc. America Bull., v. 73, p. 737-768.

Griffin, J. J., and Goldberg, E. D., 1963, Clay-mineral distribution in the Pacific Ocean, in Hill, M. N., ed., The Sea, v. 3: New York, Intersci. Pub., p. 728-741.

Grim, R. E., and Johns, W. D., 1954, Clay mineral investigation of sediments in the northern Gulf of Mexico, in Swineford, Ada, and Plummer, N. V., eds., Clays and clay minerals: Natl. Research Council Pub. 327, p. 81-103.

Grim, R. E., Dietz, R. S., and Bradley, W. F., 1949, Clay mineral composition of some sediments from the Pacific Ocean off the California coast and the Gulf of California: Geol. Soc. America Bull., v. 60, p. 1785-1808.

Holland, H. D., 1965, The history of ocean water and its effect on the chemistry of the atmosphere: Natl. Acad. Sci. Proc., v. 53, p. 1173-1182.

Hurley, P. M., Heezen, C. B., Pinson, W. H., and Fairbairn, H. W., 1963, K-Ar values in pelagic sediments of the North Atlantic: Geochim. et Cosmochim. Acta, v. 27, p. 393-399.

Junge, C. E., and Werby, R. T., 1958, The concentration of Cl^-, Na^+, Ca^{++}, and $SO_4^=$ in rain water over the U.S.: Jour. Meteorology, v. 15, p. 417-425.

Kaplan, J. R., Emery, K. O., and Rittenberg, S. C., 1963, The distribution and isotopic abundance of sulfur in recent marine sediments off southern Caliornia: Geochim. et Cosmochim. Acta, v. 27, p. 297-332.

Kennedy, V. C., 1963, Base-exchange capacity and clay mineralogy of some modern stream sediments: I.A.S.H. Comm. of Subterranean Waters Pub. 64, p. 95-105.

———— 1965, Mineralogy and cation-exchange capacity of sediments from selected streams: U.S. Geol. Survey Prof. Paper 433-D, 28 p.

Kramer, J. R., 1965, History of sea water. Constant temperature-pressure equilibrium models compared to liquid inclusion analyses: Geochim. et Cosmochim. Acta, v. 29, p. 921-946.

Kuenen, P. H., 1950, Marine Geology: New York, John Wiley and Sons, Inc., 568 p.

Livingstone, D. A., 1963a, Chemical composition of rivers and lakes, in Fleischer, Michael, ed., Data of Geochemistry: U.S. Geol. Survey Prof. Paper 440, chap. G, p. G1-G64.

———— 1963b, The sodium cycle and the age of the ocean: Geochim. et Cosmochim. Acta, v. 27, p. 1055-1069.

Lowenstam, H. A., 1961, Mineralogy, O^{18}/O^{16} ratios, and strontium and magnesium contents of recent and fossil brachiopods and their bearing on the history of the oceans: Jour. Geology, v. 69, p. 241-260.

Mackenzie, F. T., and Garrels, R. M., 1965, Silicates-reactivity with sea water: Science, v. 150, p. 57-58.

———— in press, Silica-bicarbonate balance in the ocean and early diagenesis: Jour. Sed. Petrology, in press.

Moberly, R. M., Jr., 1963, Amorphous marine muds from tropically weathered basalt: Am. Jour. Sci., v. 261, p. 767-772.

Murray, Sir John, 1887, On the total annual rainfall on the land of the globe, and the relation of rainfall to the annual discharge of rivers: Scottish Geog. Mag., v. 3, p. 65-77.

Nelson, B. W., 1963, Clay mineral diagenesis in the Rappahannock estuary: an explanation, in Bradley, W. F., ed., Clays and clay minerals: Oxford, England, Pergamon Press, New York, The Macmillan Co., p. 210.

Powers, M. C., 1954, Clay diagenesis in the Chesapeake Bay area, *in* Swineford, Ada, and Plummer, N. V., eds., Clays and clay minerals: Natl. Research Council Pub. 327, p. 68-80.
———— 1957, Adjustment of land-derived clays to the marine environment: Jour. Sed. Petrology, v. 27, p. 355-372.
———— 1959, Adjustment of clays to chemical change and the concept of the equivalence level, *in* Swineford, Ada, ed., Clays and clay minerals: Internat. Ser. Mons. Earth Sci., v. 2, p. 309-326.
Revelle, R. R., 1944, Marine bottom samples collected in the Pacific Ocean by the Carnegie on its seventh cruise: Carnegie Inst. Washington Pub. 556, p. 1-180.
Rubey, W. W., 1951, Geologic history of sea water: an attempt to state the problem: Geol. Soc. America Bull., v. 62, p. 1111-1147.
Shaw, D. B., and Weaver, C. E., 1965, The mineralogical composition of shales: Jour. Sed. Petrology, v. 35, p. 213-222.
Sillén, L. G., 1961, The physical chemistry of sea water, *in* Sears, Mary, ed., Oceanography: Am. Assoc. Adv. Sci. Pub. 67, p. 549-581.
———— 1963, How has sea water got its present composition?: Svensk Kemisk Tidsk., v. 75, p. 161-177.
———— in press, Oxidation state of earth's ocean and atmosphere. 1. A model calculation on earlier states. The myth of the "probiotic soup": Arkiv. kemi., in press.
Stefánnson, Unnsteinn, and Richards, F. A., 1963, Processes contributing to the nutrient distributions off the Columbia River and Strait of Juan de Fuca: Limnology and Oceanography, v. 4, p 394-410.
Taggart, M. S., Jr., and Kaiser, A. D., Jr., 1960, Clay mineralogy of Mississippi River deltaic sediments: Geol. Soc. America Bull., v. 71, p. 521-530.
Turekian, K. K., 1964, The geochemistry of the Atlantic Ocean Basin: New York Acad. Sci. Trans., ser. 2, v. 26, p. 312-320.
Wakeel, S. K., El, and Riley, J. P., 1961, Chemical and mineralogical studies of deepsea sediments: Geochim. et Cosmochim. Acta, v. 25, p. 110-146.
Weaver, C. E., 1959, The clay petrology of sediments, *in* Swineford, Ada, ed., Clays and clay minerals: Internat. Ser. Mons. Earth Sci., v. 2, p. 154-187.
Whitehouse, U. G., and McCarter, R. S., 1958, Diagenetic modification of clay mineral types in artificial sea water, *in* Swineford, Ada, ed., Clays and clay minerals: Natl. Research Council Pub. 566, p. 81-119.

II
Isotopes of Natural Waters: H^1, H^2, H^3, O^{16}, and O^{18}

Editor's Comments on Papers 7 Through 11

7 Friedman: *Deuterium Content of Natural Water and Other Substances*

8 Craig: *Isotopic Variations in Meteoric Waters*

9 Dansgaard: *Stable Isotopes in Precipitation*

10 Vinogradov, Devirts, and Dobkina: *The Current Tritium Contents of Natural Waters*

11 Clayton et al.: *The Origin of Saline Formation Waters: 1. Isotopic Composition*

A great deal of valuable work has been done on the paleotemperature of seawater, using measurements of the O^{18}/O^{16} ratio of calcareous shells. This volume does not include papers on paleotemperature, but it does contain representative papers discussing studies that were made on the isotopes of hydrogen and oxygen in water.

The two articles in this volume by Friedman (Paper 7) and Dansgaard (Paper 9), examining stable isotopes in natural water, are of historical importance, as they concern a subject that still requires study in great depth. The relationship between D/H and O^{18}/O^{16} in natural water was originally presented by H. Craig in 1961 (Paper 8), and in 1969 Craig published a paper on Red Sea brine (Paper 20) which demonstrates the importance of isotopic ratio data in water geochemistry. R. N. Clayton et al. (Paper 11) speculated on the origin of saline waters by collecting data on the isotopic ratios of D/H and O^{18}/O^{16} in 95 oil-field brines.

The tritium content of natural water has been examined primarily in connection with nuclear explosion tests. Included as Paper 10 is the article on tritium contents in water by A. P. Vinogradov et al. written in 1968. Papers by W. F. Libby (1961), E. A. Martell (1963), and E. Eriksson (1965) contain significant information as to the chemical behavior of tritium in the hydrosphere.

The following papers may be of interest to the reader, although they could not be included in this volume due to space limitations. In 1963 Craig explored the origins of water in geothermal areas, a study that is invaluable to the present explanations of the origin of water at the earth's surface. Craig concluded that no juvenile water exists in geothermal areas; it has been deduced from his study that water on earth may be of meteoric origin.

Papers on the isotopic ratios of snow and ice in Antarctica and Greenland contain very important data on the balance sheet of water on earth. Several significant papers in this field were written by Picciotto et al. (1960), Gonfiantini et al. (1963), and Johnsen et al. (1972).

Measurements made of the isotopic ratio of O^{18}/O^{16} in ice cores have been shown to be indicators of the climatic record by Dansgaard et al. (1971) and Fairbridge (1972). Further climatological data have been furnished by K. K. Turekian in his book *The Late Cenozoic Glacial Ages* (1971).

The D/H and O^{18}/O^{16} ratios in seawater have also been used as parameters of water mass. Horibe and Ogura (1968) reported that the deuterium content of seawater is not proportional to salinity under 100 meters in depth, and that the deuterium–salinity diagram of an individual station at various depths is characteristic. They showed that the diagram for Kuroshio is quite different from the diagrams for other stations in the open sea. The paper by Craig and Gordon (1965) on deuterium and oxygen-18 variations in the ocean and the marine atmosphere is also recommended to the reader.

Irving Friedman received his B.S. in chemistry in 1942 from Montana State College, his M.S. in chemistry in 1944 from the State College of Washington, and his Ph.D. in geochemistry in 1950 from the University of Chicago. He has received both the Congressional Antarctic Medal, and the Department of the Interior Antarctic Award. From 1950 to 1952 he worked at the Institute for Nuclear Studies in the University of Chicago as a research associate. Since 1952 he has been with the U.S. Geological Survey as a research chemist.

Harmon Craig received his Ph.D. in geology and geochemistry in 1951 from the University of Chicago. Since 1955 he has been a professor of geochemistry in the graduate department and Geological Research Division at Scripps Institution of Oceanography in San Diego. From 1951 to 1955 Craig was a research associate at the Enrico Fermi Institute for Nuclear Studies in the University of Chicago, and in 1962–1963 he was Guggenheim Fellow at the Laboratorio di Geologia Nucleare in the University of Pisa. From 1965 to 1968 Craig served as the chairman of the Department of Earth Sciences at Scripps. Since 1970 Craig has been a member of the Executive Committee of the Geochemical Oceans Sections Program of the International Decade of Ocean Exploration.

W. Dansgaard was born in 1922 and graduated in physics from the University of Copenhagen. He is presently a professor at the Geophysical Institute in the University of Copenhagen. Dansgaard measured isotopic fractionation during meterological processes in the early 1950's [see *Tellus*, **5**, 461 (1953)]. In recent years he has been occupied with paleoclimatic studies on deep ice cores from Greenland and Antarctica.

A. P. Vinogradov was born August 21, 1895; he graduated from the Medical Academy and Leningrad University in the department of chemistry in 1925. Vinogradov became a Corresponding Member of the Academy of Sciences of the U.S.S.R. in 1943 and has been an Academician since 1953. In 1949 he was named a Hero of Socialist Labor, and he received the Stalin Prize in 1949 and again in 1951. He has been a close collaborator with B. I. Vernadskii, who is remembered as the "founder" of the science of geochemistry in the U.S.S.R. In 1948 Vinogradov became the Director of the Institute of Geochemistry and Analytical Chemistry in the Academy of Sciences of the U.S.S.R.

Vinogradov's primary interests are in examining the distribution of chemical elements in the upper part of the earth's crust, in the investigation of the primary rock from which the sediments on the earth's surface were formed, and in examining the role played by volcanic materials in the formation of the upper crust. In his study on salts in ocean water, Vinogradov reached the conclusion that cations in seawater are the product of the erosion of magmatic rocks and that anions are of volcanic origin. He described geochemically more than 40 rare, widely dispersed elements for various soil zones and demonstrated their roles in various soil-forming processes. Vinogradov has investigated the association of heavy metals with bitumen. He has been active in the use of isotopes (sulfur, hydrogen, oxygen, carbon, etc.) in geochemistry, and has used the oxygen isotope O^{18} as an indicator of geochemical processes. In his work on photosynthesis, Vinogradov found that plants liberate oxygen from water and not from carbon dioxide; he has demonstrated that natural hydroxides of iron, manganese, etc., obtain oxygen from water rather than from air. From the work he has done in biogeochemistry, he has found that the majority of the elements exist in all organisms and that the chemical composition of a species is one of its characteristic features. Vinogradov developed a theoretical model for ordinary fertilizers and those which contain microelements. His investigations in biogeochemistry have gone far toward explaining the effect of the chemical environment on the evolution of flora and fauna during different geological ages.

Recently Vinogradov has been very active in the field of cosmochemistry. In the area of analytical chemistry he developed many methods of the separation of numerous stable and unstable chemical elements and introduced instrumental methods of analysis such as polarography, spectrometry, radiometry, mass spectrometry, X-ray, and luminescence.

Robert N. Clayton was born in Hamilton, Canada, in 1930. He received his B.S. and his M.S. in 1951 and 1952, respectively, from Queen's University, Ontario, and his Ph.D. in 1955 from the California Institute of Technology. Clayton was a research fellow in geochemistry at the California Institute of Technology between 1955 and 1956; an assistant professor of geochemistry at Pennsylvania State University between 1956 and 1958; an assistant professor of chemistry at the Enrico Fermi Institute and the department of chemistry and geophysical sciences in the University of Chicago from 1958 to 1962; an associate professor at the University of Chicago from 1962 to 1966. Clayton has been at the University of Chicago since 1966 as a full professor.

References

Craig, H. (1963). The isotopic geochemistry of water and carbon in geothermal areas. *Nuclear Geology on Geothermal Areas,* Consiglio Nazionale dells Ricerche, Laboratorio di Geologia Nucleare, Pisa, Italy, pp. 17–53.

Craig, H., and Gordon, L. I. (1965). Deuterium and oxygen-18 variations in the oceans and the marine atmosphere, in E. Tongiorgi (ed.), *Stable Isotopes in Oceanographic Studies and Paleotemperatures* (1950 Spoleto Conference Proceedings), Consiglio Nazionale dells Ricerche, Pisa, Italy, pp. 9–182.

Dansgaard, W., Johnsen, S. J., Clausen, H. B., and Langway, C. C., Jr. (1971). Climatic record revealed by the camp century ice core, in K. K. Turekian (ed.), *The Late Cenozoic Glacial Ages,* Yale University Press, New Haven, pp. 37–56.

Eriksson, E. (1965). An account of the major pulses of tritium and their effects in the atmosphere. *Tellus,* **17**, 118–130.

Fairbridge, R. W. (1972). Climatology of a glacial cycle. *Quaternary Res.,* **2**, 283–302.

Gonfiantini, R., Togliatti, V., Tongiorgi, E., De Breuck, W., and Picciotto, E. (1963). Snow stratigraphy and oxygen isotope variations in the glaciological pit of King Baudouin Station, Queen Maud Land, Antarctica. *J. Geophys. Res.,* **68**, 3791–3798.

Horibe, Y., and Ogura, N. (1968). Deuterium content as a parameter of water mass in the ocean. *J. Geophs. Res.,* **73**, 1239–1249.

Johnsen, S. J., Dansgaard, D., and Clausen, H. B. (1972). Oxygen isotope profiles through the Antarctic and Greenland ice sheets. *Nature,* **235**, No. 5339, 429–434.

Libby, W. F. (1961). Tritium geophysics. *J. Geophys. Res.,* **66**, 3767–3782.

Martell, E. A. (1963). On the inventory of artificial tritium and its occurence in atmospheric methane. *J. Geophys. Res.,* **68**, 3759–3769.

Picciotto, E., De Maere, X., and Friedman, I. (1960). Isotopic composition and temperature of formation of Antarctic snows. *Nature,* **187**, No. 4740, 857–859.

Turekian, K. K. (ed.). (1971). *The Late Cenozoic Glacial Ages.* Yale University Press, New Haven, 606 pp.

Deuterium content of natural waters and other substances

Irving Friedman*

Institute for Nuclear Studies, University of Chicago

(*Received* 27 *February* 1953)

ABSTRACT

A mass spectrometric method for the accurate determination of the hydrogen-deuterium ratio has been developed. It is possible to determine this ratio to $\pm 0.10\%$ using material of "normal abundance", i.e., 1 part D in 6700 parts H. Samples as small as 0.1 mg H_2 (0.001 ml H_2O) can be run. Natural evaporation and condensation that have been shown to fractionate the oxygen isotopes also fractionate the hydrogen isotopes. The ratio of these two fractionations is equal to the ratio between the ratios of the vapour pressures of H_2O/HDO and of H_2O^{16}/H_2O^{18}.

Ocean waters range from 0.0153 to 0.0156 mole % deuterium, whereas fresh waters of the United States range from 0.0133 to 0.0154 mole % deuterium. A measurement of Yellowstone Park fumarole gases gives a minimum temperature of 400° for the equilibrium $H_2O + HD = HDO + H_2$ in the gases.

The stable hydrogen isotopes, protium and deuterium, are of great interest geochemically, as they are separated or fractionated by geological or cosmological processes to a larger extent than any other pair of stable isotopes. Therefore they can serve as a useful indicator to help reconstruct the conditions under which such processes have occurred. The object of this research was to develop a method for determining the H/D ratio in milligram amounts of water or in other materials containing hydrogen. The accuracy of the method was to be better than $\pm 0.5\%$ of the ratio in water of normal deuterium abundance (1 part D in 6700 parts H); e.g., 1.000 ± 0.005 parts D in 6700 parts H.

It was decided that the only practical method of analyzing large numbers of small samples with the accuracy desired was by the use of a mass spectrometer. The first approach was the conversion of a water sample to HCl by reaction with $SiCl_4$. The HCl was then passed into a Nier-type (6-in., 60°) mass spectrometer and the ratio of the peaks at mass 38 (H^1Cl^{37}) and mass 39 (H^2Cl^{37}) was determined. Great difficulty was encountered in removing the impurities that gave a peak at mass 39—most hydrocarbons containing more than 3 carbon atoms give a large peak at mass 39 $(C_3H_3)^+$.

The next approach was the determination of the H/D ratio directly on hydrogen gas made by the reaction of a water sample with hot zinc. A new mass spectrometer tube was constructed, based on a design by Nier†, in which an additional collector tube was joined to the usual 6-in., 60° spectrometer tube in such a way that it would collect the mass 2 ion beam while the usual collector simultaneously was collecting the mass 3 ion beam.

The voltage developed by the mass-2 ion beam current flowing through a 1.5×10^{10}-ohm resistor was used to buck out the voltage developed by the mass-3 ion beam current flowing through a 1.5×10^{11}-ohm resistor. The difference

---

* Present address: U.S. Geological Survey, Washington, D.C.
† A. O. Nier, personal communication.

voltage was amplified by a vibrating reed electrometer and recorded by a Brown Electronik strip chart recorder.

In accordance with a suggestion of MARK INGRAM* the defining slits in the NIER-type ion source were both enlarged from 0·010 in. and 0·020 in. to 0·040 in. to increase the ion beam intensity, and a repeller electrode was placed in the shield box parallel to the plane of the slit plates. The voltage on the repeller electrode was variable from 0 to 10 volts negative in respect to the shield. This repeller plate established an accelerating field inside the shield box, which repelled the H_2^+ ions toward the shield, thus aiding in the removal of the H_2^+ ions by passing them through the slit in the shield. It is necessary to remove the H_2^+ from the shield box as rapidly as possible to minimize the formation of the H_3^+ ion, which will be collected on the mass-3 collector together with the HD^+ ion. A discussion of the formation of the H_3^+ ion is given by SMYTH, 1931.

H_3^+ ion Correction

In the past (A. O. NIER, 1947) the H_3^+ ion was corrected for by plotting the H/D ratio as a function of pressure of the gas in the source shield box and extrapolating to zero pressure. Since the concentration of H_3^+ is proportional to the square of the pressure (second-order reaction $H_2^+ + H = H_3^+$) and the HD^+ concentration is proportional to the pressure, the total mass-3 ion beam current will be:

$$I = ap + ap^2$$

If I is plotted vs. $1/p$, a straight line whose $p = 0$ intercept will give the true HD^+ ion current should be obtained. Since at the low gas pressures that are used the H_2^+ ion current is proportional to pressure, one usually plots H_2^+ ion current vs. $H_2^+/H_3^+ + HD^+$ ion currents.

The establishing of this curve for each sample is time-consuming and the results are inaccurate; the mass spectrometer does not remain constant for the time necessary to make such determinations, the above curve is not linear, and the extrapolation to zero pressure is fairly large. Therefore, it was decided to minimize the H_3^+, to compare sample gas with a standard gas, and to keep the H_3^+ concentration constant in both gases during measurement.

The concentration of H_3^+ ions was minimized by working at low gas pressures, by adding the repeller electrode to the source, and by using as high an accelerating voltage as possible (2200 volts with an analyzer magnetic field strength of approximately 725 gausses; the high accelerating voltage also helped draw ions out of the source shield box and by so doing reduced the probability of the H_2^+ ion colliding with a neutral molecule in the shield box. The source magnet was adjusted for maximum electron-trap current (approximately 180–220 microamperes at a total filament emission of 3 ma) rather than for maximum ion current. The electron,

---

* MARK INGRAM, personal communication.

accelerating and trap voltages were all adjusted for minimum H_3^+, as determined by the slope of the H/D ratio vs. H_2^+ beam intensity curve.

All the determinations were made at the same mass-2 ion current, $22 \cdot 2 \pm 0 \cdot 2$ volts across $1 \cdot 5 \times 10^{10}$ ohms. The repeller was then adjusted so that with the standard gas flowing into the source, the vibrating reed output was zero when the ratio dials read 00400, i.e., when the ratio, uncorrected for the difference in the resistance through which the two ion currents flowed, was $00400/111110 = HD/H_2$.

The sample gas was then switched into the source by the use of solenoid-operated valves similar to those described by McKinney et al, 1950. The pressure of the sample gas was adjusted so that the mass-2 beam of the sample gas was

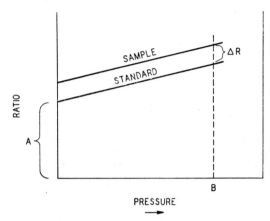

Fig. 1. Idealized plot of measured ratio HD/H_2 as a function of gas pressure in mass spectrometer ion source

equal to the mass-2 beam of the standard gas to within $\pm 0 \cdot 01$ volts (1 part in 2220). The beam intensity of mass-2 was kept constant during the course of a determination by manually adjusting the pressure of both the sample and the standard gas. This adjustment presented no problem; it had to be attended to about every 3 to 5 minutes during a 20 minute run. In order to check the beam intensity to 1 part in 5000 it was necessary to have provision to buck out the voltage as indicated by the meter with a dry cell and Helipot potentiometer arrangement, so that the difference between the actual beam voltage and the bucking-voltage could be read on a sensitive voltmeter scale. The fluctuations of the beam voltage were in the neighbourhood of $\pm 0 \cdot 005$ volts in a $22 \cdot 2$-v beam. At times it was possible to adjust the electronics so that the beam fluctuations over a period of 5 to 10 minutes were within the range $\pm 0 \cdot 002$ volts in a $22 \cdot 2$-v beam. The method of comparison of sample gas with standard gas is described elsewhere (C. R. McKinney et al, 1950).

If no H_3^+ ion was present, the percentage difference between the ratio in the standard gas and the ratio in sample gas would be:

$$\frac{100 \times (\text{ratio dial reading for sample} - \text{ratio dial reading for standard})}{\text{Ratio dial reading for standard}}$$

91

The percentage difference between the ratios in the two gases corrected for H_3^+ ion is obtained as follows:

$$\frac{(\text{Ratio dial reading for sample} - \text{ratio dial reading for standard}) \times 100}{\text{Extrapolated reading for standard}}$$

This is shown in Fig. 1. In other words the difference ΔR (Fig. 1) at a large beam intensity is determined, and this reading is divided by the extrapolated dial

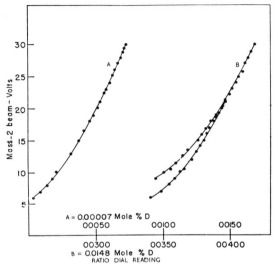

Fig. 2. Experimental plot of ratio dial reading vs. mass-2 beam intensity

reading for the standard gas. This method is applicable only to very pure H_2 gas, as impurities in the gas change the slope of the curve, so that the curves for standard gas and sample gas are no longer parallel.

This curve is not a straight line, but curves at low gas pressure. This indicates that the reactions

$$H_2^+ \rightleftharpoons H^+ + H$$
$$H_2 \rightleftharpoons H + H$$

are becoming important, and act to lower the mass 2 ion beam current.

As a check on the H_3^+ correction, two samples of "light" water containing approximately 0·00007 mole % D were analyzed. The results are shown in Fig. 2.

The curves are essentially parallel to the curves obtained with the normal water which contains about 0·0148 mole % D. The lower part of the curves were not accurately determined, since the beam intensities are too low for accurate measurements. This, plus the curvature as mentioned above made it advisable in making the H_3^+ correction to extrapolate to zero pressure from the range 15 to 25 mass-2 volts. A calibration curve was plotted after each change to the instrument, such as the removal of the source or collectors.

Sample Preparation

The vacuum line used for the preparation of the hydrogen gas from water by the reduction with hot zinc is shown in Fig. 3. A 0·01-ml sample of water to be analyzed was drawn from a pipette into a thin glass capillary tube. The capillary tube was then sealed in a flame and placed in vessel B. An iron weight was placed in the vessel and held in place by an Alnico magnet, the assembly was connected to the vacuum line by means of a ground joint, and evacuated. The vessel was sparked with a Tesla coil to help remove adsorbed gas. When the pressure fell below 10^{-4}

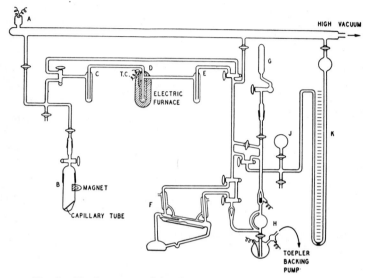

Fig. 3. Equipment used for the preparation of hydrogen gas from water samples

mm as determined by the thermocouple gauge, A, the system was isolated from the high-vacuum manifold, the trap C was cooled with liquid nitrogen, and the capillary tube broken by removing the magnet and allowing the iron weight to fall. Ten minutes was allowed for the water to be transferred to trap C. The system was then pumped to remove air that was held in the capillary tube. Trap C was allowed to warm up to room temperature while trap E was cooled with liquid nitrogen. The water vapour passed through trap D, which contained granulated zinc (about 30 grams of 10-mesh zinc) heated to about 400°C by the enclosing electric furnace. The zinc reacted with the water:

$$Zn + H_2O \rightleftharpoons H_2 + ZnO$$

producing hydrogen, which was pumped by means of the automatic Toepler pump H into the calibrated manometer K. When all the hydrogen was pumped into the manometer, the water that did not react and was condensed in trap E was made to travel back through the zinc in trap D by reversing the appropriate valves and warming E to room temperature while cooling C with liquid nitrogen.

It was found that 95 to 100% of the water reacted during the first pass over the

93

zinc if the zinc was hot enough. If the zinc was not hot enough, many passes were required and erratic results were obtained. After the hydrogen was measured, it was pumped into the sample tube G.

Care was taken to see that all the hydrogen was pumped into the tube during this transfer of the hydrogen, and during subsequent transfers, to minimize chances of isotopic fractionation.

With the line shown in Fig. 3 a sample could be analyzed completely in 45 to 70 minutes. The zinc trap D was replaced after approximately 100 runs, owing to the plugging of the trap by recrystallized zinc. All the water samples were analyzed in duplicate.

A sample of water from Lake Michigan was distilled and used as a working standard, and all the results in this paper are given in terms of this arbitrary standard.

The gas samples were placed on a vacuum manifold on the mass spectrometer and all of the gas was pumped into a reservoir behind the capillary leak, which has been described by NIER (1947). After a sample was mixed by alternate compressions and expansions, it was allowed to flow into the ion source via the capillary leak. The rate of flow of the hydrogen gas through the capillary leak into the source was about 0·05 standard cc per hour. During the course of a 10-hour day approximately 5% of the standard gas was passed through the leak. Apparently the rate of viscous flow in the forward direction through the capillary was not great enough to overcome the molecular flow in the backward direction. Therefore, a small amount of fractionation (approx. 0·05% per day) occurred and a correction had to be applied to the standard gas. This correction was determined by comparing the standard gas with two other samples of standard gas that were prepared by running triplicate samples of standard water through the line at the same time. The standard gas was replaced about every 4 to 7 days. Lengthening of the capillary tube as suggested by HALSTED and NIER (1950) should reduce this fractionation effect and lengthen the period of usefulness of each standard.

The average error in a series of 86 analyses run in duplicate is $\pm 0\cdot066\%$ ($\pm 0\cdot0000098$ mole %).

Results—Water Samples

Several water sample analyses previously made by KIRSCHENBAUM, GRAFF, and FORSTAT (1951) were checked. If the value of $0\cdot0147 \pm 0\cdot0001$ mole % D for Columbia University distilled water is accepted, our working standard is $0\cdot0148 \pm 0\cdot0001$ mole % D. With this value as a working standard, the results of the reanalyses are plotted against the data of KIRSCHENBAUM, GRAFF, and FORSTAT (Fig. 4). The differences from their standard obviously differ from our results by a factor of two. As a check, the ratio of the vapour pressures of H_2O and HDO was determined at several temperatures. The results* agree with previous published work of SCHULTZ and ZMACHINSKY† as well as with the calculated results of WAHL and UREY (1935).

---

* SAMUEL EPSTEIN and IRVING FRIEDMAN, in press.
† P. W. SCHULTZ and W. C. ZAMCHINSKY as reported by P. W. SCHULTZ, SAM Report A–595, April 12, 1943, and W. C. ZAMCHINSKY, unpublished report.

Table 1 lists the samples of natural water that were analyzed. All the results are accurate to ±0·1%. About half of the error is due to instrumental limitations and half to changes in the working standard and in the correction for H_3^+. The hydrogen results are given in terms of the percentage difference between the sample and the standard as regards the HD/H_2 ratio. Since the standard is $0·0148 \pm 0·0001$ mole % D, the conversion to mole % is given by:

$$0·0148 + (\text{hydrogen results} \times 0·0148) = \text{mole \% D in sample}$$

The oxygen results as determined by EPSTEIN (1953) are given in per mill (‰) difference in the $O^{16/18}$ ratio from the ratio in a standard CO_2 sample obtained from belemnites of the Peedee formation (Upper Cretaceous of North Carolina).

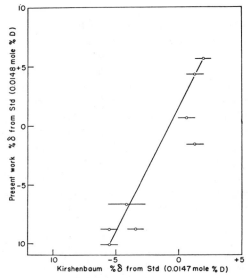

Fig. 4. A comparison of the results of the present research with the work of KIRSCHENBAUM et al

As most of the fractionation of hydrogen isotopes in the oceans is due to evaporation and condensation processes, the oxygen isotopes should be similarly separated; the ratio of the two fractionations should equal the ratio of the vapour pressures of H_2O/HDO and H_2O^{16}/H_2O^{18}. A plot both of the oxygen results as determined by EPSTEIN and MAYEDA (1953) and of the hydrogen results is given in Figs. 5. and 6 The slope of the line is approximately equal to the ratio of the vapour pressures at 30° to 40°C. The spread of the points, with the exception of the points from Great Salt Lake water, can be explained by the temperature effect.

The surface waters of the oceans have widely different isotopic compositions (see Fig. 5). In general equatorial waters are heavy, owing to the preferential loss of light hydrogen during evaporation and the concentration of the deuterium in the residue. This lighter water is deposited as rain at the higher latitudes,

Table 1. Water Samples

Ocean Water Samples

Station	Location Lat.	Location Long.	Depth Meters	Temperature °C	Hydrogen per cent*	Oxygen per mill*
Albatross Expedition						
340 #19	S 00°13′	W 18°26′	25	22·6	+5·41	+0·48
340 #20	S 00°13′	W 18°26′	500	7·1	+4·75	−0·44
340 #21	S 00°13′	W 18°26′	2500	2·95	+4·95	−0·27
340 #22	S 00°13′	W 18°26′	7500	1·4	+4·63	−0·47
373 #23	N 28°05·	W 60°49′	25	28·5	+5·64	+0·95
373 #24	N 28°05′	W 60°49′	500	16·3	+5·45	+0·48
373 #25	N 28°05′	W 60°49′	2000	3·7	+5·09	−0·05
373 #26	N 28°05′	W 60°49′	6000	2·1	+4·97	−0·19
USS Serrano						
B 105	N 48°53′	W 135°00′	602	3·8	+4·77	−0·16
B 112	N 59°24′	W 144°50′	0	5·1	+3·66	−1·0
B 117	N 57°00′	W 150°00	110	4·7	+4·27	−0·20
A 228	N 28°18′	W 117°30′	0	16·8	+4·42	−0·43
A 257	N 25°42′	W 160°20′	0	24·4	+5·15	+0·23
C 532	N 43°58′	W 133°15′	0	15·5	+4·01	−0·98
J 100	N 26°18′	W 121°40′	0	21·5	+4·38	−0·53
J 276	N 33°19′	W 128°32′	294	7·8	+4·38	−0·20
J 331	N 35°41′	W 121°50′	0	12·3	+4·17	−0·80
Snellius Expedition						
1	N 47°53′	W 125°31′	10		+3·93	−0·65
4	N 4°52·7′	W 121°30·8′	1500		+4·55	+0·30
7	S 00°4·1′	E 79°43·2′	0		+5·15	+0·95
6	S 16°00′	E 94°56′	4000		+4·72	+0·14
9	N 11°00′	W 77°40′	800		+4·72	+0·25
15	N 36°38′	E 12°56′	0		+5·42	+1·38
19	Indian Ocean		0		+4·90	+0·45
Other Samples						
Jacksonville Beach, Fla., Sept. 1951			0		+5·02	+0·65
Middle Channel, Friday Harbor, Wash., July 25, 1949			225		+4·48	−0·62
Aquarium water, La Jolla, Calif.					+4·56	−0·45
Bermuda					+5·40	+1·09
President Channel, Friday Harbor, Wash., July 16, 1949			95		+3·25	−1·62
Between Morro and Encenada, Calif., Aug. 25, 1949					+4·30	−0·28
Bering Sea off Patterson Point, Little Sitkin Island, July 1950			0		+4·07	−0·81
	N 52°8′30″	W 176°54′	0		+4·17	−0·85
Off west coast of Greenland					−4·30	−11·10

*See paragraph 2 under "Results—Water Samples"

96

Deuterium content of natural waters and other substances

Table 1.—continued

Station	Location	Depth Meters	Temperature °C	Hydrogen per cent*	Oxygen per mill*
Other Samples (cont.)					
	Off west coast of Greenland			+2·42	−3·35
	Near Mount Pleasant Island, Maine Sample 1109	29		+4·19	−1·12
	Near Mount Pleasant Island, Maine Sample 168	29		+3·75	−1·15
Fresh Water Samples					
	Snow, Chicago, Ill., 10 in. snowfall, Dec. 14, 1951			−8·28	−17·0
	Columbia River, Trail, B.C., July 17, 1943			−10·1	−17·24
	Violin Lake, Trail, B.C., June 26, 1944			−8·80	
	Violin Lake, Trail, B.C., Nov. 6, 1943			−8·83	
	Violin Lake, Trail, B.C., Sept. 22, 1943			−6·75	
	Juneau Glacier, 235 ft below surface			−6·75	−12·75
	Juneau Glacier, 155 ft below surface			−6·75	−14·90
	Grasshopper Glacier, Park County, Montana			−10·61	
	Great Salt Lake, Utah, Salt Lake boat harbor, Aug. 19, 1948			−4·92	−7·43
	Fumarole condensate from Paricutin, Jan. 1952			−0·85	+6·60
	Spring water from Paricutin region, Jan. 1948			−2·37	
	Rain water collected under Paricutin plume, 1947			−0·98	
	Standard water—Northwestern University, Dec. 17, 1942			+0·22	
	Daisy Geyser, Yellowstone Park, Wyo.			−10·03	
	Punch Bowl, Yellowstone Park, Wyo.			−10·05	
	Rain, Chicago, Ill., April 12, 1952			+0·89	−7·0
	Gullmar Fjord, west coast of Sweden; collected April 29, 1952, by Nils Jerlov during plankton outburst			+2·13	−3·57
	Mississippi River at Baton Rouge, La., May 3, 1948. Discharge about 450,000 cfs. U.S. Geol. Survey sample			+0·39	−4·90
	Mississippi River at Clinton, Iowa, June 2, 1948. Discharge 29,700 cfs. Sample collected 5 ft below surface. U.S. Geological Survey			−1·63	
	Platte River near Ashland, Nebr. May 7, 1948. Discharge 2,410 cfs. U.S. Geol. Survey sample			+0·87	
	St. Lawrence River at Ogdensburg, N.Y., June 11, 1948			+0·66	
	Susquehanna River at Marietta, Pa., May 29, 1948. Composite of 6 stations. U.S. Geol. Survey sample			−0·17	
	Apalachicola River at Chattahoochee, Fla. Discharge 16,700 cfs. U.S. Geol. Survey sample			+4·23	
	Sacramento River at Verona, Calif. Discharge 14,900 cfs. U.S. Geol. Survey sample			−1·41	
	San Joaquin River near Vernalis, Calif., July 7, 1948. Flow 1,800 cfs. U.S. Geol. Survey sample			−2·29	
	Connecticut River at Thompsonville, Conn., June 17, 1948. Flow 20,640 cfs. U.S. Geol. Survey sample			−2·15	
	Ohio River at Louisville, Ky., June 4, 1948. Flow 42,000 cfs. U.S. Geol. Survey sample			+0·19	
	Arkansas River at Van Buren, Ark., June 3, 1948. Flow 18,000 cfs. U.S. Geol. Survey sample			+3·25	

*See paragraph 2 under "Results—Water Samples"

Table 1.—*continued*

Station Location	Hydrogen per cent*	Oxygen per mill*
Fresh Water Samples (cont.)		
Rio Grande River near Mission, Tex., April 1–30, 1948. U.S. Geol. Survey sample	+3·28	
Missouri River at Kansas City, Mo., June 12, 1948. Discharge 73,000 cfs. U.S. Geol. Survey sample	−7·06	
Red River at Denison Dam near Colbert, Okla., Mar. 8–11, 14–20, 22–31; April 1–9, 11–21, 23, 25, 29–30, 1948	+3·05	
Red River of the North at Oslo, Minn., July 12, 1948. Discharge 1,750 cfs. U.S. Geol. Survey sample	−0·06	
Colorado River at Yuma, Ariz., June 4, 1948. Discharge 8,420 cfs. U.S. Geol. Survey sample	−6·06	
Snake River near Clarkston, Washington. U.S. Geol. Survey sample	−5·77	
Roanoke River near Scotland Neck, N.C., June 17, 1948. Discharge 6,500 cfs. U.S. Geol. Survey sample	+1·62	
Bermuda rain, thunderstorm, June 27, 1949	+0·53	−6·55
Monongahela River, near Morgantown, W. Va., March 26, 1943	−1·62	−9·10

making high-latitude water light. Note the analysis of Bermuda rain: +0·53% whereas surface water is +5·40%. The Bermuda rain was collected during a very heavy downpour; we can assumed that most of the water in the cloud was deposited, and that in composition the rain is a fairly good sample of the average composition of the vapour over the ocean in that region.

EPSTEIN and MAYEDA (1953) first noted that, although the surface waters of the oceans varied greatly in their oxygen isotopic composition, deep waters all tended to be light. A check of the relation between hydrogen isotopic composition and depth at two widely separated spots in the Atlantic (Albatross expedition samples) shows that the hydrogen isotopes also show this trend. As pointed out by EPSTEIN and MAYEDA, this trend may prove that polar bottom currents penetrate as far as the equator.

Chicago rain, from a large warm tropical Atlantic and Gulf of Mexico air mass, analyzed +0·89. Here also, most of the moisture in the cloud was deposited, as a 3-day rain occurred at the static juncture of a cold air mass with the warm moist tropical Gulf air. In composition, this represents a good sample of the vapour above the Gulf of Mexico region. Since much of the Great Lakes drainage rainfall is of the same origin, it is not surprising that the Great Lakes water shows about the same percentage as the rain sample. The fact that the Great Lakes water is about 1% lighter in its hydrogen deuterium ratio than the rain sample can be explained by the fact that some of the lake water originates from winter snow, which is quite light (Chicago snow −8·28%). This snow is derived from Pacific Ocean moisture that had passed over several mountain ranges and had deposited much of its moisture before arriving in the mid-west region. The explanation given by KIRSCHENBAUM (1951) for the lightness of Trail, British Columbia, water would also apply here.

"It is not unreasonable to expect that fractionation may occur as precipitation of water vapour from the air occurs. The first fraction precipitating may be higher in deuterium owing to the difference in vapour pressure of H_2O and HDO. Therefore as the winds rise to go over the mountains and much precipitation takes place, the effect may be that of Raleigh distillation in reverse, with the residual water having low deuterium content. Thus, on the basis of this hypothesis, the winds reaching the vicinity of Trail contain water somewhat depleted in deuterium. Consequently the waters which receive the last rains from these winds have a low deuterium content."

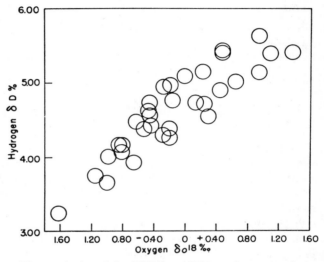

Fig. 5. A plot of the O^{16}/O^{18} and H/D results on samples of ocean water

Rivers on the Pacific coast that drain the western slope of the coast range are fairly heavy—Sacramento River -1.41%; San Joaquin River -2.29%. The same is true of east coast rivers that receive Atlantic Ocean air—Connecticut River -2.15%; Susquehanna River -0.17%. Rivers that receive Gulf air will all be of a composition close to that of Chicago rain, since there is no mountain barrier to remove much of the water before the Gulf air deposits its water in the Central Great Plains area—note the Platte River $+0.87\%$ and the Ohio River $+0.19\%$, and the Mississippi at Baton Rouge $+0.39\%$. Rivers draining the upper Plains States and the Rocky Mountain region receive much Pacific water as winter snow, and therefore they are light; for example, the Columbia River at Trail, British Columbia -10.1%. The Missouri River at Kansas City (-7.06%) has been somewhat diluted on its way south. The Mississippi River at Clinton, Iowa—mostly Gulf rain with some admixture of Pacific snow—is -1.63%. The Colorado River at Yuma, Arizona (-6.06%) and the Snake River at Clarkston, Wash.

99

(-5.77%), drain areas of high summer evaporation; which tends to make them heavier than the snow in the region (approx. -10%). Evaporation also accounts for the compositions of the Rio Grande $+3.28\%$, the Red River of Oklahoma $+3.05\%$, and the Arkansas River $+3.25\%$. If we assume that the rainfall source water is about $+0.5\%$ and that the evaporation takes place at 50°C, the fractionation during evaporation will be 5%, and about two-thirds of the water would have to evaporate to yield a residue as heavy as $+3.25\%$. The Apalachicola River of Florida yielded the heaviest river water that we determined ($+4.23\%$); and, here too, high re-evaporation of the source water is the probable explanation.

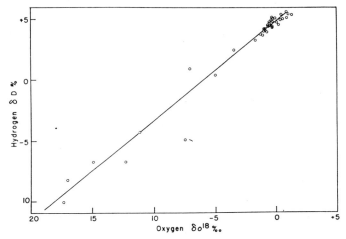

Fig. 6. A plot of O^{16}/O^{18} and H/D results on samples of ocean and fresh waters. The closely spaced group of points in the upper section of the graph represent sea water samples, and are plotted separately in Fig. 5

These samples of river water were, with a few exceptions, single samples collected during a period of a few minutes, they may have been contaminated by nearby springs or tributaries, as well as by recent rain. Therefore the conclusions drawn from the analyses of these samples must be treated as preliminary in nature.

Several melted core samples from the Juneau ice cap, Juneau, Alaska, were analyzed. The hydrogen results show no difference between the 155-ft level and the 235-ft level. The oxygen results differ greatly. This may be a case of the hydrogen ion diffusing through an immobile oxygen lattice.

The apparent seasonal variation in Violin Lake, at Trail, B.C., can be explained by summer evaporation concentrating the heavy isotope in the residue.

Miscellaneous Results

The sample of fumarole condensate from Paricutin is interesting. It is approximately 1.5% heavier in hydrogen isotopic composition than spring water from the region, a fact in itself not significant. However, the oxygen isotopic content of $+6.6‰$ indicates that the oxygen in the water equilibrated with the oxygen in the silicate rocks of the region; the rocks, according to SILVERMAN (1951) analyze +6·0‰. This equilibration must have taken place at a relatively high temperature.

Two Yellowstone Park geysers agree closely at −10·0%, a value that is probably close to the composition of the ground water and snow of the region. Samples of gas from several fumaroles in Yellowstone Park give very low results— approximately −20% (Table 2). These samples were collected by H. CRAIG and represent the hydrogen from both methane and free hydrogen in the gas. One sample from a spring issuing from Hurricane Vent, Norris Geyser Basin, showed

Table 2. Yellowstone Park Gas Samples Collected August 1950

Name Location	Hydrogen per cent*
Spring issuing from Hurricane Vent, Norris Basin	−46·3
Daisy Geyser, Upper Basin	−23·1
Punch Bowl Spring, Upper Basin	−19·2
Iron Creek, pool on bank 75 ft north of footbridge, between bridge and Cliff Geyser on west side of creek	−19·2
Kaleidoscope Group, Lower Basin. Small spring 30 ft from Kaleidoscope Geyser	−19·3

* See paragraph 2 under "Results—Water Samples"

−46·3%. According to DAY and ALLEN (1935) the gas from this spring contains 0·50% H_2 and 0·20% CH_4, the same authors claim that the other spring gases analyzed contain 0·00% H_2.

If we assume that these gases from all the springs were equilibrated at the same temperature, and subtract the methane hydrogen from the "Hurricane Vent" analysis, the hydrogen portion would give a minimum temperature of 400°C for the reaction $H_2 + HDO \rightleftharpoons HD + H_2O$ [Data from SUESS (1949)]. This conclusion is in essential agreement with the findings of BOATO and CARERI, 1952. More gases remain to be collected and separated before minimum temperatures can be assigned to these igneous activities.

In order to check the hydrogen isotopic relationship of the organic matter and the sea water in which an organism grows, two samples of abalone shell containing organic matter were run. The coarsely ground shell was mixed with powdered copper oxide previously treated with purified oxygen at 900°C; the mixture was degassed at 100°C and then heated to 800°C. The water produced by the combustion of the organic matter was processed in the usual way. The isotopic composition of the organic hydrogen was −4·5% whereas the composition of the ocean water in contact with the living organism was +4·5%. Apparently the organism fractionates its hydrogen by a large factor.

In this connection an interesting problem would be the separation and analysis of the different constituents of the organic matter in a manner that would not alter their hydrogen isotopic compositions. The problem of bound and free water in colloid systems complicates the situation.

An attempt was made to separate and analyze the hydrogen contained in an iron meteorite. As a check on the temperature necessary to remove water from the hydrated iron oxide that both coats and permeates most iron meteorite samples, a 3·5-g sample of surface scale from the Canyon Diablo meteorite was chipped off for analysis. This sample was heated in a vacuum and the evolved water collected, measured, and analyzed. Table 3 lists the results.

Table 3. Canyon Diablo meteorite

Temperature °C	Time of heating Hours	Amount of water evolved, milligrams	Analysis of water % of hydrogen*
20	12		
125	2	10	− 5·0
235	4	100	−10·5 ± 2·5
375	6	30	−11·0 ± 3
510	1	5	

* See paragraph 2 under "Results—Water Samples"

Apparently the oxide holds its water tenaciously. Heating to too high a temperature in an attempt to remove this terrestrial water causes loss of dissolved hydrogen by diffusion, as the following experiment shows.

A 10-g sample of chips of "fresh" Canyon Diablo meteorite was heated in a vacuum at 75°C, for 10 hours and at 125°C for another 10 hours. It was then heated at 450°C for 2 hours and the 10 mg of water was collected in a dry-ice trap. This fraction analyzed −9·4%. The chips were then burned in purified oxygen at 1200°C forming 0·5 mg of water, which analyzed +11·7%. Apparently while heating at 450°C hydrogen was lost by diffusion, and the deuterium was concentrated in the residue. A sample of hot-rolled iron treated in the same manner as the above meteorite sample gave essentially the same results.

In addition to loss of hydrogen by diffusion there is also the danger of contamination by exchange with hydrogen produced by the reaction of the iron with the absorbed terrestrial water during the degassing of the meteorite chips.

In view of the above results, great care must be exercised in interpreting hydrogen analysis of iron meteorites.

Acknowledgments—I would like to acknowledge the aid and encouragement of HAROLD C. UREY, who suggested the problem and who contributed greatly at every point.

I would also like to thank SAMUEL EPSTEIN and T. MAYEDA for their co-operation in making available to me hitherto unpublished figures on the oxygen isotopic analyses of the waters listed in Table I.

Thanks are also due to RICHARD BADER, HARMON CRAIG, FARRINGTON DANIELS, H. PETTERSSON, and JEROME WASSERBURG for furnishing me with samples, and to HAROLD ALLEN, DAVID LEE, and JOHN GODFREY for their assistance.

This project was financed in part by the Atomic Energy Commission under Contract No. AT (11-1)-101 and by the Office of Naval Research under Contract No. N6 ori —02028.

REFERENCES

Allen, E. T. and Day, A. L.	1935	Carnegie Inst. Wash. Pub. 466
Boata, G. and Careri, G.	1952	Nuovo Cimento **9**, 539
Epstein, S. and Mayeda, T.	1953	Geochim. Cosmochim. Acta (to be published)
Halsted, R. E. and Nier, A. O.	1950	Rev. Sci. Instruments **21**, 1019
McKinney, C. R., McCrea, J. M., Epstein, Samuel, Allen, H. A. and Urey, H. C.	1950	Rev. Sci. Instruments **21**, 724
Nier, A. O.	1947	Rev. Sci. Instruments **18**, 398
Silverman, S. R.	1951	Geochim. Cosmochim. Acta **2**, 26
Smyth, H. D.	1931	Rev. Mod. Physics **3**, 347
Suess, H. E.	1949	Z. Naturforsch. **4a**, 328
Wahl, M. H. and Urey, H. C.	1935	J. Chem. Phys. **3**, 411
	1951	"Physical Properties and Analysis of Heavy Water," Isidor Kirschenbaum, McGraw-Hill Book Co. p. 394
	1951	"Physical Properties and Analysis of Heavy Water," Isidor Kirschenbaum, McGraw-Hill Book Co. p. 398

Isotopic Variations in Meteoric Waters

Abstract. The relationship between deuterium and oxygen-18 concentrations in natural meteoric waters from many parts of the world has been determined with a mass spectrometer. The isotopic enrichments, relative to ocean water, display a linear correlation over the entire range for waters which have not undergone excessive evaporation.

Epstein and Mayeda (*1*) and Friedman (*2*) reported precise data for O^{18}/O^{16} and D/H ratios in nine nonmarine meteoric waters and found a rough linear correlation between the isotopic enrichments. In the course of research on isotopic variations in volcanic waters, I have analyzed mass spectrometrically some 400 samples of water from rivers, lakes, and precipitation in order to establish the exact nature of the isotopic relationship in meteoric waters. Gas samples were prepared by the standard CO_2–H_2O equilibration technique (*1*) and by reduction of H_2O to H_2 with uranium metal and analyzed on the McKinney-Nier type spectrometers used by the authors mentioned above as well as in my present laboratory.

The isotopic data for all samples analyzed for both isotopes (excluding detailed sets of data from Chicago and Steamboat Springs, Nev.) are shown in Fig. 1. About 40 percent of the samples are from North America, the rest being distributed all over the world. The data shown are per mil enrichments of the isotopic ratios D/H and O^{18}/O^{16} relative to a mean ocean water standard, that is,

$$\delta = [(R/R') - 1]\,1000$$

where R is either isotopic ratio and R' is the ratio in "standard mean ocean water" (SMOW) defined relative to the National Bureau of Standards isotopic water standard as described in a following report (*3*). The precision of the data is ± 0.5 per mil, or ± 1 percent of δ, for D, and ± 0.1 per mil, or ± 0.5 percent of δ, for O^{18}, the larger error applying in each case and representing ± 2 standard deviations.

The straight line in Fig. 1 represents the relationship

$$\delta D = 8\,\delta O^{18} + 10$$

(both δ values in per millage) and is seen to be an adequate fit to the data, except for waters from closed basins in which evaporation is a dominant factor governing the isotopic relationship. The samples which fit the dashed line at the high enrichment end of the curve represent rivers and lakes in East Africa. They fit a line with a slope of about 5, in contrast to the slope of 8 found for most of the data. Studies of evaporation in the laboratory, and in areas where seasonal data have been obtained, show that in free evaporation at ordinary temperatures the heavy isotope enrichment ratio $\delta D/\delta O^{18}$ consistently follows a slope of about 5 as observed in East African waters. Many of the points falling to the right of the line plotted in Fig. 1 have a similar slope of 5 when connected to points on the line which represent direct precipitation in the same area.

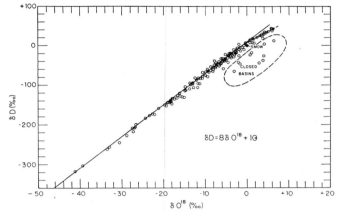

Fig. 1. Deuterium and oxygen-18 variations in rivers, lakes, rain, and snow, expressed as per millage enrichments relative to "standard mean ocean water" (SMOW). Points which fit the dashed line at upper end of the curve are rivers and lakes from East Africa.

It can be shown (*4*) that for small enrichments the slopes in Fig. 1 are the ratios of the single-stage enrichments when the isotopic concentrations are governed by vaporization or precipitation under Rayleigh conditions at constant temperature. The isotopic vapor pressure data show that slopes of 8 and 5 correspond to Rayleigh processes at liquid-vapor equilibrium at temperatures of about $-10°C$ and $+100°C$ respectively. It seems, therefore, that atmospheric precipitation follows a Rayleigh process at liquid-vapor equilibrium, as first proposed by Kirshenbaum (*5*), but that the process of free evaporation at room temperature is governed by kinetic factors. The present studies have shown that this is so up to the boiling point, and that the disequilibrium occurs principally in the O^{18}/O^{16} separation (*4*). Some of the variability along the line in Fig. 1 is certainly due to evaporation effects as well as to variations in temperature of precipitation.

All points in Fig. 1 for δD and δO^{18} lighter than -160 and -22 per mil, respectively, represent snow and ice from the Arctic and Antarctic, while tropical samples show very small depletions relative to ocean water. This distribution is expected for an atmospheric Rayleigh process as vapor is removed from poleward moving tropospheric air. However, it is actually $\log(1 + \delta)$ which should be plotted for such a process, and, in such a plot, the points in Fig. 1 fall on a curve with a continually increasing slope for lighter δ values, as would be expected (from the vapor pressure data) for precipitation at lower temperatures in high latitudes. The linear relation observed in Fig. 1 simply reflects a coincidence of the effect of the increasing difference in δ and $\log(1 + \delta)$ at high enrichments with the effect on the slope of the average temperature decrease for precipitation along a meridian from equator to poles (*6*).

HARMON CRAIG
Department of Earth Sciences,
University of California, La Jolla

References and Notes

1. S. Epstein and T. Mayeda, *Geochim. et Cosmochim. Acta* **4**, 213 (1953).
2. I. Friedman, *ibid.* **4**, 89 (1953).
3. H. Craig, *Science*, in press.
4. H. Craig, G. Boato, D. E. White, "Nuclear Processes in Geologic Settings: Proceedings of the Second Conference," *Natl. Acad. Sci.–Natl. Research Council Publ. No. 400* (1956), pp. 29–38.
5. I. Kirshenbaum, *Physical Properties and Analysis of Heavy Water* (McGraw-Hill, New York, 1951), p. 398.
6. Detailed papers on isotopic variations in meteoric and volcanic waters of specific areas will be published elsewhere. It is a pleasure to acknowledge my gratitude to Harold C. Urey in whose laboratories at the Institute for Nuclear Studies, University of Chicago, most of this work was done, to Mrs. T. Mayeda for her excellent services in the Chicago laboratory, and to G. Boato for many interesting discussions. This research has been supported by the National Science Foundation, the University of California Water Resources Commission, the Office of Naval Research, and the Atomic Energy Commission.

18 January 1961

Copyright © 1964 by the Swedish Geophysical Society

Reprinted from *Tellus*, **16**(4), 436–468 (1964)

Stable isotopes in precipitation

By W. DANSGAARD, *Phys. Lab. II, H. C. Ørsted Institute, University of Copenhagen*

(Manuscript received April 28, 1964)

ABSTRACT

In chapter 2 the isotopic fractionation of water in some simple condensation-evaporation processes are considered quantitatively on the basis of the fractionation factors given in section 1.2. The condensation temperature is an important parameter, which has got some glaciological applications. The temperature effect (the δ's decreasing with temperature) together with varying evaporation and exchange appear in the "amount effect" as high δ's in sparse rain. The relative deuterium–oxygen-18 fractionation is not quite simple. If the relative deviations from the standard water (S.M.O.W.) are called δ_D and δ_{18}, the best linear approximation is $\delta_D = 8\ \delta_{18}$.

Chapter 3 gives some qualitative considerations on non-equilibrium (fast) processes. Kinetic effects have heavy bearings upon the effective fractionation factors. Such effects have only been demonstrated clearly in evaporation processes, but may also influence condensation processes. The quantity $d = \delta_D - 8\ \delta_{18}$ is used as an index for non-equilibrium conditions.

The stable isotope data from the world wide I.A.E.A.-W.M.O. precipitation survey are discussed in chapter 4. The unweighted mean annual composition of rain at tropical island stations fits the line $\delta_D = 4.6\ \delta_{18}$ indicating a first stage equilibrium condensation from vapour evaporated in a non-equilibrium process. Regional characteristics appear in the weighted means.

The Northern hemisphere continental stations, except African and Near East, fit the line $\delta_D = 8.0\ \delta_{18} + 10$ as far as the weighted means are concerned ($\delta_D = 8.1\ \delta_{18} + 11$ for the unweighted) corresponding to an equilibrium Rayleigh condensation from vapour, evaporated in a non-equilibrium process from S.M.O.W. The departure from equilibrium vapour seems even higher in the rest of the investigated part of the world.

At most stations the δ_D and varies linearly with δ_{18} with a slope close to 8, only at two stations higher than 8, at several lower than 8 (mainly connected with relatively dry climates).

Considerable variations in the isotopic composition of monthly precipitation occur at most stations. At low latitudes the amount effect accounts for the variations, whereas seasonal variation at high latitudes is ascribed to the temperature effect. Tokyo is an example of a mid latitude station influenced by both effects.

Some possible hydrological applications are outlined in chapter 5.

1. Introduction

In the past decade an increasing number of investigators have treated the fractionation in nature of the most important isotopic components of water, H_2O^{16}, HDO^{16} and H_2O^{18}. The main results may briefly be summarized in this way:

(1) Fresh waters are poorer in heavy isotopes than sea water (GILFILLAN, 1934).
(2) The great reservoir of water, the oceans, has a fairly uniform isotopic composition disregarding those parts, which are directly mixed with fresh water (EPSTEIN & MAYEDA, 1953).
(3) The deuterium and oxygen 18 concentrations in sea water usually vary parallelly (FRIEDMAN, 1953).
(4) The heavy isotope content in precipitation decreases with the condensation temperature, which is reflected by

(*a*) variation of the composition of precipitation from individual atmospheric cooling processes—in simple cases in accordance with the Rayleigh conditions (DANSGAARD, 1953),

(*b*) the heavy isotope concentrations in fresh water decreasing with increasing latitude and altitude (DANSGAARD, 1954), and

(c) seasonal variation of the precipitation composition at high latitudes (EPSTEIN, 1956).
(5) Kinetic effect in fast evaporation can disturb the above-mentioned parallelism between the deuterium and O^{18} variations (CRAIG, BOATO & WHITE, 1956).
(6) Exchange of isotopic molecules between vapour and liquid may play an important role (FRIEDMAN, MACHTA & SOLLER, 1962).

These are the most important steps in the endeavours for clarifying the laws, which govern the fractionation processes in nature. Such developments have not yet come to an end.

Parallelly to these efforts much work has been done for utilizing the experiences in various fields, the glaciological applications being mainly represented by Epstein et al., Confiantini, Picciotto, Dansgaard and Lorius, the hydrological applications by Friedman, Craig and others. Thousands of sea water samples and ten-thousands of fresh water samples have been analysed by mass spectrometry. However, relatively few of the data have been published in spite of many of them having more than local interest. This has for several years been a handicap for the investigation of the water turnover on a global scale as well as for the solution of such local problems, which are analogous to others already investigated elsewhere. A change of practice on this point would undoubtedly be useful for many investigators and may also lead to a higher degree of systematism and standardisation of the sampling technique, which is especially important for the interpretation of precipitation data.

Most of the present investigation is based on the world wide survey of hydrogen and oxygen isotopes in precipitation organized since 1961 by the International Atomic Energy Agency (I.A.E.A.) and the World Meteorological Organization (W.M.O.). The aims of this project can be broadly stated as (1) providing a knowledge of tritium input function in various parts of the world, a parameter necessary for the application of natural tritium to hydrological problems, and (2) obtaining data on the concentration of hydrogen and oxygen isotopes, from which it should be possible to deduce some characteristics of the circulation patterns and mechanisms of the global and local movements of water.

The stable isotope data from 1961 and 1962 will be given later in this journal. In section 4 some conspicuous regularities in the data are shown, and some possible explanations are pointed out in relation to the calculations on simple processes given in chapters 2 and 3. However, the interpretation is far from completion and several of the problems are still open for discussion. Even considering ERIKSSON's (1964) important approach much work still remains to be done for interpreting the comprehensive material on the basis of meteorological studies.

1.1. MEASURING TECHNIQUE

The average occurrences of the most important isotopic components of water, H_2O^{16}, HDO^{16} and H_2O^{18}, are related as approximately

997680 : 320 : 2000 ppm (parts per million).

Nowadays, precision measurements of deuterium and O^{18} in natural waters are always made by a mass spectrometer. BOTTER & NIEF (1958) have been able to measure the absolute deuterium content in water, whereas no means are available for an accurate determination of the absolute O^{18} content. It is much easier to measure relative or absolute differences between two samples. Fortunately, the greatest interest is attached to the variations in isotopic composition, the measurement of which is, consequently, a common feature of all techniques hitherto applied in this field.

In this work all data will be given as the relative deviation, δ, of the heavy isotope content of a sample from that of a standard. If the absolute content is denoted by a, δ may be considered as

$$\delta = \frac{a_{\text{sample}} - a_{\text{standard}}}{a_{\text{standard}}} \cdot 10^3 \text{\textperthousand}.$$

The reference standard is SMOW (Standard Mean Ocean Water, CRAIG, 1961b). The measuring accuracy is ± 2‰ on δ_D for deuterium and ± 0.2‰ on δ_{18} for O^{18}. For details of the measuring technique reference is made to DANSGAARD (1961).

If a sample composition is given by δ' relative to a secondary standard, which deviates δ_{st}‰ from SMOW, the δ for the sample relative to SMOW will be

$$\delta = \delta' + \delta_{st} + \delta'\delta_{st} \qquad (1)$$

(BOATO, 1960). Thus, the δ function is not additive in a simple sense.

1.2. ISOTOPE FRACTIONATION FACTORS

The isotopic composition of natural waters covers a large range, which amounts to more than 400‰ for δ_D and 40‰ for δ_{18}, i.e. at least 200 times the measuring accuracy used here, or 400 times the best accuracy obtainable.

Isotopic fractionation of water is caused by several processes in nature, e.g. biological activity and exchange with other materials; of greatest interest from meteorological, hydrological and glaciological points of view is the fact that the volatility of H_2O^{16} is higher than those of the heavy isotopic components. This causes fractionation in all condensation processes and also in the evaporation of well mixed liquid water.

The fractionation factors for such processes depend upon the temperature and the rate of reaction. If the process proceeds so slowly that the equilibrium conditions are practically realized at the boundary between the phases, the fractionation factor for evaporation of liquid water becomes simply the ratio between the vapour pressure of the light component (p) and that of a heavy one ($p' < p$):

$$\alpha = \frac{p}{p'}.$$

At normal temperature the α's for HDO and H_2O^{18} are approx. 1.08 and 1.009, respectively. This means that vapour in equilibrium with water is depleted some 80‰ in deuterium and 9‰ in O^{18} relative to the water:

$$a_{vapour} = \frac{p'}{p} a_{water} = \frac{a_{water}}{\alpha},$$

or, when abbreviating the indices and assuming the water to be SMOW ($\delta = 0$):

$$\delta_v = \frac{a_v - a_w}{a_w} = \frac{1}{\alpha} - 1. \qquad (2)$$

Conversely, the first small amount of condensate from such vapour will have a composition, which is

TABLE 1

$t\,^\circ C$	α_D	α_{18}
100	1.029	1.0033_0
80	1.037	1.0045_2
60	1.046	1.0058_7
40	1.060	1.0074_0
20	1.079_1	1.0091_5
0	1.106_0	1.0111_9
-10	1.123_9	1.0123_0
-20	1.146_9	1.0135_0

$$a_c = \frac{p}{p'} a_v = \alpha \frac{a_w}{\alpha} = a_w, \qquad (3)$$

i.e. the same as that of SMOW. Using δ the corresponding calculation is the following: In analogy with (2) the condensate composition relative to the vapour is

$$\delta_c' = \alpha - 1. \qquad (4)$$

Inserting (2) and (4) into (1) gives

$$\delta_c = \left(\frac{1}{\alpha} - 1\right) + (\alpha - 1) + \left(\frac{1}{\alpha} - 1\right)(\alpha - 1) = 0. \qquad (5)$$

The α_D and α_{18} values used in this work for temperatures $t > 0°C$ are those measured by MERLIVAT et al. (1963) and ZHAVORONKOV et al. (1955), respectively. The α values below the freezing point are difficult to measure but very important for condensation processes in nature, so we must use extrapolated values in the temperature range $-20 < t < 0°C$. As to α_{18}, the formula of Zhavoronkov et al.,

$$\alpha_{18} = 0.9822 \exp(15.788/RT),$$

has been used down to $-20°C$, whereas the α_D values have been chosen by a second order extrapolation from the curve found by Merlivat et al. (Table 1).

In section 2 a few equilibrium processes will be treated quantitatively.

If we turn to non-equilibrium processes, i.e. fast reactions, the situation becomes more complicated. The values given above cannot be applied in such cases, because of the existence of a kinetic effect on the fractionation during the change of phase. As pointed out by CRAIG, BOATO & WHITE (1956) the observations may be understood qualitatively by assuming the rate of reaction, c, of the light component to be the fastest one:

$$c(H_2O^{16}) > c(HDO) > c(H_2O^{18}).$$

Considering an evaporation of water the relatively fast escape of H_2O^{16} corresponds to an increase of its vapour pressure and, thus, to effective fractionation factors, α^K, higher than α. DANSGAARD (1961) reported values of α_{18}^K up to 1.019, corresponding to a 100‰ increase of $\alpha_{18} - 1$, which governs the simple evaporation-condensation processes (cf. eq. (2) and (4)).

The physical explanation for the kinetic effect is not known (not even the equilibrium case is fully explained), but we shall frequently, especially in section 3, touch upon some of its important consequences for the global movement of water.

Since any evaporation of water in nature takes place into environments, already containing some vapour, one must also consider the influence of exchange of isotopic molecules between the vapour and the evaporating water (DANSGAARD 1953; FRIEDMAN & MACHTA 1962; CRAIG et al., 1963; ERIKSSON 1964). This influence, of course, depends upon the vapour content and its composition. The consequences for precipitation will be discussed in sections 3.3 and 4.2.3.

2. Equilibrium processes

In this chapter we shall treat the isotopic fractionation of water caused by some simple equilibrium processes, i.e. kinetic effects are neglected. The considerations are limited to two-phase systems with no supply of material from the environments, such systems not necessarily being identical with closed two-phase systems; e.g. Rayleigh processes are considered in sections 2.1.2 and 2.2. The results will later be compared with the actual findings.

2.1. CONDENSATION

Most important for the isotopic composition of precipitation is the condensation. As mentioned in section 1.2 the first small amount of water condensed from vapour in equilibrium with SMOW will have the same composition as SMOW (eq. (5)). By further condensation the vapour preferentially looses the heavy components, i.e. δ_v for the remaining vapour and, consequently, δ_c for newly formed condensate both get more and more negative.

2.1.1. Condensation in a closed two-phase system

If the condensation takes place in a closed two phase system with *equilibrium between the total liquid and vapour phases* at any stage, and if δ_c for the first small amount of condensate is supposed to be zero, δ_c^* for the total liquid phase can be shown to change by further condensation as

$$\delta_c^* = \frac{1}{\alpha_0} \cdot \frac{1}{\varepsilon F_v + 1} - 1, \quad \varepsilon = \frac{1}{\alpha} - 1,$$

α_0 being the value of α at the beginning of the processes, and F_v the remaining fraction of the vapour phase.

Similarly, for the composition of the vapour phase

$$\delta_v^* = \frac{1}{\alpha_0 \alpha} \cdot \frac{1}{\varepsilon F_v + 1} - 1.$$

For $F_v \to 0$: $\delta_c^* \to \frac{1}{\alpha_0} - 1$,

which is equal to δ_v^* for $F_v = 1$.

By *isothermal condensation* $\alpha = \alpha_0$, and $\delta_v^* \to 1/\alpha_0^2 - 1$ for $F_v \to 0$. The two practically straight lines in Fig. 1 show δ_c^* and δ_v^* as functions of F_v.

If the condensation is caused by *cooling*, α increases as the process proceeds. The compositions of the liquid and the vapour phases, denoted by δ_c' and δ_v', are shown by the dashed curves in Fig. 1. The limit of δ_v' for $F_v \to 0$ cannot be stated, since α is not known at very low temperatures.

When the temperature is low enough to allow *sublimation*, no exchange occurs between the solid and vapour phases. The composition of the newly formed solid material and that of the vapour will change as under Rayleigh conditions (curves δ_c and δ_v in Fig. 1, cf. p. 440), and the mean composition of the total solid phase will change as

$$\delta_s = \frac{1}{\alpha_0} \cdot \frac{1 - F_v^{\alpha_m}}{1 - F_v}$$

(cf. DANSGAARD, 1961, pp. 43–45), α_m being the value of α at the mean temperature. δ_s

corresponding to an isobaric cooling of saturated air from $t_0 = 20°C$ is shown by the thin, upper curve in Fig. 1.

2.1.2. *Rayleigh processes*

If the condensation of the vapour proceeds under Rayleigh conditions (i.e. a slow process with immediate removal of the condensate from the vapour after formation), δ_c for the liquid or solid phase and δ_v for the vapour phase will change as

$$\left. \begin{array}{l} \delta_c = \dfrac{\alpha}{\alpha_0} F_v^{\alpha_m - 1} - 1, \\[1em] \delta_v = \dfrac{1}{\alpha_0} F_v^{\alpha_m - 1} - 1, \end{array} \right\} \quad (6)$$

α, α_0 and α_m referring to the momentary condensation temperature t, to the initial temperature t_0 and to $(t + t_0)/2$, respectively (cf. DANSGAARD, 1961, pp. 45–46). In case of isothermal condensation, $\alpha = \alpha_0 = \alpha_m$. In any case, δ_c and $\delta_v \to -\infty$ for $F_v \to 0$ as indicated by the δ_c and δ_v curves in Fig. 1, which have been calculated by assuming an isobaric cooling of a saturated air mass from $t_0 = 20°C$.

The Rayleigh process thus leads to much higher fractionation than processes, in which the two phases are allowed to equilibrate by exchange. In nature, exchange will, more or less, smooth out the phenomenon. The curves δ_c' and δ_c, consequently, represent the two extremes to be expected.

In the following are given some calculations, based upon equation (6), on the variations of the deuterium and O^{18} components of the condensate in Rayleigh cooling processes. F_v has throughout been calculated as the mixing ratio (gram vapour per kg dry air) at the temperature t and the pressure p, divided by the initial mixing ratio at temperature t_0 and pressure p_0 (1000 mb). This procedure accounts for the variation of F_v due to changes in both temperature and pressure (or volume).

For the initial temperature t_0 between 0° and 80° the cooling processes have been supposed to be isobaric, for t_0 equal to 100°C isochoric (constant volume).

In the upper left part of Table 2, columns 2–7 give the remaining fractions F_v of the vapour as well as δ_D and δ_{18} for new condensate

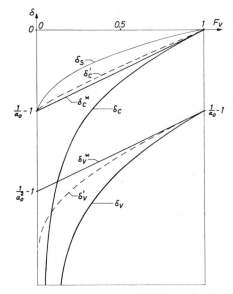

FIG. 1. Isotopic fractionation of the remaining vapour and of the condensate as a function of the remaining fraction, F_v, of vapour.

$\delta_v^*(\delta_c^*)$ equilibrium between the total liquid and vapour phases during isothermal condensation.

$\delta_v'(\delta_c')$ same as above during condensation by cooling.

$\delta_v(\delta_s)$ sublimation by cooling. δ_s is the mean composition of the solid phase.

$\delta_v(\delta_c)$ sublimation or condensation by cooling under Rayleigh conditions. δ_c is the composition of newly formed condensate.

after 20° and after 40° isobaric cooling below t_0. Calculated data on moist-adiabatic cooling from $t_0 = 0°$ and $t_0 = 20°$ are given in the lower left part of Table 2. Due to the expansion, and the consequent less condensation, in such processes the δ's are numerically lower than the corresponding values in the isobaric case.

2.2. EVAPORATION

During their fall from the cloud to the ground the rain drops are subjected to evaporation and exchange with the environmental vapour. These processes are, thus, important for the final composition of liquid precipitation when it reaches the ground. The evaporation is relatively high in dry air, in which case the process probably proceeds under non-equilibrium conditions (EHHALT et al., 1963). On the other hand,

TABLE 2

t_0 °C	Condensation by isobaric (for $t_0=100$°C isochoric) cooling						Evaporation (isothermal)			
	$t=(t_0-20)$ °C			$t=(t_0-40)$ °C			$F_w=0.3$		$F_w=0.1$	
	F_v %	δ_D ‰	δ_{18} ‰	F_v %	δ_D ‰	δ_{18} ‰	δ_D ‰	δ_{18} ‰	δ_D ‰	δ_{18} ‰
(1)	(2)	(3)	(4)	(5)	(6)	(7)	(8)	(9)	(10)	(11)
0	20	−150	−17.1				122	13.4	247	25.8
20	26	−95	−11.7	5.2	−223	−28.2	92	11.1	184	21.5
40	30	−63	−8.2	7.7	−153	−19.8	71	8.9	140	17.1
60	32	−46	−6.0	9.6	−104	−14.0	54	7.1	107	13.6
80	27	−44	−5.4	8.7	−87	−11.4	44	5.4	86	10.0
100	49	−14	−1.6	2.1	−39	−4.3	34	4.0	67	7.6
	Condensation by adiabatic cooling ($p_0=1000$ mb)									
0	29	−107	−12.9							
20	43	−52	−6.6	13	−144	−18.3				

in case of high humidity the exchange will possibly be the dominating factor (FRIEDMAN & MACHTA, 1962).

Thus, evaporation under Rayleigh conditions may very well be a process of negligible importance in nature. Nevertheless, for comparison this process has been treated quantitatively in the isothermal case using the usual formula

$$\delta_w = F_w^\varepsilon - 1, \quad \varepsilon = \frac{1}{\alpha} - 1, \quad (7)$$

in which δ_w is the composition of the remaining fraction F_w of the water reservoir, if the initial composition is supposed to be $\delta_w^0 = 0$. In the right part of Table 2 are listed the calculated δ_D and δ_{18} for $F_w = 0.3$ and 0.1 at various temperatures. All δ's are, of course, positive due to the preferential escape of H_2O^{16}.

2.3. TEMPERATURE EFFECT

Isothermal condensation never happens in the atmosphere. Any formation of precipitation is caused by some kind of cooling process. However, we cannot in general use the isotopic composition of a given amount of precipitation as an indication of the condensation temperature (not even in case of no kinetic, exchange or evaporation effects), because the observed composition of the individual rain is a function of several parameters, e.g. the thermodynamic conditions during the cooling, the initial composition etc. The influence of the condensation temperature is the easiest to calculate, at least under simplified conditions. Cooling processes under Rayleigh conditions, have already been considered in section 2.1.2., and some of the results are listed in Table 2. Fig. 2 shows how δ_{18} of the condensate decrease with temperature.

Logarithmic differentiation of eq. (6) gives

$$\log(\delta+1) = \log \alpha - \log \alpha_0 + (\alpha_m - 1) \log F_v,$$

$$\frac{d\delta}{\delta+1} = \frac{d\alpha}{\alpha} + (\alpha_m - 1)\frac{dF_v}{F_v} + \log F_v \, d\alpha_m,$$

$$\frac{d\delta}{dt} = \left(\left(\frac{1}{\alpha} + \frac{1}{2} \log F_v\right)\frac{d\alpha}{dt}\right.$$

$$\left. + \frac{\alpha_m - 1}{F_v}\frac{dF_v}{dt}\right)(\delta+1), \quad (8)$$

$d\alpha_m$ being put equal to $\frac{1}{2}d\alpha$.

Table 3 gives the isotopic change per degree cooling, $d\delta/dt$, for various initial dew points t_0 and also the average values, $(d\delta/dt)_a$, in the temperature ranges $t_0 \to (t_0-20)$ and $(t_0-20) \to (t_0-40)$. The figures in parenthesis refer to vapour-ice equilibrium. Apparently, one can easily measure a change corresponding to cooling 1°C at normal and low temperature.

Two attempts have been made to correlate the composition of precipitation with the temperature of formation. DANSGAARD (1953) measured rain from a warmfront, in which the air was moist-adiabatically cooled from $t_0 = 12°C$ to approx. $-8°C$. The observed change in O^{18} content was $8‰$, or $11‰$ when applying an estimated correction for evaporation (and exchange) from the drops. The latter figure corresponds to $0.50‰$ per centigree in agreement with Table 3; linear interpolation to $t_0 = 12°C$ in column 13 gives, namely,

$$\left(\frac{d\delta_{18}}{dt}\right)_a = 0.45 ‰/°C.$$

In the low temperature range (-18 to $-30°$) PICCIOTTO et al. (1960) found 8 and $0.9‰/°C$ for D and O^{18} in Antarctic snow. These mean values are compared with $d\delta/dt$ at $-20°C$ after an isobaric cooling from $20°C$ (vapour-ice equilibrium from $0°C$). Eq. (8) gives:

$$\frac{d\delta_D}{dt} = 7.7 ‰/°C \quad \text{and} \quad \frac{d\delta_{18}}{dt} = 0.95 ‰/°C.$$

In the following subsection Fig. 3 shows a linear correlation between the annual means of

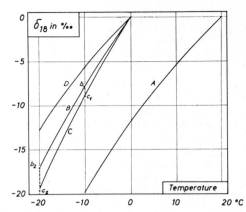

FIG. 2. δ_{18} of newly formed condensate as a function of temperature in Rayleigh cooling processes. For the first stage condensate $\delta_{18} = 0$:

A. Isobaric cooling. Vapour-water equilibrium. $t_0 = 20°C$.
B. Isobaric cooling. Vapour-water equilibrium. $t_0 = 0°C$.
C. Isobaric cooling. Vapour-ice equilibrium. $t_0 = 0°C$.
D. Moist-adiabatic cooling. Vapour-water equilibrium. $t_0 = 0°C$.

The dashed lines reflect turnover from vapour-liquid (b) to vapour-ice equilibrium (c). Further sublimation by cooling will, practically speaking, follow the curve C (DANSGAARD, 1961).

TABLE 3. *Variation (in ‰ per °C) of the isotopic composition of newly formed condensate in cooling processes.*

Columns 2, 5, 9 and 12: slope of the δ-t curve at the initial temperature t_0. The figures in parenthesis refer to vapour-ice equilibrium. *Columns 3, 6, 10 and 13*: the average slope for the first 20 centigrees cooling. *Columns 4, 7, 11 and 14*: the average slope for the next 20 centigrees cooling.

Isobaric cooling

t_0 (1)	$d\delta_D/dt$ (2)	$(d\delta_D/dt)_a$ $t_0 \to t_0 - 20$ (3)	$t_0 - 20 \to t_0 - 40$ (4)	$d\delta_{18}/dt$ (5)	$(d\delta_{18}/dt)_a$ $t_0 \to t_0 - 20$ (6)	$t_0 - 20 \to t_0 - 40$ (7)
0	6.3 (6.9)	7.0 (7.7)		0.71 (0.82)	0.85 (0.97)	
20	4.0	4.8	6.4 (7.0)	0.48	0.58	0.78 (0.88)
40	2.7	3.2	4.5	0.35	0.41	0.58

Moist-adiabatic cooling

	(9)	(10)	(11)	(12)	(13)	(14)
0	4.3 (4.6)	5.6 (6.2)		0.51 (0.58)	0.64 (0.73)	
20	2.0	2.6	4.6 (5.3)	0.24	0.33	0.58 (0.67)
40	0.5	0.82	1.8	0.06	0.105	0.23

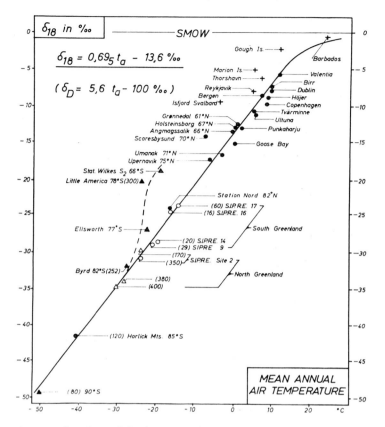

FIG. 3. The annual mean δ_{18} of precipitation as a function of the annual mean air temperature at surface. The figures in parenthesis indicate the total thickness (in cm) of the investigated snow layers.

the O^{18} content in precipitation and the surface air temperature, t_a,

$$\delta_{18m} = 0.69_5 t_a - 13.6\%_0 \qquad (9)$$

over a very wide temperature range including North Atlantic coast stations and Greenland ice cap stations. If the mean surface temperature is supposed to vary parallel to the mean condensation temperature from one place to the other, equation (9) gives:

$$\left(\frac{d\delta_{18}}{dt}\right) \simeq 0.70 \%_0/°C. \qquad (10)$$

In the high temperature range ($0 < t_a < 10°C$) this figure should be compared to $d\delta_{18}/dt = 0.66\%_0/°C$, calculated from equation (8) with $t_0 = 20°C$ and $t = 5°C$ after an isobaric cooling. All the Greenland ice cap stations ($-30 < t_a <$ $-10°C$) are high altitude stations. The precipitating air masses are moist-adiabatically cooled, at least from $t = 0°C$. Furthermore, they are always in vapour-ice equilibrium at the sites in question. Thus, we assume an isobaric cooling from $t_0 = 20°C$ to $t = 0°C$ followed by a moist-adiabatic cooling (vapour-ice) from $t = 0°$ to $t = -20°C$. Equation (8) then gives $d\delta_{18}/dt = 0.67\%_0/°C$ at $t = -20°C$ ($=$ the mid temperature in the interval considered) in fairly good agreement with (10).

Thus, the observed temperature effects are in agreement with those calculated under the assumption of the condensation part of the water cycle in nature being a simple Rayleigh process starting at $20°C$. This is, of course, not the same as to state that all precipitation at high latitudes originates from those parts of the oceans, where the temperature is approx.

20°C. But as a simplified model the Rayleigh condensation is, apparently, sufficient for the interpretation of most observations (cp. p. 456).

2.3.1. Glaciological applications

As pointed out in the beginning of section 2.3., one cannot use the composition of the individual rain as a direct measure of the condensation temperature. Nevertheless, it has been possible to show a simple linear correlation between the annual mean values of the surface temperature and the O^{18} content in high latitude, non-continental precipitation (DANSGAARD, 1961). The main reason is that the scattering of the individual precipitation compositions, caused by the influence of numerous meteorological parameters, is smoothed out when comparing average compositions at various locations over a sufficiently long period of time (a whole number of years).

The somewhat revised and extended correlation is shown in Fig. 3. The mean annual δ_{18} for the precipitation is plotted against the mean annual air temperature at surface, t_a, which varies approx. parallely with the mean annual condensation temperature from one place to the other. The 38 considered stations are the following:

(1) 17 continental stations from the North Atlantic region (filled circles) in the temperature range $-17 < t_a < 11°C$.
(2) 6 island stations with $-4 < t_a < 25°C$ (crosses).
(3) Greenland ice cap stations and 1 Antarctica ice cap station (open circles) with $-41 < t_a < -14°C$ measured by the present author (the sample from Horlich Mtns. was collected by Dr. Bjørn Andersen, Norsk Polarinstitutt).
(4) 4 Greenland and 5 Antarctica ice cap stations (triangles) with $-50 < t_a < -19°C$ measured by EPSTEIN & BENSON, (1959) and by EPSTEIN & SHARP (1962).

It is quite obvious that the curve should not exceed $\delta_{18} = 0$ (SMOW), since positive δ's occur only as a result of special processes like accidental evaporation from liquid precipitation.

In the high temperature range ($t_a > -5°C$) a continental effect appears, the δ_{18} decreasing for a given t_a when going inland. The reasons are probably (1) increasing participation of isotopically light re-evaporated fresh water, and (2) increasing formation of convective rain; the difference between the surface temperature and the actual condensation temperature is, namely, greater in showers than in other types of precipitation processes.

For all Greenland stations (except Scoresbysund) there is a very close correlation in the range $-30 < t_a < 0°C$:

$$\delta_{18m} = 0.69_5 t_a - 13.6\%_0. \qquad (9)$$

Owing to the smooth course of the mean annual isotherms on the Greenland ice cap the sites of formation of icebergs extruded from the West coast glaciers can be estimated using the $\delta_{18}-t_a$ correlation. The equation used by DANSGAARD (1961) and SCHOLANDER et al. (1962) was not quite identical with (9), but the difference is small, and none of the conclusions drawn has to be changed.

The δ_D-t_a correlation analogous to (9) may be derived by introducing the $\delta_D-\delta_{18}$ relation (16) for continental stations (cf. subsection 4.1.2., p. 454):

$$\delta_D = 5.6\, t_a - 100\%_0. \qquad (11)$$

The glaciological application of the equations (9) and (11) mentioned above, and also the isotope-stratigraphic application of the seasonal change in composition (EPSTEIN & BENSON, 1959), are based upon the close correlation between the δ's and the condensation temperature. Other conditions for successful glaciological applications are, however, that (1) considerable accumulation occurs both in summer and winter; (2) the trajectories of precipitating air masses are roughly the same for the whole area considered and all the year round; (3) no considerable ablation takes place (by evaporation, melting or drift); and (4) the topography of the considered area is simple (cp. LORIUS, 1963).

If we look at the accumulation zone of the Greenland ice cap West of the ice shed, the above mentioned conditions are mainly fulfilled, e.g. precipitating air masses usually come from directions between South and West.

Antarctica, however, is surrounded by a relatively warm ocean, and a given location on the ice cap may receive precipitation from various directions and from air masses with essentially different pre-histories. Furthermore, in many parts of the continent the topography is not simple, and the precipitation is so sparse

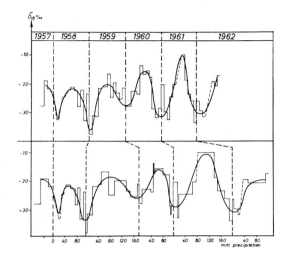

FIG. 4. δ_{18} in monthly precipitation at station Nord (82° N).

that ablation may disturb the stratigraphy. These may all be contributory reasons for the more complex correlation (if any) between δ_{18} and t_a indicated by the 4 filled triangles (Fig. 3) for $t_a > -25°C$ in Antarctica.

Furthermore, the above may also explain why the isotope-stratigraphic method has met serious difficulties in Antarctica (EPSTEIN & SHARP, 1962), which never occurred in Greenland.

In this connection it may be useful to consider the δ_{18} variation in precipitation at the Greenland station Nord (82° N). The δ_{18} record of monthly precipitation goes back to Sept. 1957 (Fig. 4). The mean annual air temperature is $-17°C$ and, consequently, the δ_{18} is very low. The δ_{18} curve shows approx. the same width of the minima and the maxima.

The method of determining annual accumulation on polar glaciers is based on the seasonal variation. Even though this variation broadly appears from Fig. 4, one can easily find δ_{18} values, which are unusual for the season (cf. Sept. 57, Dec. 58 and Jan. 61).

In the lower section of Fig. 4 the horizontal lines are plotted proportional to the actual amount of precipitation. This curve could be the result of O^{18} analysis of samples collected on a nearby glacier, if no melting or drift had disturbed the stratigraphy. Such disturbances are sources of error in the glaciological method

just mentioned; evidently, on the hypothetic Station Nord glacier a snow drift of 25 mm water equivalent could erase the 1957-58 minimum. Unusual isotopic composition of a relatively large amount of precipitation constitutes another source of error; thus, it seems possible that the low Sept. 57 could be misinterpreted as a winter minimum.

2.4. AMOUNT EFFECT I

In chapter 4, and especially section 4.2.3., a negative correlation will be demonstrated between δ and the amount of monthly precipitation, i.e. low δ's in rainy months and high δ's in months with sparce rain. This "amount effect" is found all the year round at most tropical stations, and in the summer time at mid latitudes, but never at polar stations, where the temperature effect is the dominating factor.

When looking for reasons for the amount effect we are confronted with difficulties, mainly because of the extreme complexity of the isotopic turnover in the processes forming convective rain. The air moves in vertical direction, and the condensate formed at any stage falls down through all the foregoing ones. Thereby, it is mixed up with the droplets and takes up new vapour at the lower stages. On top of that, further complications are added by exchange in the cloud and, during the fall from the cloud to the ground, by exchange and evaporation in the non-saturated air (the situation is somewhat simpler in case of hail, cf. FACY et al. (1963)).

However, we may consider qualitatively several possible, and probably contributing, reasons for the amount effect: (1) Looking at a given mass of condensing vapour the δ of newly formed condensate and, thereby, δ_{mean} of the total amount of condensate, n, decrease as the cooling proceeds (cp. Fig. 1, p. 440). Since, furthermore, n increases with the degree of cooling, one may expect an amount effect with a negative correlation between δ_{mean} and n in cases, where a given location is passed by storms of various degrees of development. (2) The fractionation by isotopic exchange between the falling drops and the environmental vapour (cf. section 3.3.) is most pronounced for light rain causing relatively high δ's in such rain, because the vapour below the cloud has not yet been exposed to cooling processes (in heavy rain the vapour composition may, more or less, be determined by the liquid phase). (3) The same tendency would seem to appear due to evaporation from falling drops. Low humidity of the low altitude air causing a great relative loss of liquid material (i.e. small values of n) and, thereby, considerable enrichment of the rain (i.e. high δ's).

Thus, both evaporation and exchange usually tend to enrich small amounts of rain in heavy isotopes. Low δ's in tropical rain must, therefore, be due to deep cooling of the air followed by only little enrichment by the other processes (e.g. in case of heavy and/or long lasting rain and high humidity), whereas high δ's could be due to either the rain being a first stage product and/or a later stage product enriched by exchange and evaporation (light rain and low humidity). Direct evidence of enrichment by evaporation from falling drops have been reported by DANSGAARD (1953, 1961, p. 71) and by EHHALT et al. (1963).

When going towards higher latitudes the amount effect becomes less pronounced, partly because of the lower degree of evaporation from falling drops. In the polar regions the isotopic composition is closely connected to the low altitude temperature, because most precipitation is formed at lower altitudes, and precipitating clouds consist usually of ice crystals, which are exposed to negligible isotopic fractionation by evaporation or exchange with environmental vapour (cf. section 3.3.).

One might expect that a parameter like the mean intensity of precipitation, i, would show some correlation with δ in the tropics, because low intensity would seem to correspond to relatively (1) low degree of cooling and, at the same time, (2) high degree of evaporation and (3) exchange with low altitude vapour. However, such a correlation between i and δ does not seem to be a general feature, at least not for continental stations.

Let us, as an example, consider Binza, Congo. For each period of precipitation the Congo Meteorological Service has observed the amount of precipitation (n mm) and the duration of the period (T hours). For 16 monthly samples the correlation coefficient between δ_{18} and Σn (= the total monthly amount of precipitation) is 0.82. However, the Binza data also enable us to calculate the weighted mean intensity of rain,

TABLE 4

Sample	Σn mm	i mm/hour	δ_{18} ‰	Number of rain periods
Sept. 61	84	5.3	−1.75	7
Nov. 61	255	4.5	−8.48	34

$$i = \frac{\Sigma(n^2/T)}{\Sigma n} \text{ mm/hour}$$

for each month (i varied between 1.2 and 19.8 mm/hour, δ_{18} between −0.35 and −8.48‰). The correlation coefficient between δ_{18} and i is only 0.07. Thus, δ has no correlation with i.

In Sep. and Nov. 1961 the rain fell with approx. the same mean intensity, but Nov. was dominated by frequent periods of rain, and the total amount was 3 times greater than in Sep. (Table 4). The very low δ_{18} in Nov. is due to a combination of (1) deep cooling by high lifting and (2) generally high humidity and, consequently, relatively little enrichment by evaporation from falling drops.

These data point to the humidity (or rather the precipitative water, i.e. the total amount of water in a 1 cm² column) as being a very important factor, at least at continental stations. The same feature appears in the record on Copenhagen precipitation covering 8 years. In the relatively dry spring the δ_{18}'s are generally almost 2‰ higher than in the autumn, in spite of the condensation temperature being generally lower in spring.

2.5. THE δ_D–δ_{18} RELATION

Whereas a F_v–δ diagram, e.g. Fig. 1, reflects only the fractionation of one heavy component relative to the light one (H_2O^{16}), the plotting of δ_D against δ_{18} opens up the possibility of following the important processes, in which the two heavy components behave in different ways, e.g. kinetic effects. However, it is worth noticing that even the simple equilibrium processes considered in this chapter lead to δ_D–δ_{18} relations, which are not quite simple.

The condensation curves C_0, C_{20} and C_{30} shown in Fig. 5 are obtained (partly from Table 2) by plotting the calculated δ_D against the calculated δ_{18} for the initial temperature t_0 equal

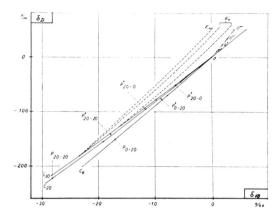

FIG. 5. Calculated δ_D–δ_{18} relations. The C-curves reflect the composition of newly formed condensate in isobaric condensation processes starting at 0°, 20° and 30°C. The composition of the first stage product is supposed to be $\delta_D = \delta_{18} = 0$. The E-curves reflect the composition of the remaining water in isothermal evaporation processes.

to 0°, 20° and 30°C, respectively. Point O represents the composition of the first small amount of condensate released at t_0, i.e. SMOW. As the cooling proceeds, the composition of newly formed condensate changes so that its characteristic (δ_D, δ_{18}) point moves down one of the curves depending on t_0. Temperature marks are given on the C_0 and C_{30} curves for each 10 degrees cooling. E.g. the point P_{0-20} on C_0 indicates the composition of new condensate released at $t = -20°$ in an isobaric process starting at $t_0 = 0°$. Had the process been moist-adiabatic, the corresponding composition would have been that indicated by P'_{0-20}. Similarly, the "moist-adiabatic" points P'_{20-0} and P'_{20-20} correspond to the "isobaric" points P_{20-0} ($t_0 = 20°$, $t = 0°$) and P_{20-20} ($t_0 = 20°$, $t = -20°$) on C_{20}.

Already at this stage it is evident that, even under the very simplest assumptions, the position of a given rain in the δ_D–δ_{18} diagram is a function of several parameters, such as (1) the initial vapour composition, (2) the initial dew point t_0, (3) the degree of cooling, $t_0 - t$, and (4) the way of cooling. Furthermore, it is clear that all rains cannot be expected to lie along one single curve in the δ_D–δ_{18} diagram; the way of cooling as well as t_0 influence the slope of the curve, which reflects the proceeding condensation from a given vapour reservoir.

Other, even more important factors, e.g. kinetic effects, will be mentioned in section 3.1.

The two dashed evaporation curves, E_0 in Fig. 5, show the variation of a water reservoir evaporating at 0°, the initial compositions being given by two points on C_{30}. The curve E_{10} from the lower point reflects an evaporation at 10°.

The slopes of such curves have shown to be an important characteristic of natural fractionation processes. A mathematical expression for the slope, s_C, of the simple condensation processes is evaluated from equation (8), p. 441. Since this equation is valid for both the deuterium and the O^{18} component, s_C is calculated as

$$s_C = \frac{d\delta_D}{d\delta_{18}} = \frac{d\delta_D}{dt} \bigg/ \frac{d\delta_{18}}{dt}.$$

Similarly, the slope s_E of the curves for isothermal evaporation is derived from (7):

$$d\delta_w = \varepsilon F_w^{\varepsilon - 1} dF_w$$

valid for both components, i.e.

$$s_E = \frac{d\delta_D}{d\delta_{18}} = \frac{\varepsilon_D}{\varepsilon_{18}} F_w^{\varepsilon_D - \varepsilon_{18}}. \quad (12)$$

Values of s_C and s_E are listed in Table 5 for all stages of the processes considered in Table 2. The columns 2 and 5 give s_C and s_E in the starting point ($\delta_D = \delta_{18} = 0$). The figures in parenthesis denote mean values of the slopes from the beginning of the process to the stage in question.

Disregarding the initial value at $t_0 = 100°$, s_C is 8.0 ± 0.9, with a minimum at $t_0 = 60°$. In the very important case C_{20} ($t_0 = 20°$), the mean value of s_C is 8.0 ± 0.2, over a range of 40 degrees cooling both under isobaric and moist adiabatic cooling. As stated by DANSGAARD (1961) a change from vapour-water to vapour-ice equilibrium will have no significant effect.

The value 8.0 ± 0.2 is in accordance with earlier findings of 8.1 ± 0.4 (sea water, FRIEDMAN, 1953), 8.1 ± 0.4 (Greenland ice, DANSGAARD, NIEF & ROTH, 1960), and 8 (CRAIG, 1961). The latter author claims a slope of 5 to correspond to Rayleigh processes at liquid-vapour equilibrium at about 100°C. Such low slopes are certainly not found here, when using

TABLE 5. Slope of $\delta_D - \delta_{18}$ curves.

t_0	s_C. Condensation by isobaric cooling (for $t_0 = 100°C$ isochoric)			s_E. Isothermal evaporation		
	$t = t_0$	$t = t_0 - 20$	$t = t_0 - 40$	$F_w = 1$	$F_w = 0.3$	$F_w = 0.1$
(1)	(2)	(3)	(4)	(5)	(6)	(7)
0	8.9	8.5 (8.8)		8.6	9.6 (9.1)	10.5 (9.6)
20	8.2	7.9 (8.1)	7.5 (7.9)	8.1	8.7 (8.3)	9.4 (8.6)
40	7.8	7.6	7.2	7.7	8.2 (8.0)	8.6 (8.2)
60	7.7	7.5	7.2	7.5	7.9	8.2
80	8.3	7.6	7.2	8.0	8.3	8.6
100	11.6	8.5	7.9	8.6	8.8	9.1

	s_C. Condensation by adiabatic coling ($p_0 = 1000$ mb)		
0	8.7	(8.3)	
20	8.0	(7.9)	(7.8)

the α-values given in Table 1, which are the most reliable known to the present author.

It should be noted that whereas for isothermal evaporation

$$s_E \cong \varepsilon_D/\varepsilon_{18},$$

according to eq. (12), because the exponent $\varepsilon_D - \varepsilon_{18}$ is small and constant, the slope of the condensation curves cannot be calculated approximately as

$$s_C = \frac{\alpha_D - 1}{\alpha_{18} - 1}, \quad (13)$$

mainly because the term containing $d\alpha/dt$ in eq. 8 is not negligible; e.g. for $t_0 = 20°$ and $t = -20°$, eq. (13) gives $s_C = 10.9$, whereas (8) gives 7.5 (Table 5).

A most interesting possibility has been pointed out by ERIKSSON (1964), namely, that $\alpha - 1$ in the Rayleigh formula (6) should be replaced by $\sqrt{\alpha} - 1$, when considering average values, due to eddy diffusion processes in the atmosphere. This procedure gives the values of δ_D, δ_{18} and s_C listed in Table 6 for the isobaric process from $t_0 = 20°$, to $t = -20°$ i.e. only approx. 40 % of the fractionation calculated before, and an almost linear $\delta_D - \delta_{18}$ curve with a slope close to 7.3.

3. Non-equilibrium processes

Any use of the α factors (Table 1, p. 438) involves the assumption of equilibrium, which is only true in practically infinitely slow processes. Many observations show that lack of equilibrium has important consequences for several *evaporation* processes in nature. However, up till now, only weak indications of kinetic isotope effect have been found in the investigated *condensation* processes. All reported examples (CRAIG, BOATO & WHITE, 1956; BOATO, 1960; CRAIG, 1961; EHHALT et al., 1963; CRAIG, GORDON & HORIBE, 1963) have been concerned with local evaporation phenomena, whereas the process of highest importance for the global movement of water, namely the evaporation from the oceans, has never been the object of a systematic investigation.

3.1. KINETIC EFFECT

The theory behind the non-equilibrium processes is very complicated (not even the α values at equilibrium are fully explained). Qualitatively, however, many observed deviations from the

TABLE 6. *Isobaric cooling from $t_0 = 20°C$. Exponent $\sqrt{\alpha_m} - 1$ in eq. (6).*

$t°C$	δ_D ‰	δ_{18} ‰	s_C
20	0	0	7.3
−20	−88	−12.1	7.5

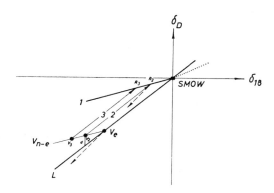

FIG. 6. The approximate δ_D–δ_{18} relation in equilibrium processes (slope $\cong 8$) and in non-equilibrium processes (slope < 8 for evaporation, slope > 8 for condensation). See text for details.

simple equilibrium processes can be interpreted as consequences of the various isotopic components having different rates of reaction, c:

$$c(H_2O^{16}) > c(HDO^{16}) > c(H_2O^{18}). \quad (14)$$

By extremely fast evaporation (into vacuum) one may, as a rough, approximation consider the flux of a given isotopic component from the water surface as being proportional, not only to the equilibrium vapour pressure of the component in question, but also to its rate of diffusion from the deeper layers. This would cause an effective fractionation factor of $\alpha^K = \mu\alpha$ in stead of α, μ being the ratio between the diffusion rates of the light and heavy components, i.e. $\alpha_D^K = \sqrt{19/18} \cdot 1.08 = 1.11$ for HDO^{16}, and $\alpha_{18}^K = \sqrt{20/18} \cdot 1.009 = 1.063$ for H_2O^{18} in case of mono-molecular movement in the liquid phase. However, according to the cell-cluster model, the water molecules move in clusters, consisting of several molecules. If, for example, six molecules move together, the mass numbers of the isotopic components become 108, 109 and 110, the latter two clusters containing one deuterium and one O^{18} atom, respectively. Hence, $\alpha_D^K = 1.09$, $\alpha_{18}^K = 1.028$. This is not far from the experimental results, $\alpha_D^K = 1.075 - 1.090$, $\alpha_{18}^K = 1.025$, obtained by evaporation from a thin water jet into a fastly passing stream of dry air. However, in this case the presence of air highly reduced the rate of evaproration, so that the diffusion of the vapour into the air may be the critical point. Anyhow, the experiment shows

the deuterium component to be much less sensitive to kinetic effects than the O^{18} component, since $\varepsilon_D(=1/\alpha_D - 1)$ deviated no more than 15 % from the equilibrium value, whereas ε_{18} deviated 200 %.

According to eqt. (12) the slope of the first part of the curve for a fast evaporation at 20°C becomes 3.2, when using the above mentioned values of α^K, in stead of 8 in case of equilibrium. In Fig. 6 the line L is drawn with a slope of 8. By fast evaporation of a limited amount of SMOW, the composition of the remaining liquid would change along some line like the dotted one.

If vapour in equilibrium with SMOW is denoted by V_e on line L (slope = 8, Fig. 6), fastly evaporated vapour must have lower δ_D and δ_{18}, because H_2O^{16} is the fastest reacting component. The (δ_D, δ_{18}) point, V_{n-e}, of such non-equilibrium vapour is laying above L, because $c(HDO^{16}) > c(H_2O^{18})$. The slope $s_v (<8)$ and length of $V_e V_{n-e}$ depends upon factors like the humidity, the temperature and the wind speed component perpendicular to the water surface, the turbulence in the water etc.

The first stage of an equilibrium condensation from any vapour on the line $V_e V_{n-e}$ will cause rain laying on line No. 1 ($\delta_D = s_v \cdot \delta_{18}$) through SMOW, e.g. vapour V_2 gives a first stage rain R_2, V_3 gives R_3 etc. By further equilibrium condensation later stage rains will follow C-curves starting at R_2, R_3 or another point on line No. 1. According to Fig. 5 and Table 5 we do not commit any considerable error, at least not in the high δ range, by approximating the C-curves to lines with slope 8. Consequently, in Fig. 6 the composition of later stage rains may be considered to move from the first stage point down along some line parallel to L, e.g. from SMOW along L, from R_2 along line No. 2 etc. (at the same time the remaining vapour in question moves in the same direction from the V point).

The experiments show that evaporation from a limited water surface goes on under equilibrium conditions only by very slow removal of vapour. On the other hand, no distinct indication of non-quilibrium condensation has been reported till now, as far as the present author knows. Nevertheless, the relation (14) might also influence the condensation process in extreme cases, but in the opposite direction, because the relatively high rate of reaction of the light component would counteract the preferential condensation of the less volatile heavy components, i.e. both of the α^K's would be lower than under equilibrium. If, in analogy with the evaporation, $\alpha_D - 1$ is supposed to be less sensitive to kinetic effects than $\alpha_{18} - 1$, a hypothetic non-equilibrium condensation of vapour V_e, would give a first stage condensate deviating from SMOW along a line like No. 1. Let R_2 be the composition of such first stage condensate. At later stages (continental rains?) the new condensate and the remaining vapour would follow a curve with a slope > 8 (the line $R_2 V_e$ shown dashed in Fig. 6).

The above is only a rough, qualitative consideration which, nevertheless, may be useful, when trying to interpret the numerous data in chapter 4. In order to simplify the terminology we shall now introduce a new parameter:

The surplus of deuterium relative to L is in the following denoted by d (e.g. SMOW and V_e in Fig. 6 have $d = 0$, whereas V_2 has a positive d). This parameter can apparently be used for indicating non-equilibrium conditions.

Equilibrium processes do not change the d-index for any of the phases, e.g. SMOW ($d = 0$) gives equilibrium vapour V_e ($d = 0$). Furthermore, vapour V_2 is condensed into R_2 with the same d as V_2. By further equilibrium condensation both R_2 and V_2 move down line 2 parallel to L, i.e. their d remains unchanged.

Non-equilibrium evaporation from a limited amount of water reduces the d-index of the water as long as exchange is not a dominating factor, e.g. fast evaporation of a limited amount of SMOW would make the water moving upwards along a line with a slope < 8 (dotted in Fig. 6), i.e. the d-index would get more and more negative. Similarly, n–e condensation from a limited amount of vapour would decrease the d-index of the vapour (cf. the dashed curve from V_e in Fig. 6), but this process does not seem to be common in nature.

By Non-equilibrium evaporation from an infinitely large and well mixed reservoir, d of the water will, of course, remain constant and, disregarding exchange, the d-index of the vapour will be positive and it will increase with the rate of reaction (cf. $V_e \to V_{n-e}$ in Fig. 6; as usual, exchange between the phases will reduce the effect). In this very important case d is, thus, not only a non-equilibrium indicator, but even a "rate of evaporation" index. Furthermore,

TABLE 7. *Mean δ_{18} in ocean samples* (EPSTEIN & MAYEDA, 1953).

Number of samples	Depth, m	Mean δ_{18} ‰
21	0– 25	+ 0.38
13	500–2000	– 0.19
8	4000–7500	– 0.41

supposing condensation to be an equilibrium process in nature, the averaged d of precipitation at a given locality reflects the rate of evaporation in the source area. This would seem to be a powerful tool in chemical meteorology.

3.2. THE COMPOSITION OF EVAPORATING OCEAN WATER

Unfortunately, relatively few measurements have been published on ocean water. The definition of SMOW (CRAIG, 1961 b) is based upon a NBS standard water and upon EPSTEIN & MAYEDA's (1953) data on 6 Atlantic, 11 Pacific and 2 Indian Ocean samples taken from 500 to 2000 m depth.

Owing to the evaporation from the surface layer of the oceans, which is not completely mixed with the rest, the upper layer has a composition with positive δ's relative to SMOW. This appears clearly from Epstein and Mayeda's data as seen in Table 7. Of these samples only one is from a latitude higher than 45°. In the first group 5 surface samples from the Atlantic subtropical high pressure region have a mean of + 0.69 ‰, whereas the mean of 9 surface samples from the East Pacific is + 0.15 ‰. Thus, the surface water shows measurable regional variations, but usually positive δ_{18} because of the evaporation. If the evaporation takes place under non-equilibrium conditions, the evaporating surface water probably has a composition somewhere on the dotted line in Fig. 6, i.e. a negative d-index. The line L is, nevertheless, considered most rational to use as a reference, but it is worth noticing that SMOW represents evaporating ocean water only as far as first order effects are concerned.

3.3. EXCHANGE

A drop of water falling through saturated air of the same temperature will, practically speaking, be in quantitative equilibrium with

TABLE 8. *Adjustment time* (τ_a) *and distance of fall* (f_a) *for the HDO content in falling drops of radius r.*

r cm	τ_a sec	f_a m
0.01	7.1	5.1
0.05	9.2	370
0.075	164	890
0.10	242	1600
0.15	360	2900

the vapour, i.e. per unit of time the number of water molecules leaving the drop is the same as that condensing on the surface of the drop. However, a further condition for isotopic equilibrium, i.e. for the above being true for each isotopic component, is that the heavy isotope content of the water phase, a_w, is α times that of the vapour phase, a_v, or δ'_w (relative to the vapour composition) being equal to $\alpha - 1$ (cf. eq. (4)). Otherwise, the exchange of molecules between the phases will cause a successive change of a_w until it reaches the value αa_v (or $\delta'_w = \alpha - 1$).

The consequences of the process for the condensation in a cloud was discussed by DANSGAARD (1953, p. 465), who assumed a fast equilibration between the cloud droplets and the environmental vapour.

BOLIN (1958) has derived a formula for the adjustment time, i.e. the time it takes for a falling drop to reach to a composition, which deviates $1/e$ of ($\alpha a_v - a_w$) from the equilibrium value, αa_v. Bolin's formula was essentially verified experimentally by FRIEDMAN, MACHTA and SOLLER (1962). Some of their results for the HDO component are listed in Table 8.

Apparently, τ_a and, consequently, the corresponding distance of fall, f_a, increase rapidly with r. Thus, cloud droplets equilibrate almost instantly. Even large drops may undergo considerable change in composition. However, the composition of the liquid phase must be expected to be the dominating one in cases of heavy and long lasting rainfall, especially if the drops travel through relatively stagnant air.

Evidently, attempts to interpret the isotopic composition of rain should involve the consideration of the vapour composition, but, unfortunately, this is not technically feasible in a world wide project like that described in chapter 4.

If the humidity is less than 100 %, a net evaporation will diminish the drop, and a kinetic effect will complicate the situation. Unlike the case treated by CRAIG et al. (1963) the isotopic composition of the drop does not reach to a constant value before the drop disappears, because the diminishing mass of the drop will cause a decreasing velocity of fall and, thereby, a decreasing rate of evaporation.

Exchange between snow and vapour can only affect the outer layers of molecules appreciably and is, therefore, considered negligible.

4. The I.A.E.A.-W.M.O. precipitation survey

As mentioned in the introduction a world wide precipitation survey has been organized by I.A.E.A. and W.M.O. since 1961. Samples of monthly precipitation have been collected for isotope analysis at an increasing number of stations (at present over one hundred) spread over all continents (Fig. 7).

The stable isotope data on samples from 1961 and 1962 will be given later in this journal.

4.1. The mean annual isotopic composition

Up till now, a complete record of stable isotope measurements covering at least one year has been obtained from some 70 stations. The data indicated on the five following figures are mean values, δ_m, of samples of monthly precipitation.

Whenever possible, δ_m has been calculated over all months of a whole number of years. Due to changing weather conditions from one year to the other, the δ_m can, of course, vary as well, especially at dry stations, where it may be determined by a few rain storms. Nevertheless, with this remark in mind, the δ_m's have been interpreted tentatively as normals in the following, but if the composition at a single station is in contradistinction to that of otherwise apparently similar stations, one should probably wait a couple of years to decide whether this is accidental or not.

The simple mean value, e.g. over 12 months,

$$\delta_m = \frac{1}{12} \sum_{i=1}^{12} \delta_i$$

reflects the gross climatic conditions the year round, whereas the weighted mean value

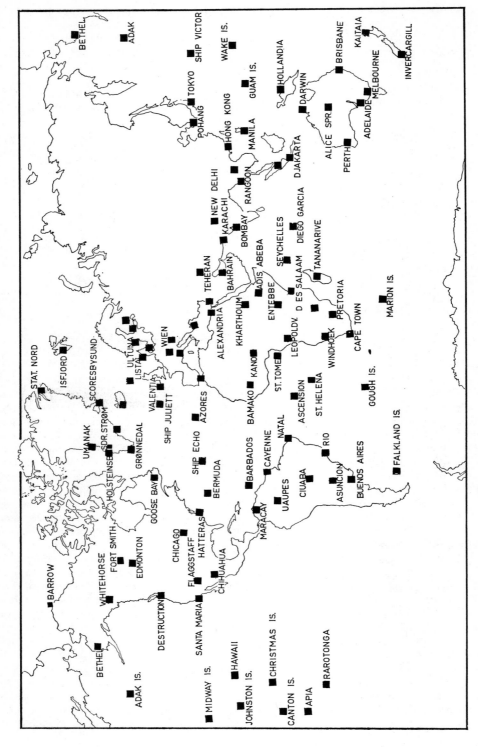

Fig. 7. The I.A.E.A.-W.M.O. precipitation network.

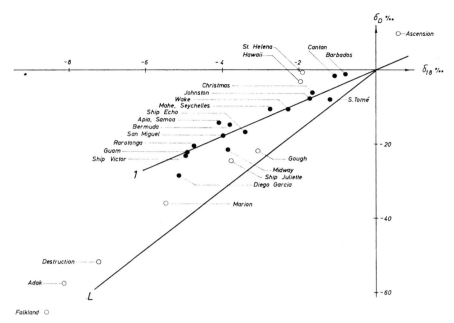

FIG. 8. Unweighted means for island stations.

$$\delta_m^w = \frac{1}{P} \sum_{i=1}^{12} (p_i \cdot \delta_i)$$

p_i and P being the monthly and annual amount of precipitation, is rather determined by the conditions in the rainy periods.

In some cases with occasionally very sparse precipitation the unweighted mean, δ_m, may be heavily influenced by a few months, in which the more or less accidental evaporation from falling drops causes δ's extremely deviating from normal. In regions with dry seasons, the weighted δ_m^w is, therefore, preferred as a basis for comparison.

4.1.1. Island stations

In Fig. 8 the dots indicate the unweighted mean composition $(\delta_{Dm}, \delta_{18m})$, for tropical and subtropical islands (filled circles) and for other islands (open circles). Let us, for a moment, look apart from Ascension, Hawaii and St. Helena for reasons given below, and from Canton, because at this latter station the rainfall was rather unusual in the period considered; thus, 4 out of 12 months had less than 10 % the normal precipitation; with the amount effect in mind the unweighted Canton composition may, therefore, be interpreted at being unusually high.

The $(\delta_{Dm}, \delta_{18m})$ points for the other 15 tropical islands and ships fit reasonably well to the heavy line No. 1 through SMOW:

$$\delta_D = (4.6 \pm 0.4) \delta_{18} + (0.1 \pm 1.6) \quad (15)$$

calculated by the method of the least squares. The correlation coefficient is 0.95 ± 0.09.

The heavy isotope content is not connected with the surface temperature. However, there seems to be some correlation with the mean monthly amount of precipitation, which is below 100 mm at each of the stations Barbados, Canton, S. Tome, Johnston, Christmas and Wake, but from 120 to 245 mm at the lower 10 stations. This effect of low δ's in abundant rain was discussed in section 2.4. and will be shown more clearly later (section 4.2.3.).

The slope of 4.6 indicates that the simple Rayleigh condensation process under equilibrium conditions cannot account for the isotopic fractionation, since the calculation on such processes gives 7.5 as the lowest slope in the temperature range $20 > t > -20°C$. Neither can a kinetic effect in the condensation be responsible for the phenomenon, because this effect would be reflected by a slope higher than 8 (cf. section 3.1., p. 449).

It is more profitable to consider the rain at

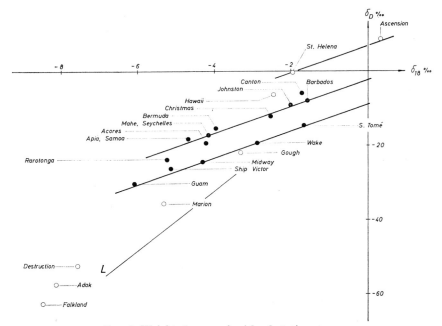

FIG. 9. Weighted means for island stations.

the tropical island stations as being representative for the first stage of the condensation of ocean vapour. A tropical cloud over the ocean has an influx of air containing vapour formed by fast evaporation of sea water. If the vapour composition deviates (along the line V_e–V_{n-e}, Fig. 6) from the equilibrium composition (V_e) to a degree, which increases with the cloud activity, the first stage of equilibrium condensation should produce rains along a line through SMOW and with a slope < 8, as described in section 3.1., p. 449. This model apparently fits the observations quite well (cf. lines No. 1 in Figs. 6 and 8).

St. Helena, Hawaii and Ascension seem to deviate significantly from the rest. As to Hawaii the precipitation in 1962 was only half the normal, so the normal composition is probably closer to line No. 1. This is supported by the fact that the unweighted mean composition for the more normal year 1963 was $\delta_{Dm} = -6.9‰$, $\delta_{18m} = -2.51$. A possible explanation for the odd composition at St. Helena, (Hawaii?) and Ascension is this: They are all three located on the equatorward side of a subtropical high pressure. The prevailing winds have an equatorward component and come from regions with cool sea currents. The air masses are unstable, and they have not reached to build up the high amount of precipitable water (total amount of water vapour in a column of air), which is typical for the low latitude islands in the Western parts of the oceans. Therefore, the first stage condensation temperature, t_0, is considerably lower than the evaporation temperature, t_e (though the precipitation may still represent a first stage condensation). If the corresponding α-values are denoted by α_0 and α_e, the equations (2), (4) and (5) on p. 438 give for the first condensate from equilibrium vapour:

$$\delta_c^0 = \frac{1}{\alpha_e} - 1 + \alpha_0 - 1 + \left(\frac{1}{\alpha_e} - 1\right)(\alpha_0 - 1) = \frac{\alpha_0}{\alpha_e} - 1.$$

Since $\alpha_0 > \alpha_e$, the first stage condensate will be enriched relative to SMOW, corresponding to the points O', O'' etc. (Fig. 5, p. 447) for $t_e - t_0 = 10°$, $20°$ etc. Furthermore, a kinetic effect in the evaporation will cause first stage condensate along a line with slope < 8 through one of the displaced O-point, i.e. a pattern into which St. Helena, Hawaii and Ascension fit (cf. Fig. 8).

Of the 6 remaining island stations, Marion, Ship Juliette and Gough are lying close to the polar fronts, and Reykjavik, Adak and Falkland on higher latitudes. They all fit—more or less—

STABLE ISOTOPES IN PRECIPITATION 455

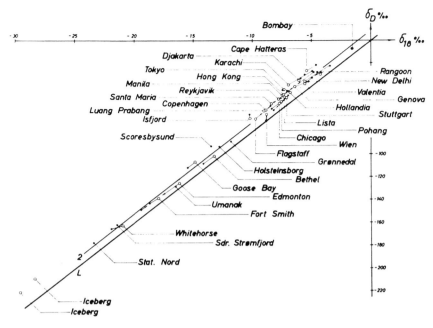

Fig. 10. Weighted (open circles) and unweighted (dots) means for Northern Hemisphere continental stations, disregarding Africa and the Near East.

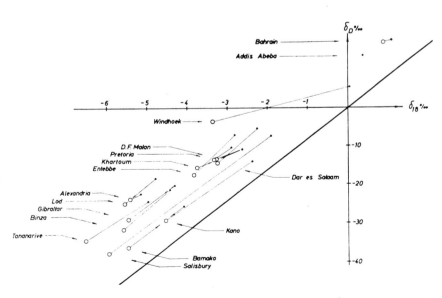

Fig. 11. Weighted (open circles) and unweighted (dots) means for African and Near East stations.

179

the line for continental stations. Precipitation occurs mainly in connection with cyclonic activity and frontal lifting.

If we consider the weighted means ($\delta^w_{18m}, \delta^w_{Dm}$) shown as large circles in Fig. 9 (filled for tropical and subtropical stations, otherwise open), it seems as the low latitude stations split up into three groups along three (heavy) lines, all with a slope close to 3.5:

$$\delta_D = 3.5\, \delta_{18} + d_0.$$

(1) $d_0 = +6.5\%_0$: Ascension, and St. Helena. Their common climatic features are outlined above.

(2) $d_0 = -2.5\%_0$: Canton, Christmas, Johnston, Apia, Barbados, Seychelles, Ship E., Bermuda. The 4 former stations are all located in the mid Pacific.

(3) $d_0 = -9.0\%_0$: S. Tome, Wake, Midway, Ship V. and Guam. The 4 latter stations are all located in the Western part of the Pacific. They have all more than 1000 mm of annual precipitation.

The slope (3.5) being lower than that of line No. 1 for the unweighted means (4.6) may indicate that the intense rain, which dominates the weighted means, originates from vapour released from sea water in a very fast evaporation.

Furthermore, the low value of d_0 in group (3) could be due to the mean condensation temperature for the heaviest rain being relatively low (very high clouds), i.e. such rain does not correspond to a first stage condensate.

Practically all δ^w_m's are lower than the corresponding δ_m's. This is, of course, a result of the amount effect. At Hawaii, no amount effect is observed, because the precipitation is uniformly distributed over all months.

4.1.2. Northern hemisphere continental stations, except Africa and the Near East

The unweighted means for 34 stations shown as dots in Fig. 10 and 1 Antarctic station, which falls outside the frames of the figure, determine a line (No. 2) with the equation

$$\delta_{Dm} = (8.04 \pm 0.14)\delta_{18m} + (9.5 \pm 1.5), \quad (16)$$

the correlation coefficient being 0.995 ± 0.018. For clearness, only fractions of this line are shown in Fig. 10.

The equation for the weighted means is

$$\delta^w_{Dm} = (8.1 \pm 0.1)\delta^w_{18m} + (11 \pm 1). \quad (17)$$

Both of the equations (16) and (17) are in essential agreement with that given by CRAIG (1961 a).

The position of a given station in the diagram is first and foremost determined by the condensation temperature: all tropical stations lie along the upper end of the line, all polar stations along the mid and lower end. However, Fig. 10 also demonstrates the importance of loss of moisture from an air mass during its passage of extensive territories (cf. the continental effect, Fig. 3 and p. 444) and, especially, high mountains; cf. Bombay →New Delhi, Valentia →Stuttgart → Vienna, Holsteinsborg (West coast of Greenland) →Sdr. Strømsfjord (on the other side of high mountains, at the same altitude and with the same mean air temperature as Holsteinsborg).

Line No. 2 in Fig. 10 (eq. 16) fits very well in with a C_{20} curve displaced $+9.5\%_0$ in deuterium. A comparison with line No. 2 in Fig. 6 and the explanation on p. 449 leads to the conclusion that most continental precipitation on the Northern hemisphere is formed on later stages of an equilibrium condensation from ocean vapour, which has been evaporated in a non-equilibrium process fast enough to give a surplus of deuterium, $d = +10\%_0$.

Another (weaker) indication of the evaporation from the ocean being a non-equilibrium process is deduced from the odd position of Bombay in Fig. 10 with considerably higher δ's than the other stations: In the rainy season the prevailing monsoun winds reach only 700–1000 m above the ground level, and the enormous amounts of rain (550 mm per month from June through Aug.) cannot originate exclusively from this layer. Most of the rain is rather released by convective activity above the monsoun air. On the way down through this air, the isotopic composition of the drops is more or less adjusted by exchange to that of the water in equilibrium with the monsoun vapour. If we assume the exchange to be complete, and that we are dealing with the first stage of the condensation, the composition of Bombay rain reflects that of the monsoun vapour and the deviation of the rain from SMOW is equal to the deviation of the monsoon vapour from vapour in equilibrium with SMOW. The former deviation is $-7‰$ and $-1.4‰$ for HDO and H_2O^{18} with a ratio of 5.0.

4.1.3. Other continental stations

The *African stations* (plus Teheran, Lod and Bahrain) are plotted in Fig. 11, the unweighted means as dots, the weighted means as open circles with lower δ's than the former due to the amount effect. A low slope of the line connecting a dot with its corresponding circle (e.g. Windhoek) indicates non-equilibrium evaporation from falling rain of low intensity. The stations Kano (Nigeria) and Bamako (Mali) both have a monsoon climate like that of New Delhi, and their positions in the diagram are, accordingly, close to line No. 2 in Fig. 10 ($d = 6‰$). The other African and Near East stations, however, have generally a higher d than the European, North American and Asian stations (cf. Table 9).

If we turn to *South America*, the Northeastern part of the continent is steadily dominated by moist trade winds from the Atlantic Ocean. Fig. 12 shows that both the weighted and the unweighted means at the coast stations Cayenne, Natal, Rio de Janeiro and Buenos Aires are close to the line No. 1 for tropical island stations. A continental effect is responsible for the low δ's at the inland station Ciuaba. The apparent lack of amount effect at Asuncion is due to the mean values being calculated from only 4 samples collected in rainy months.

Of the 8 investigated *Australian* and *New Zealand* stations only the tropical Darwin has a $d < 10‰$ (Fig. 13).

From *Antarctica* we have only one representative sample collected at Horlick Mountains (85° S) by Dr. Bjørn Andersen. The heavy isotope content is very low ($\delta_D = -338.3‰$, $\delta_{18} = -41.94‰$) and corresponds to the very low

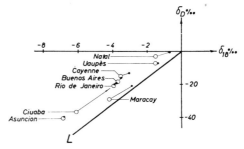

FIG. 12. Weighted (open circles) and unweighted (dots) means for South American stations.

mean air temperature (cf. Fig. 3, p. 443). Of the more than hundred investigated stations, this is the only one with a negative d ($-2.8‰$), which may indicate that at least part of the precipitation originates from considerably colder regions of the ocean than in other parts of the world, including the Greenland ice cap (cf. the positive d's for the two icebergs). Referring to Fig. 5, a condensation starting at $t_0 = 0°C$ would proceed along C_0 with a slope > 8 and, thus, lead to negative d's at very low temperatures, even if the initial vapour had a minor positive d. (cf. FRIEDMAN et al., 1964, p. 197).

In conclusion, a region consisting of (1) Africa and the Near East, (2) South America and (3) Australia except the moist, tropical regions, has precipitation with $d \cong 14‰$ (approx. 5‰ higher than Northern Hemisphere continental stations, cf. Table 9). On the lines of the conclusion in the foregoing subsection this is taken as an indication of such precipitation originating from maritime vapour (V_3 in Fig. 6) formed under conditions, which are farther from equilibrium than in other parts of the world. The

TABLE 9. *d for weighted means.*

Stations	$d‰$	d mean
Northern hemisphere	+ 9.4	+ 9.4
Africa and Near East	+ 14.8	
South America	+ 12.6	+ 14.4
Australia	+ 14.1	

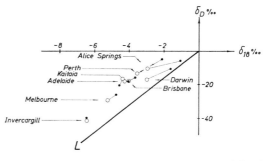

FIG. 13. Weighted (open circles) and unweighted (dots) means for Australian and New Zealand stations.

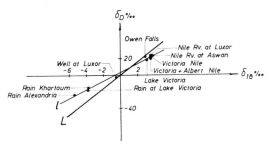

Fig. 14. Weighted means for precipitation at East African stations and February composition of the Nile River.

existence of dry climates prevailing in extensive areas of the above mentioned region supports this suggestion, as such climatic conditions would seem to lead to an increase in the isotopic departure of the vapour from the equilibrium vapour.

4.1.4. *Isotopic fractionation along the Nile River*

A number of river and ground water samples were collected from North and East Africa along the Nile River in Feb. 1964. Although the following considerations are made on a slender basis, they give some indication of the isotopic turnover in this interesting region.

In Fig. 14 the weighted mean composition of rain at Entebbe has been plotted as representative for the supply of water to Lake Victoria. The Lake Victoria water (sampled at Entebbe) and also the Owen Falls (20 km from the lake) both deviate along the line 1 with a slope of approx. 5 indicating fast evaporation from the lake. In addition to the weighted mean composition of rain at Entebbe Fig. 14 shows those at Khartoum and Alexandria, which also fit line 1 through the February composition of the Nile.

Shortly after having passed Lake Albert, the (Albert) Nile water has a composition (with negative *d*), which is essentially the same as the composition two weeks later at Aswan and Luxor, some 3000 km further North. Thus, if the sparse winter supply of water (and exchanged vapour) to the Nile from the winter dry regions North of 5° N can be neglected, the evaporation from the river cannot be considerable in winter (cp. FRIEDMAN et al., 1964).

Whereas the drainage of Lake Victoria is roughly unchanged during the year, the additional water from the tributary rivers is considerable in the summer time. The composition of the Nile during high water must, consequently, be much closer to that of Khartoum rain. This is supported by the sample collected from a deep well at Luxor (Fig. 14), approx. 3 km from the Nile, because the ground water at this location probably represents a weighted mean of the Nile river.

Thus, it is reasonable to assume both the tributary water and the Nile water far North to be somewhere along line 1 and, since possible seeping does not cause any fractionation, this is also true for the evaporated material. Consequently, the evaporation from the Nile takes place under non-equilibrium conditions corresponding to a slope of 5, as would also be expected considering the dry climate in question.

4.2. The mean monthly isotopic composition

The hitherto considered annual mean values are most suitable for the investigation of the global movement of water. If we want to look at the local isotope effects, we must, of course, consider the monthly samples. They give some more detailed informations and enable us to investigate possible correlations between the isotopic composition and parameters like the condensation temperature, the origin of the vapour and the amount of precipitation.

A $\delta_D - \delta_{18}$ plot of the monthly samples from the individual station is one useful mean for studying the local isotope turnover and its fitting into a large pattern on a global scale. Unfortunately, it is not feasible to present a $\delta_D - \delta_{18}$ diagram for each of the many stations. Neither is it the purpose of this paper to look into local details, such as water balance problems, though complete stable isotope records as those given in the appendix must constitute a good basis for such studies (cf. section 4.1.4.).

Instead, some of the most conspicuous features have been listed in Table 10 for those stations from which a reasonably representative number of data are available up till now. Furthermore, in section 4.2.2. six groups of typical $\delta_D - \delta_{18}$ correlations will be defined.

Table 10 shows in

Column 1: name and country of the station, alphabetic order.

Column 2: number of samples considered (from 1961–62, a few from 1963).

STABLE ISOTOPES IN PRECIPITATION

TABLE 10

s = summer, w = winter, a = autumn. *Column 2*: n = number of monthly samples. *Column 5*: d = surplus of deuterium relative to line L, cf. Fig. 6. *Column 6*: Temperature effect, cf. section 4.2.4. *Column 7*: Amount effect in ‰ change of δ_{18} per 100 mm precipitation. Scattering in ‰, cf. section 4.2.3.

Station (1)	n (2)	δ_{18} max (3)	δ_{18} min (3)	Slope (4)	d max (5)	d min (5)	Temp. effect (6)	Amount effect (7)
Acores (see San Miguel)								
Adelaide, Australia	12	−1 exc. Jan.	−5 −8	8 ± 1.5	15		−	−
Addis Abeba, Ethiopia	4	+3	−3	(10)	20	11	−	−
Adak, N. Pacific	12	−5	−11	8 ± 1	8		−	−
Alexandria, Egypt	7	−4	−6	−	18		−	+
Alice Springs, Austr.	10	+5 exc. Jan.	−4 −8	5 ± 1 exc. Jan.	18	−10	−	+ ?
Apia, Samoa, S. Pac.	10	−2	−7	−	19		−	+
Ascension Isl., S. Atl.	13	+2	−1	−	5		−	+ ?
Athens, Greece	13	−2	−8	8 ± 1.5	19		−	+
Bahrain	8	+6	−2	3 ± 1	23	−12	−	−
Barbados Isl., N. Atl.	22	+2	−4	8 ± 1.5	5		−	−1.2 2
Bermuda Isl., N. Atl.	12	−2	−6	−	16		−	−
Bethel, Alaska	12	−9	−21	7.5 ± 1	5		+	−
Binza, Congo	16	0	−9	8 ± 1	14		−	−2.2 4
Bombay, India	10	0	−3	8 ± 1.5	5		−	−
Brisbane, Australia	12	−2 exc. Jul.	−4 −11	8 ± 2	14		−	+
Buenos Aires, Argentine	18	−2	−8	8 ± 1	13		−	−
Canton, S. Pac.	23	+2 exc. Dec. 62:	−3 −10	−	7		−	−1.5 2
Cape Hatteras, U.S.A.	12	−3	−7	4.5 ± 1.5 exc. Jan.	22	4	−	+
Cape Town (see D. F. Malan)								
Cayenne, Fr. Guiana	8	−2	−5	−	11		−	+ ?
Chicago, U.S.A.	12	−2	−19	7 ± 1 (w) 5 ± 1 (s)	22	−15	0.35	+ (s)
Chihuahua, Mexico	7	−4	−9	−	12		−	−
Christmas Isl., N. Pac.	13	0	−3	−	7		−	−1.3 2.5
Ciuaba, Brazil	13	+3	−10	8.0 ± 0.5	11		−	−3.6 6
Dar es Salaam, Tangan.	20	0	−7	8 ± 2	13		−	−1.2 4
Darwin, Australia	7	+2 exc. Jan.	−4 −9	3.5 ± 1 exc. Jan.	16	−9	−	−
Destruction Isl., E. Pac.	9	−6	−9	4 ± 2	21	7	−	−
D. F. Malan, S. Africa	17	−1	−7	6 ± 1.5	24	−2	+	−
Diego Garcia, Ind. Oc.	8	−3	−6	6 ± 2	13		−	−
Djakarta, Indonesia	12	−3	−9	8.0 ± 0.5	15		−	+
Edmonton, Canada	17	−11	−26	8.0 ± 0.5	7		0.4	−
Entebbe, Uganda	12	0	−8	10 ± 1	11	−28	−	−2.2 3
Falkland (see Stanley)								
Flagstaff, Arizona	13	0	−16	8 ± 1	10		+	− exc. for $p < 15$
Fort Smith, Canada	9	−13	−26	8 ± 1.5 exc. Aug. & May	3		+	−
Genoa, Italy	13	−3 exc. Jul.	−9 +2	8 ± 1 (w) 4.3 ± 1 (s)	16	−25	−	−
Gibraltar (see North Front)								
Goose Bay, Canada	20	−10	−28	8.0 ± 0.3	12		+	−
Gough Isl., S. Atl.	21	−1	−5	−	3		+	−

Table 10 (continued)

Station (1)	n (2)	δ_{18} max (3)	δ_{18} min	Slope (4)	d max (5)	d min	Temp. effect (6)	Amount effect (7)
Guam Isl., Pac.	13	−1	−8	8±1.5	17	−	−	−2.0 1.6 exc. Sep. $p = 550$
Hawaii (Hilo), N. Pac.	23	−1	−4	−	12	−	−	−
Hollandia, New Guinea	22	−3	−9	8±1.5	11	−	−	+
Hong Kong (Kings Park)	20	0	−12	7±1	8	−	−	−1.4 5
Isfjord, Svalbard	15	−6	−15	7.5±1	14	−	−	−
Johnston Isl., N. Pac.	12	0	−3	5±2	6	−	−	−1.5 2
Kano, Nigeria	11	+10	−8	5.2±0.5 8 in rainy season	19	−20	−	−2.2 5
Karachi, Pakistan	7	+1	−13	8±1	12	−	?	+
Lista, Norway	20	−4	−9	8±1	7	−	+	+ (s)
Lod (Beth Dagon), Israel	13	0	−8	5.5±1.5	31	4	−?	+?
Luang Prabang, Laos	13	0	−13	8.0±0.5	12	−	−	+
Mahe, Ind. Ocean	15	+4	−8	5.5±1 exc. Sep. ($p = 350$)	20	−8	−	−1.8 5 exc. Feb.
Manila, Philippines	19	0	−10	8.0±0.5	14	−	−	−1.4 5 for $p < 200$
Maracay, Venezuela	12	+1	−7	8±1.5	2	−	−	−3.5 ?
Marion Isl., S. Atl.	21	−3	−6	−	8	−	−	−
Melbourne, Australia	12	−1	−9	7.0±0.3	16	6	+	+ (s and a)
Midway, N. Pac.	13	−1	−6	8±1	9	−	−	+ Feb.–Aug.
Natal, Brazil	9	+1	−3	−	5	−	−	?
New Delhi, India	15	+3	−11	8±1	5	−	−	(+)
North Front, Gibraltar	13	−4 exc. Jun. & Sep.	−7	−	23	−4	−	+
Nord, Greenland	17	−9	−33	7.4±0.5	10	−	+	−
Perth, Australia	12	0 exc. Jan. (+6, $p = 2$)	−5	4±1	17	−13	+	+
Pohang, Korea	21	−4	−9	−	7	−	(+)	(+)
Pretoria, S. Africa	17	0	−5	5±1.5	20	−6	−	−2.0 3
Rangoon, Burma	12	−2	−8	7±1.5	11	−	−	+
Rarotonga, S. Pac.	6	−2	−6	−	18	−	−	+
Reykjavik, Iceland	14	−5	−11	8±1	10	−	+?	+?
Rio de Jan., Brazil	13	−2	−6	8±1.5	12	−	−	+
Salisbury, S. Africa	22	+8	−10	8±0.5 6.5±1 (for pos. δ_{18})	8	−	−	+
San Miguel Isl., N. Atl.	11	−2	−6	8±1.5	14	−	−	+
Santa Maria, Cal., U.S.A.	6	+1	−10	?	6	−	−	+
S. Tomé Isl., S. Atlantic	9	0 exc. May +14	−4	−	2	−	−	+
St. Helena, S. Atl.	8	−1	−2	−	15	−	−	−
Seychelles, (see Mahe)								
Stanley, S. Atl.	14	−6	−12	8±1	4	−	?	?
Stuttgart, Germany	23	−4	−16	8±1 6.5±1 (s)	4	−	+	−
Tananarive, Madagaskar	11	−2	−10	8±1	15	−	−	(+)
Teheran, Iran	14	+4	−7	5±1.5 exc. Sep.	22	−11	(+)	(+)
Tokyo, Japan	22	−4	−12	4.7±1	21	0	+	+ (exc. w)
Uaupès, Brazil	7	+4	−5	−	4	−	−	−

Table 10 (continued)

Station (1)	n (2)	δ_{18} max (3)	δ_{18} min	Slope (4)	d max (5)	d min	Temp. effect (6)	Amount effect (7)
Waco, Texas, U.S.A.	11	−2	−6	?	11	−		+
Wake Isl., Pac.	12	0	−5	8 ± 1	8	−	−	−1.6 2
Valentia, Eire	12	−2	−7	8 ± 1	7		+	?
Weather Ship E, N. Atl.	12	0	−6	3 ± 1.5	22	−5	−	?
Weather Ship J, N. Atl.	13	−1	−7	8 ± 1.5	6		+	?
Veracruz, Mexico	9	−4	−12	8 ± 1	10		−	−
Whitehorse, Canada	24	−16	−28	5.5 ± 1 exc. Apr.	22	−18	+	−
Vienna, Austria	19	−1	−16	(w)8 ± 1.5 (s)6 ± 1	17	−10	+	+ (s)
Windhoek, S. Africa	14	+9	−8	8 ± 1.5 3.5 ± 1.5 for pos. δ_{18}	28	−28	−	+

Column 3: range of δ_{18}.
Column 4: the slope (with uncertainty) of the line representing the best approximation to the apparent $\delta_D - \delta_{18}$ correlation.
Column 5: the d-index of this line (cf. p. 450) if it has a slope close to 8; otherwise the maximum and minimum of d for the individual samples. In case of no $\delta_D - \delta_{18}$ correlation the mean d for the individual samples is given.
Column 6: distinct temperature effect (in ‰ δ_{18} per °C) usually indicated by a +; lack of distinct temperature effect is denoted by −.
Column 7: distinct amount effect denoted by + or, in some cases by the slope (in ‰ change of δ_{18} per 100 mm precipitation) and the width of the band obtained by plotting the individual monthly δ_{18} as a function of the monthly precipitation (cf. section 4.2.3.).

4.2.1. The range of δ_{18}

Referring to the sections 2.3. and 2.4. the δ_{18} values at a given station vary due to the influence of many factors. Three of the most important reasons are (1) varying condensation temperature (most pronounced at high latitudes, especially under continental conditions), (2) varying degree of evaporation from falling drops (most pronounced in regions with alternating rainy and dry seasons, especially in hot climates), and (3) seasonal shift of source area of the precipitating vapour.

Table 10, column 3, gives some impression of the δ_{18} range for the individual station. To facilitate a general survey mean values have been calculated for stations with common climatic features. If some station is missing in this connection, and if no special remark is made, the reason is incomplete knowledge to one or more of the parameters in question.

In Table 11 the ocean islands and weather ships (column 2) have simply been arranged according to their latitude (column 1). Furthermore, in

Column 3: t_a = the mean annual air temperature at surface.
Column 4: Δt_m = the mean difference between the highest and lowest monthly mean air temperature at surface.
Column 5 and 6: $p_{max}(p_{min})$ = the average amount of precipitation fallen in the rainiest (driest) month, from which samples have been analysed.
Column 7: δ_{18m} = the average of the mean annual δ_{18}.
Column 8: $\Delta\delta_{18m}$ = the mean difference between the highest and lowest δ_{18} measured for the individual station.

$\Delta\delta_{18m}$ is practically the same for the 3 groups of stations. However, the reason for the variation reflected by $\Delta\delta_{18m}$ is not the same. At the low latitude stations Δt_m is only 2°C, whereas the low p_{min} indicates the occurrence of periods with sparse precipitation; consequently, the $\Delta\delta_{18m}$ should be ascribed to the amount effect (section 2.4). Conversely, at the mid and high

TABLE 11. *Ocean islands and weather ships.*

Latitude (1)	Stations (2)	t_a °C (3)	Δt_m °C (4)	p_{max} mm (5)	p_{min} mm (6)	δ_{18m} ‰ (7)	$\Delta \delta_{18m}$ ‰ (8)
0°–23°	Apia, Ascension, Barbados, Canton, Christmas, Gaum, Hawaii, Johnston, S. Tome, Wake	26.2	2.1	237	16	−1.6	4.0
23°–45°	Acores, Bermuda, Gough, Midway, Ship E.	18.9	8.1	361	96	−3.4	4.4
45°–90°	Adak, Falkland, Marion, Ship J.	6.8	7.8	261	52	−6.6	5.2

latitude stations the relatively high p_{min} indicates a more uniform distribution of the precipitation the year round; the high Δt_m (= 8°C) points to a simple seasonal temperature variation as being responsible for the δ_{18} variation.

The temperature effect is also reflected by δ_{18m} decreasing with t_a. However, the δ_{18m} and t_a values are hardly relevant for calculating the temperature effect, if only for the reason that the altitude of the formation of precipitation is higher at low latitudes than at high latitudes. What should be compared is the mean condensation temperature and δ_{18m}.

4.2.2. *Slope of* $\delta_D = f(\delta_{18})$ *Table 10, Column 4*

A $\delta_D - \delta_{18}$ plot of the individual monthly samples from a given station shows the characteristics of the relative $\delta_D - \delta_{18}$ fractionation. Only a few typical examples of such plots will be given below. However, a study of all diagrams shows distinct common features for many stations.

In the following are shown six types and degrees of correlation. Practically all stations fit into one of these six groups:

I. All samples relatively close together. No distinct correlation between δ_D and δ_{18} (Fig. 15). To this group belong:

Acores	Hawaii
Apia	Johnston (IV)
Ascension	Marion
Bermuda	Natal
Canton	San Tome
Destruction	Ship E. (IV)
Gough	Ship J.

i.e. disregarding Natal (cf. p. 456), only island stations and ships with no dry season and little seasonal temperature variation. Ship E. and Johnston may also be incorporated into group IV.

A few stations are very much like the above, except for a few samples with either (*a*) considerably lower δ's than the rest and displaced from these by a slope ~8 (due to heavy rain; e.g. point P_1 in Fig. 15; this is true for Dar es Salaam, Brisbane and Adelaide) or (*b*) considerably higher δ's than the rest and displaced from these by a slope <8 (due to fast evaporation from falling drops; this is true for the periodically dry Perth).

II. Wider δ-range than group I. Distinct linear correlation with slope≅8 as shown in Fig. 16 ($\delta_D = 8\delta_{18} + d$). To this group belong:

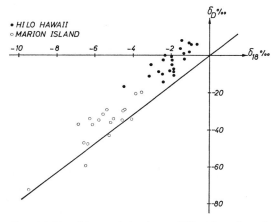

FIG. 15. Monthly precipitation at Hilo, Hawaii, and Marion Island.

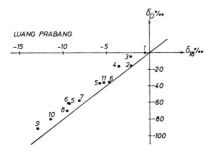

FIG. 16. Monthly precipitation at Luang Prabang, Laos, Slope = 8. The figures indicate the number of the month.

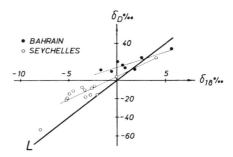

FIG. 18. Monthly precipitation at Bahrain and Mahe, Seychelles Islands. Slope < 8.

Adak Island*	Karachi
Barbados Isl.	Leopoldville
Bethel*	Lista*
Bombay	Luang Prabang
Buenos Aires	Manila
Djakarta	Maracay
Edmonton*	Midway
Falkland Isl.*	New Delhi
Flagstaff	Rangoon
Goose Bay*	Reykjavik*
Guam Isl.	Rio de Jan.
Hollandia	Valentia*
Isfjord*	Wake Isl.

The wide δ-range is caused by either (1) considerable amount effect or (2) considerable seasonal temperature variation (denoted by *).

It is interesting that all South and East Asian stations under the influence of the monsoun are included in this group.

All stations but Flagstaff, New Delhi and Edmonton are situated close to an ocean.

III. Fig. 17: Slope $\cong 8$ for low δ's (variations due to amount effect and/or temperature effect). Slope < 8 for high δ's in relatively dry and/or warm season (fast evaporation from falling drops). To this group belong:

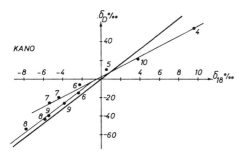

FIG. 17. Monthly precipitation at Kano, Nigeria. The figures indicate the number of the month.

Chicago	Stuttgart
Fort Smith	Tananarive
Genoa	Windhoek
Kano	Vienna
Salisbury	

All these stations have a continental climate. For Chicago, Fort Smith, Stuttgart and Vienna the low slope part of the relation is a pronounced summer phenomenon with intense evaporation under relatively dry and warm (hot) conditions. For Salisbury and Windhoek the low slope occurs in the Southern hemisphere winter, which is the relatively dry season in South Africa.

For Kano and Tananarive some samples are situated along the prolongation of the low slope line. The whole low slope line represents the precipitation at the beginning and at the end of the rainy season, whereas the samples from the rainy season are along the line with slope $\cong 8$.

IV. Fig. 18: Slope < 8 (found in South African rains by EHHALT et al., 1963).
To this group belong:

Alice Springs	Melbourne
Bahrain	Nord
Darwin	Pretoria
Gibraltar	Seychelles Isl.
Lod	Whitehorse

and possibly

Cape Town (D.F. Malan)	Hong Kong
Diego Garcia Isl.	Johnston

Of the island stations Mahe, Seychelles, is located in the Western part of the Indian Ocean, i.e. in one of the source areas of the monsoon vapour. The net evaporation is extremely high, and the vapour cannot be expected to be in equilibrium with the ocean water. This

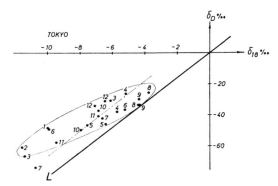

FIG. 19. Monthly precipitation at Tokyo. Seasonal variation of d: All winter months are above the thin, dashed line, which corresponds to $d = +14‰$.

may be the reason why 10 out of 11 monthly samples are close to the line (Fig. 18)

$$\delta_D = (4.8 \pm 1)\delta_{18} + (4 \pm 8),$$

which is practically identical with eq. (15); cf. also the suggested explanations for line No. 1 in Figs. 6 and 8 (p. 449 and p. 453).

Several of the above-mentioned stations at low latitudes have a generally dry climate (Alice Springs and Bahrain) or a periodically dry climate (Darwin, Gibraltar, Lod and Pretoria). The amount effect accounts for the δ variations and non-equilibrium evaporation from falling drops probably accounts for the low slope.

At the two high latitude stations Nord and Whitehorse the precipitation is sparse. The temperature effect is responsible for the δ variations. The summer (rain) samples have generally higher δ's but lower d's than the winter (snow) samples, and non-equilibrium evaporation from the liquid precipitation probably accounts for the low slope.

V. Fig. 19: Seasonal variation of d, high in winter, low in summer. To this group belong Pohang and Tokyo.

As to Tokyo the δ's are dominated by the temperature effect in the winter and by the amount effect the rest of the year (cf. section 4.2.3., p. 464). In winter the prevailing wind direction is Northwest (i.e. from the Asian continent). The dry continental winds collect moisture from the Sea of Japan in a fast evaporation (high d). Much of this moisture is precipitated on the Northeastern slope of the Island. The rest is possibly mixed up with Pacific vapour causing precipitation in Tokyo with not too low δ's, but relatively high d. In the summer time the prevailing Southeasterly winds produce occasionally heavy rainfall of the monsoun type with low d.

VI. Fig. 20: Slope higher than 8.

In spite of the limited number of samples this seems to be the case for Entebbe and Adis Abeba. If so, the reason might be that light rains (with high δ's in accordance with the amount effect) exchange with fastly evaporated fresh water (with high d in accordance with the kinetic effect) to a higher degree than heavy rain.

4.2.3. Amount effect II

The amount effect has been discussed in section 2.4. and demonstrated in sections 4.1.1. (pp. 452 and 454, cf. also Fig. 9), 4.1.2. (cf. the low latitude stations in Fig. 10), 4.2.1. (p. 461) and 4.2.4. (p. 465).

However, the clearest indication of the δ's and the amount of rain varying in antiphase at low latitude stations appears by considering the individual monthly samples, or rather the unweighted means for two neighbouring months. Such means of δ_{18} are shown as dots in the upper parts of Figs. 21, 22 and 23 for Wake, Binza (Leopoldville) and Tokyo respectively. The open circles indicate δ_{18} in a single month. The dashed curve shows the corresponding variation of δ_D. Under the δ-diagrams is plotted the average precipitation, $\bar{p}$, of two neighbouring months.

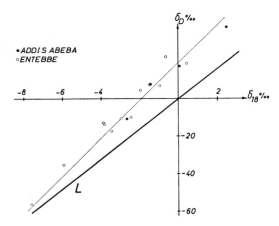

FIG. 20. Monthly precipitation at Entebbe, Uganda, and Addis Abeba, Ethiopia. Slope > 8.

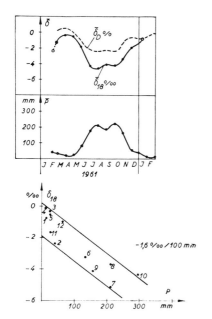

FIG. 21. Amount effect at Wake Island. Upper section: $\bar{\delta}$ = mean δ of two neighbouring months. Mid section: $\bar{p}$ = the corresponding amount of precipitation. Lower section: δ_{18} plotted against p for the individual month.

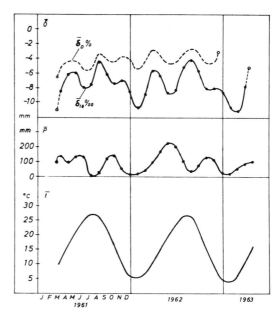

FIG. 23. Combined temperature effect and amount effect at Tokyo. Upper section: $\bar{\delta}$ = mean δ of two neighbouring months. Mid section: $\bar{p}$ = the corresponding amount of precipitation. Lower section: The corresponding mean air temperature at surface.

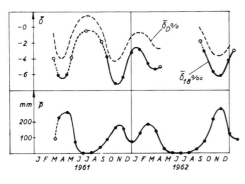

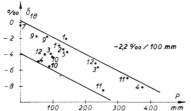

FIG. 22. Amount effect at Binza (Leopoldville), Congo. See legend to Fig. 21.

For the tropical Wake and Binza $\bar{\delta}$ and $\bar{p}$ are, obviously, varying with a negative correlation factor. This is also true for Tokyo (Fig. 23) except in the winter time, when the monthly mean temperature is only 5°C, i.e. when the climate of the Japanese Islands is more like that of high latitude regions, in which the temperature effect dominates the isotopic variation. The δ_{18} curve for Tokyo reflects a seasonal variation due to strongly varying temperature (the $\bar{\delta}_{18}$ has an absolute maximum in summer and an absolute minimum in winter) superimposed by a short periodic variation in antiphase with $\bar{p}$ (except in mid winter).

In the lower parts of Figs. 21 and 22 δ_{18} for the individual months have been plotted against p. The amount effect may be characterised by the slope and width of the band covering most of the points, for Wake: $-1.6\permil$ per 100 mm, width $2\permil$, for Binza: $-2.2\permil$ per 100 mm, width $4\permil$. These figures are listed in Table 10, column 7, together with the corresponding ones for other low latitude stations. However, in many cases, e.g. Tokyo, a correlation clearly

Table 12

Group	Station	$\delta_s - \delta_w$ ‰
A	Apia, Samoa	−2.1
	Guam	−2.4
	Seychelles	−2.5
	Wake	−2.2
B	Bamako	−4.3
	Binza	−2.2
	Ciuaba	−5.4
	Hong Kong	−5.1
	Manila	−3.5
	Maracay	−3.6
	New Delhi	−2.4
	Rio de Janeiro	−2.3
	Salisbury	−6.5
	Windhoek	−2.6
C	Genova	+3.0
	Gibraltar	+4.0
	Perth	+3.0
D	Chicago	+8.6
	Copenhagen	+2.8
	Edmonton	+7.6
	Flagstaff	+5.6
	Fort Smith	+4.0
	Goose Bay	+5.6
	Nord	+9.7
	Sdr. Strømfjord	+7.9
	Stuttgart	+5.5
	Whitehorse	+3.1
	Vienna	+5.2
E	Cape Hatteras	−3.3
	Isfjord	−2.7

appears in a $\bar{\delta} - \bar{p}$ plot (upper parts of Figs. 21–23), but not in a $\delta_{18} - p$ plot (lower parts of Figs. 21 and 22). In such cases presence of a $\bar{\delta} - \bar{p}$ correlation is indicated by a + in Table 10, column 7.

The amount effect is distinct for practically all low latitude stations with varying p. For the 6 tropical islands: Canton, Christmas, Guam, Johnston, Seychelles and Wake it averages in δ_{18} to −1.6‰ per 100 mm (width 2.5‰) and for the 8 partly continental stations: Binza, Ciuaba, Dar es Salaam, Entebbe, Hong Kong, Kano, Manila and Pretoria the average is −2.0‰ per 100 mm (width 4.5‰), i.e. a little higher than for the islands.

At many mid latitude stations the amount effect is found in the summer time, e.g. Chicago, Copenhagen and Melbourne, whereas the lower part of Fig. 4, p. 445, shows no amount effect at the high latitude station Nord, probably because of negligible convective activity and little evaporation from liquid precipitation.

4.2.4. Seasonal variation

In this section δ_s and δ_w denote the unweighted means of the summer and winter months, respectively. On the Northern hemisphere this means May–Oct. and Nov.–Apr., and vice versa on the Southern hemisphere. All the stations for which $|\delta_s - \delta_w| > 2$ are listed in Table 12.

The islands in group A and the more or less continental stations in group B are all located at low latitudes. They have all rainy summers and relatively dry winters. The negative values of $\delta_s - \delta_w$ reflect the amount effect.

The mid latitude stations in group C have all relatively dry summers. Both the amount effect and the temperature effect contribute to the positive values of $\delta_s - \delta_w$.

Most of the stations in group D are high latitude continental stations. The seasonal variation of the temperature at surface is more than 16°C (more than 25°C, if we disregard the 3 European stations). None of these stations has a dry season. The generally high values of $\delta_s - \delta_w$ are mainly due to the temperature effect.

The two stations in group E are both lying in the Gulf Stream and they should be compared with the high latitude island stations, none of which shows distinct seasonal variation. The negative $\delta_s - \delta_w$ at Isfjord, Svalbard, is interesting, the more so as (1) the monthly mean temperature varies from +5°C in July to −12°C in March, (2) the highest $\delta_s - \delta_w$ (+10‰) is found at Station Nord only 800 km West of Isfjord. The reason may be that the climate at Isfjord is dominated by moist air masses from mid latitudes, whereas that of Station Nord is frequently influenced by air masses from the Arctic Ocean.

5. Hydrological applications

The well known glaciological applications of isotope variations are based upon the consequences of the temperature effect, which appear as

(1) seasonal variation,
(2) variation with the altitude.

The latter variation has also been demonstrated in rather dry and warm climates (e.g. in Greece;

PAYNE, personal communication, 1963), and it may, under favourable conditions, be used as a hydrological tool, e.g. for investigating local sources of ground water.

A third consequence of the temperature effect (often combined with the evaporation effect) is

(3) the variation with the amount of rain in the Tropics and the Subtropics, which has been demonstrated in the above. This variation may also involve some hydrological scopes, especially in nonmountainous zones of the Tropics, where (1) and (2) are negligible. Thus, one might think of using the low δ's of abundant rain for clarifying parts of the water turnover at a given locality, e.g. the contributions of such rain to the ground water and to the run-off water.

Finally, it should be mentioned that the fractionation connected to the evaporation

(4) increase of δ with the degree of evaporation (cf. 2.2.),

(5) variation of $d\delta_D/d\delta_{18}$ with the rate of evaporation (cf. 3.1.),

involve the possibility of elucidating the sides of a local hydrological pattern, especially in dry areas, in which evaporation from surface water is an important factor.

In principle, any natural tracing of water can be used for hydrological studies. This tracing is weak (usually less than 100 times the measuring accuracy) relative to artificial tracing of minor amounts of water. On the other hand, the natural labelling of water occurs in a much larger scale than in any hydrological experiment with artificially induced tracers, e.g. a heavy thunderstorm may release enormous amounts of isotopically light water over a large area, which counteracts the weak labelling of the water.

Acknowledgements

The author is indebted to Prof., Dr. phil. Jørgen Koch and to Dr. Bryan R. Payne, I.A.E.A., for helpful support in several phases of the work, to Mr. Jørgen Møller, Chem. engineer, Mr. John H. Jensen, Miss Ellen M. Olsen, Mrs. Ingrid Svendsen, Mr. N. Henriksen and Mr. Jørgen Madsen for excellent technical assistance, and to the numerous people at the meteorological stations, who have made an invaluable effort in collecting and shipping thousands of samples.

Thanks are also due to Prof., Dr. phil. Mogens Pihl, Dr. tech. Bent Buchmann and Phil. Dr. E. Eriksson for reading the manuscript.

The procurement of the Tomson-Houston deuterium mass spectrometer was made possible by a grant from The Carlsberg Foundation, and financial and organizing support was given by the International Atomic Energy Agency.

REFRENCES

BOATO, G., 1960, Isotope fractionation processes in nature. Summer Course on Nucl. Geol., Pisa.
BOLIN, B., 1958, On the use of tritium as a tracer for water in nature. Proceedings of Second United Nations International Conference on the Peaceful Uses of Atomic Energy, Vol. 18, pp. 336–343.
BOTTER, R., and NIEF, G., 1958, Joint Conference on Mass Spectrometry, Sept. 1958. Pergamon Press, London.
CRAIG, H., 1961a, Isotopic variations in meteoric waters. Science, 133, pp. 1702–03.
CRAIG, H., 1961b, Standard for reporting concentrations of deuterium and oxygen-18 in natural waters. Science, 133, pp. 1833–34.
CRAIG, H., BOATO, G., and WHITE, D. E., 1956, Isotopic geochemistry of thermal waters. Nat. Acad. Sci., Nucl. Sci. Ser., Rep. No. 19, pp. 29–36.
CRAIG, H., GORDON, L. J., and HORIBE, Y., 1963, Isotopic exchange effects in the evaporation of water. Journ. Geophys. Res., 68, pp. 5079–87.
DANSGAARD, W., 1953, The abundance of O^{18} in atmospheric water and water vapour. Tellus, 5, pp. 461–69.
DANSGAARD, W., 1954, The O^{18} abundance in fresh water. Geochim. et Cosmochim. Acta, 6, pp. 241–60.
DANSGAARD, W., 1961, The isotopic composition of natural waters. Medd. om Grønland, 165, Nr. 2, pp. 1–120.
DANSGAARD, W., NIEF, G., and ROTH, E., 1960, Isotopic distribution in a Greenland iceberg. Nature, 185, pp. 232–33.
EHHALT, D., KNOT, K., NAGEL, J. F., and VOGEL, J. C., 1963, Deuterium and oxygen 18 in rain water. Journ. Geophys. Research, 68, pp. 3775–80.
EPSTEIN, S., 1956, Variations of the O^{18}/O^{16} ratios of fresh water and ice. Nat. Acad. Sci., Nucl. Sci. Ser., Rep. No. 19, pp. 20–25.
EPSTEIN, S., and BENSON, C., 1959, Oxygen isotope studies. Trans. Am. Geophys. Union, 40, pp. 81–84.
EPSTEIN, S., and MAYEDA, T., 1953, Variations of the O^{18} content of waters from natural sources. Geochim. et Cosmochim. Acta 4, pp. 213–24.
EPSTEIN, S., and SHARP, R., 1962, Comments on annual rates of accumulation in West Antarctica. Publ. 58 of the I.A.S.H. Commission of Snow and Ice (Symposium of Obergurgl), pp. 273–85.

ERIKSSON, E., 1964, The history of the major pulses of tritium in the stratosphere. *Tellus* (in press).

FACY, L., MERLIVAT, L., NIEF, G., and ROTH, E., 1963, The study of the formation of a hailstone by means of isotopic analysis. *Journ. Geophys. Res.*, **68**, pp. 3841–48.

FRIEDMAN, I., 1953, Deuterium content of natural waters and other substances. *Geochim. et Cosmochim. Acta*, **4**, pp. 89–103.

FRIEDMAN, I., MACHTA, L., and SOLLER, R., 1962, Water vapor exchange between a water droplet and its environment. *Journ. Geophys. Res.*, **67**, pp. 2761–70.

FRIEDMAN, I., REDFIELD, A. C., SCHOEN, B. and HARRIS, J., 1964, The variation of the deuterium content of natural waters in the hydrologic cycle, *Rev. of Geophys.*, **2**, pp. 177–224.

GILFILLAN, E. S. JR., 1934, The isotopic composition of sea water. *Journ. Am. Chem. Soc.*, **56**, pp. 406–08.

GONFIANTINI, R., and PICCIOTTO, E., 1959, Oxygen isotope variations in antarctic snow samples. *Nature* **184**, pp. 1557–58.

GONFIANTINI, R., TOGLIATTI, V., TONGIORGI, E., DE BREUCK, W., and PICCIOTTO, E., 1963, Snow stratigraphy and oxygen isotope variations in the glaciological pit of King Baudouin Station, Queen Maud Land, Antarctica. *Journ. Geophys. Res.*, **68**, pp. 3791–98.

LORIUS, C., 1961, Concentration en deutérium des couches de névé dans l'Antarctique. *Ann. de Geophys.*, **17**, pp. 378–87.

LORIUS, C., 1963, Le deutérium possibilités d'application aux problèmes de recherche concernant la neige, le névé et la glace dans l'Antarctique. *Comité National Français des Recherches Antarctiques*. No. 8, pp. 1–102.

MERLIVAT, L., BOTTER, R., and NIEF, G., 1963, Fractionnement isotopique au cours de la distillation de l'eau. *Journ. Chim. Phys.*, **60**, pp. 56–61.

PICCIOTTO, E., DE MAERE, X., and FRIEDMAN, I., 1960, Isotopic composition and temperature of formation of antarctic snows. *Nature*, **187**, pp. 857–59.

ROTH, E., 1963, L'utilation des mesures de teneur en deutérium pour l'étude des phénomènes meteorologiques et géologiques, *J. Chimie Physique*, **60**, pp. 339-350.

ZHAVORONKOV, UVAROF and SEVRYUGOVA, 1955, Primenenie Mechenykh. *Atomov Anal. Khim. Akad. Nauk. USSR*, pp. 223–33.

Copyright © 1968 by the American Geological Institute

Reprinted from *Geochem. Internat.*, 5(5), 952–966 (1968)

THE CURRENT TRITIUM CONTENTS OF NATURAL WATERS*

A. P. Vinogradov, A. L. Devirts and E. I. Dobkina
Vernadskiy Institute of Geochemistry
and Analytical Chemistry,
Academy of Sciences of the USSR,
Moscow

(ABSTRACT)

Tritium analyses have been made on a number of water and plant samples collected primarily during 1966-1968 in the western USSR. The analyses were made by gas-proportional counting either of C_2H_2, made by reacting sample H_2O with C^{14} free CaC_2, or of C_6H_6 made by hydrogenerating "dead" or sample C_2H_2 with H_2 from the sample H_2O. The counters were steel or copper in conventional steel shields with a ring of geiger guard counters in anticoincidence with the main counter. 2 σ detection limits for 1000 minute counts were 3 TU for C_6H_6 at 2 atmospheres and 18 TU for C_2H_2 at 1 atmosphere. No enrichment was done.

The tritium content of atmospheric precipitation and water vapor near Moscow ranged from 400 to 800 TU in 1966 and from 40 to 300 TU in 1967. Lakes in the central and NW USSR ranged from 170 to 370 TU in 1967, while the Pakhva and Oka Rivers near Moscow in September 1967 had 65 and 34 TU, respectively. --F. J. Pearson, Jr.

I. INTRODUCTION

There has recently been much interest in the heavy isotope of hydrogen, tritium (T), because nuclear weapons tests have led to a considerable increase in the abundance of this isotope since 1952-4, while the subsequent cessation of large-scale testing has made T a convenient tracer for many geologic processes that formerly could not be followed with natural T [1-5].

1. Tritium in Nature

The isotope of hydrogen have the following natural abundances: $_1H^1$ 99.984%, $_1H^2$ (D) 0.0156%, and $_1H^3$ (T) ~10^{-16}%. The first two isotopes are stable, while T is a β-emitter of maximum energy 18 keV and half-life 12.26 years, the decay scheme being
$_1H^3 \rightarrow _2H^3 + \beta^-$.

Tritium is produced naturally in the atmosphere, lithosphere, and hydrosphere. It is produced in the atmosphere when cosmic particles (protons and neutrons) interact with nitrogen, oxygen, and argon. In the lithosphere it is produced by neutrons, from spontaneous fission of uranium and from (α, n) reactions, with fission less important than bombardment by α-particles from the decay of U and Th in rocks. These neutrons interact via Li^6 (n, T) He^4 to produce tritium. The amount of tritium produced in the lithosphere is small relative to that produced in the atmosphere. The reaction Li^6 (n, T) He^4 also occurs in the hydrosphere, but the amount produced is even smaller, since the

―――――――
*Trans. from: Geokhimiya, No. 10, pp. 1147-1162, 1968.

Editor's Note: This article was translated from *Geochimiya*, **10**, 1147–1162 (1968)

light isotope of hydrogen captures the neutrons. The main source of tritium is thus the atmosphere, the main reaction being $N^{14} + n \rightarrow C^{12} + T$ for $E_n > 4.4$ MeV. The rate of production of tritium is highest at a height of ~10 km and for neutrons of energy 15-20 MeV [6]. Other data [7] indicate that 2/3 of all cosmic-ray tritium may be formed in the stratosphere, mainly between the tropopause and 25 km. Table 1 gives some estimates of the mean rate of formation Q.

It is clear from Table 1 that the various estimates of Q range over nearly a factor of 10, so the natural tritium level cannot be stated with certainty. If we take 0.12 atom/cm² per sec (the least value of Q) as the rate, this corresponds to a total of 1800 g on the planet, of which ~11 g remains in the atmosphere, 13 g is in groundwater, and the rest is in the ocean [9].

Tritium is also produced in meteorites by high-energy protons, the amounts in iron meteorites ranging from 11 to 290 and in stony meteorites from 100-600 disintegrations per minute per kg [18].

About 99% of the tritium is converted to HTO and participates in the normal cycle of water. Tritium also occurs as HT and CH_3T. The mean lifetime of tritium in the stratosphere is about 2 years; the mean lifetime of an HTO molecule in the troposphere, or the dwell time in the troposphere, is about 1 month [19, 20].

Rutherford's group made the first search for natural tritium in the 1930s; not knowing that it was radioactive, they searched for it with a mass spectrograph and did not observe it. Some years later, Alvarez showed [21] that T is radioactive. Faltings and Harteck discovered natural T in atmospheric hydrogen [22], while Gross et al. [23] found it in rainwater. Since then, many specimens of natural objects have been examined for T to establish the distribution of this isotope on the Earth.

Little is known about tritium concentrations before nuclear tests were begun. The natural T level in water is only about 1 atom per 10^{18} atoms of protium, and a quantity of this order is very difficult to measure. Tritium concentrations are expressed in tritium units (TU), with 1 TU = T/H = 10^{-18}. The T content of tropospheric water ranges from somewhat above 0.1 TU (water vapor over the sea) to 80 TU (snow in Greenland). The concentration in stratospheric water is larger by a factor of 10^5, mainly because of the low water content of the stratosphere and the long half-life of water there. Rivers and lakes have up to 10 TU, while ocean water has ~1 TU. The concentration in

Table 1

Rates of Formation of Tritium in Nature

Sphere	rate Q*	Year published	Reference
Atmosphere	0.10—0.20	1953	[8]
	0.12	1954	[9]
	0.14	1955	[10]
	1.2	1957	[11]
	1.2	1957	[12]
	1.06	1958	[13]
	0.9	1958	[5]
	0.75	1958	[14]
	0.6—1.3	1960	[15]
	0.25—0.35	1961	[16]
	0.6	1962	[7]
	0.20±0.05	1967	[17]
Lithosphere	10^{-3}	1954	[9]
Hydrosphere	10^{-6}	1953	[8]

*Q rate of formation of T atoms per sec per cm² of the Earth's surface.

molecular hydrogen at ground level before 1954 was 3800-35,000 TU, i.e., 1000 times that in tropospheric water vapor. The level in hydrogen in the stratosphere is [19, 24] twice that in atmospheric HT. Tritiomethane is formed via $CH_4 + HT \rightarrow CH_3T + H_2$ in the troposphere, where HT from the stratosphere mixes with atmospheric methane. The upper limit for T in tropospheric methane before 1953 is given [19, 25] as 870 TU.

The above data relate to tritium of natural origin. Testing of nuclear devices began in the early 1950s, which began to raise the tritium contents of natural objects, especially the atmosphere. Figure 1a shows the global production of tritium; up to 1963, the planet received 200 kg. Figure 1b shows the effect of this on precipitation [26]. The first peak represents the summer of 1954 and constitutes 700 TU (mean for the quarter), so 1954 represents the boundary between the cosmic era of T formation and the nuclear one.

Many laboratories made observations on the global distribution of tritium in subsequent years. Specimens were collected at over 100 points spread over the Earth. The results show that tritium is identically distributed throughout the northern hemisphere and tends to diminish towards the equator and over oceans at a given latitude (the latter effect due to dilution with water vapor from the ocean). The level over continental areas rose to a mean maximum of 1000 TU [27] in 1958-9 and to 5000 TU in 1963 [26], individual samples of precipitation giving up to

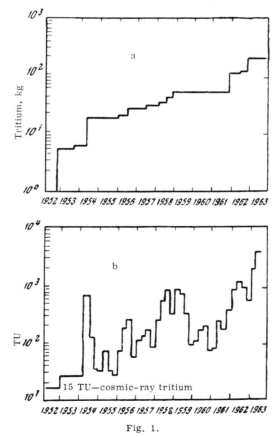

Fig. 1.

a) Global production of artificial tritium; b) tritium content of rain at Ottawa (Canada).

10,000 TU (Northern Canada). The rate of decrease in T after 1963 was less than that after 1959 (period of moratorium), which may be due to differences in the tests in 1958 and 1961-3. Precipitation in the southern hemisphere in 1963 averaged 15 TU, or less than that in the northern hemisphere by a factor of 300-400.

The literature does not yield much evidence on tritium in the atmosphere and hydrosphere after 1964.*

2. Methods of Assaying Tritium

The isotope emits low-energy β-rays, and this, with the low T level before nuclear tests, made it difficult to assay. All T assays were performed by preliminary enrichment, but any such process involves an additional source of error in determining the enrichment factor. Two methods of enrichment have been found successful: electrolysis [9, 28-30] and thermal diffusion [31], with enrichment monitored via deuterium [31, 32].

The introduction of artificial T into the atmosphere allowed assays to be made with little or no enrichment [33, 34].

The decay of tritium is such that the only possible counters are ones enclosing the sample to give 4π geometry. Measurements have thus been made with gas counters and with liquid scintillation counters. Tritium as hydrogen has been assayed in Geiger counters with the addition of ethylene [29], ethylene and argon [33], or propane [35]. Hydrogen has also been used [36] in a proportional counter with CO_2. Tritium as hydrocarbons has been measured in proportional counters, the gases (ethane and propane) containing 2 or 4 atoms of the sample hydrogen [26, 37, 38]. Tritium has been used in scintillation counters directly as water [39, 40] and as benzene [34, 41].

As in the case of C^{14}, measurement of T requires minimization of the background due to cosmic radiation and environmental radioactivity [42].

---

*Editor's note: probably reflects publishing time lags. Data are certainly being collected.

955

II. EXPERIMENTAL

The aims of the present work were to record tritium contents in the atmosphere, hydrosphere, and biosphere in the USSR, and to compare the results with those for other parts of the world. Preliminary calculations were made using the characteristics of available radiocarbon dating equipment, and it was decided to assay T as hydrocarbons in a proportional counter without preliminary enrichment [43, 44].

1. Preparation of Hydrocarbons

Acetylene and ethane were chosen as the counting gases, since both have good counting characteristics. In the case of ethane, it is possible to introduce 2, 4, or 6 hydrogen atoms from the sample. We examined only water specimens.

Acetylene: This was produced via $CaC_2 + 2H_2O = C_2H_2 + Ca(OH)_2$. The carbide was originally synthesized from ancient carbon (marble or anthracite) by the reaction between magnesium and carbonate. Subsequently we used tested technical carbide. The acetylene was produced in a vacuum apparatus, and the gas was stored before measurement.

Ethane: This was produced by hydrogenation of acetylene in the presence of Raney nickel [42, 44], the synthesis being accelerated by inclusion of a circulating pump. The hydrogen was generated by reaction of water with magnesium in a steel tube 35 mm in diameter and ~1 m long. The lower part of the tube had a branch with a ground joint to a glass dropping funnel. The water entered the bottom of the tube drop by drop and was evaporated there by a gas burner. The steam passed over Mg turnings heated to 600°, a fresh batch of Mg being used for each water specimen in order to avoid memory effects. The system was evacuated before the reaction was started. Production of 1 liter of acetylene requires about 4 g of technical carbide and 15 ml of water. Hydrogenation of 1 liter of acetylene requires 2 liters of hydrogen, which is provided by about 2 ml of water. This means that, if the water sample is small, acetylene from ancient water may be employed, the water sample being used only to produce hydrogen. The resulting ethane is then $C_2H_2H_4^*$, in which H* is hydrogen containing the tritium. The ethane is also stored in glass vessels to permit decay of the radon present in the freshly prepared gas.

2. Counting Apparatus

The detectors for tritium β-rays were proportional counters made of stainless steel (volume 1 liter) or electrolytic copper (volume 1.6 liters), the anode wire being tungsten (diameter 0.028 mm) and the counter ends being of PTFE. The counters allowed of evacuation and filling with gas at high pressure. The shielding (Fig. 2) consisted of steel enclosing blocks of wax saturated with boric acid, a layer of mercury, and a ring of Geiger counters in providing an anticoincidence (AC) circuit with the proportional counter. Figure 3 shows the block diagram of the apparatus. The pulses from the proportional counter pass via a linear amplifier to a single-channel pulse-height analyzer whose window sidth can be adjusted within wide limits. This width was chosen with its lower limit such as to exclude noise and interference, while the upper limit excluded pulses whose energy exceeded the maximum energy of the tritium β-rays. The pulses transmitted by the analyzer are shaped to rectangular form and pass to circuit AC-2, which selects pulses not coincident with pulses from the Geiger counters,

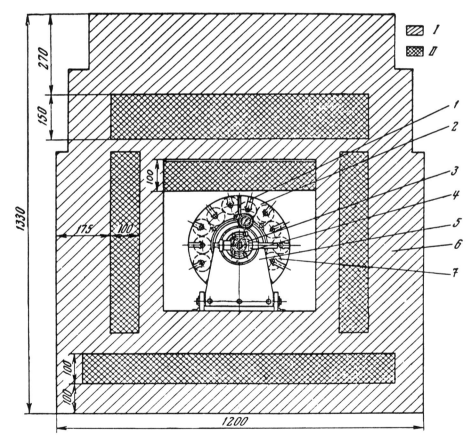

Fig. 2. Location of counter in shield:
1) and 2) supports for guard counters, 3) shield for proportional counter, 4) proportional counter, 5) mercury shield, 6) steel shield, 7) Geiger counters; I) steel body, II) wax containing boric acid.

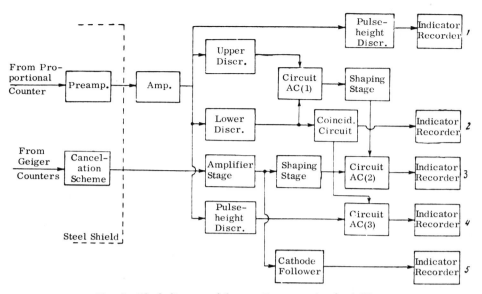

Fig. 3. Block diagram of the counting apparatus for tritium.

which are recorded in channel 3 (scalar 3) and serve to indicate the tritium activity. The other four channels are accessory ones and are as follows: channel 1) integral count from the proportional counter, channel 2) coincidences between the proportional and guard counters, channel 4) large pulses, and channel 5) guard counters. The integral background of the steel counter was 3.6 min^{-1}, as against an unscreened background of 400 min^{-1}, or a background inside the steel but without the Geiger counters of 180 min^{-1}. The exact window width is chosen in accordance with gas pressure, detector background, and detector efficiency.

The acetylene was used in the steel counters at pressures up to 1 atm. The C_2H_2 was used with windows of 5-31 or 5-45 V, which are equivalent to energy ranges of 1.4-8.5 and 1.4-12 keV. In the first case, the counter background was 0.56 min^{-1}; in the second, 0.90 min^{-1}. Ethane was used at 2 atm in these same counters, with the window accpeting the range 3 to 19 keV. The background was then 1.52 min^{-1}.

The standard water in all these cases was artesian water from a depth of 160 m, but subsequent studies showed that this water could not be considered as tritium-free. A zero standard was used on deep-water mineral springs. Table 2 gives the results. The artesian water (T-11) showed the highest activity, which indicates a connection with current water. The lowest value attained was for Smirnovskaya water taken directly from the spring (at Zheleznovodsk). The Dzermuk water was that commercially available. The Smirnovskaya water gives 0.70 min^{-1} and has been taken as the zero standard (counter background).

Ethane has also been used in the copper counters: the background given in the same energy range by Smirnovskaya water was 1.28 min^{-1}.

The efficiency was monitored with artificial tritium in both gases giving 30 min^{-1} per liter.

All background and specimen measurements were made over periods of ~20 hr, which gave acceptable counting statistics and a relative error in the result between 1.5 and 30% in accordance with the activity level. The error was larger when acetylene was used, and this means that samples with over 100 TU should be used with acetylene in the 1 liter counter if the error is not to exceed 10%. This is quite adequate for atmospheric T in recent years.

Ethane provides higher sensitivity and accuracy. The gas can be safely used in

957

Table 2

Comparison of Groundwater Activities with 1 Liter
of C_2H_6 at 2 Atm

Laboratory number	Specimen	Counts/min
T_{E_6}-11	Artesian water used in making acetylene and hydrogen	1.52 ± 0.02
T_{E_2}-11	C_2H_2 from artesian water, hydrogen from Kipp's apparatus via $Zn + 2HCl_2 + H_2$	1.17 ± 0.02
T_{E_6}-70	Dzhermuk mineral water	0.95 ± 0.03
T_{E_6}-75	Smirnovskaya mineral water	0.70 ± 0.02

the 1.6 liter copper counter at 2 atm, which gives the high performance factor $P = H\epsilon/\overline{n_b} = 26.6$, in which H is the number of gram-moles of hydrogen introduced into the counter, n_b is the background rate (min^{-1}), and ϵ is efficiency (%). Table 3 gives the P for various methods and equipment, while the last column gives the minimum amounts of tritium that can be determined on the 2 σ criterion (σ is standard deviation). In measuring low activities, it can be assumed that $\sigma = \sqrt{2n_b t}$, in which t is measurement time [42]. This t has been taken as 1000 min in all cases in Table 3, so that the results are comparable. The present method gives the lowest K_{min}, at 3 TU, but the 4 σ criterion was applied in the acetylene measurements (T_A) at low levels in Table 4, so the limit of observation becomes 34 TU there.

III. MEASUREMENTS OF NATURAL SUBSTANCES

We examined precipitation and atmospheric water vapor from the Moscow region, rivers and lakes in the European part of the USSR, and surface waters of the oceans. Atmospheric precipitation was collected regularly between November 1966 and February 1968 at a height of 2 m from the ground. Collections were not made as systematically in earlier years, so the specimens do not give an exact picture of the monthly T, but the results are of interest for comparison with other parts of the world. Table 4 gives all the results. All specimens analyzed for tritium were given a laboratory number prefixed by T, to distinguish them from specimens assayed in the laboratory for C^{14} dating; subscript A denotes acetylene, while subscript E denotes ethane, with 2, 4, or 6 indicating the number of hydrogen atoms from the specimen. For instance, T_{E_2}-3 indicates use of ethane with only 2 hydrogen atoms derived from sample T-3, i.e., the gas was $C_2H_2^*H_4$.

The data in section I of Table 4 are in chronological order from 1961. The highest peak (~1000 TU) occurs in the specimen of 18 May 1962 (T-10), other individual collections in this period (T-7 and T-9) giving lower values. Samples over longer periods (T-8 from end of 1961 to early 1962; T-29, mean sample for 1962) give the lower values of 425 and 325 TU respectively. Year-average values for other points near the latitude of Moscow are 428 TU for Lista in Norway (59°N, 07°E) and 1408 TU for Edmonton in Canada (54°N, 113°W) [26]. Results for points well to the north and south of the latitude of Moscow in 1962 are as follows: 2027 TU for North Greenland (81°N, 17°W) and 12 TU for Malan in South Africa (34°S, 18°W). In 1963, the levels for the latter two places rose by factors of 2-3, but at all times from the start of nuclear tests the level in the southern hemisphere remained much below that in the north. The level at our station at the start of 1963 was 845 TU (T-30), i.e., twice that for the start of 1962, or more than 2.5 times the mean for 1962, which is close to the increase at various latitudes. The activity level began to fall in 1964-6, as is clear from samples

Table 3

Comparison of Counter Systems Used in Measuring Low Levels of Tritium Activity [20, 29, 34, 38]

Method and source	H_2(g-mole) in specimen	n_b, c/min	ε, %	$H_2 \times \varepsilon/\sqrt{n_b}$	K_{min}, TU
Liquid scintillation counters					
[34]	1.55	52	29	6	11
[34]	0.39	15	34	3.5	20
[40]	1.75	52	6.9	1.7	41
[39]	0.90	61	14	1.6	42
[41]	0.08	19	26	0.47	46
Geiger counters					
[33]	0.1155	4	90	5.2	9
O'Brien	0.070	2.7	~100	4.3	11
[35]	0.040	2.25	94	2.5	19
[24]	0.068	20	~100?	1.5	45
[9]	0.0097	8	~100?	0.34	137
[28]	0.0023	1.5	~100?	0.19	250
[29][1]	0.0045	17	~100?	0.11	627
Proportional counters					
[38]	0.45	5.3	95	18.5	4
[37]	0.196	1.6	80	16.5	5.5
[36]	0.0166	5	90	0.66	102
Heys et al.	0.0041	0.58	~100?	0.54	127
Present work, copper counter (1.6 liters, C_2H_6*, 2 atm)	0.429	1.28	70	26.6	2.5
Present work, steel counter (1 liter, C_2H_6*, 2 atm)	0.267	0.70	70	22	3
Present work, steel counter (1 liter, C_2H_2*, 1 atm)	0.045	0.56	67	3.9	17
Same	0.045	0.90	82	3.8	18

[1] A standard SBM-7 counter was used.

collected over long periods (T-31 and T-18). Increased levels occurred in individual rainfalls (T-17, T-21), and also in T-19.

Figure 4 gives a fuller picture of the T level near the ground from November 1966. The level in rain in 1967 was less than that in 1966 (40-300 TU as against 400-800). The level should continue to fall in future years if no fresh artificial T enters the atmosphere. Our results for April and May 1967 show a slight rise, with a distinct peak in August; various reasons might be given for the latter, one of which is the correlation between T in rain and Sr^{90} in air [45, 46], the increase in the T in precipitation starting two months after that in Sr^{90}. This delay may be due to secondary evaporation of T from the continent, which thus affects the T in precipitation. It has been found [47] that the maximum fallout of Sr^{90} in 1961-3 in the Moscow area occurred in June, the same occurring [48] in 1965-6. The August 1967 peak may thus be ascribed to secondary evaporation. Figure 4 shows individual values (T-33, T-34, T-46) as well as monthly means, the two sets agreeing closely. Section II of Table 4 gives the T content of water vapor near the ground. These samples were taken on days when no precipitation occurred and showed the same levels as precipitation in the same month, except that T-42 was somewhat high, this being collected in the period 1-10 July 1967, when there was no rain. The broken line in the figure shows the amount of precipitation in the area. The year-average level, including precipitation, for 1967 was 183 TU.

Table 4

Tritium Contents of Various Objects

Laboratory number	Time and place of collection	TU
	I. Atmospheric precipitation	
T_{E2}-3	1961 13—14.III	287 ± 10
T_{E2}-8	1961—1962 1.XII—10.IV	425 ± 20
T_{E2}-9	1962 20.I	490 ± 27
T_A-50	1962 21—22.II	480 ± 20
T_{E2}-10	1962 18.V	1030 ± 17
T_A-29	1962 Aver. sp. from 11 specimens of rain in the following periods: 20.I, 15.II, 21—22.II, 30—31.V; 1—10.VI, 11—20.VI, 21—30.VI, 1—10.VII, 11.VII—20.VIII, 18.X—4.XI, 4.XI—4.XII	324 ± 11
T_A-28	1962 4.XI—4.XII	128 ± 11
T_A-30	1962-1963 Average sample from two specimens collected from 5 Dec. to 1 March and 1-30 April	845 ± 16
T_A-31	1963-1964 Average sample from three specimens collected 1 Dec. to 21 Feb., 3 May, and 10 Nov.	653 ± 11
T_A-15	1965 7.I	178 ± 11
T_A-18	1965—1966 1.XII—17.II	420 ± 13
T_A-19	1966 17.II—23.III	746 ± 18
T_A-20	1966 5.VII	492 ± 23
T_A-21	1966 21.VII	770 ± 40
T_A-14	1966 23.VII—3.VIII	430 ± 4
T_A-17	1966 4.VIII	600 ± 17
T_A-16	1966 1—30.XI	60 ± 11
T_A-22	1966 1—31.XII	125 ± 11
T_A-24	1967 1—31.I	198 ± 16
T_A-26	1967 1—28.II	109 ± 11
T_A-35	1967 24.III	196 ± 11
T_{E4}-35	1967 24.III	195 ± 11
T_{E6}-35	1967 24.III	202 ± 16
T_A-27	1967 1—31.III	114 ± 11
T_A-32	1967 1—30.IV	238 ± 16
T_A-34	1967 1—31.V	232 ± 16
T_A-33	1967 16.V	225 ± 16
T_A-38	1967 1—30.VI	165 ± 11
T_A-44	1967 11.VII (sole rain in period 1-14 July)	158 ± 11
T_A-45	1967, 20—31.VII	178 ± 11
T_A-51	1967 1—31.VIII	290 ± 11
T_A-51A	1967 9—31.VIII	335 ± 16
T_A-46	1967 6.VIII *	286 ± 11
T_A-57	1967 1—30.IX	98 ± 11
T_A-58	1967 1—31.X	81 ± 7
T_A-1	1967 1—30.XI	35 ± 11
T_A-63	1967 6.XI*	49 ± 12
T_A-65	1967 1—31.XII	186 ± 11
T_A-67	1968 1—31.I	111 ± 11
T_A-68	1968 1—29.II	47 ± 11
	II. Atmospheric water vapor at ground level	
T_A-36	1967 1—30.VI	171 ± 16
T_A-42	1967 1—10.VII	256 ± 16
T_A-43	1967 11—13.VII	204 ± 16
T_A-54	1967 14—28.VIII	312 ± 20

Table 4 (Continued)

Laboratory number	Time and place of collection	TU
	III. Surface waters	
T_{E_2}-4	1960, August, Black Sea, to south of Yalta	<34
T_A-62	1966, August, Pacific Ocean, Kuril-Kamchatka trench	<34
T_A-39	1967, April, Atlantic, Bay of Biscay	<34
T_{E_6}-39	Same	29 ± 6
T_A-40	1967, April, Atlantic, near equator, 30°W	<34
T_{E_6}-40	Same	31 ± 6
T_A-48	1967, July, Lake Ladoga	173 ± 11
T_A-49	1967, August, Lake Onega	266 ± 16
T_A-47	1967, August, Lake Zarasay, Lithuania	373 ± 11
T_A-41	1967, August, Lake Krugloye, Moscow oblast	256 ± 16
T_A-73	1967, August, Lake Dolgoye, Moscow oblast	204 ± 12
T_A-52	1967, September, Pakhra R., Moscow oblast	65 ± 11
T_A-56	1967, September, Oka R., Moscow oblast, near Serpukhov	34 ± 11
T_A-64	1967, November, Black Sea near Batumi	<34
	IV. Moscow water	
T_{E_2}-2	1960	287 ± 10
T_A-23	1967, 1-31. I	106 ± 16
T_A 25	1967, 1-28. II	109 ± 11
T_A-66	1968, 1-31. I	68 ± 12
	V. Plant material	
T_A-55	1967, fall harvest, tomatoes from Moscow oblast	199 ± 16
T_A-59	1967, fall harvest, apples from Moscow oblast	140 ± 16
T_A-53	1967, fall harvest, tomatoes from Kherson oblast, Ukraine	133 ± 7
T_A-60	1967, fall harvest, grapes from North Caucasus	65 ± 11

Note: All tritium contents calculated with allowance for isotope fractionation in the preparation of acetylene. Our preliminary measurements indicate that the isotope effect is about 30%.
*Precipitation collected in Moscow and Batumi respectively.

Section III of Table 4 gives data for rivers, lakes, and oceans. Lakes Ladoga, Onega, and Zarasay (all in the NW of the USSR) give 170, 286, and 373 TU respectively (T-48, T-49, and T-47). In the center of the European part of the USSR, T-41 and T-73 gave 256 and 204 TU. The rivers in central regions in 1967 gave lower levels: 65 TU (T-52) for the R. Pakhra and 34 TU (T-56) for the R. Oka. The mean value for water from the Moscow R. (section IV) at the start of 1967 was ~100 TU (T-23, T-25), which is about 1/3 of the value for 1960 (T-2). Moscow water at the start of 1968 had ~70 TU (T-66).

Ethane with 6 hydrogen atoms from the sample was used with the lower levels encountered with ocean water, since C_2H_2 and $C_2H_4H_2$ did not produce useful results (T_{E_2} 4, T_A-62, T_A-39, T_A-40, T_A-64). In 1967, the surface waters of the Atlantic had ~30 TU (T_{E_6}-39, T_{E_6}-40) which is much above the level of 1953-4 (mean values of 0.24 to 2.1 TU for a series of samples) [10, 49].

Section V of Table 4 gives results for fruit and vegetables of the 1967 harvest from various latitudes in the European part of the USSR. Specimens T-53, T-55, and T-59 reflect the levels in precipitation; the lower value for T-60 evidently reflects the closeness of the sea and the latitude effect.

Table 5 illustrates the latitude effect for precipitation via year-average means for

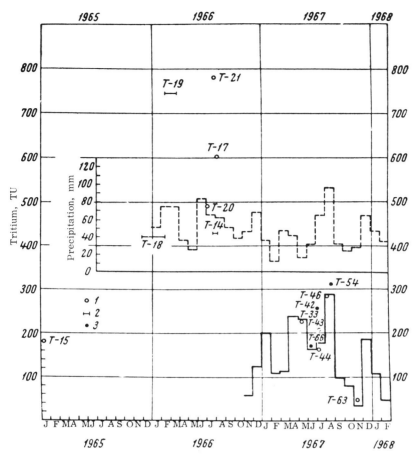

Fig. 4. T content of the troposphere as HTO in 1965-8:
1) single specimens, 2) specimens taken over interval shown, 3) atmospheric water vapor at ground level.

continental, coastal, and island stations [26, 27, 50]. The table includes our values of 1966 and 1967 for Moscow, the 1966 figure being from the available samples (Table 4). Earlier years are not represented in the table, as the annual collections were not reasonably complete.

Summary

Nuclear tests up to 1962 introduced about 200 kg of T, which is far more than the preexisting level. This artificial tritium upset natural equilibria in the geospheres. The time elapsed since the cessation of tests (Moscow treaty) is about half the half-life, and the tritium activity in the atmosphere has fallen because the amount in the stratosphere has been reduced (τ about 2 years). The levels at the end of 1967 and in early 1968 have fallen so far that it has become difficult to use acetylene counters, so $C_2H_2H_4^*$ or $C_2H_6^*$ has had to be used even for the atmosphere.

The present results are for HTO in the atmosphere and hydrosphere in the USSR; they are not very extensive, but they add to the global data on tritium in natural waters. For convenience of comparison we have taken the contents before testing began as 15 TU for precipitation and 1 TU for ocean surface water, as representing natural equilibrium. Some conclusions can then be

Table 5

Annual Mean Tritium Concentrations in Precipitation in Various Parts of the World [26, 27, 50]

Site and country	Position		TU (10^{-18} T/H)												
	Lat.	Long.	1953	1954	1955	1956	1957	1958	1959	1960	1961	1962	1963	1964	1965	1966	1967
North Greenland	81° N	17° W	—	—	—	—	—	—	—	—	—	2030	5800	6390	—	—	—
Is Fjord, Spitzbergen	78° N	14° E	—	—	—	—	—	—	—	—	574	512	1850	—	—	—	—
Reyjavik, Iceland	64° N	22° W	—	—	—	—	—	—	—	—	84	336	1540	—	—	—	—
Bethel, Alaska, USA	61° N	162° W	—	—	—	—	—	—	—	—	—	629	2070	—	—	—	—
Juddinge, Sweden	59° N	18° E	—	—	—	—	—	290	476	140	89	650	—	—	—	—	—
Lista, Norway	58° N	7° E	—	—	—	—	—	—	—	—	59	428	1240	—	—	—	—
Moscow, USSR	56° N	37° E	—	—	—	—	—	—	—	—	—	—	—	—	—	—	183
Edmonton, Canada	54° N	113° W	—	—	—	—	—	147	128	55	224	1410	4020	—	—	—	—
Valencia, Ireland	52° N	10° W	—	—	—	76	71	254	242	73	34	250	1080	—	—	—	—
Vancouver, Canada	49° N	123° W	—	77	—	—	—	—	—	—	64	670	2930	2310	—	—	—
Stuttgart, Germany	48.5° N	9° E	—	—	—	—	—	—	—	—	75	910	3020	—	—	—	—
Vienna, Austria	48° N	17° E	—	—	—	—	—	—	—	—	97	—	—	—	—	—	—
Ottawa, Canada	45° N	76° W	27	144	43	148	122	515	542	161	143	890	1510	770	—	456	—
Flagstaff, USA	35° N	112° W	—	—	—	—	—	—	—	—	450	480	3660	1670	—	—	—
Srinagar, India	34° N	75° E	—	—	—	—	—	—	—	—	—	—	—	—	—	—	—
Hsiangan	22° N	114° E	—	—	—	—	—	—	—	—	24	55	167	—	—	—	—
Hilo, Hawaian Islands	20° N	155° W	—	—	—	—	—	—	—	—	—	—	—	—	—	—	—
Manila, Phillipines	15° N	121° E	—	—	—	—	—	—	—	—	20	72	180	150	—	—	—
Trivandrum, India	8° N	77° W	—	—	—	—	—	—	—	—	—	7	170	98	—	—	—
Minikoi, India	8° N	73° E	—	—	—	—	—	—	—	—	10	38	125	—	—	—	—
Entebbe, Uganda	0° N	32° W	—	—	—	—	—	—	—	—	19	13	64	—	—	—	—
Kinshasa, Congo	4° S	15° W	—	—	—	—	—	—	—	—	12	13	18	—	—	—	—
Dar-es-Salam, Tanganyika	7° S	39° W	—	—	—	—	—	—	—	—	19	44	54	—	—	—	—
Windhoek, SW Africa	23° S	17° W	—	—	—	—	—	—	—	—	—	—	—	—	—	—	—
Pretoria, South Africa	26° S	28° E	—	—	—	—	—	—	—	—	—	30	42	—	—	—	—
Malan, South Africa	34° S	18° E	—	—	—	—	—	—	—	—	11	12	18	—	—	—	—
Melbourne, Australia	38° S	145° E	—	—	—	—	—	—	—	—	—	22	—	—	—	—	—

drawn. The year-average tritium content for 1967 at Moscow was 183 TU, which exceeds the level before 1952 by more than an order of magnitude. The levels of 1967 (40-300 TU) are less than those of 1966 (400-800 TU). Precipitation has shown a clear tendency to reduction in tritium content since 1963. The increase at the end of the summer in 1967 was due to reevaporation from the continent. The T concentrations at coastal and island stations should be correspondingly less than those for the continental station, since these areas are more influenced by ocean water, which is in molecular exchange with atmospheric HTO.

The T level in Atlantic surface water in spring 1967 was 30 units above the level of 1952. The data show that the dwell time in the surface layer of the Atlantic exceeds the dwell time in the stratosphere. The dwell time of T in the surface (mixing) layer of the ocean is at present debatable.

Lakes in NW and central areas in the USSR contained somewhat more T than the rivers in 1967, evidently from accumulation of precipitation from previous years, although the levels in the lakes were of the same order as for precipitation in 1967. Plant material of 1967 also reflects the T content of precipitation in that year.

We are indebted to M. M. Senyavin and V. D. Vilenskiy for providing some of the samples and to N. I. Prokofyeva and D. F. Frantsuzov for assistance.

REFERENCES

1. Vinogradov, A. P. Vvedeniye v Geokhimiyu Okeana (Introduction to the geochemistry of the ocean), Nauka, Moscow, 1967.

2. Libby, W. F. Significance of tritium in hydrology and meteorology. V sb.: Geokhimicheskiye Issledovaniya (In: Geochemical researches). Inos. Lit., Moscow, 1961.

3. Libby, W. F. Tritium geophysics: recent data and results. Tritium in the Physical and Biological Sciences, Vol. 1, 5, IAEA, Vienna, 1962.

4. Giletti, B. J. and J. L. Kulp. Application of tritium measurements to oceanography and meteorology. Trans. Amer. Geophys. Union, Vol. 37, 345, 1956.

5. Bolin, B. On the use of tritium as a tracer for water in nature. Second United Nations International Conference Peaceful Uses of Atomic Energy, Geneva, 1958, Vol. 18, 336, 1959.

6. Luyanas, V. Yu. K Voprosu o Vozniknovenii Tritiya v Atmosfere (Origin of tritium in the atmosphere), Trudy AN Lit. SSR, Ser. B, 1 (32), 21, 1963.

7. Begemann, F. Der natürliche Tritiumhaushalt der Erde und die Frage seiner zeitlichen Variation. Chimia, Vol. 16, No. 1, 1962.

8. Fireman, E. L. Measurement of the (n, H^3) cross section in nitrogen and its relationship to the tritium production in the atmosphere. Phys. Rev., Vol. 91, No. 4, 1953.

9. Kaufman, S. and W. F. Libby. The natural distribution of tritium. Phys. Rev., Vol. 93, No. 6, 1954.

10. Buttlar, H. and W. F. Libby. Natural distributions of cosmic-ray produced tritium. J. Inorg. Nucl. Chem., Vol. 1, No. 1/2, 1955.

11. Begemann, F. and W. F. Libby. Continental water balance, ground water inventory and storage time, surface ocean mixing rates and world-wide water circulation patterns from cosmic ray and bomb tritium. Geochim. et cosmochim. Acta, Vol. 12, 1957.

12. Craig, H. Distribution, production rate, and possible solar origin of natural tritium. Phys. Rev., Vol. 105, No. 3, 1957.

13. Begemann, F. New measurements of the world-wide distribution of natural and artificially produced tritium. Second United Nations International Conference Peaceful Uses of Atomic Energy, Geneva, 1958, Vol. 18, 545, 1959.

14. Giletti, B. J., F. Bazan and J. L. Kulp. The geochemistry of tritium. Trans. Amer. Geophys. Union, Vol. 39, No. 5, 1958.

15. Wilson, A. T. and G. J. Fergusson. Origin of terrestrial tritium. Geochim. et cosmochim. Acta, Vol. 18, No. 3/4, 1960.

16. Craig, H. and D. Lal. The production rate of natural tritium. Tellus, Vol. 13, 1961.

17. Teegarden, B. I. Cosmic-ray production of deuterium and tritium in the earth's atmosphere. J. Geophys. Res., Vol. 72, No. 19, 1967.

18. Lavrukhina, A. K. and G. M. Kolesov. Izotopy vo Vselennoy (Isotopes in the universe). Atomizdat, Moscow, 1965.

19. Begemann, F. The tritium content of atmospheric hydrogen and atmospheric methane. Earth Science and Meteoritics, p. 169, 1963.

20. Buttlar, H. Tritium in rainwater. Earth Science and Meteoritics, p. 188, 1963.

21. Alvarez, L. W. and R. Cornog. Helium and hydrogen of mass 3. Phys. Rev., Vol. 56, No. 6, 1939.

22. Faltings, V. and P. Harteck. Der Tritiumgehalt der Atmosphäre. Z. Naturforsch., Vol. 5a, No. 8, 1950.

23. Grosse, A. V., W. M. Johnston, R. L. Wolfgang and W. F. Libby. Tritium in nature. Science, Vol. 113, No. 2923, 1951.

24. Begemann, F. The tritium content of atmospheric hydrogen and atmospheric methane. J. Geophys. Res., Vol. 68, No. 13, 1963.

25. Martell, E. A. On the inventory of artificial tritium and its occurrence in atmospheric methane. J. Geophys. Res., Vol. 68, No. 13, 1963.

26. Thatcher, L. L., B. R. Payne and J. F. Cameron. Trends in the global distribution of tritium since 1961. Radioactive fallout from Nuclear Weapons Tests: Proceedings of the Second Conference, 1965.

27. Eriksson, E. On account of the major pulses of tritium and their effects in the atmosphere. Tellus, Vol. 17, No. 1, 1965.

28. Brown, R. M. and W. E. Grummitt. The determination of tritium in natural waters. Canad. J. Chem., Vol. 34, 1956.

29. Romanov, V. V. and V. N. Soyfer. Apparatura i Metodika Izmereniya Pirodnogo Tritiya (Apparatus and methods for measuring natural tritium). V sb.: Yadernaya Geofizika (Nuclear geophysics). Gostoptekhizdat, Moscow, 1962.

30. Ostlund, G. and L. B. Lundgren. Stockholm natural tritium measurements. Tellus, Vol. 16, No. 1, 1964.

31. Buttlar, H. and B. Wiik. Enrichment of tritium by thermal diffusion and measurement of dated Antarctic snow. Science, Vol. 149, No. 3690, 1965.

32. Ostlund, H. G., R. M. Brown and A. E. Bainbridge. Standardization of natural tritium measurements. Tellus, Vol. 16, No. 1, 1964.

33. Buttlar, H., W. Stahl and B. Wiik. Tritiummessungen an Regenwasser ohne Isotopenanreicherung. Z. Naturforsch., Vol. 17a, No. 1, 1962.

34. Tamers, M. A. and R. Bibron. Benzene method measures tritium in rain without isotope enrichment. Nucleonics, Vol. 21, No. 6, 1963.

35. Ostlund, G. A hydrogen gas counting system for natural tritium measurements. Tritium in the Physical and Biological Sciences, Vol. 1, 333, IAEA, Vienna, 1962.

36. Gonsior, B. Tritium-Anstieg im atmosphärischen Wasserstoff. Naturwissenschaften, Vol. 46, No. 6, 1959.

37. Bainbridge, A. E., P. Sandoval and H. E. Suess. Natural tritium measurements by ethane counting. Science, Vol. 134, No. 3478, 1961.

38. Eulitz, G. W. Sensitive tritium counting with a propane proportional counting system. Rev. Sci. Inst., Vol. 34, No. 9, 1963.

39. Boyce, I. S. and J. F. Cameron. A low-background liquid-scintillation counter for the assay of low-specific-activity tritiated water. Tritium in the Physical and Biological Sciences, Vol. 1, 231, IAEA, Vienna, 1962.

40. Kaufman, W. J., A. Nir, G. Parks and R. M. Hours. Recent advances in low-level scintillation counting of tritium. Tritium in the Physical and Biological Sciences, Vol. 1, 249, IAEA, Vienna, 1962.

41. Tamers, M., R. Bibron and G. Delibrias. A new method for measuring low level tritium using a benzene liquid scintillator. Tritium in the Physical and Biological Sciences, Vol. 1, 303, IAEA, Vienna, 1962.

42. Vinogradov, A. P., A. L. Devirts, E. I. Dobkina, N. G. Markova and L. G. Martishchenko. Opredeleniye Absolyutnogo Vozrasta po C^{14} pri Pomoshchi Proportsional'nogo Schetchika (C^{14} Age determination with a proportional counter). Akad. Nauk SSSR, Moscow, 1961.

43. Vinogradov, A. P., A. L. Devirts, E. I. Dobkina and N. G. Markova. Opredeleniye Absolyutnogo Vozratsa po C^{14}: Soobshcheniye III (C^{14} Age determination: Pt. III), Geokhimiya, No. 5, 1962.

44. Devirts, A. L., E. I. Dobkina and N. G. Markova. Metodika Opredeleniya Radiouglerodnym Metodom Absolyutnogo Vozrasta Organicheskikh Ostaktov (Age determination on organic remains by the radiocarbon method). V sb.: Paleogeografiya i Khronologiya Verkhnego Pleystotsena i Golotsena po Dannym Radiouglerodnogo Metoda (Paleogeography and chronology of the late Pleistocene and Holocene from radiocarbon data). Papers for the 7th INQUA Congress, Nauka, Moscow, 1965.

45. Libby, W. F. Geophysics of tritium, J. Geophys. Res., Vol. 66, 1961.

46. Israel, G., W. Roether and G. Schumann. Seasonal variations of bomb-produced tritium in rain. J. Geophys. Res., Vol. 68, No. 13, 1963.

47. Malakhov, S. G., Ye. N. Davydov and M. P. Nekhorosheva. Vremennyye Kolebaniya Kontsentratsii Produktov Deleniya v Prizemnom Sloye Atmosfery v Podmoskov'ye i na Ostrove Kheysa Zemli Frantsa Iosifa v 1956-1963 gg (Temporal variations in the concentration of fission products in the ground-level of the atmosphere in the Moscow region and at Heis Island, Franz Josef Land, in 1956-1963). V sb.: Radioaktivnyye Izotopy v Atmosfere i Ikh Ispol'zovaniye v Meteorologii (Radioactive isotopes in the atmosphere and their use in meteorology). Atomizdat, Moscow, 1965.

48. Zykova, A. S., Ye. L. Telushkina, G. P. Yefremova, G. A. Kuznetsova, V. P. Rublevskiy and V. I. Shumakov. Radioaktivnost' Atmosfernogo Vozdukha i Nekotorykh Produktov Pitaniya v g. Moskve v 1965 i 1966 gg (Radioactivity of the atmosphere and certain foods in Moscow in 1965 and 1966). Atomizdat, Moscow, 1967.

49. Giletti, B. J. and F. Bazan. Rates of movement and mixing of water masses in the Atlantic ocean. Bull. Geol. Soc. America, Vol. 67, No. 12, 1956.

50. Athale, R. N., D. Lal and Rama. The measurement of tritium activity in natural waters. Part II. Proc. Indian Acad. Sci., Vol. 65A, No. 2, 1967.

Received for publication July 11, 1968

UDC 539.163

Copyright © 1966 by the American Geophysical Union

Reprinted from *J. Geophys. Res.*, **71**(16), 3869–3882 (1966)

The Origin of Saline Formation Waters
1. Isotopic Composition

R. N. Clayton,[1] I. Friedman,[2] D. L. Graf,[3] T. K. Mayeda,[1]
W. F. Meents,[4] and N. F. Shimp[4]

The content of total dissolved solids and δD and δO^{18} values are given for 95 oil-field brines from the Illinois, Michigan, and Alberta basins and the Gulf Coast. The variation in deuterium content among basins is found to be much greater than that within each basin. Oxygen isotopic composition, on the other hand, shows a large range in each basin, strongly correlated with salinity. The relationships between isotopic and chemical compositions of the brines lead to the following conclusions: (a) the water is predominantly of local meteoric origin; (b) the deuterium content has not been greatly altered by exchange or fractionation processes; (c) extensive oxygen exchange has taken place between water and reservoir rocks. Several samples contain water which appears to have originated as precipitation during Pleistocene glacial periods.

Introduction

The processes involved in the development of saline formation waters (brines) are of major hydrologic and geochemical interest. However, as *Chave* [1960] pointed out, attempts to relate brine compositions to the compositions of ancient oceans are thwarted by the extensive chemical changes that have taken place in the brines after sediment deposition. Clearly, there have been more unknowns than measurables in classic studies of brine origin.

The relations governing the distribution of stable hydrogen and oxygen isotopes in surface waters are by now reasonably well understood after the research of the last fifteen years. We attempt to use these relations, together with isotopic and chemical analyses, to reach conclusions about brine histories. The Illinois and Michigan basins were initially chosen for study because of relative structural simplicity, extensive geologic documentation, and the presence of numerous oil wells from which to sample. The Michigan basin contains great thicknesses of salt and anhydrite, whereas the Illinois basin has only anhydrite of limited thickness and extent. When results from these two basins suggested that meteoric waters were involved, additional samples from Alberta and the Gulf Coast were added to the study.

Our approach has been initially to consider all the samples from a given basin or geographical area as a group, regardless of their depth or lateral position. Such an approach seeks chemical and physical mechanisms operative throughout the whole sedimentary rock mass that may be detected before proceeding to the hydrologic and petrologic details around individual sampling points.

Most of the samples come from the Illinois and Michigan basins, and the location of these sampling points is shown in Figure 1. Table 1 gives the total dissolved solids content and isotopic composition of all samples, as well as the in situ temperatures needed for testing temperature dependence of these isotopic compositions.

A second paper in this series [*Graf et al.*, 1965] deals with isotope effects in shale ultrafiltration. In the third paper [*Graf et al.*, 1966] the chemical compositions and the findings of the first two papers are used, and mechanisms are suggested for the formation of brines rich in calcium chloride.

Deuterium concentrations in this paper are reported relative to standard mean ocean water

---

[1] The Enrico Fermi Institute for Nuclear Studies, The University of Chicago, Chicago, Illinois.
[2] Branch of Isotope Geology, U. S. Geological Survey, Denver Federal Center, Denver, Colorado.
[3] Illinois State Geological Survey, Urbana; now at Department of Geology and Geophysics, University of Minnesota, Minneapolis.
[4] Illinois State Geological Survey, Urbana.

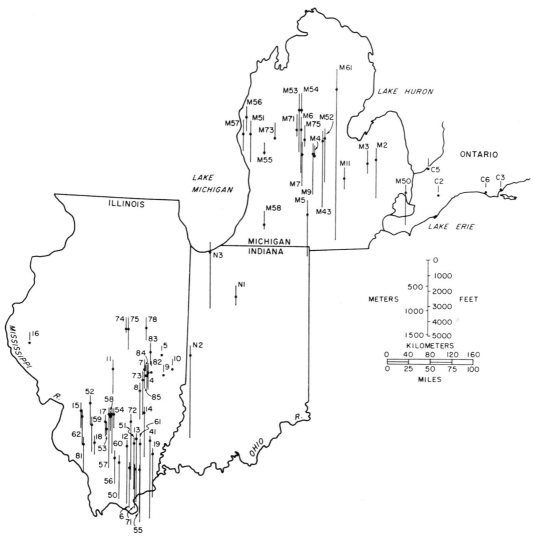

Fig. 1. The geographic locations of samples from the Illinois and Michigan basins, shown as dots projected onto the plane of sea level, with the lengths of the well above and below sea level drawn to scale.

(SMOW) [*Craig*, 1961b] in units of δD, defined as

$$\delta D(\%) = \frac{(D/H)_{\text{sample}} - (D/H)_{\text{SMOW}}}{(D/H)_{\text{SMOW}}} \times 100$$

Oxygen isotopic concentrations are also given relative to SMOW, but in parts per thousand:

$$\delta O^{18}(\text{‰})$$
$$= \frac{(O^{18}/O^{16})_{\text{sample}} - (O^{18}/O^{16})_{\text{SMOW}}}{(O^{18}/O^{16})_{\text{SMOW}}} \times 1000$$

In Figures 2–5 are shown the isotopic compositions of hydrogen and oxygen plotted versus total salinity, expressed as gram-equivalents per liter. No significant change is seen if salinity is expressed in other ways, e.g., as grams per liter. Nor is there a better correlation between isotopic abundance and the concentration of any individual solute species. The isotopic compositions of mean ocean water [*Epstein and Mayeda*, 1953] and the appropriate local present-day meteoric water [*Friedman et al.*, 1964; *Dans-*

TABLE 1. The Isotopic Compositions, in Situ Temperatures, and Total Dissolved Solids Contents of Brine Samples

Sample Number*	δD, ‰	δO^{18}, ‰	Temperature, °C	Total Dissolved Solids, gram-equiv./liter
Illinois Basin				
4		−1.82	28.3	1.88
5		−6.65	14.8	0.45
6		−0.94		
6A	−1.7		38.0	2.32
7		−2.12		
7A	−1.5	−2.32	28.1	2.05
8		−4.79	32.6	1.97
9		−5.78		0.67
9A	−3.9	−6.21	16.0	0.58
10		−5.72	15.6	0.87
11	−1.60	+0.49	31.5	2.16
12	−1.55	+1.51	50.7	2.66
13	−1.55	+2.74	48.5	2.85
14	−4.66	−3.20	39.1	2.34
15		−2.96		1.03
15A	−3.7		27.1	0.96
16	−8.23	−10.15	16.6	0.20
16A	−8.53	−10.21		
17		−2.17		1.42
17A	−2.0		26.5	
18	−2.57	−2.27	23.9	1.13
19	−1.20	−0.65	37.0	2.20
41	−1.8	+5.10	56.2	3.40
50	−3.1	−2.41	37.1	2.43
51	−3.5	−3.35	40.0	2.54
52	−6.0, −5.6	−7.30	28.3	0.18
53	−3.2	−4.34	20.8	1.54
54	−1.8	−2.49	25.1	1.94
55	−0.2	+0.26	39.0	2.25 or 2.86*
56	−3.0	−3.56	31.2	2.04
57	−2.8	−1.47	45.8	1.34
58	−1.6	−2.36	25.2	1.94
59	−2.2	−2.07	34.5	1.76
60	−2.4	−2.86	40.0	1.48
61	−0.4	+3.76	55.2	3.45
62	−5.4	−7.47	32.4	0.32
71	−1.8	−0.80	38.0	2.01
72	−2.2	−0.76	48.1	2.18
73	−2.5	−1.58	37.5	2.28
74	−5.4	−8.12	19.0	0.069
75	−5.1	−7.62	26.1	0.024
78	−5.0	−6.90	22.1	0.040
81	−2.5	−1.13	32.2	1.78
82	−4.3	−6.30	13.7	0.103
83	−4.9	−6.98	21.3	0.106
84	−2.5	−3.40	20.9	1.63
85	−3.8	−5.34	13.9	0.42
N1	−5.8	−8.22	18.2	0.23
N2	+0.7	−2.42	60.0	3.52
N3	−5.0	−6.50	24.5†	1.17
Michigan Basin				
M2	−4.3	+4.34	35.9	7.32
M3	−4.95	−4.64	20.8	4.39

210

TABLE 1. (Continued)

Sample Number*	δD, ‰	δO¹⁸, ‰	Temperature, °C	Total Dissolved Solids, gram-equiv./liter
M4	−3.72	−1.63	15.4	5.36
M5	−4.60	−1.38	44.2	5.39
M6	−4.00	−2.00	38.0	4.55
M7	−4.30	−1.39	31.7	5.63
M9	−3.05	−1.32	31.1	5.81
M11	−4.27	−3.95	20.8	4.50
M43	−5.3	+2.58	52.2	7.69
M50	−4.6	+0.16	30.0	6.08
M51	−4.0	−2.90	30.2	5.19
M52	−3.8	−0.06	36.8	5.46
M53	−5.2	−2.12	38.4	4.94
M54	−2.5	+0.02	18.2	5.23
M55	−5.2	−4.94	13.4	2.76
M56	−10.9	−13.11	20.9	1.90
M57	−5.5	−4.09	24.2	3.95
M58	−5.2	−5.64	19.1	3.74
M61	−6.0	−1.95	87.8	5.36
M71	−3.7	+0.55	19.0	5.54
M73	−3.0	+0.78	16.3	5.04
M75	−3.7	+0.57	18.6	5.66
C2	−5.8	−7.79	9.1	0.55
C3	−6.5	−8.79	9.2	0.38
C5	−4.2	−6.58	9.5	0.48
C6	−10.6	−13.31	8.7	0.064
Gulf Coast‡				
G3	+1.44	+8.76	72.5	2.50
G6	+1.45	+4.56	59.9	1.64
G7A	+1.27		85.8	2.13
G10	−0.40	+4.05	93.9	3.00
G13	+0.03	+4.60	83.3	3.14
G14	−0.60	+2.96	63.5	1.06
G15	−0.11	+8.65	142.8	5.75
G24	+2.94		79.1	1.46
G25	+0.69		89.2	3.05
G30	−0.53	+3.35	104.3	3.65
G38	−1.23	+2.54	77.3	1.66
G39	−1.00	+5.19	129.5	2.33
G79	−0.66		115.1	2.03
G79A	+0.28		117.4	2.12
G81	−0.86	+2.05	67.3	1.32
Alberta Basin				
A	−4.9	+7.80	82.3	3.62
B	−5.3	+3.61	77.8	2.90
C	−9.8	−5.07	32.2	0.39
D	−12.0	−12.81	26.7	0.093
E	−8.0	−4.58	35.0	0.39
F	−5.6	+4.03	96.7	3.03
G	−9.0	−3.01	55.6	2.05
H	−7.0	+2.31	56.7	3.48

* The detailed geologic descriptions of brine sampling points are given in *Graf et al.* [1966].
† At 960 meters.
‡ Deuterium analyses quoted from unpublished measurements by Harmon Craig.

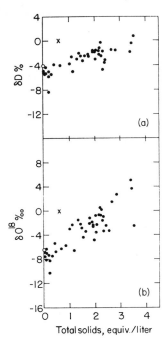

Fig. 2. Isotopic compositions of Illinois basin brines: (a) hydrogen, (b) oxygen. Ocean water indicated by cross, present-day meteoric water by open circle.

gaard, 1964] are shown for comparison. The reference points for δD and δO^{18} contents of meteoric water in the Alberta basin are weighted values at Edmonton as shown in Dansgaard's Figure 10. The δD reference values for the other three areas are rough estimates from Figure 14 of *Friedman et al.* [1964], and the corresponding δO^{18} values have been calculated from the relationship given between these two isotopic quantities by *Craig* [1961a].

Hydrogen Isotopic Composition

The δD values from each basin vary about 2‰ on either side of a mean. In the Michigan and Gulf Coast basins, there is no systematic variation of δD with salinity; in Illinois and Alberta, there is some suggestion of an increase of deuterium content with salinity.

There are large variations of mean δD from basin to basin, the Gulf Coast samples having a deuterium content very similar to that of ocean water and the Alberta samples having δD values about 10‰ lower. It is seen in Figures 2a, 3a, 4a, and 5a that for each basin the δD value obtained by extrapolating the brine data to low salinity corresponds to the δD value of local meteoric water and differs considerably from the value for ocean water, except for the Gulf Coast where the two values nearly coincide. This relationship eliminates the possibility that the brines are simple marine evaporite waters, for in that event each basin should show a series of points passing through the ocean water value, increasing in deuterium content with increasing salinity. Neither can the brine water have originated in a well-mixed oceanic reservoir, even though a very high percentage of the sedimentary rocks in the areas sampled are known from petrologic and paleontologic evidence to be marine. The most obvious conclusion is that the original water from the depositional basin has been lost during compaction and subsequent flushing and that the formation water now encountered originated as precipitation over land, under climatic conditions not greatly different from those prevailing today.

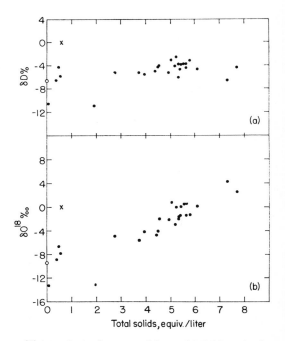

Fig. 3. Isotopic compositions of Michigan basin brines: (a) hydrogen, (b) oxygen. Ocean water indicated by cross, present-day meteoric water by open circle.

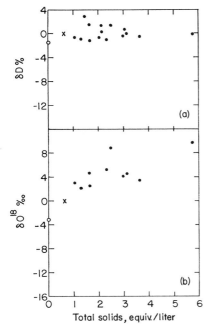

Fig. 4. Isotopic compositions of Gulf Coast brines: (a) hydrogen, (b) oxygen. Ocean water indicated by cross, present-day meteoric water by open circle.

they attributed to isotopic exchange between water and petroleum hydrocarbons. This conclusion is supported by their laboratory observation of isotopic exchange between petroleum and heavy water. However, the importance of exchange with hydrocarbons cannot be evaluated at this time because nothing is known about the isotopic fractionation between water and hydrocarbons. Even the direction of the fractionation is unknown.

Deuterium exchange between H_2S and H_2O is utilized commercially in the production of heavy water, and it will take place in nature where H_2S-rich natural gases occur. *Hitchon* [1963] reported natural gases from western Canada with as much as 54 volume % H_2S, although 82% of the analyses had less than 5% H_2S. Even in the extreme case, however, the reservoir of hydrogens in water would dominate unless there were extensive local recycling near an oil pool. *Roth* [1957] noted that the 85 ppm deuterium in H_2S from the natural gas at Lacq, southwestern France, corresponds to equilibrium with an infinite supply of water containing 150 ppm deuterium (the local value) at the

Those processes which might alter D/H ratios within a basin are (1) climatic variation with time, giving rise to changes in the isotopic composition of the precipitation, (2) isotopic exchange between water and other hydrogen compounds such as hydrocarbons, hydrogen sulfide, and hydrated minerals, (3) evaporation, and (4) diffusion and related effects.

We believe the effect of climatic variation can be demonstrated for a few extreme samples of unusually low δD and δO^{18} contents, which are interpreted as originating during periods of Pleistocene glaciation. Some of the scatter of the data may be the result of earlier, less extreme, climatic changes. A change of 4°C in mean annual temperature will cause a change of 2.2% in δD of mean precipitation [*Dansgaard*, 1964].

Very little information is available on exchange rates or equilibrium isotope fractionations between water and other geologically important hydrogen compounds. *Alekseev and Kontsova* [1962] found an increase of deuterium content of formation waters with depth, which

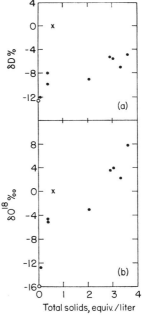

Fig. 5. Isotopic compositions of Alberta basin brines: (a) hydrogen, (b) oxygen. Ocean water indicated by cross, present-day meteoric water by open circle.

213

estimated reservoir temperature of 140°C.

There remains the possibility of isotopic exchange with the water of hydration in sedimentary minerals, principally gypsum and clays, together with addition of water from these minerals to the brine whenever conditions of pressure, temperature, and chemical concentration lead to dehydration. Gypsum dehydration takes place within the range of in situ conditions at which our samples are found. At 1 atm pressure and $a_{H_2O} = 1$, the gypsum-anhydrite transition is at $57 \pm 2°C$ [*Hardie*, 1964]. For lower activities of water, anhydrite is stable to lower temperatures, but the metastable persistence of gypsum to 70°C and higher in some solutions is well known [*Zen*, 1962].

We have found no published values of the hydrogen isotopic composition of water and hydroxyl contained in clay minerals. The deuterium exchange experiments of *Moum and Rosenqvist* [1958] [see also, *Faucher and Thomas*, 1955; *Romo*, 1956] suggest that at 100°C even the least accessible hydrogens, those of hydroxyl ions, should exchange within 10 years.

Without fundamental data on hydrogen isotope exchange among water, hydrocarbons, and minerals, it is not possible to rule out such exchange as a factor in determining D/H ratios in oil-field brines. However, the small variation observed among D/H ratios within each basin suggests that the effects of exchange are small.

Some correlation can be found between δD values and positions of the samples in any one basin. A mechanism to account for this observation, isotopic fractionation by filtration through shale micropore systems, is postulated in the second paper of the series [*Graf et al.*, 1965].

OXYGEN ISOTOPE COMPOSITION

The oxygen isotope data can be examined to see whether they support the conclusion that the water of the oil-field brines is of meteoric origin. In each basin there is a large range of O^{18}/O^{16} ratios for the waters, and the heavy isotope is enriched in the waters of greatest salinity. The extrapolations to zero salinity of the relationships implied in Figures 2b, 3b, 4b, and 5b again coincide with the local meteoric waters. This coincidence may be fortuitous, however, as indicated below.

There is a striking resemblance between the patterns of isotopic behavior seen here and those described by *Craig et al.* [1956] for normal chloride or for near-neutral hot springs from a large number of thermal areas: 'Isotopi-

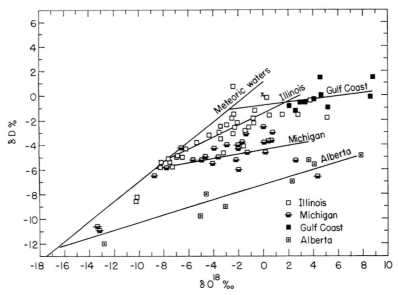

Fig. 6. Isotopic compositions of all brines in this study. The solid line is the locus of values for meteoric waters throughout the world [*Craig*, 1961a; *Dansgaard*, 1964]. Cross represents mean ocean water.

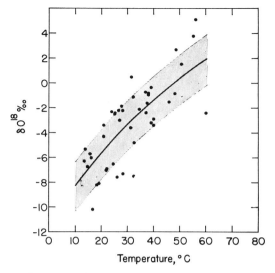

Fig. 7. Variation of brine O^{18}/O^{16} ratio with well temperature for the Illinois basin. The equilibrium curve is drawn for calcite of $\delta = +24.2‰$ (see text).

cally, these springs are either identical in composition with the associated surface river or lake waters, or they have about the same deuterium concentration as the local meteoric waters, but are enriched in O^{18} with respect to these waters by amounts ranging up to about 0.4%. This enrichment is generally observed in the higher temperature, deeper circulating springs.' These authors suggested that the observed enrichment of O^{18} in the waters results from isotopic exchange between water and rock at elevated temperatures.

Our data can be presented on a plot of δD versus δO^{18} (Figure 6) similar to those used by Craig et al. to discuss hot spring systems. The least saline waters in each basin approach the line for meteoric waters, and the more saline ones depart from that line in the direction of O^{18} enrichment. No simple relationship between the isotopic compositions of brines and ocean water is evident. Rather, the picture for oil-field brines is qualitatively like that for hot springs: series of points for each area extend from the value of local meteoric water and show a large enrichment in O^{18} but a relatively small enrichment in deuterium. However, the 'oxygen shift' is very large compared with that observed in most hot spring systems, probably because the ratio of water to rock is smaller for the brine exchange system.

Recent studies on oxygen isotopic variation of rocks in geothermal areas [*Clayton and White*, in preparation; *Clayton and Steiner*, in preparation] indicate that extensive exchange with the reservoir rocks is common. Hydrothermally altered calcite, for example, is found to be in isotopic equilibrium with the surrounding water at temperatures at least as low as 70°C. It has been shown in the laboratory that isotopic exchange between water and calcite is much more rapid than that between water and silicates, equilibrium being attained within 1 month at 190°C [*Clayton*, 1959]. Hence it might be expected that the exchange of oxygen between water and limestones would be the dominant factor in determining the isotopic composition of oxygen in oil-field brines. The δO^{18} of these waters should depend on (1) the δO^{18} value of initial, unexchanged water, presumably of meteoric origin, (2) the δO^{18} value of the solid with which most of the exchange takes place, presumably calcitic limestone, (3) the extent of equilibration between water and rock, and (4) the temperature of equilibration. Our measurements suggest that the major variations result from the temperature effect.

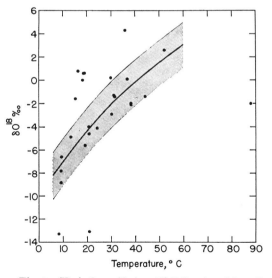

Fig. 8. Variation of brine O^{18}/O^{16} ratio with well temperature for the Michigan basin. The equilibrium curve is drawn for calcite of $\delta = +25.3‰$ (see text).

Figures 7–10 show the measured δO^{18} values plotted versus the bottom-hole temperatures of the wells (see appendix for a discussion of temperature estimates). On each graph there is superimposed a curve drawn as follows: For that basin a single value of δO^{18} for calcite, chosen to give the best fit to the data, was assumed. Then, using the equilibrium constants [*Clayton*, 1961] for the calcite-water system, we calculated values of δO^{18} over the observed temperature range for water equilibrated with calcite of the assumed composition. The shaded band indicates that most of the points fall within 2‰ of the equilibrium line.

Two important features are evident: (1) In all basins and at all temperatures, the gross trend of δO^{18} versus temperature follows the equilibrium line; (2) the value of δO^{18} assumed for calcite is practically the same from basin to basin. This value of $\delta = 23.8 \pm 2$‰, equivalent to $\delta = -6.5 \pm 2$‰ relative to the PDB standard, is just the value widely observed for old, unmetamorphosed limestones. The range of δ from -4.5 to -8.5‰ relative to PDB includes 80% of the Paleozoic limestones analyzed by *Keith and Weber* [1964].

The scatter of the data around the equilibrium line may result from several factors: (1) lack of equilibrium with carbonates (especially likely for the lowest-temperature waters and for those from carbonate-poor rocks); (2) error in temperature assignment (most probable in the deepest wells); and (3) local deviations in the isotopic composition of reservoir carbonates from the mean value. The points which fall far from the equilibrium line include groups of

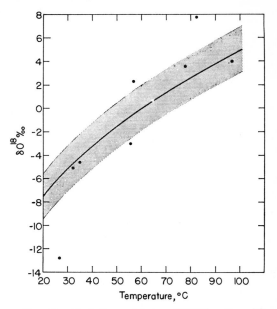

Fig. 10. Variation of brine O^{18}/O^{16} ratio with well temperature for the Alberta basin. The equilibrium curve is drawn for calcite of $\delta = +22.3$‰ (see text).

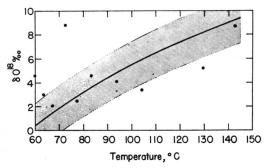

Fig. 9. Variations of brine O^{18}/O^{16} ratio with well temperature for the Gulf Coast samples. The equilibrium curve is drawn for calcite of $\delta = +22.8$‰ (see text).

samples from the same rock unit, such as M4, M54, M71, M73, and M75 from the 'Stray' sandstone member of the Michigan formation in Michigan, and 8, 14, and 51 from the Fredonia limestone member of the Ste. Genevieve formation in Illinois.

Treating pre-Tertiary carbonate wall rock as an oxygen isotopic reservoir of uniform composition of $+22$ to $+26$‰ raises some difficult questions. Exchange with a large amount of ocean water raised to rather high temperature would have been needed to bring about the shift to this range from the $+29$ to $+30$‰ of marine sediments. If the shift were a result of exchange with fresh water, the limestones themselves should show a latitude effect. Obviously a more comprehensive sampling of limestones with latitude and depth is needed, together with serial isotopic analyses of successive layers from pore walls. We report in Table 2 the bulk oxygen and carbon isotopic compositions of pieces of limestone core taken close to three of the Illinois brine sampling points lying at depths of 700 to 1250 meters. The δC^{13} values are typical of marine limestones, the δO^{18} values are typical of those for marine limestones al-

TABLE 2. Isotopic Analyses of Limestones Lying Close to Brine Sampling Points

Core Sample Number	Well from Which Core Was Taken	Depth of Core Sample below Surface, meters	δO^{18}, ‰	δC^{13},† ‰
11C	Natl. Assoc. Petrol., No. 2 Sincox, 29-13N-1E, Christian Co., Illinois	713	+25.31	+2.47
12C	Sun Oil, No. 1, Glenn-Cates-Orrick Unit, 1-4S-4E, Jefferson Co., Illinois	1247	+26.36	+3.85
51C	Texas, No. 1, Louis, 5-3S-6E, Wayne Co., Illinois	1031–1033	+26.50	+2.52

† C^{13} relative to PDB standard.

tered by exposure to fresh water, and both are toward the heavy end of the range for the category. The average oxygen isotopic composition, $\delta = +24$‰, calculated for calcite from the brine isotopic analyses is applicable only to that portion of the rock which is undergoing exchange with water. If the interiors of grains are unexchanged marine carbonate, bulk carbonate analyses like those in Table 2 should show samples richer in O^{18} than the calculated values.

Samples Containing Water of Pleistocene Age

Some low-salinity samples are depleted in D and O^{18} relative to present-day precipitation in the general geographic areas from which they come. The best examples are sample 16 from Illinois and sample C6 from Ontario, although we shall also discuss samples 52 and 62. Sample M56, similar isotopically to the other four, is a moderately saline brine.

Sample 52 comes from a Devonian sandstone that crops out along the Mississippi and lowermost Illinois rivers some 80 km northwest of the sampling point. The water in the present-day Mississippi at the latitude of these outcrops is, of course, depleted in the heavy isotopes relative to precipitation there because of an extensive drainage basin to the north; *Friedman et al.* [1964, Table 11] gave $\delta D = -5.6$ and -6.2‰ for the river at St. Louis, Missouri, and at Clinton, Iowa. These are precisely the measured values for sample 52, which could thus be infiltrated river water of present-day isotopic composition. The same reasoning applies to sample 62, from an Ordovician limestone that crops out in the Mississippi gorge under Pleistocene alluvium some 24 to 40 km to the west of the sampling points.

The present-day Illinois (ancient Mississippi) and Mississippi rivers are about 40 km to the east and west, respectively, of the sampling point for sample 16, and the partially gravel-filled valleys of these rivers are cut within about 75 meters of the level of the bottom of the well [*Horberg*, 1950]. Infiltration of river water might, therefore, be suspected. However, the δD value of -8.2‰ for sample 16 is too small to be explained in this way, and we are left with the possibility that the water is from Pleistocene glacial time. A check sample from a nearby well, 16A, gives $\delta D = -8.5$‰.

Parallel arguments for the three Illinois samples may be made from δO^{18} analyses. The measured value of Mississippi River water at St. Louis (-8.67‰ [*Epstein and Mayeda*, 1953], corrected to SMOW) is more than adequately light to explain the -7.3‰ measured for sample 52 and the -7.47‰ of sample 62, but not the -10.15 and -10.21‰ of samples 16 and 16A.

Sample C6 is from a 79-meter well on the north shore of Lake Erie, in Late Silurian rocks that crop out in a broad belt 40 km to the northeast. Sample M56 was taken from Middle Devonian rocks (Traverse group) encountered in a well 11 km east of Lake Michigan. These beds crop out under the lake, probably about 40 km from the east shore. The lightest water in the Great Lakes chain, that in Lake Superior, has $\delta D = -7.5\%$ [*Friedman et al.*, 1964], so that infiltrated lake water is inadequate to explain the $\delta D = -10.6$ for C6 and that of -10.9 for M56. These two samples also have unusually low O^{18} contents, which suggests that water entered the aquifer at a time when the climate was colder than at present.

The appreciable salinity of M56 could have resulted from glacial-age water dissolving bedded salt, or the sample could be a mixture of glacial-age water and pre-existing saline formation water. This sample is discussed further by *Graf et al.* [1965].

The calculation of maximum possible flow rates for samples 16, M56, and C6 involves estimating the time at which glacial-age climate returned to normal in the area between central Illinois and the Canadian border. Depending upon the distance from the ice front over which climate is believed to be modified, and upon the relative importance one attaches to evidence from pollen studies and to estimates of the return of sea level to normal, the outside limits for this time point are 14,000 and 5000 years ago. Maximum possible flow rates then become 3 to 8 m/yr for samples 16 and C6 and 4 to 11 m/yr for M56. These values are toward the low end of the normal range of flow rates, 1.2 to 540 m/yr, reported by *Todd* [1959, p. 53], and should in reality be even lower to the extent that these waters originated within rather than at the end of the Pleistocene glaciation.

Discussion

The arguments in this paper have dealt only with the gross features of isotopic composition, variations greater than $\pm 2\%$ in D/H and greater than $\pm 2\%_o$ in O^{18}/O^{16}. The subtle consequences of lithologic and hydrologic patterns on isotopic composition within an individual basin are discussed by *Graf et al.* [1965]. Because saline formation waters apparently do not form by simple concentration of marine water, a more complex explanation is required for the origin of the dissolved solids in these brines [see *Graf et al.*, 1966].

Our conclusion regarding the origin of oil-field brine water differs from that of *Degens et al.* [1964], which was based upon the oxygen isotopic composition of brines from Oklahoma, Colorado, Texas, Florida, and Utah. These authors obtained patterns qualitatively like our Figures 2b, 3b, 4b, and 5b but interpreted them as resulting from mixing of marine and fresh waters. Their oxygen isotopic variations might also be explained by exchange with reservoir rocks. Deuterium analyses should discriminate between these alternatives. It is of interest that some of the brines of Degens et al., from Oklahoma, have O^{18}/O^{16} ratios much less than that of present-day local precipitation and were explained as originating during an earlier, colder climate.

The correspondence in deuterium content between oil-field brines and present-day local precipitation suggests that these waters entered the rocks when the climate was not greatly different from that of the present. If our correlation of small δ values with precipitation during Pleistocene glacial periods is correct, the other samples must be pre-Pleistocene precipitation because the waters of Pleistocene age are from shallow wells relatively near aquifer outcrops.

Appendix

Illinois samples were usually obtained from cased wells with brine in-flow from levels considered to be uncontaminated by brine disposal or water flooding. Isocon maps available for many formations [*Meents et al.*, 1952, and unpublished data] made it possible to monitor brine concentration with time to make certain that drilling mud, dissolved solids from acidizing, and other contaminants were flushed out after completion of a well and to guard against major casing leaks. A few samples were taken by bailer-sampling the bottom of the water column during cable-tool drilling or a drill-stem test, if there was rapid in-flow of a large amount of brine. The usual practice was to sample such a water column at a number of points, at least in the lower part, to establish constancy of composition. In sampling at the surface, care was taken to ensure that the brine samples had not undergone evaporation or chemical or heat treatment; this meant sampling at the well head, or from a flowing discharge pipe directly connected to the well head, or from untreated brine in a storage tank under an oil layer. We played no part in the collection of Gulf Coast samples in 1947, but have attempted to ensure that samples collected for us in Alberta and the Michigan Basin were taken carefully. Sampling is clearly still the major limitation in this study.

Unfiltered aliquots of the brine samples were distilled to dryness before isotopic analysis. In general, repeated mass spectrometer runs were made until agreement was reached to ±0.1‰ for oxygen and to ±0.1% for hydrogen.

The Gulf Coast samples were collected in 1947 and analyzed for oxygen in 1950 and for deuterium at an unspecified time between 1947 and 1952. The possibility of evaporation because of poor sealing during this extended storage period makes it unwise to place much emphasis on any individual value, but it is quite unlikely that the general isotopic composition of the group of samples is erroneous. The same comment applies to the oxygen isotopic analyses, made in 1963, of samples 50–60 inclusive, which were collected and analyzed for deuterium in 1957.

The isotopic and chemical values as tabulated for Michigan basin sample M2 have been adjusted to eliminate the effect of 2% local fresh water added to the brine wells to prevent clogging of the well by precipitated salts. Sample M43 has been similarly adjusted for an estimated addition of 5% of fresh water.

Most temperature measurements in oil wells are made during or very shortly after completion of drilling, before thermal equilibrium with wall rock has been established. If there is concern about thermal equilibrium it is likely to be the short-term, dynamic equilibrium for a constant rate of fluid movement, as in water injection [see *Bullard et al.*, 1964]. The best compilation of static equilibrium values is that of *Moses* [1961], and we used his contoured temperature gradient map in computing in situ temperatures of Gulf Coast brine samples, taking as an assumed mean annual air temperature the value of 23°C to which the depth-temperature curves of the area extrapolate [*Nichols*, 1947]. In the other areas studied, mean annual temperatures for the period 1899–1938 [*U. S. Dept. of Agriculture*, 1941, pp. 672–673, 703] were used in similar calculations. The uncertainty in surface temperature is probably within ±2°C. Moses did not state the distribution of depths of the wells upon which his map is based, and we may have introduced an error in using his gradients for those of our Gulf Coast samples that come from depths of 2100 to 3300 meters.

Temperature logs and flowing-brine temperatures for shallow wells in southwestern Ontario typically show an increase from 8°C at a depth of 30 meters to 10°C at 240 meters (W. D. Brittain, personal communication), in good agreement with a mean annual temperature of 8 or 9°C at the nearby outcrop areas. Temperature measurements have been made by the Oil and Gas Conservation Board of Alberta in most of the Alberta wells sampled, and temperature gradients calculated from these measurements are in excellent agreement with those derived from OGCB temperatures for nearby Alberta fields listed by *Moses* [1961].

Averages of scattered values given by Moses for nine counties in the Illinois basin range from 2.4 to 3.1°C per 100 meters of depth and show a rude zoning adequate for estimating in situ temperatures for our samples from those counties. M. G. Simmons (personal communication) measured a gradient of about 2.09°C per 100 meters at depths to 1050 meters in three counties between Urbana and Chicago. For samples from the rest of Illinois, we have assumed a gradient value of 2.7. Scattered temperature measurements of uncertain validity in the Michigan basin suggest gradients ranging from 2.0 to 3.6°C per 100 meters, and we have computed in situ temperatures for all but one of the Michigan samples from an assumed gradient of 2.7. This assumption yields for deep-lying sample M61 a temperature 24°C higher than the measured value, and the average of these two temperatures is given in Table 1. For M43 we use the temperature measured by the Dow Chemical Company. Our value of N3 is calculated from a bottom-hole temperature measurement made in that well by M. G. Simmons.

Uncertainties in chemical analytical values are discussed in the third paper of the series [*Graf et al.*, 1966].

Acknowledgments. We owe a major debt to the members of the Oil and Gas Unit of the Michigan Geological Survey, in particular to B. F. Ackerman, S. L. Alguire, Robert Breed, V. F. Sargent, John Snider, F. G. Terwilliger, and Russell Wiles, for helping us to sample in Michigan. We are equally grateful to Brian Hitchon of the Research Council of Alberta and to Donald Shaw and other members of the Oil and Gas Conservation Board of Alberta, who between them planned and carried out the Alberta sampling program. Dr. Hitchon also supplied us with information from his personal files on the geochemistry of formation fluids in Alberta. It is a pleasure to acknowledge additional help given us in sampling outside Illinois by R. J. Anderson, R. C. Hultin, and R. David Matthews of the Dow Chemical Company, W. D. Brittain of the Department of Energy and Resources Management of Ontario, T. A. Dawson and R. K. Leininger of the Indiana State Geological Survey, and C. L. Lundy and Harry Middleton of the Michigan Consolidated Gas Company. The Gulf Coast samples were collected by the Humble Oil Company in 1947 for Harold C. Urey.

M. G. Simmons of the Dallas Seismological Observatory furnished us bottom-hole temperature measurements made in Illinois, Indiana, and Michigan in 1964. Translations from the hydrogeological literature were made by P. X. Sarapuka and S. C. Csallany. Among our colleagues at the Illinois State Geological Survey who contributed to geologic and hydrologic interpretation, Elwood Atherton, R. E. Bergstrom, R. F. Mast, D. H. Swann, W. Arthur White, and H. B. Willman deserve particular mention. We have benefitted from the interest shown in this work by Harmon Craig, K. S. Deffeyes, B. B. Hanshaw, G. E. Hendrickson, Gordon Rittenhouse, K. E. Vanleir, J. D. Bredehoeft, D. E. White, R. O. Rye, M. D. Mifflin, Raymond Siever, and B. F. Jones. These acknowledgments apply to the complete set of three papers on brine origin.

The oxygen isotopic analyses were supported by National Science Foundation grants NSF-G-19930 and NSF-GP-2019 to Robert N. Clayton. Parts of the paper were written by D. L. Graf during the tenure of a Research Fellowship at Harvard University under the auspices of the Committee on Experimental Geology and Geophysics.

References

Alekseev, F. A., and V. V. Kontsova, Distribution of deuterium in interstitial waters of oil pools, *Yadern. Geofiz., Sb. Statei, 1961,* pp. 196–201, 1962.

American Public Health Association, *Standard Methods for the Examination of Water and Wastewater,* Boyd Printing Company, Albany, New York, 1960.

Bullard, H. M., R. D. Clarke, and D. H. Rush, Primary logging as applied to past primary production, 2, *Producers Monthly, 28* (10), 27–31, 1964.

Chave, K. E., Evidence on history of sea water from chemistry of deeper subsurface waters of ancient basins, *Bull. Am. Assoc. Petrol. Geologists, 44,* 357–370, 1960.

Clayton, R. N., Oxygen isotope fractionation in the system calcium carbonate-water, *J. Chem. Phys., 30,* 1246–1250, 1959.

Clayton, R. N., Oxygen isotope fractionation between calcium carbonate and water, *J. Chem. Phys., 34,* 724–726, 1961.

Clayton, R. N., and A. Steiner, Oxygen isotope studies in New Zealand geothermal areas, in preparation.

Clayton, R. N., and D. E. White, Oxygen isotope studies in the Salton Sea geothermal area, in preparation.

Craig, H., Isotopic variations in meteoric waters, *Science, 133,* 1702–1703, 1961a.

Craig, H., Standard for reporting concentration of deuterium and oxygen-18 in natural waters, *Science, 133,* 1833–1834, 1961b.

Craig, H., G. Boato, and D. E. White, Isotopic chemistry of thermal waters, *Natl. Acad. Sci., Nucl. Sci. Ser., Rept. 19,* 29–36, 1956.

Dansgaard, W., Stable isotopes in precipitation, *Tellus, 16,* 436–468, 1964.

Degens, Egon, J. M. Hunt, J. H. Reuter, and W. E. Reed, Data on the distribution of amino acids and oxygen isotopes in petroleum brine waters of various geologic ages, *Sedimentology, 3,* 199–225, 1964.

Epstein, S., and T. Mayeda, Variation of O^{18} content of waters from natural sources, *Geochim. Cosmochim. Acta, 4,* 213–224, 1953.

Faucher, J. A., and H. C. Thomas, Exchange between heavy water and clay minerals, *J. Phys. Chem., 59,* 189–191, 1955.

Friedman, I., A. C. Redfield, B. Schoen, and J. Harris, The variation of the deuterium content of natural waters in the hydrologic cycle, *Rev. Geophys. 2*(1), 177–224, 1964.

Graf, D. L., I. Friedman, and W. F. Meents, The origin of saline formation waters, 2, Isotopic fractionation by shale micropore systems, *Illinois State Geol. Surv. Circ. 393,* 1965.

Graf, D. L., W. F. Meents, I. Friedman, and N. F. Shimp, The origin of saline formation waters, 3, Calcium chloride waters, *Illinois State Geol. Surv. Circ.,* in press, 1966.

Hardie, L. A., Gypsum-anhydrite equilibrium at 1 atm pressure, paper presented at Geological Society of America annual meeting, Miami Beach, Florida, 1964.

Hawkins, M. E., W. D. Dietzman, and C. A. Pearson, Chemical analyses and electrical resistivities of oilfield brines from fields in east Texas, *U. S. Bur. Mines Rept. Invest. 6422,* 20 pp., 1964.

Hawkins, M. E., O. W. Jones, and C. Pearson, Analyses of brines from oil-productive formations in Mississippi and Alabama, *U. S. Bur. Mines Rept. Invest. 6167,* pp. 1–22, 1963.

Hillebrand, W. F., G. E. F. Lundell, H. A. Bright, and J. I. Hoffman, *Applied Inorganic Analysis,* John Wiley & Sons, New York, 1953.

Hitchon, Brian, Geochemical studies of natural gas, 2, Acid gases in western Canadian natural gases, *J. Can. Petrol. Technol., 2*(3), 100–116, 1963.

Horberg, L., Bedrock topography of Illinois, *Illinois State Geol. Surv. Bull. 73*, 111 pp., 1950.

Keith, M. L., and J. N. Weber, Carbon and oxygen isotopic composition of selected limestones and fossils, *Geochim. Cosmochim. Acta, 28,* 1787–1816, 1964.

Meents, W. F., A. H. Bell, O. W. Rees, and W. G. Tilbury, Illinois oilfield brines, their geologic occurrence and chemical composition, *Illinois State Geol. Surv., Illinois Petrol. No. 66,* pp. 1–38, 1952.

Moses, P. L., Geothermal gradients now known in greater detail, *World Oil, 152*(6), 79–82, 1961.

Moum, J., and I. Th. Rosenqvist, Hydrogen (protium)–deuterium exchange in clays, *Geochim. Cosmochim. Acta, 14,* 250–252, 1958.

Nichols, E. A., Geothermal gradients in mid-continent and Gulf Coast oil fields, *Am. Inst. Mining Met. Engrs. Trans., 170,* 44–47, 1947.

Romo, L. A., The exchange of hydrogen by deuterium in hydroxyls of kaolinite, *J. Phys. Chem., 60,* 987–989, 1956.

Roth, Etienne, Problemes relatifs à la production d'eau lourde en France, *Rappt. Comm. Energie Atomique No. 657,* pp. 339–352, 1957.

Todd, D. K., *Ground Water Hydrology,* John Wiley & Sons, New York, 1959.

United States Department of Agriculture, Yearbook (*Climate and Man*), 1941.

Zen, E-an, Phase-equilibrium studies of the system $CaSO_4$-NaCl-H_2O at low temperatures and 1 atmosphere pressure, *Geol. Soc. Am. Spec. Paper 68,* p. 306, 1962.

(Manuscript received December 2, 1965.)

III
Chemical Composition of Rain and River Waters

Editor's Comments on Papers 12, 13, and 14

12 **Sugawara:** *Migration of Elements Through Phases of the Hydrosphere and Atmosphere*

13 **Junge and Werby:** *The Concentration of Chloride, Sodium, Potassium, Calcium, and Sulfate in Rain Water over the United States*

14 **Garrels and Mackenzie:** *Origin of the Chemical Compositions of Some Springs and Lakes*

The papers included in this part contain information on the chemical compositions of rain and river waters, and on the factors controlling their chemical compositions. In his article of 1967 (Paper 12), K. Sugawara catagorized the compositions of rain and river waters in Japan, and discussed the sources of chemical species in rainwater by the use of the enrichment factor. Junge and Werby in their paper of 1958 (Paper 13) discussed their observations of the chemical compositions of rainwater in the United States. The chemistry of rainwater can be classified according to origin into three divisions: marinogenic, lithogenic, and a group of industrial origin. Sea spray and smoke from industrial activity have been suggested as sources for sulfate in rainwater; it has also been suggested that sulfate is produced in polluted coastal areas, where biogenic H_2S enters the atmosphere to be oxidized to SO_2 and further to SO_3 (Koyama et al., 1965; Kellogg et al., 1972).

Lead in snow and ice in Greenland is thought to have originated from smelteries prior to 1940 and, after 1940, from burned gasoline (Patterson, 1971). The major source of arsenic content in the earth's rainfall was considered by Kanamori and Sugawara (1965) to be waste gases of sulfide ores and burning fuels. Miyake (1971) discussed the sources of artificial radioactive substances in the ocean.

The data contained in these papers indicate that atmospheric and hydrospheric influences on the distribution and migration of chemical species within the water cycle are virtually inseparable. *Impingement of Man on the Oceans*, edited by D. W. Hood in 1971, is an excellent book which treats in depth the subject of pollution.

The chemical compositions of river and lake waters have been discussed by D. A. Livingstone (1963), and the factors controlling the chemical composition of river water were reviewed by Sugawara. The primary source of river water is atmospheric precipitation, which is modified in its chemical composition by several factors: the addition of elements from dry fallout, from thermal and mineral springs, or from chemicals distributed for domestic, farming, and industrial purposes; the gain or loss of a variety of chemical elements through ionic exchange and through the dissolution of soil and rock materials; and adsorption and fixation by soil materials. The contribution of a particular element from soil and rock material to river water may be estimated by obtaining the total contribution of that element from all other sources and subtracting that sum from the content of the element in river water. The contribution from soil and rock materials to river water is significant in terms of the chemical weathering of rocks. Garrels and Mackenzie (Paper 14) and Bricker et al. (1968) presented the chemical reactions for weathering processes of rocks with water by supposing that primary rock-forming silicates are altered to clay minerals plus dissolved chemical species supplied by the dissolution of rocks. Such a treatment of the topic of the chemical composition of natural water makes it possible to estimate the degree of fragmentation of parent rocks, to forecast rockslides and landslides (Kitano et al., 1967), and to estimate the amounts of rocks weathered and clay minerals formed during a given period.

Ken Sugawara was born in 1899 in Tokyo, Japan. He graduated from the department of chemistry in the Faculty of Science at Tokyo Imperial University in 1923. Sugawara served as a professor in the department of chemistry at Nagoya University from 1941 to 1963, and was the Dean of the Faculty of Science at Nagoya University between 1948 and 1952; he is presently a Professor Emeritus of Nagoya University. Upon the establishment of the Geochemical Society of Japan in 1953, Sugawara was appointed chairman of the board of the society, and between 1964 and 1966 he served as the president of the society. Since 1965 he has been the chairman of the Joint Committee for Geochemistry and Cosmochemistry of the Science Council of Japan. He served as the vice-president of the Sagami Chemical Research Center between 1963 and 1972, and has since been a supervisor of the Research Center. Sugawara received the Sakurai Prize in 1942 for his work on chemical studies in lake metabolism from the Chemical Society of Japan, and in 1958 he received the Prize of the Academy of Japan for his geochemical studies on migration of elements through phases of the hydrosphere and atmosphere.

Christian Junge was born at Elmshorn/Holstein, Germany, in 1912. He studied meteorology and geophysics at the University of Graz (Austria), and at the universities of Hamburg and Frankfort in Germany. Junge worked as a meteorologist in Germany between 1935 and 1953, and as a senior scientist at the Cambridge Air Force Research Center in Bedford, Massachusetts, from 1953 to 1961. Junge returned to Germany following a call to the professional Chair for Meteorology at the University of Mainz, and he was simultaneously installed as the director of the Meteorological Institute. He is presently the director of the Department for Chemistry of the Atmosphere and Physical Chemistry of Isotopes at the Max-Planck Institute in Mainz. Junge received the Alfred Wegener Medal of the Society of the Deutsche Meterologische Gesellschaften in 1968, and the Carl Gustaf Rossby Research Medal of the AMS in 1973.

Robert M. Garrels was born in 1916. He received his B.S. from the University of Michigan in 1937, and both his M.S. (1939) and his Ph.D. (1941) from Northwestern University; he was presented with an honorary M.S. from Harvard University in 1952, and an honorary Doctor of Science degree from the University of Brussels in 1969. Garrels served as an instructor, an assistant professor, and an associate professor at Northwestern University between 1941 and 1952; he has been chief of the solid-state group at the U.S. Geological Survey from 1952 to 1955. At Harvard University Garrels was an associate professor from 1955 to 1957, a professor form 1957 to 1963, and chairman of the Department of Geological Sciences from 1963 to 1965. He was the president of the Geochemical Society in 1962. Garrels served as a professor of geology at Northwestern University between 1965 and 1969, a professor of geology at Scripps Institution of Oceanography between 1969 and 1971, and James Cook Professor of Oceanography at the University of Hawaii. Since 1972 Garrels has been a professor of geology at Northwestern University.

References

Bricker, O. P., Godfrey, A. E., and Cleaues, E. T. (1968). *Mineral-Water Interaction During the Chemical Weathering of Silicates.* American Chemical Society, Washington, pp. 296–307.

Hood, D. W. (ed.) (1971). *Impingement of Man on the Oceans.* Wiley-Interscience, New York, 314 pp.

Kanamori, S., and Sugawara, K. (1965). Geochemical study of arsenic in natural waters: I. Arsenic in rain and snow. *J. Earth Sci., Nagoya Univ.,* **13**, 23–35.

Kellogg, W. W., Cadle, R. D., Allen, E. R., Lazrus, A. L., and Martell, E. A. (1972). The sulfur cycle. *Science,* **175**(4022), 587–596.

Kitano, Y., Kato, K., Kanamori, S., Kanamori, N., and Yoshioka, R. (1967). Rockslides resulting from the geochemical weathering of rocks. *Annals Disas. Prev. Res. Inst. Kyoto Univ.,* No. 10-A, pp. 557–587 (in Japanese).

Koyama, T., Nakai, N., and Kamata, E. (1965). Possible discharge rate of hydrogen sulfide from polluted coastal belts in Japan. *J. Earth Sci., Nagoya Univ.,* **13**, 1–11.

Livingstone, D. A. (1963). Chemical composition of rivers and lakes, in M. Fleischer (ed.), *Data of Geochemistry,* 6th ed., *U.S. Geol. Survey Prof. Paper 440-G,* 64 pp.

Miyake, Y. (1971). Radioactive models, in D. W. Hood (ed.), *Impingement of Man on the Oceans,* Wiley-Interscience, New York, pp. 565–588.

Patterson, C. (1971). Lead, in D. W. Hood (ed.), *Impingement of Man on the Oceans,* Wiley-Interscience, pp. 245–258.

Reprinted from *Chemistry of the Earth's Crust*, Vol. 2, A. P. Vinogradov, ed., Israel Program for Scientific Translation, Ltd., Jerusalem, 1967, pp. 501–510

MIGRATION OF ELEMENTS THROUGH PHASES OF THE HYDROSPHERE AND ATMOSPHERE

Ken Sugawara

I. INTRODUCTION

The present paper is a part of the brief summary of the progress of a research program, carried out over the past ten years at the author's laboratories, aimed at determining the migration of elements through phases of the hydrosphere and atmosphere, the processes involved in the migration and at the quantitative estimation of the amounts of different elements which migrate.

The deviations in composition of the salt in atmospheric precipitation from sea salt are outlined for individual elements by using a term "enrichment coefficient".

Elements can be classified into various groups: the marinogenic group, the major source of which is sea spray; the lithogenic group, the major source of which is soil dust blown off the earth's surface; and another group to which fumes and gases from the combustion of fuels and other industrial activities contribute greatly.

Possible factors affecting the chemical composition of land waters are discussed. On the basis of the fundamental idea that the primary source of land waters is atmospheric precipitation, discussions are held on the manner and extent of participation of these factors in modifying the original composition of precipitation, finally forming river water.

The total yearly material discharge, as well as the net supply of elements from the Japanese Islands to the ocean, was estimated.

The effect of this net contribution of elements on the composition of sea water is also discussed.

The composition of atmospheric precipitation is discussed using the data of chemical analysis of approximately 300 samples of rain water and snow collected at various sites in Japan.

The composition of land water is discussed on the basis of the average composition of 43 representative rivers in Japan.

For the calculation of the yearly material discharge and the net contribution of elements to the ocean from the Japanese Islands, the total outflow of thermal and mineral springs with their average composition, amounts of chemicals consumed for domestic, farming and other purposes in Japan, and the possible extent of erosion of soil and rocky materials were taken into account.

II. COMPOSITION OF PRECIPITATION AND THE ANNUAL SUPPLY OF ELEMENTS FROM THE SKY TO THE JAPANESE ISLANDS

Table 1 provides data on the average composition of atmospheric precipitation in Japan.

TABLE 1

Average composition of atmospheric precipitation in Japan and enrichment coefficients of elements

Element	Chemical composition mg/l	Enrichment coefficient	Element	Chemical composition mg/l	Enrichment coefficient
Na	1.1	1.95	Si	0.83	4,700
K	0.26	12	Fe	0.23	400,000
Ca	0.97	39	Al	0.11	200,000
Mg	0.36	5.3	P	0.014	3,600
Sr	0.011	24	As	0.0016	30,000
Cl	1.1	1.0	Cu	0.00083	7,100
F	0.089	1,100	Zn	0.0042	14,000
I	0.0018	2,400	Mo	0.000060	2,100
$SO_4 - S$	1.5	29.5	V	0.0014	80,000

The elements found in precipitation can classified into three groups: the marinogenic group, the lithogenic group, and a group of industrial origin.

The elements of the marinogenic group are those whose source is primarily the sea. They are emitted mainly as spray and in other forms from the sea surface. They are supplemented by soil materials blown off the surface of the Earth and from wastes of industrial combustion. Sodium, potassium, magnesium, calcium, strontium, chlorine, iodine, fluorine and sulfur belong to this group. They are characterized by not very high values of the enrichment coefficient, a term which was proposed by the present author /1/ to compare the content of an element in atmospheric precipitation with that in sea water by reference to chlorine.

$$(M/Cl)_{sample} : (M/Cl)_{sea\ water},$$

where $(M/Cl)_{sample}$ and $(M/Cl)_{sea\ water}$ are, respectively, the ratios in equivalents of the element to chlorine in the sample and in sea water (see Table 1).

Silicon is a representative of the lithogenic group. In fact, the presence of silicon in rain water is detectable only after the sample is evaporated and subjected to alkali fusion, a fact indicating that the silicon in rain water is not in a dissolved state but is present as debris of silica or silicate minerals blown off the Earth's surface. Iron and aluminum are assumed to belong to the same group and to be directly or indirectly associated with silicon. They are characterized by high values of the enrichment coefficient, ranging in the order of ten to hundreds of thousands as compared with the highest value of 39 for Ca of the marinogenic group. It is unimaginable that these elements are supplied from the sea at such a high degree of enrichment. Vanadium, molybdenum, copper and zinc also show high enrichment coefficients. Though their origin is not clear, they may tentatively be classified in the lithogenic group.

Elements of the third group, the source of which is thought to be largely industrial wastes, are expected to increase in number in the future with increase in industrial activities. Arsenic can be mentioned as a member in view of its concentration in atmospheric precipitation and other findings.

Putting aside the details of the disucssion on these topics, the total annual supply of elements from the sky to the Japanese Islands and possible contributions from various sources to this supply, as calculated from available data, are listed in Table 2, Columns 3 — 6.

The total fallout, including precipitation and dry fallout, was calculated on the basis of the average composition of atmospheric precipitation, taking the average annual rainfall as 1,800 mm and assuming that the composition of dry fallout is approximately equal to that of precipitation.

Calculation of the elements supplied from deflated soil was based on the assumption that the silicon in both atmospheric precipitation and dry fallout is derived completely from the soil, and that the composition of the soil is identical to that of Norway loam as determined by M. Goldschmidt /2/. The abundance of elements excluded from his analysis was assumed to be equal to that in average igneous rock.

The contribution of industrial activity was calculated only for those elements for which a calculation can be made using sufficiently reliable data: calcium, sulfur and arsenic.

The contribution of smoke and other chimney dusts of industrial activities in the case of other elements was omitted for the following reasons: their contribution, if any, is thought to be small as compared with the contributions of other sources, and reliable data are not available.

The net contribution of the sea, as spray and in other forms, was obtained by subtracting the contributions of deflated soil and chimney wastes from the total supply in Column 3.

III. AVERAGE COMPOSITION OF RIVER WATER OF JAPAN AND FACTORS DETERMINING IT

The primary source of land water is atmospheric precipitation which is subject to the action of various factors ultimately determining the composition of river water. Such factors include conceivable additions of material from dry fallout, thermal and mineral springs and chemicals distributed for domestic, farming and other purposes, gain through ionic exchange and dissolution of soil and rock materials, and loss through ionic exchange, adsorption and fixation by soil materials.

In order to assess the effectiveness of these factors, the above-cited data on the average composition of the rain water in Japan, on one hand, and values of the average composition of 43 representative rivers in Japan, on the other, were adopted as bases (Table 3). The rivers were so selected that their drainage area covers 14 % of the entire country and is, moreover, representative of the different geological conditions in Japan. Furthermore, the area was chosen so as to include the rivers as far downstream as the point where the water is still practically free of contamination by industrial wastes. The water was sampled at points where the rivers cross the border of this area, the samples thus reflecting overall changes in water composition brought about by factors acting in the area. The population

Migration of Elements Through Phases of the Hydrosphere and Atmosphere

TABLE 2

Material balance of the Japanese Islands

Group	Element	Total supply from the sky in the forms of precipitation and dry fallout 10⁴ mg/yr	Direct supply from the sea as spray and in other forms 10⁴ mg/yr	Supply from blown-off soil 10⁴ mg/yr	Supply from smoke and other industrial dust 10⁴ mg/yr	Total discharge to the sea 10⁴ mg/yr	Addition to land waters — Industrial products 10⁴ mg/yr	Addition to land waters — Thermal and mineral springs 10⁴ mg/yr	Net contribution to the sea 10⁴ mg/yr	Net contribution to the sea 10⁻⁴ mg/l/yr
Marinogenic	Na	23.8	22.8	0.98	—	41.5	11.8*	3.7	6.9	2.0
	K	5.7	3.6	2.1	—	5.6	4.6*	0.2	-2.6	-0.76
	Mg	7.7	6.4	1.3	—	13.4	0	0.24	7.0	2.1
	Ca	19.8	18.3	1.4	0.10	39.2	4.9	1.1	20.9	6.1
	Sr	0.26	0.25	0.01	—	0.32	0	0.0023	0.07	0.021
	Cl	25.7	25.5	0.15	—	51.3	13.1*	6.0	12.7	3.7
	I	0.033	0.033	0.00002	—	0.013	0	0.0035	-0.018	-0.0053
	F	1.65	1.63	0.02	—	0.84	0	0.0091	-0.79	-0.23
	S	33.0	28.6	0.069	4.3	29.7	9.9	2.63	1.1	0.32
Lithogenic	Si	17.8	0.0	17.8	—	47.0	0	0.18	47.0	13.8
	Fe	4.95	—	3.3	—	3.7	0	1.5	3.7	1.1
	Al	2.38	—	4.5	—	2.4	0	0.42	2.4	0.70
	Mo	0.001	—	0.001	—	0.0034	0	0.00004	0.0034	0.0010
	V	0.0026	—	0.0012	—	0.0056	0	—	0.0056	0.0016
	Cu	0.015	—	0.0027	—	0.011	0	0.004	0.011	0.0032
	Zn	0.073	—	0.0060	—	0.29	0	0.0006	0.029	0.0085
Industr.	As	0.035	—	0.00071	0.014	0.030	0.027	0.0090	0.030	0.0088

* Elements supplied from the sea.

density in the area is $100/km^2$ against $240/km^2$ over the entire country. Out of the total annual flow (4.8×10^{11} l) of thermal and mineral springs in Japan, 1.6×10^{11} l comes from the area in question. From the annual rainfall and annual runoff of rivers in this area, the rate of evaporation was calculated as 17.8 %, a value which is identical to the average for the entire country. Table 5 provides a comparison of the major chemicals which are consumed annually in this area for domestic, farming and other purposes with the entire consumption in Japan.

TABLE 3

Average composition of 43 representative rivers in Japan and contributions from various sources for determining the composition

Element	Composition of the rivers	Contribution from various sources				Balance of contributions and river values
		precipitation	dry fallout	industrial products	thermal and mineral springs	
Na	5.1	1.3	2.3	0.29	0.22	1.0
K	1.0	0.31	0.55	0.81	0.012	—0.70
Mg	2.4	0.42	0.74	0	0.016	1.2
Ca	6.3	1.1	1.9	0.31	0.066	2.7
Sr	0.057	0.014	0.025	0	0.00014	0.018
Cl	5.2	1.4	2.47	0.90	0.37	0.06
I	0.0022	0.0018	0.0032	—	0.00021	—0.0030
F	0.15	0.089	0.16	—	0.00053	—0.10
S	3.5	1.8	3.2	0.32	0.17	—2.0
Si	8.1	0.98	1.7	0	0.016	5.4
Fe	0.48	0.27	0.48	—	0.09	—0.36
Al	0.36	0.13	0.23	—	0.028	—0.028
Mo	0.00060	0.000060	0.00011	—	0.0000022	0.00043
V	0.0010	0.0014	0.025	—	—	—0.0029
Cu	0.0014	0.00083	0.015	—	0.00023	—0.0011
Zn	0.0050	0.0042	0.074	—	0.000037	—0.0070
As	0.0017	0.0019	0.034	0.0017	0.00053	—0.0058

The assessment is made starting with chlorine (Table 3). The chlorine contribution of atmospheric precipitation to river water is calculated as 1.4 mg/l by multiplying the original concentration in the precipitation, 1.2 mg/l, by 1.78, the concentration factor by evaporation. The contribution of thermal and mineral springs is assessed as 3.7 mg/l from the average content of chlorine and the volume of water in the thermal and mineral springs in Japan (Table 4) which feed the rivers in question. The contribution of soil and rocky materials is calculated as 0.06 mg/l on the basis of the possible contribution of silicon of the same source; this will be discussed further on. The contribution of chlorine from industrial products which are distributed in the area for domestic, farming and other purposes is 0.09 mg/l as itemized as follows: 0.45 mg/l for sodium chloride consumed for domestic use as calculated on the assumption that the salt is evenly consumed per capita throughout the entire nation. The chlorine from ammonium chloride and potassium chloride consumed as fertilizer is calculated as 0.21 mg/l and 0.24 mg/l, respectively, on the assumption that they are distributed in the proportion of the farm coverage of this area to that of the whole country (Table 5). In the final count, the difference between the total of these contributions from different sources and the average of the actually observed values of river chlorine is taken as indicating the contribution from dry fallout. This difference is 2.47 mg/l, a value 1.77 times the contribution of atmospheric precipitation.

TABLE 4

Average compositions of thermal and mineral spring waters in Japan*

Element	Thermal and mineral springs mg/l
Na	770
K	42
Mg	56
Ca	221
Sr	0.5
Cl	1,250
F	1.9
I	0.73
SO_4–S	570
Si	38
Fe	313
Al	95
As	0.43
Mo	0.0077
Cu	0.8
Zn	0.13

* Calculated by Y. Kitano.

TABLE 5

Industrial products consumed for domestic, farming and other purposes

Element	Consumed for	Whole country 10^{14} mg/yr	The tested area 10^{14} mg/yr
Na	Domestic purpose as NaCl	3.9	1.6
	Industrial purpose as NaCl	7.9	0
K	Farming as KCl	4.2	4.2
NH_4	Farming as NH_4Cl	0.42	0.42
Ca	Farming as superphosphate	2.6	0.94
	Farming as cyanamide	2.3	0.81
Cl	Domestic purpose as NaCl	6.1	2.5
	Farming as KCl	3.8	1.4
	Farming as NH_4Cl	3.3	1.2
SO_4–S	Farming as superphosphate	1.0	0.36
	Farming as $(NH_4)_2SO_4$	3.9	1.4
	Farming as K_2SO_4	0.1	0.036
	Industrial processes	4.9	0
As	Agriculture as insecticide	0.027	0.7796

Assessment for other elements can be made in a similar but partly modified way. Through the assessments the salt composition of fallout is tentatively assumed to be identical to that of the atmospheric precipitation. Thus, the amount of each element from dry fallout is taken to be 1.77 times the corresponding contribution of atmospheric precipitation as given in the preceding paragraph.

Thus, for sodium the contributions of atmospheric precipitation, dry fallout, thermal and mineral springs, and the consumption of chemicals are calculated as 1.3 mg, 2.3 mg, 0.22 mg and 0.29 mg, respectively, per liter. In this case, the balance of the river value and the total of these contributions, 1.00 mg/l, is the net contribution of sodium from soil and rock materials during the passage of water through soil strata, because this element is least prone to adsorption or fixation during passage through the ground.

Potassium is processed similarly. Of interest is the fact that its balance is negative. The situation, however, is understandable when the adsorbability of the ion of this element is taken into account. A considerable proportion of potassium may escape by adsorption in some ground layer, and even if a certain amount is added to the permeating water from other layers, the gain does not cover the loss and the balance remains negative.

Among the tested elements, calcium shows the greatest gain in balance. Limestone and calcareous soils are undoubtedly responsible for this large value.

The sulfate balance is of special interest. In view of the big contribution of atmospheric precipitation, the consumption of sulfate manure and thermal and mineral springs, the negative balance of this element is exceedingly high. Reduction to sulfide in an anaerobic ground layer causes escape of sulfate from the water and its accumulation as insoluble metallic sulfides. Surely a counterreaction is also possible whereby solid sulfur compounds are oxidized to soluble sulfates which are liberated into aerated water when they come into contact with the latter. However, the gain does not cover the loss, leading to a big negative balance.

With regard to its high content, silicon parallels calcium and sulfate both in atmospheric precipitation and in river water. However, while in precipitation silicon, as distinct from calcium and sulfate, is not in a dissolved state but in the form of solid particles, in river water it is almost completely dissolved as are calcium and sulfate. Naturally, it is not likely that the silicon supplied by atmospheric precipitation and dry fallout constitutes a direct source of the silicon in river water. Thus, the total silicon in river water is concluded to have been supplied through the dissolution of soil and rock materials during the passage of the water through ground layers and to constitute the net contribution of these materials, including the supply from thermal and mineral springs.

One is confronted with the question whether the iron and aluminum in atmospheric precipitation are in the dissolved state. They are determined after the sample has been evaporated on addition of hydrochloric acid. In view of this and other circumstances, it is likely that the values from analysis nearly cover the total iron and aluminum in the sample, and that both elements are present as solid minerals. If this be the case, the iron and aluminum from atmospheric precipitation and dry fallout are not expected to constitute the direct sources of both elements in river water. They are likely to have been supplied by dissolution during the passage of the water through ground strata. Furthermore, the oxide-reductive nature of iron complicates its behavior underground. In anaerobic environs the iron tends to be reduced not only to form soluble ferrous ions, but further to separate an insoluble sulfide, while the anaerobic water laden with ferrous ions tends to precipitate ferric hydroxide when aerated. In effect, the balance is negative for both elements.

For minor elements, with the exception of molybdenum, it is remarkable that the balance of the river water value and the value for atmospheric precipitation and dry fallout is always negative. Evidently, this deficit is accounted for by adsorption, fixation by chemical changes, etc., during filtration of water through soil strata. In fact, the usual groundwater issuing from deeper strata has been proved to be greatly impoverished in copper and iodine as compared with atmospheric precipitation and surface water /3, 4/.

IV. CYCLIC, SEMICYCLIC AND NONCYCLIC ELEMENTS

If the process of migration of an element, starting with its emission in the form of spray from the sea to air, followed by deposition on the ground in the form of precipitation and dry fallout, up to its return to the sea in the form of river and other land discharges, is referred to as a "cycle," the elements so far tested can be classified into two categories: "cyclic" and "noncyclic."

Silicon, aluminum, iron, arsenic, molybdenum, vanadium, copper and zinc apparently fall within the noncyclic group, because their enrichment coefficients are too great for one to assume that a considerable portion of the amounts of these elements in atmospheric precipitation has been supplied in the form of spray from the sea, whereas the contribution of deflated soil materials or industrial wastes appears to account satisfactorily enough for the contents of these elements in atmospheric precipitation.

On the other hand, many of the so-called marinogenic elements are cyclic, as evidence shows that a substantial portion of these elements in atmospheric precipitation is derived from the sea, and their values of balance in precipitation and river water are clearly positive. Sodium, calcium, magnesium, strontium /9/, and chlorine are examples of such elements.

Not all the marinogenic elements are cyclic. Thus, considerable portions of the iodine, fluorine /10/, sulfur and potassium supplied from the sea are retained in the ground and only fractions are returned to the sea. They are classified as "semicyclic elements."

It is needless to say that the "cycle" is not a cycle in a strict sense. In other words, the term "cycle" does not imply that the supplied elements return in their original amount; they are subjected to various changes on their path of migration and are affected by ionic exchange, dilution effect, fixation, redissolution and other changes, especially during the course of passage through ground strata. Those elements which are eventually discharged into the sea in an amount more than equivalent to that supplied from the sea are termed "cyclic". Among the cyclic elements those which are least prone to adsorption, fixation and other such changes and rather readily find their way back to the sea, may well be called "more perfectly cyclic elements." Sodium and chlorine are examples of such elements.*

V. TOTAL ANNUAL DISCHARGE OF ELEMENTS FROM THE JAPANESE ISLANDS

The total annual discharge of elements from the Japanese Islands is assessed as follows:

The first approximation is made by multiplying the river water values cited above by the total annual outflow from the entire country. Values thus obtained are corrected by taking into account the local differences in the consumption of chemicals and in the outflow of thermal and mineral springs. The final results of the calculation are listed in Table 2, Column 7. For example, the first approximation for chlorine is 29.1×10^{14} mg which

---

* Even in the case of these elements, the ion individuals in the original rain water are nearly completely replaced by ones from the "pool" via the so-called " dilution effect" during the passage of water through soil layers. Indeed, this was clearly shown recently by K. Kodaira in his investigation of the behavior of strontium in rain water during its permeation through soil layers, using Sr^{85} as a tracer. (K. Kodaira. Studies on the Behavior of Calcium and Strontium in a Rainwater-Soil System (Japanese).— National Institute of Agricultural Sciences. 1963).

is corrected to 41×10^{14} mg on addition of chlorine from consumed chlorides and from thermal and mineral springs.

VI. NET CONTRIBUTION OF ELEMENTS TO THE SEA AND SECULAR CHANGE IN THE SALT COMPOSITION OF SEA WATER

The annual net contribution of individual elements to the sea is calculated by subtracting the annual supply from the sea from annual discharge calculated above (Table 2, Columns 10, 11). The annual supply includes both the supply via the air and the supply by consumed industrial products the materials of which come from sea water.

The results of the calculation provide us with clues to the nature of the secular change in composition of sea water which occurred in the past and is expected to continue to occur in the future.

Such a change is very small in terms of a year. If the net runoff from all continents and islands of the world is assumed to be evenly distributed throughout the ocean water, the Japanese Islands are found to contribute 3.38×10^{18} l water, this figure being based on the fact that the total volume of ocean water of the world is $1,372 \times 10^{21}$ l and that the Japanese Islands cover 369,766 km^2 of the entire land area of the world (148,698,000 km^2). The net change of the elements in sea water per year is calculated in Table 2, Column 11. The change only becomes remarkable after millions of years. Despite this it is very important to note that many elements tend to increase, while other elements, e.g., iodine and fluorine, obviously decrease in concentration with time.

It must be further added that not all the elements which apparently should increase in concentration in sea water with time either did so or do so to a noticeable degree. Some elements constantly escape from the water by forming precipitates or by being associated with such precipitates. Thus, calcareous organisms are constantly assimilating calcium from water in order to form their calcareous shells. Diatoms and other siliceous organisms take up silicon, thus lowering its concentration in the sea. Iron and other elements are separated from the sea by forming coagula of hydrated metallic oxides. Some other elements tend to be captured by these coagula, sinking and settling to the ocean floor. Thus, the actual increase of these elements in the sea does not follow the course predicted simply on the basis of the calculated amounts of net increase. The rates of precipitation and redissolution of elements in the sea itself are important topics requiring further extensive and quantitative investigation.

In this connection, the finds on molybdenum are of special interest.

While in land discharge five minor elements — molybdenum, vanadium, copper, zinc and arsenic — show nearly the same order of concentration, with the highest value for zinc, 0.005 mg/l, and the lowest value for molybdenum, 0.006 mg/l, the usual concentration of these elements in sea water is more uniform, with 0.0019 mg V, 0.002 mg Zn, 0.001 mg Cu and 0.0018 mg As per liter. Molybdenum is exceptional in that it is abundant in sea water with a concentration of 0.010 mg/l (one order of magnitude higher than the other minor elements)/4, 5, 6, 7, 8/.

The situation becomes understandable when the different tendencies of these elements to co-precipitation in the pH range of sea water are taken into account. An experiment in natural and artificial sea water at pH 8.0 showed that coprecipitation by ferric hydroxide coagulum is 100 % in the case of vanadium and 10 % in the case of molybdenum /11/. This indicates that vanadium is more likely to be separated from sea water than molybdenum. Actually, deep-sea sediments are enriched in vanadium which has a high abundance ranging from 10 — 300 ppm with an average value of 186 ppm, while the abundance of molybdenum varies from 2 — 32 ppm, with an average of 8 ppm /12, 13/.

Thus, the higher content of molybdenum in sea water as compared with other minor elements is taken as being the result of the accumulation of this element in the past ages. This accumulation can be expected to continue in the future, while the concentrations of other elements will remain little changed even if the net contribution from the land is not small.

Institute of Chemistry and Water Research Laboratory,
Faculty of Science, Nagoya University, Nagoya, Japan

REFERENCES

1. SUGAWARA, K. Chemistry of Ice, Snow and Other Water Substances in Antarctica. — The Antarctic Record, Vol. 11:836; Extrait de la publication, 55 de l'A.I.H.S. Colloque sur la glaciologie antarctique, Vol. 44. 1961.
2. GOLDSCHMIDT, V.M. Grundlage der quantitativen Geochemie. — Fortschr. Mineral, Kryst. Petrog., Vol. 17:112. 1933.
3. MORITA, Y. Distribution of Copper and Zinc in Various Phases of the Earth Materials. — J. Earth Sciences, Vol. 3:40, Nagoya Univ. 1955.
4. TERADA, K. Distribution and Migration of Iodine in and through Atmosphere and Hydrosphere (Japanese). — Water Research Laboratory, Nagoya Univ. 1959.
5. SUGAWARA, K., K. TERADA, S. KANAMORI, N. KANAMORI and S. OKABE. On Different Distribution of Calcium, Strontium, Iodine, Arsenic and Molybdenum in the Northwestern Pacific, Indian, and Antarctic Oceans. — J. Earth Sciences, Vol. 10:34-50, Nagoya Univ. 1962.
6. SUGAWARA, K., H. NAITO and S. YAMADA. Geochemistry of Vanadium in Natural Waters. — J. Earth Sciences, Vol. 4:46, Nagoya Univ. 1956.
7. SUGAWARA, K., M. TANAKA and S. OKABE. Geochemistry of Molybdenum in Natural Waters, Vol. II. — J. Earth Sciences, Vol. 8:114. Nagoya Univ. 1961.
8. KANAMORI, S. Microdetermination of Arsenic and Distribution and Migration of Arsenic in and through the Sea, Air and Land Waters with Processes Involved (Japanese). — Institute of Chemistry, Faculty of Science, Nagoya Univ. 1962.
9. KANAMORI, N. Distribution and Migration of Calcium and Especially Strontium in and through the Atmosphere and Hydrosphere (Japanese). — Water Research Laboratory, Nagoya Univ. 1962.
10. KOBAYASHI, S. Geochemistry of Fluorine in Natural Waters (Japanese). — Sugiyama Women's College, Nagoya Univ. 1961.
11. KUWAMOTO, T. Co-Precipitation of Vanadium (V) Together with Ferric Hydroxide (Japanese). — J. Chem. Soc. Japan, Vol. 81:1669. 1960.
12. OANA, S. Contents of Vanadium, Chromium and Molybdenum in Deep-Sea Deposits (Japanese), Vol. I and II. — J. Chem. Soc. Japan, Vol. 59:1234. 1938; Vol. 61:1060. 1941.
13. KURODA, K. Contents of Vanadium, Chromium and Molybdenum in Deep-Sea Deposits (Japanese), Vol. III. — J. Chem. Soc. Japan, Vol. 63:496. 1943.

Errata

Page 231, Table 3, should read

Table 3

	Dry fallout
V	0.0025
Cu	0.0015
Zn	0.0074
As	0.0034

Page 231, line 4 from Table 3 should read: by 1.178, the
Page 231, line 5 from Table 3 should read: as 0.37 mg/l from
Page 231, line 12 from Table 3 should read: is 0.90 mg/l as
Page 232, Table 5, should read

Table 5

	The tested area, 10^{14} mg/yr
K	1.5
NH_4	0.15
As	0.0096

THE CONCENTRATION OF CHLORIDE, SODIUM, POTASSIUM, CALCIUM, AND SULFATE IN RAIN WATER OVER THE UNITED STATES

By Christian E. Junge

Geophysics Research Directorate, Air Force Cambridge Research Center

and

R. T. Werby

Werby Laboratories, Inc.

(Manuscript received 4 January 1958)

ABSTRACT

The distribution of yearly averages of the concentration of various inorganic ions in rain water over the United States is discussed. The major source of Cl⁻ is the ocean. The Cl⁻/Na⁺ ratio, however, is considerably less than that in sea water. It is very likely that this is caused by excess Na⁺ from the soil. A similar distribution of excess material from the soil is observed with K⁺. In contrast to Na⁺ and K⁺, which are rather uniformly distributed over the United States, Ca⁺⁺ shows highest values over the Southwest, in agreement with the occurrence of dust storms.

Most of the SO₄⁻⁻ over the ocean originates from sea spray. The source of additional SO₄⁻⁻ is the land. Budget considerations indicate that about 30 per cent of this additional SO₄⁻⁻ on a global scale is due to human activities.

On the basis of the data presented, the average global residence time of SO₂ is estimated to be 40 days. This value is compared with available data on residence times of other constituents which are also primarily controlled by washout.

1. Introduction

From July 1955 through July 1956, a network of rain sampling stations for chemical analysis was operated in the United States. Results on sea salt, ammonia, and nitrate have already been published (Junge and Gustafson, 1957; Junge, 1957a). In this paper the rest of the data will be discussed.

The precipitation samples were collected and analyzed on a monthly basis. The monthly average concentrations of the various ions obtained in this way were plotted on maps and compared with the general weather pattern. These maps showed considerable fluctuations, but no obvious correlations with the circulation could be detected. We decided, therefore, to present here only maps of yearly averages. Such maps can be regarded as a fairly good representation of the "chemical climate" and may prove most valuable for general geochemical considerations, some of which will be discussed here shortly.

The presentation of data is restricted further to maps of *concentration*. The concentration in rain water reflects to a large extent the concentration of the same constituent in the air and is, therefore, of special importance. This is particularly true for constituents which are fairly uniformly distributed throughout the troposphere (Junge, 1957a). If necessary, the total amounts of material brought down by rain can easily be obtained with the help of the annual rainfall of the same period or, more generally, of the average annual rainfall.

2. Analytical procedures

Details about the network and the procedures of collection are given in previous papers (Junge and Gustafson, 1956 and 1957).

Sodium, potassium, and calcium were determined by flame photometry (Beckman DU, flame attachment, photomultiplier). Sodium and potassium determinations were made on specimens as collected with an accuracy of ±5 per cent. Calcium was determined, in most cases, on samples previously concentrated, with an accuracy of ±10 per cent.

Sulfates were determined nephelometrically by the method of Keily and Rogers (1955). This method, by the use of ethanol as a stabilizer, solid barium chloride, and controlled agitation, is a great improvement over previous turbidimetric and nephelometric techniques and, with care, gives an accuracy of ±10 per cent.

Chlorides were determined by a nephelometric method developed by the authors and similar to that used for sulfates. It is vital, for stabilization of the silver chloride "precipitate" formed, that the sample be diluted with an equal volume of highest purity ethanol. With the use of great care and replicate determinations, we could attain an accuracy of ±10 per cent.

In certain coastal and maritime stations where greatest possible accuracy in establishing the Cl⁻/Na⁺ ratio was desired, a null method was used. Appropriate dilutions were made of actual sea water until the "synthetic" sample had a sodium value (flame photometer) exactly equal to that of the actual sample being investigated. With the sodiums "balanced" in this manner, replicate chloride comparisons were made between both "synthetic" and sample and the actual ratio Cl⁻/Na⁺ calculated from the

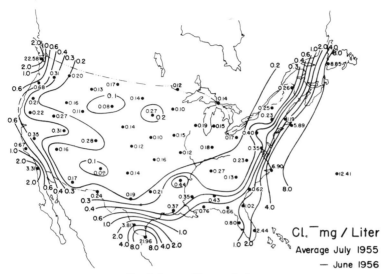

FIG. 1. Average Cl⁻ concentration.

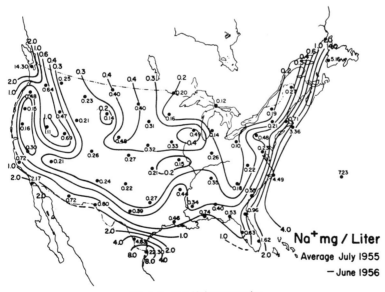

FIG. 2. Average Na⁺ concentration

theoretical of 1.88. In this manner, relatively small deviations from theoretical could be established with a considerably higher accuracy.

3. Chloride and sodium

In fig. 1 and 2 the chloride and sodium distributions are plotted. Both components are characterized by a rapid drop from the coast inland. Then, about 500 mi from the coast there is a leveling off so that fairly constant values are maintained throughout the continent. It was shown (Junge and Gustafson, 1957) that this distribution can be explained by different vertical distributions of the sea-salt particles over sea and land as a result of vertical mixing and that washout by precipitation is involved only to a minor degree. According to this, the rapid drop reflects primarily the dilution of sea spray particles in the lowest 2 to 3 km of the troposphere, where most rain is formed. This dilution occurs when maritime air masses, with the highest concentration of sea-spray particles in the ground layers, move inland as vertical mixing tends to produce a uniform vertical distribution. The constant level inland corresponds to the area where, on the average, this vertical mixing is completed.

An interesting irregularity can be noticed in the area of Brownsville and Laredo, Texas. Large salt flats in the coastal districts are known to produce fairly high concentrations of salt in the air. Due to the same mixing process that occurs in the case of ocean sources, the influence of this source becomes negligible beyond a distance of approximately 500 mi. This area of natural mineral pollution will be disregarded in our further considerations.

Inasmuch as a previous paper (Junge and Gustafson, 1957) has discussed the geographical distribution of sea-salt components, we will here be concerned primarily with their ratios. Fig. 3 is a plot of the ratios of chloride to sodium, values for which are shown in figs. 1 and 2. The ratios of the major constituents of sea water itself are given in table 1.

The ratio of Cl^-/Na^+ is only slightly lower than that of sea water along the coast, but drops considerably further inland. Only a few isolated local monthly values were higher than 1.88, and were located mainly

TABLE 1. Average ratios of the major constituents in sea water (Sverdrup et al, 1946).

Ratio	Value
Cl^-/Na^+	1.800
K^+/Na^+	0.360
Ca^{++}/Na^+	0.379
SO_4^{--}/Cl^-	0.140

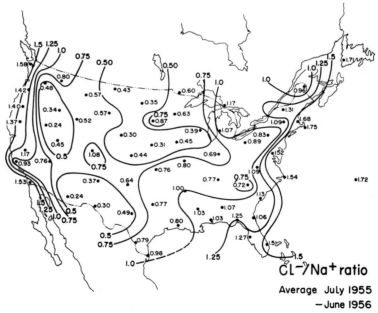

FIG. 3. Average ratio of Cl^-/Na^+.

in industrial areas. Otherwise, all the values were lower than that of sea water.

To give some impression of the yearly fluctuations of the ratio Cl^-/Na^+, the monthly averages were plotted in fig. 4. To make these averages more representative for continental conditions, the stations along the coasts were excluded. A second curve gives the average of the eastern half of the continental stations only. It can readily be seen from these curves that the variations of the Cl^-/Na^+ ratio in time and space are not very considerable. Slightly higher values during most of the winter may be related to a more maritime climate during this period of increased west wind circulation.

This pattern of the Cl^-/Na^+ ratio over the United States is somewhat different from that encountered over North Europe. Rossby and Egnér (1955) found ratios as high as 3.0 during advection of air masses from the south, while with air masses arriving from other directions, especially from the northeast, the values dropped to 0.2. The comparison with the U. S. data suggests that these strong fluctuations must be a peculiarity of the chemical climate in Europe, the reason for which is not known.

The deviation of the Cl^-/Na^+ ratio from 1.88 may be caused either by a loss of Cl^- caused by a decomposition of sea-spray particles or by additional Na^+ (mineral dust from the soil). The distribution of this excess Na^+ can easily be calculated by subtracting from the observed Na^+ value the theoretical net value (based on the composition of sea water) corre-

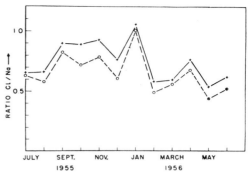

FIG. 4. Circles: monthly average ratios of Cl^-/Na^+ for the United States with the exclusion of the coastal stations. Crosses: the same but only for the eastern half of the United States.

sponding to the observed Cl^- value (fig. 5). Over the ocean and along the coasts the values in this map cannot be considered very accurate because they appear as small differences of two relatively large figures. But the values inland are fairly reliable. The distribution of these excess Na^+ values is comparatively uniform in contrast to what one might expect on the basis that the dry Southwest is a major source of mineral dust. If we disregard the Brownsville area, there is only one area of rather high values inland, namely that between the eastern and western Rocky Mountains. Before further discussion, however, the results of the K^+ and Ca^{++} analyses should be presented.

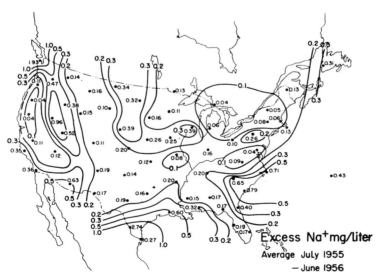

FIG. 5. Average excess Na^+ concentration, calculated as the difference between fig. 2 and the expected Na^+ concentration according to fig. 1 and the composition of sea water.

4. Potassium and calcium

The K^+ distribution is given in fig. 6. Its general features are very similar to those of the Na^+ map. Whereas in sea water the ratio K^+/Na^+ is 0.041, in rain water this ratio is found to be in the order of magnitude of 1.0. The only reasonable explanation for this fact is an additional source of K^+ from the soil, because chemical decomposition of sea-spray particles or, as sometimes is assumed, a mechanical separation of their constituents is very unlikely to increase this ratio by a factor of about 20. With the exception of the coastal stations, the amount of K^+ due to sea water is so small that fig. 6 can, for all practical purposes,

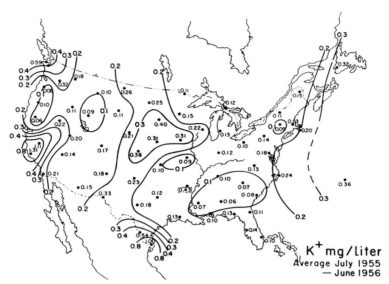

FIG. 6. Average K^+ concentration.

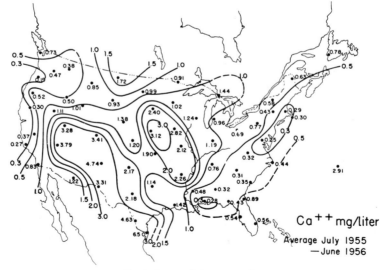

FIG. 7. Average Ca^{++} concentration.

be regarded as an excess K^+ map and be compared with fig. 5. It is of interest to note that figs. 5 and 6 show some common features, especially in the center of the continent. This should be expected if the source of the excess of both ions is the soil, since both ions are chemically so closely related.

The excess of Ca^{++} over sea water is even higher than for K^+ (fig. 7). But in contrast to the excess Na^+ and K^+, the Ca^{++} concentration shows a large and pronounced maximum in the arid Southwest, as would be expected from the occurrence of dust storms in that area. It should be mentioned here that the funnels of the rain collectors were opened during rainfalls only. The Ca^{++} values refer, therefore, exclusively to the soluble Ca^{++} components in rain water and do not include dry fallout between rainfalls, which may be considerable in the arid areas.

$CaCO_3$ is a major constituent of the soil in the arid areas. It is converted to soluble salts by the action of the CO_2 and SO_2 in the air either at the ground or, as aerosol, in the air. Na and K, on the other hand, occur in the soil primarily as silicates and

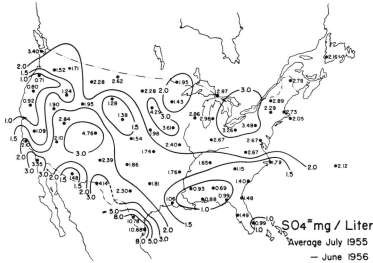

FIG. 8. Average SO_4^{--} concentration.

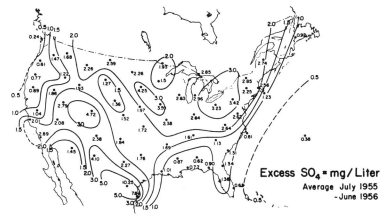

FIG. 9. Average excess SO_4^{--} concentration calculated as the difference between fig. 8 and the expected SO_4^{--} concentration according to fig. 1 and the composition of sea water.

will be converted into soluble salts only within the soil and predominantly in moist areas. Different weathering processes of this kind are, therefore, offered here as a tentative explanation for this striking difference between the Ca^{++} map and those for Na^+ and K^+. The rain samples were also analyzed for magnesium. The analytical method used was the Hunter modification of the Titon yellow method (Hunter, 1950). It gave good and reliable results, but needed considerable amounts of sample so that large areas with low rainfall could not be covered by data. For this reason, the data are not included here.

To our knowledge, Kalle (1953–54) was the first to point out in a careful study that the deviations of the ratios Cl^-/Na^+, K^+/Na^+, etc., in rain from those of sea water are due to cyclic mineral dust. He used data from the Scandinavian rain collection network. Others claimed that the decrease of the ratio Cl^-/Na^+ over land was caused by the release of Cl from the sea spray particles by various chemical processes (Rossby and Egnér, 1955). In fact, a gaseous chlorine component was found in our atmosphere (Junge, 1956). Since there can be little doubt, however, that the excess K^+ must come from the soil, we see no reason why a similar process should not be primarily responsible for Na^+ enrichment also. Furthermore, direct micro-analysis of sea-salt particles collected in the atmosphere did not indicate such decomposition (Junge, 1956; Twomey, 1954). It is even questionable if over such large areas as represented by the United States a decomposition of the sea-salt particles can result in a deviation of the rain-water composition from that of sea water. Since sooner or later all material will be washed out, it is unlikely that under equilibrium conditions there will be a net gain of one component on a continental scale.

Summarizing, we believe that there is little evidence that decomposition of sea-salt particles is of importance in the chemistry of rain water. However, further observations are needed to resolve this question.

5. Sulfate

For the sulfate concentration, we again plotted the observed (fig. 8) and the calculated excess distribution (fig. 9). It becomes immediately apparent that most of the sulfate over the sea comes from sea salt. But even over the sea a small additional amount is present. The accuracy of these excess values over the sea (e.g., the Bermudas) and near the coast is, of course, not very great because the SO_4^{--} here appears as a small difference between two fairly large figures. This excess SO_4^{--} increases rapidly over land, which indicates that its source is located here. Even the outline of Florida is reflected by the iso-lines. Fig. 9 also shows that industry is an important but not predominant source of excess sulfate. The relatively high values of SO_4^{--} in the Southwest and other areas with little or no industry lead to the conclusion that the earth surface must be a very efficient source of sulfur. It is very likely that this sulfur leaves the ground partly as particulate material, as in the cases of Ca^{++}, K^+, and Na^+, but that the major source is the exhaustion of H_2S formed by the decomposition of organic matter. This H_2S will be oxidized to SO_2 which in turn is converted to SO_4^{--} by photochemical processes (Gerhard, 1953) and by oxidation in cloud droplets (Junge and Ryan, 1957). Since only a fraction of these gases will be converted in each condensation process, the gases can spread fairly uniformly over the whole globe. This will result in an apparently uniform distribution of the excess SO_4^{--} in rain water over the oceans and remote places, which is indicated by the data given below. A few unpublished measurements of the exhaustion of soil gases and considerations of the chemistry of the soil indicate that no SO_2 is released directly from the soil.

We think that enough data are now available to discuss the sulfur cycle on a global basis. We exclude from this discussion the component due to sea salt, because it is of little general interest. To estimate the total amount of excess SO_4^{--} brought down by rain over the entire earth, we separate land and sea. The average concentration of excess SO_4^{--} over the United States (excluding the locally disturbed values of Brownsville and Laredo), according to fig. 9, is 2.15 mg per 1. Another fairly representative figure can be derived from the Scandinavian rain analysis data (Egnér and Eriksson, 1957). For example, the concentration of excess SO_4^{--} for the year August 1956 to July 1957 for sixty stations was 2.9 mg per 1. Both the United States and the Northern European values may be a little too high if applied to the large continental areas of Asia, Africa, etc., because the former are influenced by industry to a higher degree. Only a few data are available from Russia, mostly in or near towns (Eriksson, 1952). They give an average of 2.5 mg per 1, but again they are likely to be too high to represent the whole country. Considering these values and the vast unpopulated areas of the earth, we think that 2.2 mg per 1 may be a fairly good average for the concentration of excess SO_4^{--} over land for the whole earth.

It is a little more difficult to find a corresponding figure for the oceans. Very few data are available for pure maritime locations and they are only useful with simultaneous Cl^- data to calculate the excess. In table 2 we compiled those data which could be found to satisfy these conditions. Although Gorham (1955) made his measurements in Scotland, the minimum of excess SO_4^{--} for pure maritime conditions can be derived from his data with fair reliability. We added one value from ice samples from the interior of Greenland (unpublished) which can also be considered as representative for the level of places remote from any continental sources. The average of all these samples is close to 0.5 mg per 1. The average can, of course, only be regarded as tentative, but some minimum values from continental places seem to confirm that there is a fairly defined and constant lower level of excess SO_4^{--} of the same magnitude in an undisturbed atmosphere. With these data and average figures for precipitation over land and ocean (Möller, 1951) we obtain table 3.

Estimated figures for the total amount of sulfur emitted to the atmosphere by industrial and human activities in the world are given by Magill and others (1956) and partly reproduced in table 4.

We see immediately from these figures the predominant role of the combustion of coal. Emitted sulfur has not increased very greatly in recent years because increasing fuel consumption is partly compensated by better methods of sulfur recovery in industrial exhaust gases. We think that for 1956 a total of 4×10^7 tons is a conservative estimate. This value is, of course, a kind of lower limit because it does not include the combustion of wood and other non-industrial sources.

The figure on sulfur emission into the atmosphere by combustion (crude oil and coal, table 4) can be checked in the following way: according to Revelle and Suess (1957) the average total annual production of CO_2 by combustion between 1940 and 1949 is 6.3×10^9 tons, corresponding to 1.7×10^9 tons carbon, which is approximately equivalent to the weight of the fuels. The estimated sulfur emission of fuels alone for the year 1943 is 3.0×10^7 tons. This gives an average sulfur content of the fuels of 1.76 per cent, a value which is in good agreement with data on coal and oil.

The estimated emission of SO_4^{--} into the air for 1956 is $4.0 \times 10^7 \times 96/32 = 1.2 \times 10^8$ tons. This is 33 per cent of the total amount of excess SO_4^{--} brought down by precipitation. The major uncertainty of this figure enters with the calculation of the average concentration of SO_4 in rain over land and sea, which may be off by about 30 per cent. The other figures used can be considered accurate within 10 to 20 per cent. But these inaccuracies do not alter the interesting conclusion that the sulfur emission by human activities has reached a considerable fraction of the total global sulfur budget. Since most of this sulfur is produced in the northern hemisphere, the percentage of industrial sulfur may be even higher here because of long delay times of lateral mixing across the equator.

The importance of this may be illustrated by the observation that SO_4^{--} is an important constituent of the atmospheric condensation nuclei (Junge, 1956). A possible increase of condensation nuclei may, in turn, have an effect on the rain formation processes over oceans and other remote places, where a partial lack of condensation nuclei may be assumed.

It may be of interest in this connection to give some estimation of the average residence time of sulfur in the atmosphere. The average sulfur-dioxide concentration in the atmosphere is very likely to be around 10 μg per m^3 (Junge, 1957b). Assuming that this concentration is constant throughout the tropo-

TABLE 2. Excess SO_4^{--} concentrations in rain water over sea and in places not influenced by continents.

Place	SO_4^{--} mg per l
Bermudas } United States network	0.38
Azores	1.04
Scotland (Gorham, 1955)	0.40
Greenland, inland ice	0.30
Average	0.53

TABLE 3. Total amounts of precipitation and excess SO_4^{--} per year.

Area	Precipitation tons	Aver. SO_4^{--} conc. mg per l	Amount of SO_4^{--} tons
Land	10×10^{13}	2.2	2.2×10^8
Sea	34×10^{13}	0.5	1.7×10^8
Earth	44×10^{13}	—	3.9×10^8

TABLE 4. World estimated sulfur emitted to the atmosphere per year in 10^6 tons, according to Magill and others, 1956.

Year	Zinc plants	Lead smelters	Copper smelters	Nickel smelters	Crude oil	Coal	Total
1937	1.300	0.269	4.690	0.776	4.078	23.250	34.363
1943	1.462	0.252	5.512	1.238	4.513	25.700	38.677

TABLE 5. Average world-wide residence times of some tropospheric constituents controlled by washout.

Constituent	Method	Reference	Residence times Days	Relative to water
Water	Budget considerations	E.g., Lettau 1954	10	1.0
Tropospheric radioactive debris, Nevada tests	Direct observations	Stewart, Crooks and Fisher 1955	31	3.1
Natural aerosols	Derived from measurements of natural radioactivity	Blifford, Lockhart, and Rosenstock, 1952	17	1.7
		Haxel and Schumann, 1955	6	0.6
Sea salt particles	Washout considerations	Junge and Gustafson, 1957	6	0.6
SO_2	Budget considerations		43	4.3

sphere, the total amount of SO_2 in the atmosphere would be 3.1×10^7 tons or 4.6×10^7 tons SO_4^{--}. With a total amount of annual SO_4^{--} precipitation of 3.9×10^8 tons, the average residence time is 43 days. This appears to be a little larger than that of particulate matter. We give some data for comparison in table 5. All the given residence times are very tentative. But it appears that SO_2 has a comparatively long one as compared to particulate matter. Since washout is the most important process of removal from the atmosphere, it is reasonable to express the figures in terms of the residence time of water. This value is a good indicator of the mechanisms of washout and cleaning in our atmosphere, and seems to us of special importance to characterize the removal of a particular constituent from the atmosphere. It may be called the *washout index*.

The small values in Column 5 indicate how efficiently the natural cleaning process works. Only one or a few turnovers of the water content in the atmosphere are necessary to remove all pollution. The importance of this fact for the global aspects of air pollution cannot be overemphasized.

REFERENCES

Blifford, I. H., L. B. Lockhart, Jr., and H. B. Rosenstock, 1952: On the natural radioactivity in the air. *J. Geophys. Res.*, **57**, 499–509.

Egnér, H., and E. Eriksson, 1957: Current data on the chemical composition of air and precipitation. *Tellus*, **9** (and previous volumes), 140–143.

Eriksson, E., 1952: Composition of atmospheric precipitation, Part II. *Tellus*, **4**, 280–303.

Gerhard, E. R., 1953: *The photochemical oxidation of sulphur dioxide to sulphur trioxide and its effect on fog formation.* Tech. Rep. No. 1, 1–101, Contract No. SF-9, Eng. Exper. Sta., Univ. of Illinois.

Gorham, E., 1955: On the acidity and salinity of rain. *Geochim. et Cosmochim. Acta*, **7**, 231–239.

Haxel, O., and G. Schumann, 1955: Selbstreinigung der Atmosphäre. *Z. Physik*, **142**, 127–132.

Hunter, J. G., 1950: An absorptiometric method for the determination of magnesium. *The Analyst*, **75**, 91–99.

Junge, C. E., 1956: Recent investigations in air chemistry. *Tellus*, **8**, 127–139.

—, and P. E. Gustafson, 1956: Precipitation sampling for chemical analysis. *Bull. Amer. meteor. Soc.*, **37**, 244.

—, and ——, 1957: On the distribution of sea salt over the United States and its removal by precipitation. *Tellus* **9**, 164–173.

—, and T. Ryan, 1957: *The oxidation of sulfur dioxide in dilute solutions.* Submitted to Quart. J. r. meteor. Soc.

—, 1957a: *The distribution of ammonia and nitrate in rain water over the United States.* Submitted to Trans. Amer. geophys. Union.

—, 1957b: Air chemistry. To appear in Vol. IV, *Advances in geophysics.* New York, Academic Press.

Kalle, K., 1953–1954: Zur Frage des "cyklischen salzes." *Annal. Meteor.*, **6**, 305–314.

Keily, H. J., and L. B. Rogers, 1955: Nephelometric determination of sulfate impurities in certain reagent grade salts. *Anal. Chem.*, **27**, 759–762.

Lettau, H., 1954: A study of the mass, momentum, and energy budget of the atmosphere. *Arch. Meteor. Geophys. Biokl.*, A, **7**, 133–157.

Magill, P. L., F. R. Holden, and C. Ackley, (editors), 1956: *Air pollution handbook.* New York, McGraw-Hill. 2–45.

Möller, F., 1951: Die Verdunstung als geophysikalisches problem. *Naturwissenschaftliche Rundschau*, 45–50.

Revelle, R., and H. E. Suess, 1957: Carbon dioxide exchange between atmosphere and ocean, and the question of an increase of atmospheric CO_2 during the past decades. *Tellus*, **9**, 18–27.

Rossby, C. G., and H. Egnér, 1955: On the chemical climate and its variation with the atmospheric circulation pattern. *Tellus*, **7**, 118–133.

Stewart, N. G., R. N. Crooks, and E. M. R. Fisher, 1955: *The radiological dose to persons in the United Kingdom due to debris from nuclear test explosions.* Rep. for the Meteor. Res. Pap. Comm. on the medical aspects of nuclear radiation, Atomic Energy Res. Estab., Harwell, June 1955, 1–22.

Sverdrup, H. V., M. W. Johnson, and R. H. Fleming, 1946: *The oceans.* New York, Prentice-Hall, 166.

Twomey, S., 1954: The composition of hygroscopic particles in the atmosphere. *J. Meteor.*, **11**, 334–338.

14

Copyright © 1967 by the American Chemical Society

Reprinted from *Equilibrium Concepts in Natural Water Systems*, American Chemical Society, Washington, D.C., 1967, pp. 222–242

Origin of the Chemical Compositions of Some Springs and Lakes

ROBERT M. GARRELS

Northwestern University, Evanston, Ill.

FRED T. MACKENZIE

Bermuda Biological Station for Research, St. George's West, Bermuda

The spring waters of the Sierra Nevada result from the attack of high CO_2 soil waters on typical igneous rocks and hence can be regarded as nearly ideal samples of a major water type. Their compositions are consistent with a model in which the primary rock-forming silicates are altered in a closed system to soil minerals plus a solution in steady-state equilibrium with these minerals. Isolation of Sierra waters from the solid alteration products followed by isothermal evaporation in equilibrium with the earth's atmosphere should produce a highly alkaline Na-HCO_3-CO_3 water; a soda lake with calcium carbonate, magnesium hydroxysilicate, and amorphous silica as precipitates.

As natural waters circulate in the water cycle, their compositions change continuously. Evaporation occurs, solid materials react, gases are gained or lost, differential diffusion of components occurs through permeable and semipermeable media, and organisms absorb or lose constituents. Our only hope of gaining some insight into the genesis of various types of water bodies is to study some carefully chosen systems in which relatively few processes are involved and which can be isolated sufficiently from their surroundings to make some approximate mass balances.

We have chosen to study the genesis of the spring waters of the Sierra Nevada because of the availability of a careful set of analyses of the waters plus determinations of the primary igneous rock minerals and of the soil minerals derived from them (3). As we will demonstrate, the Sierra system emerges as one in which a few primary igneous rock minerals are being attacked by soil waters high in dissolved CO_2, but otherwise nearly pure, to yield soil minerals plus spring water. The system is apparently "closed"; there is little loss or gain of H_2O or CO_2 during the interaction of soil water and primary silicates. Furthermore, the chemical composition of the igneous rocks of the Sierra is reasonably representative of rocks of the continental crust; consequently, relations in the Sierra system may have widespread application to rock-water systems important in space and time.

Weathering Relations

Feth *et al.* (*3*) carefully studied the Sierra Nevada spring waters. We will summarize their results before extending some of their interpretations.

The granitic rocks from which the springs issue range from quartz diorite to quartz microcline gneiss. Feldspars and quartz are the major minerals in the rocks, with accessory hornblende and biotite. The K-feldspar and plagioclase feldspars, although differing widely from place to place, are about equally abundant; the plagioclase ranges in composition from oligoclase (An_{26} minimum) to andesine (An_{40} maximum). Andesine is the dominant plagioclase. The dissolved content of the springs comes almost entirely from attack of CO_2-rich soil water on these silicates, especially on the plagioclase. An aluminosilicate residue, stripped of alkali and alkaline earth metals, is left behind. Traces of gibbsite are found as well as some mica and montmorillonite. Although kaolinite was identified as an important alteration product in almost all instances, the bulk of the residue apparently averages out near the composition of kaolinite. Table I gives the mean values for the compositions of ephemeral and perennial springs.

The ephemeral springs are on the average aggressive waters with a calculated CO_2 pressure of about $10^{-1.8}$ atm. (as compared with $10^{-3.5}$ atm. for ordinary air) and have reacted sufficiently with the rock minerals to use up about half of the original dissolved CO_2 picked up while the waters pass through the soil zone. The perennial springs apparently average about the same initial CO_2 pressure, but about three-fourths of the CO_2 has been changed into HCO_3^- by reaction. Figure 1 (*1*) shows that the sodium content and the pH of the ephemeral springs are consistent with the reaction of CO_2-containing water with a plagioclase feldspar to form kaolinite in a closed system.

Reconstruction of Original Minerals. To test the conclusions of Feth (*3*) concerning the weathering reaction, allow an average spring water to back-react with kaolinite, the chief weathering product, and see if the original rock minerals can be formed. The reactions and products

Table I. **Mean Values for Compositions of Ephemeral and Perennial Springs of the Sierra Nevada**[a]

	Ephemeral Springs		Perennial Springs	
	p.p.m.	molality $\times 10^4$	p.p.m.	molality $\times 10^4$
SiO_2	16.4	2.73	24.6	4.1
Al	0.03	—	0.018	—
Fe	0.03	—	0.031	—
Ca	3.11	0.78	10.4	2.6
Mg	0.70	0.29	1.70	0.71
Na	3.03	1.34	5.95	2.59
K	1.09	0.28	1.57	0.40
HCO_3	20.0	3.28	54.6	8.95
SO_4	1.00	0.10	2.38	0.25
Cl	0.50	0.14	1.06	0.30
F	0.07	—	0.09	—
NO_3	0.02	—	0.28	—
Dissolved solids	36.0		75.0	
pH	6.2[b]		6.8[b]	

[a] Ref. 3, p. 16.   [b] Median Value.

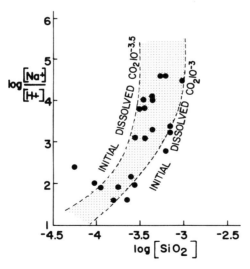

Figure 1. Logarithm of the ratio of Na^+ to H^+ in Sierra spring waters vs. logarithm of silica content. Solid circles are based on water analyses; dashed lines are theoretical compositions that should result from the attack of CO_2-bearing water on plagioclase to yield solution plus kaolinite

are shown in Table II. First the cations and anions in snow water were subtracted from the spring water solutes to determine the materials derived from the rock. Because mean compositions were used, a slight deficiency of anions resulted; this was corrected by giving HCO_3^- a slightly higher value than that realized by subtracting the mean concentration of HCO_3^- in snow from that of the springs. Otherwise no corrections were made. Then enough Na^+, Ca^{2+}, HCO_3^-, and SiO_2 reacted with kaolinite to make plagioclase, using up all the Na^+ and Ca^{2+}. The resultant calculated plagioclase does in fact have a composition similar to that found in the rocks.

Next all Mg^{2+} and enough K^+, HCO_3^-, and SiO_2 reacted with kaolinite to make biotite, leaving a small residue of K^+, HCO_3^-, and SiO_2 to form K-feldspar. Some silica remained—about 4% of the original concentration. Thus, the mass balance is probably within the limits of error of the original values of concentration used for the median composition of spring waters.

The reactions work out too well to leave much doubt that we are indeed dealing with a closed system reacting with CO_2 and that the weathering product is kaolinite or a material with a composition close to kaolinite in terms of the Al to Si ratio and the balance of alkali or alkaline earth cations.

Some Quantitative Aspects of the Weathering Process. The success of reconstructing the original rock minerals leads to several important conclusions.

First, there is at most a slight excess of silica over that needed to form the original silicates. Thus, the conclusion of Feth *et al.* (3) is re-emphasized—*i.e.*, the silica in the water came from the breakdown of the silicates and an insignificant amount from direct solution of quartz. The silica represents CO_2, changed into HCO_3^- by the weathering process.

Second, Feth *et al.* (3) observe that the waters gain much of their silica in a few feet of travel, showing that it is the action of the high CO_2 water that produces "kaolinite." The rock minerals react, forming "kaolinite" continuously in the system, and the "kaolinite" controls the water composition by its presence. If the aluminum analyses were not so low, and hence analytically suspect, an attempt could be made to calculate an equilibrium constant for the substance formed. All that can be said at the moment is that the values of SiO_2 and Al concentrations and of pH are reasonable for those controlled by an aluminosilicate of the approximate stability of kaolinite.

Third, Table II illustrates that about 80% of the rock-derived dissolved constituents in the ephemeral springs can be accounted for by the breakdown of plagioclase alone. Even though K-feldspar is abundant in the rocks, little breakdown occurs; the high Na to K ratio in these waters apparently is related to the differential rate of weathering of the feldspars rather than to K^+ adsorption after release upon clay minerals, as is often assumed. In the Sierra, the residual material left after initial attack of the aggressive waters on the rock minerals should be chiefly a mixture of quartz, K-feldspar, and kaolinite, and perhaps small amounts of aluminum oxide hydrates.

Some insight into what happens after the original strong attack on the silicates can be gained by subtracting the ephemeral spring analyses from those of the perennial springs. The perennial springs circulate deeper and have higher pH values when they encounter the rocks. Table III shows the "pick up" of constituents by continued circulation.

It is immediately apparent that the constituents added by deeper circulation are different from those derived by initial attack in some important ways. The ratio of SiO_2 to Na^+ in the increment is nearly 1 to 1. Because weathering of plagioclase to kaolinite releases dissolved SiO_2 and Na^+ in a ratio of 2 to 1 (see Table II) and plagioclase is the only reasonable source of Na^+, a solid other than kaolinite is being produced. Table IV is an attempt to deduce the reactions that occur during deeper

Table II. Source Minerals
Ephemeral Springs

Reaction (coefficients $\times 10^4$)	Na^+	Ca^{2+}
Initial concentrations in spring water	1.34	0.78
Minus concentrations in snow water	1.10	0.68
Change kaolinite back into plagioclase Kaolinite $1.23\ Al_2Si_2O_5(OH)_4 + 1.10\ Na^+ + 0.68\ Ca^{2+}$ $+ 2.44\ HCO_3^- + 2.20\ SiO_2 =$ Plagioclase $1.77\ Na_{0.62}Ca_{0.38}Al_{1.38}Si_{2.62}O_8 + 2.44\ CO_2 +$ $3.67\ H_2O$	0.00	0.00
Change kaolinite back into biotite Kaolinite $0.037\ Al_2Si_2O_5(OH)_4 + 0.073\ K^+ + 0.22\ Mg^{2+}$ $+ 0.15\ SiO_2 + 0.51\ HCO_3^- = 0.073$ Biotite $KMg_3AlSi_3O_{10}(OH)_2 + 0.51\ CO_2 + 0.26\ H_2O$	0.00	0.00
Change kaolinite back into K-feldspar $0.065\ Al_2Si_2O_5(OH)_4 + 0.13\ K^+ + 0.13\ HCO_3^-$ $+ 0.26\ SiO_2 =$ K-feldspar $0.13\ KAlSi_3O_8 + 0.13\ CO_2 + 0.195\ H_2O$	0.00	0.00

circulation. With the thought that Cl⁻ is derived from NaCl, and SO_4^{2-} from $CaSO_4$, enough Na^+ and Ca^{2+} are removed to balance these anions. This assumption is weak, but fortunately the concentrations of SO_4^{2-} and Cl⁻ are so low that this step is not important.

Next it seems reasonable to use up Mg^{2+} and K^+ along with kaolinite to make biotite, inasmuch as the ratio of K^+ to Mg^{2+} in the waters is nearly that of a K-Mg biotite. This leaves Na^+ equal to SiO_2, so the remaining SiO_2 was apportioned between the reconstruction of kaolinite to plagioclase and that of montmorillonite to plagioclase. The montmorillonite composition chosen was that of a calcium beidellite of average cation exchange capacity. Beidellites seem to be prevalent under weathering conditions, and at such low concentrations of Na^+ and K^+ it is almost certain that the exchange positions are primarily occupied by Ca^{2+}.

After these steps, a considerable amount of Ca^{2+} and HCO_3^- remain, and we believe that with deep circulation, the waters which have lost much of their original aggressiveness, pick up Ca^{2+} (and perhaps some Mg^{2+}) from minor amounts of carbonates encountered en route.

for Sierra Nevada Springs

(concentrations moles/liter × 10⁴)

Mg^{2+}	K^+	HCO_3^-	SO_4^{2-}	Cl^-	SiO_2	Products moles/liter × 10⁴
0.29	0.28	3.28	0.10	0.14	2.73	
0.22	0.20	3.10	—	—	2.70	Derived from rock
	minus plagioclase					
0.22	0.20	0.64	0.00	0.00	0.50	1.77 $Na_{0.62}Ca_{0.38}$ feldspar
	minus biotite					
0.00	0.13	0.13	0.00	0.00	0.35	0.073 biotite
	minus K-feldspar					
0.00	0.00	0.00	0.00	0.00	0.12	0.13 K-feldspar

Table III. Effects of Deeper Circulation on Spring Waters[a]

	SiO_2	Ca^{2+}	Mg^{2+}	Na^+	K^+	HCO_3^-	SO_4^{2-}	Cl^-
Perennial Spring	4.10	2.60	0.71	2.59	0.40	8.95	0.25	0.30
Ephemeral Springs	2.73	0.78	0.29	1.34	0.28	3.28	0.10	0.14
Increment	1.37	1.82	0.42	1.25	0.12	5.67	0.15	0.16

[a] Analyses are mean values in moles/liter $\times 10^4$.

The over-all picture of what happens to the soil waters, as illustrated by Tables II and IV, is that initially they rapidly attack the rocks, kaolinizing chiefly plagioclase plus biotite and K-spar. As they penetrate more deeply, the reaction rate slows down, and both kaolinite and montmorillonite are weathering products. Also, an important part of the Ca^{2+} comes from solution of small amounts of carbonate minerals.

Before attempting to document the formation of montmorillonite, a few remarks on weathering rates are of interest. Roughly one-half the initial CO_2 seems to be expended in the soil zone, and most of the rest alters rock minerals at deeper levels. The total reacting capacity is about

Table IV. Source Minerals and Concentrations in

Reaction (coefficients $\times 10^4$)	Na^+	Ca^{2+}
Initial concentrations (perennial minus ephemeral)	1.25	1.82
Remove $Ca^{2+} = SO_4^{2-}$, and $Na^+ = Cl^-$	1.09	1.67
Adjust HCO_3^- = total electrical charge of cations	1.09	1.67
Change kaolinite back into biotite 0.07 $Al_2Si_2O_5(OH)_4$ + 0.42 Mg^{2+} + 0.14 K^+ + 0.28 SiO_2 + 0.98 HCO_3^- = 0.14 $KMg_3AlSi_3O_{10}(OH)_2$ + 0.98 CO_2 + 0.49 H_2O	1.09	1.67
Change kaolinite back into plagioclase 0.26 $Al_2Si_2O_5(OH)_4$ + 0.235 Na^+ + 0.144 Ca^{2+} + 0.47 SiO_2 + 0.52 HCO_3^- = 0.38 $Na_{0.62}Ca_{0.38}Al_{1.38}Si_{2.62}O_8$ + 0.52 CO_2 + 0.78 H_2O	0.85	1.53
Change montmorillonite back into plagioclase 0.81 $Ca_{0.17}Al_{2.33}Si_{3.67}O_{10}(OH)_2$ + 0.85 Na^+ + 0.38 Ca^{2+} + 0.61 SiO_2 + 1.62 HCO_3^- = 1.37 $Na_{0.62}Ca_{0.38}Al_{1.38}Si_{2.62}O_8$ + 1.62 CO_2 + 1.62 H_2O	0.00	1.15
Precipitate $CaCO_3$ 1.15 Ca^{2+} + 2.30 HCO_3^- = 1.15 $CaCO_3$ + 1.15 CO_2 + 1.15 H_2O	0.00	0.00

9×10^{-4} equivalents per liter and corresponds to the destruction of about 3.5×10^{-4} moles of plagioclase, 0.2×10^{-4} moles of biotite, and about 0.2×10^{-4} moles of K-spar. The annual precipitation in this part of the Sierra averages about 100 cm./year. Therefore, the rate of chemical weathering is about 3.6×10^{-5} moles/year/sq. cm. If the rock consists of one-third plagioclase by volume, the rock should be disintegrated to an average depth of one meter in about 9000 years, and the residue would be chiefly a rubble of quartz, K-feldspar, and kaolinite.

Phase Control of Water Composition. The conclusion that kaolinite forms quickly and continuously during weathering, plus the strong suggestion that the deeper circulating waters are forming montmorillonite as well, can be tested qualitatively by predicting the compositional genesis of the waters, especially in terms of Ca^{2+} and Na^+. If one assumes that the CO_2-bearing waters continuously react with plagioclase to form first "kaolinite" and then "montmorillonite" and that these phases maintain equilibrium with the waters as they are continuously fed from the feldspar, then observed compositions can be compared with those based on these assumptions.

Weathering Products of Deeper Circulation

moles/liter $\times 10^4$

Mg^{2+}	K^+	HCO_3^-	SO_4^{2-}	Cl^-	SiO_2	Mineral altered and product (moles/liter $\times 10^4$)
0.42	0.12	5.67	0.15	0.16	1.37	
0.42	0.12	5.67	0.00	0.00	1.37	
0.42	0.12	5.39	0.00	0.00	1.37	
0.00	0.02	4.41	0.00	0.00	1.09	0.14 biotite 0.07 kaolinite
0.00	0.00	3.89	0.00	0.00	0.62	0.38 plagioclase 0.26 kaolinite
0.00	0.00	2.27	0.00	0.00	0.01	1.37 plagioclase 0.81 montmorillonite
0.00	0.00	0.03	0.00	0.00	0.01	1.15 calcite

For the reaction of montmorillonite to kaolinite,

$$2H^+ + 3Ca_{0.33}Al_{4.67}Si_{7.33}O_{20}(OH)_4 + 7H_2O =$$
$$7Al_2Si_2O_5(OH)_4 + 8SiO_{2aq} + Ca^{2+}$$

the equilibrium constant is

$$K = \frac{[Ca^{2+}][SiO_2]^8}{[H^+]^2}$$

Thus, for a montmorillonite of the average composition shown and neglecting changes in the activity of water, a water undersaturated with respect to montmorillonite but saturated with respect to kaolinite should have a value for the quotient less than K, and all waters saturated with respect to both phases should have a constant value of the quotient. Figure 2 shows a plot of the compositions of Sierra springs, in which the

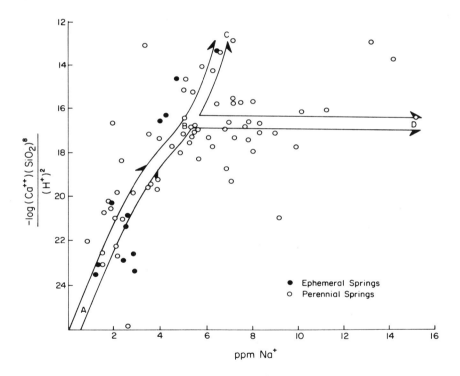

Figure 2. pK' for equilibrium between Ca-beidellite and kaolinite plotted as a function of Na content of Sierra Nevada spring waters. Arrow ABC is the path of water evolution calculated for the reaction from $Na_{0.62}Ca_{0.38}$ plagioclase to kaolinite in a closed system with an initial dissolved CO_2 of 0.0006 moles per liter. Arrow ABD is the expected path if evolution is also controlled by the two-phase equilibrium kaolinite-montmorillonite

logarithm of the quotient $\dfrac{m_{Ca^{2+}} m^8_{SiO_2}}{m^2_{H^+}}$ (pK') is plotted against p.p.m. Na$^+$. The arrow ABC is calculated assuming that only kaolinite results from decomposition of plagioclase, whereas the arrow ABD is the predicted evolution if kaolinite is formed first, but that montmorillonite also forms when the waters gain sufficient SiO$_2$, Ca^{2+}, and a high enough pH. The position of the arrow for kaolinite-montmorillonite equilibrium has been estimated from the boundary on Figure 3 and the exchange constants for Na-Ca montmorillonites. Despite the scatter, it is clear that the ephemeral springs alter feldspar chiefly to kaolinite and that many of the perennial springs, although their evolution is to kaolinite, have compositions that suggest a halt in that path of evolution and possible control by equilibration with both phases.

If this is true, and it must be regarded as a highly tentative conclusion, then the upper limits of silica content of many natural waters, which are far less than saturation with amorphous silica ($\cong$ 115 p.p.m.), may well be controlled by equilibrium between the waters and various silicate phases. This does not mean that the controlling solids are well-crystallized, clearly distinguishable substances, but there is definitely an interplay between the waters and solid aluminosilicates. Furthermore, the aluminosilicates apparently differ from each other in important compositional steps, and are not simply continuous gradations resulting from progressive adsorption and alteration as water compositions change. This generalization from the Sierra studies is more strikingly shown by Bricker and Garrels (1). Figure 3 is adapted from their work and shows that the compositions of dilute ground and surface waters in silicate-bearing rocks are contained within a set of phase boundaries derived from equilibrium relations among silicate phases.

An idealized water, derived from the attack of CO$_2$-bearing water on a typical felsic rock to produce only kaolinite, should have the following characteristics, expressed in terms of molar concentrations:

(1) HCO$_3^-$ should be the only anion, except for small concentrations of Cl$^-$ and SO$_4^{2-}$ from fluid inclusions in the minerals, oxidation of pyrite, and other minor sources.

(2) Na$^+$ and Ca^{2+} should be the chief cations, and the ratio of Na$^+$ to Ca^{2+} should be the same as that in the plagioclase of the rock.

(3) The total of Mg^{2+} and K$^+$ should be less than about 20% of the total of Na$^+$ and Ca^{2+}. The ratio of Mg^{2+} to K$^+$ should range around 1 to 1, with higher values related to higher percentages of mafic minerals, and lower values to higher ratios of K-feldspar to mafic minerals.

(4) The ratio of SiO$_2$ to Na$^+$ should be about 2 to 1, with somewhat higher ratios from rocks unusually high in K-spar and/or mafic minerals.

(5) The ratio of Na$^+$ to K$^+$ should be 5 to 1 or higher because of the abundance and high weathering rate of plagioclase, as opposed to the

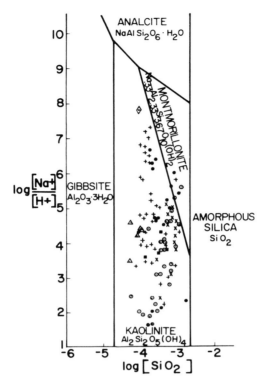

Figure 3. Stability fields of some minerals in the Na_2O-Al_2O_3-SiO_2-H_2O system at 25°C. as a function of Na^+, H^+, and dissolved silica. Points are from ground water analyses of siliceous rocks

⊙ = granite, rhyolite　■ = andesite, diorite
△ = Jamaican bauxites　× = basalt, gabbro
◇ = sea water　● = siltstone, clay, shale
\+ = sandstone, arkose, graywacke

low weathering rate of K-feldspar and the generally low abundance of micas or other K-bearing phases.

Table V compares the Sierra waters with some waters from other felsic rocks. The fundamental similarities are clearly apparent as well as the minor variations that show the imprint of the details of rock mineralogy and small additions of constituents from various other sources.

Evaporative Concentration of Sierra Waters

The alteration of the igneous rocks of the Sierra was treated as an example of the genesis of waters in an essentially closed system. There

Table V. Compositions of Waters from Various Igneous Rocks (mole %)

	Sierra Ephemeral Springs[a]	Sierra Perennial Springs[a]	Springs from Sierra Volcanic Rocks[a]	Ground Water from Granite, R.I.[b]	Ground Water from Rhyolite, N. M.[b]
SiO_2	33.8	21.0	19.5	19.9	38.6
Ca^{2+}	8.5	13.0	10.3	9.6	4.6
Na^+	13.8	13.0	10.3	15.6	20.0
Mg^{2+}	2.8	3.5	8.0	6.6	2.5
K^+	2.5	2.0	1.4	1.2	1.3
HCO_3^-	38.7	45.0	48.5	38.0	28.0
Cl^-	—	1.5	1.4	8.4	2.5
SO_4^{2-}	—	1.5	0.2	0.6	0.8
pH	6.2	6.8	7.0	7.6	7.2

[a] Data from Ref. 3. [b] Data from Ref. 8.

was no indication from an analysis of the processes involved that a "final" equilibrium had been attained. Although the rates of altering the primary silicates diminished as CO_2 was used up and the concentration of dissolved solids increased, primary rock-forming silicates definitely are unstable in any waters produced. It would be interesting to attempt to deduce the final composition of waters and solids if a given quantity of CO_2-charged water were permitted to react indefinitely with a felsic rock, but the real situation seems to be that the waters emerge from the closed system when they have reached compositions similar to those of the Sierra and begin to lose and gain constituents from other sources.

Perhaps the next step in trying to understand the complexities of the genesis of natural waters is to see what might happen if the Sierra springs were isolated from the parent rock and solid alteration products and permitted to evaporate isothermally in equilibrium with the CO_2 of the earth's present atmosphere. In terms of a natural situation, this process might resemble the fate of waters discharged to the east of the Sierra and evaporated in a playa of the California desert. It is, of course, impossible to have a real situation without adding reactive solids from some source, but such interference could be treated afterwards in terms of additional variables. Also, this particular situation might shed some light on the effects of the "igneous rock component" of natural waters during the evaporative processes that change stream waters into oceans.

Table VI gives the mean composition of the perennial springs of the Sierras. We have chosen to ignore Fe, Al, NO_3, and F and restricted ourselves to the following analysis (Table VI).

Calculations and Procedures. The general procedure we have used is to calculate the effects of concentrating the waters by various factors

Table VI. Composition of Sierra Spring Water Used in Evaporation Study (3)

	p.p.m.	moles/liter $\times 10^4$
SiO_2	24.6	4.10
Ca^{2+}	10.4	2.60
Mg^{2+}	1.70	0.71
Na^+	5.95	2.59
K^+	1.57	0.40
HCO_3^-	54.6	8.95
SO_4^{2-}	2.38	0.25
Cl^-	1.06	0.16 [a]

pH (median) = 6.8; ionic strength = 0.0013

[a] Cl^- has been diminished by 0.14 from the calculated 0.30 derived from 1.06 p.p.m. given in the analysis to correct initial electrical imbalance between anions and cations. This change is important in the compositions calculated but not in the general pattern of change to be presented.

up to 1000-fold, determining the stages at which various solids should precipitate from the system, and determining the effects of removing solids on composition and pH.

The conditions imposed are that the water remains in equilibrium with a CO_2 pressure of $10^{-3.5}$ atm., that the temperature remains constant at 25°C., and that pure water (except for a little CO_2) is continuously removed from the system. It is further assumed that any solids formed remain in equilibrium. As it turns out, the question as to whether precipitated solids are isolated after formation or continue to react is not important here, although it is important in many comparable systems.

To keep track of the minerals that might form, we tried to assess the possible combinations of dissolved species to form solids. Table VII

Table VII. Solids Considered as Possible Precipitates in Concentrated Spring Water

Species	Solid	Equilibrium Constant
Na^{2+}	None	None
K^+	None	None
Ca^{2+}		
gypsum	$CaSO_4 \cdot 2H_2O$	$a_{Ca^{2+}} a_{SO_4^{2-}} = 10^{-4.62}$
calcite	$CaCO_3$	$a_{Ca^{2+}} a_{CO_3^{2-}} = 10^{-8.35}$
Mg^{2+}		
brucite	$Mg(OH)_2$	$a_{Mg^{2+}} a^2_{OH^-} = 10^{-11.15}$
magnesite	$MgCO_3$	$a_{Mg^{2+}} a_{CO_3^{2-}} = 10^{-8}$
hydromagnesite	$Mg_4(CO_3)_3(OH)_2$	$a^4_{Mg^{2+}} a^3_{CO_3^{2-}} a^2_{OH^-} = 10^{-34.9}$
sepiolite	$MgSi_3O_6(OH)_2$	$a_{Mg^{2+}} a^3_{SiO_2} a^2_{OH^-} = 10^{-24}$
SiO_2		
amorphous silica	$SiO_2 \cdot 2H_2O$	$a_{SiO_2} = 10^{-2.7}$

shows the species we considered and some data concerning them. The values for the equilibrium constants, except for sepiolite, have been calculated from the free energy values given by Garrels and Christ (4). The value for sepiolite is discussed in the text.

A few notes on our selections and on some of the other mineral possibilities may help to clarify the use of these particular species. Gypsum seems to be the most likely sulfate phase. It is stable with respect to anhydrite, is a common precipitate from low temperature natural waters, and except in highly concentrated brines, double sulfates would not be expected. Calcite was chosen in preference to aragonite or possibly a whole variety of magnesian calcites because it is the stable phase and commonly precipitates directly from dilute solution. Also, the difference in solubility between calcite and aragonite is not great enough to influence the general picture. Dolomite is certainly a possibility in the system, as is the metastable mineral huntite ($CaMg_3(CO_3)_4$), but even at $1000\times$ concentration of these waters the conditions apparently necessary in nature to precipitate dolomite (evaporation of sea water to a dense brine, gypsum precipitation, and $\frac{Mg^{2+}}{Ca^{2+}}$ ratio ≥ 30 (2) are not approached). The low Al (Table I) precludes more than traces of aluminosilicates, and Mg is the only cation likely to make a silicate species. Hostetler (5), after reviewing the occurrence of the Mg-silicates, concluded that sepiolite tends to form in most natural solutions at room temperature in preference to serpentine or talc, especially in a high silica environment, and he cites many occurrences of sepiolite plus magnesite and amorphous silica. Furthermore, Siffert (7) synthesized sepiolite within a day at room temperature at pH values of about 8.8 in high SiO_2/MgO solutions. Our greatest difficulty was in assigning an equilibrium constant for sepiolite. From Siffert's work and using the formula in Table VII, we get a K value of about 10^{-22} for the reaction:

$$3H_4SiO_4 + Mg^{2+} + 2OH^- = MgSi_3O_6(OH)_2 + 6H_2O$$

On the other hand, we precipitated a magnesium silicate from sea water by adding sodium metasilicate that gave an approximate K of 10^{-24} after aging. Siffert's material was freshly precipitated and could be expected to age significantly, like other poorly crystalline silicates. Consequently, the value chosen is open to considerable doubt.

Some other phases considered were nesquehonite ($MgCO_3 \cdot 3H_2O$), artinite ($Mg_2CO_3(OH)_2 \cdot 2H_2O$), and lansfordite ($MgCO_3 \cdot 5H_2O$), but all are rare and according to Hostetler (5) are not stable.

Our procedure was to follow the changes in concentrations and ionic strength as the water is evaporated, and by correcting ion molalities by activity coefficients, keep track of the ion activity products of the various

Table VIII. Calculated Ion Activity Products of Various Concentration

	Calcite $CaCO_3$	Gypsum $CaSO_4 \cdot 2H_2O$	Brucite $Mg(OH)_2$
Initial water in equilibrium with atmosphere	8.8	8.3	15.7
Concentrated by a factor of			
1.25	Saturated	8.1	15.5
2.0	Saturated	8.0	15.1
5.0	Saturated	7.7	15.9
10.0	Saturated	7.9	15.9
100.0	Saturated	8.3	15.9
1000.0	Saturated	8.8	15.4
Equilibrium Constant	8.35	4.62	11.15

^a Products are expressed as negative logarithms (pK').

solids as shown in Table VIII. If an ion activity product reached the equilibrium constant value, we then maintained equilibrium with that solid during continuing concentration. Sample calculations show the procedure used.

Sample Calculations. The first step was to determine if the water analysis represents a solution in equilibrium with the atmosphere. For the reaction

$$CO_2 + H_2O = HCO_3^- + H^+$$

the equilibrium constant is

$$\frac{a_{HCO_3^-} \, a_{H^+}}{a_{CO_2} a_{H_2O}} = K_{CO_2} = 10^{-7.82} \qquad (1)$$

Rearranging, and writing in terms of activity coefficients and molalities and assuming the pH is a measure of a_{H^+}

$$a_{CO_2} = \frac{\gamma_{HCO_3^-} m_{HCO_3^-} a_{H^+}}{K_{CO_2} a_{H_2O}} \qquad (2)$$

In all solutions considered, the activity of water is greater than 0.99 and was considered unity throughout. The ionic strength (I) of the original water is only 0.0013 and that of the final water 0.408, so that individual ion activity coefficients calculated from Debye-Hückel equations (4) were used throughout to estimate ion activities. The calculated a_{CO_2}, or P_{CO_2}, using the median pH of 6.8 is shown in Equation 3.

Possible Solid Compounds in Sierra Water as a Function of by Evaporation

Magnesite $MgCO_3$	Hydromagnesite $Mg_4(CO_3)_3(OH)_2$	Sepiolite $Mg(SiO_2)_3(OH)_2$	Silica Gel SiO_2
9.4	43.6	25.9	3.4
8.8	42.5	25.3	3.3
8.7	40.2	24.3	3.1
9.6	44.6	Saturated	3.0
9.6	48.9	Saturated	Saturated
9.6	44.0	Saturated	Saturated
9.0	42.5	Saturated	Saturated
8.0	34.9	24.0	2.7

$$P_{CO_2} = \frac{0.97 \times 10^{-3.04} 10^{-6.8}}{10^{-7.82}} = 10^{-2.0} \text{ atm.} \qquad (3)$$

The CO_2 pressure of the water is considerably above the atmospheric value of $10^{-3.5}$ and therefore the initial water is not in equilibrium with the atmosphere.

The next step, which is representative of all pH calculations, is to find the pH and the accompanying changes in dissolved carbonate species when the water comes to equilibrium with the atmosphere with respect to P_{CO_2}. As CO_2 is lost, the pH will rise.

The equation for electrical neutrality is:

$$2m_{Ca^{2+}} + 2m_{Mg^{2+}} + m_{Na^+} + m_{K^+} + m_{H^+} = \\ 2m_{SO_4^{2-}} + m_{Cl^-} + m_{HCO_3^-} + 2m_{CO_3^{2-}} + m_{OH^-} \qquad (4)$$

Rewriting to put pH-dependent species on the right

$$2m_{Ca^{2+}} + 2m_{Mg^{2+}} + m_{Na^+} + m_{K^+} - 2m_{SO_4^{2-}} - m_{Cl^-} = \\ m_{HCO_3^-} + 2m_{CO_3^{2-}} + m_{OH^-} - m_{H^+} \qquad (5)$$

The terms on the right of Equation 5 can be expressed in terms of pH, equilibrium constants, P_{CO_2}, and ionic activity coefficients; the terms on the left are known from the water analysis (Table VI). Rearranging Equation 2 gives Equation 6.

$$m_{HCO_3^-} = \frac{K_{CO_2}P_{CO_2}}{\gamma_{HCO_3^-}a_{H^+}} = \frac{10^{-7.82}10^{-3.5}}{\gamma_{HCO_3^-}a_{H^+}} = \frac{10^{-11.32}}{\gamma_{HCO_3^-}a_{H^+}} \quad (6)$$

The parallel equation for CO_3^{2-} is

$$m_{CO_3^=} = \frac{K_{HCO_3}K_{CO_2}P_{CO_2}}{\gamma_{CO_3^{2-}}a_{H^+}^2} = \frac{10^{-10.33}10^{-7.82}10^{-3.50}}{\gamma_{CO_3^{2-}}a_{H^+}^2} = \frac{10^{-21.65}}{\gamma_{CO_3^{2-}}a_{H^+}^2} \quad (7)$$

and for OH^-

$$m_{OH^-} = \frac{K_{H_2O}}{\gamma_{OH^-}a_{H^+}} = \frac{10^{-14}}{\gamma_{OH^-}a_{H^+}} \quad (8)$$

Substituting analytical values of molality for the terms on the left of Equation 5, Debye-Hückel activity coefficients in the right-hand terms, and neglecting m_{H^+}

$$0.00052 + 0.000142 + 0.00026 + 0.00004 - 0.000050 - 0.000016 =$$

$$\frac{10^{-11.32}}{0.97a_{H^+}} + \frac{2(10^{-21.65})}{0.86a_{H^+}^2} + \frac{10^{-14}}{0.97a_{H^+}}$$

collecting terms,

$$0.000896 = \frac{10^{-11.32}}{a_{H^+}} + \frac{10^{-21.28}}{a_{H^+}^2} + \frac{10^{-14.0}}{a_{H^+}}$$

solving for a_{H^+},

$$a_{H^+} = 10^{-8.26}; \text{pH} = 8.26.$$

From this value of pH, the values for $m_{HCO_3^-}$, $m_{CO_3^{2-}}$, and m_{OH^-} are obtained from Equations 6, 7, and 8, and the total water analysis can be determined. The only important change is in pH, which rises from 6.8 to 8.26; the change is accompanied by a loss in dissolved CO_2 and a slight increase in CO_3^{2-}—actually not enough to be worth showing in the analysis.

The next step is to determine whether or not the water as a result of the pH change has become saturated with respect to any of the solid phases considered (Table VII). For this, ion activity products are computed for each phase and compared with the equilibrium value (Table VIII). The calculation for calcite is given to illustrate the procedure. The ion activity product is

$$\gamma_{Ca^{2+}}m_{Ca^{2+}}\gamma_{CO_3^{2-}}m_{CO_3^{2-}} = K'$$

The value of $\gamma_{CO_3^{2-}}m_{CO_3^{2-}}$ is obtained from the pH and Equation 7:

$$\gamma_{CO_3^{2-}}m_{CO_3^{2-}} = \frac{10^{-21.65}}{(10^{-8.26})^2} = 10^{-5.13}$$

The Debye-Hückel equation yields 0.88 for $\gamma_{Ca^{2+}}$, and $m_{Ca^{2+}}$ is 2.60×10^{-4} or $10^{-3.59}$ (Table VI). Thus K' is $(0.88)(10^{-3.59})(10^{-5.13}) = 10^{-8.78}$.

Consequently, because the equilibrium value of the ion activity product is $10^{-6.35}$, the water is slightly undersaturated with respect to calcite, its Ca^{2+} content is fixed by CO_2 pressure and pH; in other words, because the product of $a_{Ca^{2+}}$ and $a_{CO_3^{2-}}$ is a constant, $a_{Ca^{2+}}$ and hence $m_{Ca^{2+}}$ can be expressed as a negative term on the right side of the electrical balance equation. After sepiolite precipitates, it can be handled similarly. The

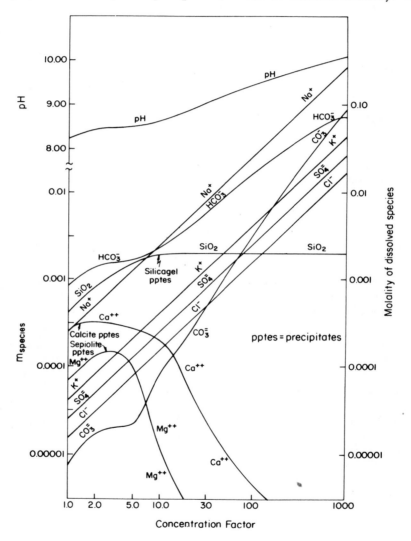

Figure 4. Calculated results of evaporation of typical Sierra Nevada spring water at constant temperature in equilibrium with atmospheric CO_2

analytical expression, after the water becomes saturated with calcite, sepiolite, and silica, is

$$m_{Na^+} + m_{K^+} - 2m_{SO_4^{2-}} - m_{Cl^-} = m_{HCO_3^-} + 2m_{CO_3^{2-}}$$
$$+ m_{OH^-} - 2m_{Ca^{2+}} - 2m_{Mg^{2+}},$$

or

$$m_{Na^+} + m_{K^+} - 2m_{SO_4^{2-}} - m_{Cl^-}$$
$$= \frac{10^{-11.32}}{\gamma_{HCO_3^-} a_{H^+}} + \frac{2(10^{-21.65})}{\gamma_{CO_3^{2-}} a_{H^+}^2} + \frac{10^{-14}}{\gamma_{OH^-} a_{H^+}} - \frac{2(10^{13.3} a_{H^+}^2)}{\gamma_{Ca^{2+}}} - \frac{2(10^{12.1} a_{H^+}^2)}{\gamma_{Mg^{2+}}}$$

In calculating for a given degree of water concentration, the ionic strength is estimated, and tentative activity coefficients for the ions are obtained. After an approximate pH is obtained and values can be assigned for the molalities of the pH-dependent species, new values for the ionic strength and activity coefficients are obtained, and the electrical balance equation is solved again. One such iteration usually suffices to provide satisfactory results. Using the Debye-Hückel activity coefficients obviously leads to an increasing uncertainty in γ_i values with increasing ionic strength, and the values calculated for 1000× concentration should be regarded as rough approximations.

Discussion. A summary of the calculated changes in water composition during evaporation is shown in Figure 4. In essence, evaporation changes the water from a nearly neutral Na-Ca-HCO$_3$ water to a highly alkaline Na-HCO$_3$-CO$_3$ water. Calcium and magnesium are removed by early precipitation of calcite and sepiolite, and their concentrations diminish to small values as the values of CO$_3^{2-}$ and OH$^-$ rise. Because of the initial high silica, there is enough silica to permit the water to precipitate silica gel at about 10× concentration. Na$^+$, K$^+$, Cl$^-$, and SO$_4^{2-}$ concentrate without forming solids.

Table IX. Analyses of Various

	Na^+	K^+	Ca^{2+}
Sierra 100 ×	595	159	0.2
Lower Alkali Lake, Eagleville, Calif.[a]	370	11	6.9
Amargosa River, Nev.[a]	423	17	2.0
Kurusch Gol, Iran[a]	730	29	11
Keene Wonder Spring, Calif.[b]	1040	25	23
Sierra 1000 ×	5950	1590	0
Soap Lake, Wash.[a]	12,500	12,500	3.9
Lenore Lake, Wash.[a]	5360	5360	3.0

[a] Analyses from Ref. 6, Chapter G.
[b] Analyses from Ref. 8, Chapter F.

Buffering of pH during the early heavier precipitation of calcite and sepiolite is clear and is reflected in a near constancy of HCO_3^- and CO_3^{2-}. However, after Ca^{2+} and Mg^{2+} are substantially reduced, the pH again rises with further concentration. Because the Sierra waters are so low in sulfate, gypsum does not precipitate; abstraction of Ca^{2+} as calcite never permits the solubility product of gypsum to be exceeded.

The 100× and 1000× waters resemble some natural waters. Table IX compares, in p.p.m., the hypothetical Sierra waters with several natural water bodies.

Summary of Genesis and Evaporation of Sierra Waters

The preceding study of the origin of the composition of spring waters in felsic rocks and the results of evaporating the waters in the absence of the solid weathering residues show that the processes are highly asymmetric. The weathering study showed that dissolved silica comes chiefly from the incongruent solution of silicate minerals rather than from congruent solution of quartz, and as a result the univalent and bivalent cations in solution derived from these silicate minerals are balanced by HCO_3^-. During evaporation in equilibrium with the atmosphere, the bivalent ions tend to form insoluble carbonates or hydroxysilicates, and the pH of the evaporating solution remains relatively constant until the silicates have been largely removed. The univalent ions, on the other hand, do not form similar compounds of low solubility, and concentrate continuously with a concomitant increase in HCO_3^-, CO_3^{2-}, and pH. Therefore, waters maintained in equilibrium with the atmosphere and separated from decomposition products, with univalent ions originally derived from incongruent solution of silicates will inevitably become highly alkaline if concentrated greatly.

Saline Waters, p.p.m.

Mg^{2+}	HCO_3^-	CO_3^{2-}	SO_4^{2-}	Cl^-	SiO_2	pH
0	990	212	240	56	120	9.45
0.9	1200	1200	307	1160	63	?
0.2	639	639	257	109	70	?
25	1200	1200	306	342	?	?
38	1070	29	796	567	57	8.4
0	4250	4500	2400	560	120	10.02
23	11,270	5130	6020	4680	101	?
20	6090	3020	2180	1360	22	?

It is obvious that the silica present in these waters, if denied the opportunity to react with aluminous minerals, must precipitate as amorphous silica or as silicates of the bivalent cations, as illustrated by sepiolite in the example chosen. In nature the silica may be removed by organisms, such as diatoms, but they do not change the gross chemistry.

Acknowledgments

We have benefitted greatly from discussions with many colleagues in many places. We especially want to thank Roland Wollast of the University of Brussels, Harold Helgeson of Northwestern University, Charles Christ of the United States Geological Survey, and Owen Bricker of The Johns Hopkins University for reading the manuscript and for offering many useful suggestions and criticisms.

Literature Cited

(1) Bricker, O. P., Garrels, R. M., *Proc. Ann. Rudolphs Conf.*, 4th, 1965, Rutgers Univ., New Brunswick, 1967.
(2) Deffeyes, D. K., Lucia, F. J., Weyl, P. K., *Science* **143**, 678 (1963).
(3) Feth, J. H., Roberson, C. E., Polzer, W. L., *U.S. Geol. Surv. Water Supply Paper* **1535-I**, 170 (1964).
(4) Garrels, R. M., Christ, C. L., "Solutions, Minerals, and Equilibria," Appendix 2, p. 450, Harper and Row, New York, 1965.
(5) Hostetler, P. B., Ph.D. thesis, Harvard University, 1960.
(6) Livingstone, D. A., *U.S. Geol. Surv. Profess. Paper* **440-G**, 61 (1963a).
(7) Siffert, B., *Geol. Alsace Lorraine* **21**, 32 (1962).
(8) White, D. E., Hem, J. D., Waring, G. A., *U.S. Geol. Surv. Profess. Paper* **440-F**, 14 (1963).

RECEIVED April 28, 1966. Work supported by grants from the National Science Foundation (GP-4140) and the Petroleum Research Fund of the American Chemical Society. Contribution No. 390, Bermuda Biological Station for Research.

IV
Chemical Composition of Connate and Thermal Waters

Editor's Comments on Papers 15 Through 19

15 White: *Magmatic, Connate, and Metamorphic Waters*

16 Chave: *Evidence of History of Sea Water from Chemistry of Deeper Subsurface Waters of Ancient Basins*

17 Roedder: *Studies of Fluid Inclusions: I. Low Temperature Application of a Dual-Purpose Freezing and Heating Stage*

18 Ivanov: *Principal Geochemical Environments and Processes of the Formation of Hydrothermal Waters in Regions of Recent Volcanic Activity*

19 Ellis and Mahon: *Natural Hydrothermal Systems and Experimental Hot-Water/Rock Interactions*

Included in this part are selected papers dealing with the change in chemical composition of natural water confined for a length of time in sediments. Several papers, including an excellent review by D. E. White (Paper 15), discuss the geochemistry of thermal and mineral waters. Subsurface waters underlying ancient basins are thought to be remnants of seawater trapped in sediments at the time of their deposition. K. E. Chave (Paper 16) studied these brines to evaluate the possibility of using their chemical compositions as indicators of ancient seawater chemistry; the magnitude of diagenetic change is so great, however, that it is unlikely that evidence on ancient seawater chemistry can be obtained from this line of study.

Original ocean water in contact with sediments evolves to connate or fossil water having a very high calcium chloride concentration, which differs greatly from its initial composition. The mechanisms of the diagenetic process in subsurface waters have been studied with the aid of stable and radioactive isotopes, as well as with common chemical constituents (White et al., 1963; White, 1965; Valyashko and Vlasova, 1965; Graf et al., 1966; Degens and Chilingar, 1967; Billings et al., 1969; Hitchon et al., 1971). Calcium chloride water has been observed in Don Juan Pond in Antarctica, where the mineral antarcticite ($CaCl_2 \cdot 6H_2O$) was discovered (Meyer et al., 1962; Torii and Ossaka, 1965); calcium chloride water has also been recorded in the groundwaters discharged from the swarm earthquake area in central Japan (Yoshioka et al., 1970).

Reviews on thermal and mineral waters were published by both White (1957a) and by V. V. Ivanov (Paper 18). Natural hot water compositions have been compared by Ellis and Mahon with the chemical compositions resulting from the experimental interaction at 150–350°C with volcanic rocks and graywacke (Paper 19; Ellis and Mahon, 1967). Underground temperatures have been estimated from the silica, sodium, potassium, and calcium contents of waters from hot springs and wet-steam wells by Fournier and Rowe (1966), Ellis (1970), and Fournier and Truesdell (1973).

The chemical composition of fluid inclusions in minerals has been studied to discern the nature of reactions between minerals and pore liquids. E. Roedder's model (Paper 17) for the method by which chemical composition in a very small volume of liquid fluid is readily measured is presented here. Interaction between water and sediment necessarily affects the chemical composition of interstitial water; papers on interstitial waters are not presented in this volume, owing to space limitations.

Donald E. White was born in Dinuba, California, in 1914. He received his B.A. in geology from Stanford University in 1936, and his Ph.D. in geology from Princeton University in 1939. He has been with the U.S. Geological Survey since August 1939. White is primarily involved with research on natural hydrothermal systems with applications to the origin of mineral deposits and to the utilization of geothermal energy. He has served as an informal geothermal energy consultant to New Zealand since 1948, to Iceland since 1951, to the U.S. industrial exploration teams since 1949, and in recent years White has served many countries in this capacity, including Japan, Mexico, Nicaragua, El Salvador, Taiwan, the Philippines, Turkey, Ethiopia, Kenya, and India. White received the Department of the Interior Distinguished Service Award from the United States in 1971.

Keith E. Chave was born in Chicago in 1928. He received his B.A. in 1948, his M.S. in 1951, and his Ph.D. in 1952 from the University of Chicago. Chave was a research geologist at the California Research Corporation in La Habra from 1952 to 1959. He held the position of an assistant professor between 1959 and 1961, an associate professor from 1961 to 1964, and a full professor from 1964 to 1967 at Lehigh University. Since 1967 Chave has been a professor with the Department of Oceanography at the University of Hawaii and has served as chairman of the department since 1970; he has been an associate of the Woods Hole Oceanographic Institution since 1962.

A. J. Ellis, a graduate of Otago University, has specialized in the field of hydrothermal geochemistry, undertaking both laboratory experimental work on the physical chemistry of waters at high temperatures and pressures, and field studies in the geothermal areas of New Zealand. Ellis is the director of the chemistry division of the New Zealand Department of Scientific and Industrial Research.

Edwin Woods Roedder was born in Monsey, New York, in 1919. He received his B.A. from Lehigh University in 1941, and both his M.A. (1947) and his Ph.D. (1950) from Columbia University. Roedder was an assistant professor from 1950 to 1952, and an associate professor from 1952 to 1955 at the University of Utah. Roedder has been with the U.S. Geological Survey since 1955, and is presently with the Experimental Geochemistry and Mineralogy Branch of the U.S. Geological Survey in Washington, D.C.

V. V. Ivanov works at the Central Institute of Resorts in Moscow, where he is in charge of the Hydrogeological Department, which studies mineral and thermal waters. Ivanov is also a professor of the chair of hydrogeology at the Moscow State University.

References

Billings, G. K., Hitchon, B., and Shaw, D. R. (1969). Geochemistry and origin of formation waters in the western Canada sedimentary basin, 2. Alkali metals. *Chem. Geol.*, **4**, 211–223.

Degens, E. T., and Chilingar, G. V. (1967). Diagenesis of subsurface waters, in G. Larsen and G. V. Chilingar (eds.), *Diagenesis in Sediments, Developments in Sedimentology*, Vol. 8, American Elsevier, New York, pp. 477–502.

Ellis, A. J. (1970). Quantitative interpretation of chemical characteristics of hydrothermal systems. Proceedings of the United Nations Symposium on the Development and Utilization of Geothermal Resources, Pisa, Vol. 2, Part 1, *Geothermics* (Special Issue), **2**, 516–528.

Ellis, A. J., and Mahon, W. A. J. (1967). Natural hydrothermal systems and experimental hot water/rock interactions (Part II). *Geochim. Cosmochim. Acta*, **31**, 519–538.

Fournier, R. O., and Rowe, J. J. (1966). Estimation of underground temperatures from the silica content of water from hot springs and wet-stream wells. *Amer. J. Sci.* **264**, 685–697.

Fournier, R. O., and Truesdell, A. H. (1973). An empirical Na-K-Ca geothermometer for natural waters. *Geochim. Cosmochim. Acta*, **37**, 1255–1275.

Graf, D. L., Meets, W. F., Friedman, I., and Shimp, N. E. (1966). The origin of saline formation waters: III. Calcium chloride waters. *Illinois State Geol. Survey Circ. 397*, 60 pp.

Hitchon, B., Billings, G. K., and Klovan, J. E. (1971). Geochemistry and origin of formation waters in the western Canada sedimentary basin—III. Factors controlling chemical composition. *Geochim. Cosmochim. Acta,* **35**, 567–598.

Meyer, G. H., Morrow, M. B., Wyss, O., Berg, T. E., and Littlepage, J. L. (1962). Antarctica: The microbiology of an unfrozen saline pond. *Science,* **138**, 1103–1104.

Torii, T., Ossaka, J. (1965). Antarcticite: A new mineral, calcium chloride hexahydrate. *Science,* **149**, 675–977.

Valyashko, M. G., and Vlasova, N. K. (1965). On the formation of calcium–chloride brines, *Geochem. Internat.,* **2**, 25–36.

White, D. E. (1957). Thermal waters of volcanic origin. *Geol. Soc. America Bull.,* **68**, 1637–1658.

White, D. E. (1965). Saline waters of sedimentary rocks. Fluids in Subsurface Environments—A Symposium, *Amer. Assoc. Petrol. Geol. Mem. 4,* pp. 342–366.

White, D. E., Hem, J. D., and Waring, G. A. (1963). Chemical composition of subsurface waters, in M. Fleischer (ed.), *Data of Geochemistry,* 6th ed., *U.S. Geol. Survey Prof. Paper 440-F,* 67 pp.

Yoshioka, R., Okuda, S., and Kitano, Y. (1970). Calcium chloride type water discharged from the Matsushiro area in connection with swarm earthquakes. *Geochem. J.,* **4**, 61–74.

Reprinted from *Geol. Soc. America Bull.*, **68**, 1659–1682 (Dec. 1957)

MAGMATIC, CONNATE, AND METAMORPHIC WATERS

By Donald E. White

Abstract

Some major types of water of "deep" origin are believed to be recognizable from their chemical and isotopic compositions. Oil-field brines dominated by sodium and calcium chlorides differ markedly from average ocean water. In general, the brines are believed to be connate in origin ("fossil" sea water) with a negligible to high proportion of meteoric water. Many brines, particularly in pre-Tertiary rocks, are much higher in salinity than sea water and are greatly enriched in calcium as well as sodium chloride. Brines near the salinity of sea water are generally higher, relative to sea water, in bicarbonate, iodine, boron, lithium, silica, ammonium, and water-soluble organic compounds, and lower in sulfate, potassium, and magnesium.

Many changes take place after sea water is entrapped in newly deposited marine sediments: (1) Iodine, silicon, boron, nitrogen, and other elements have been selectively concentrated in organisms that decompose during and after burial in sediments. Many of the elements may redissolve in the interstitial water. (2) Bacteria are active in the sediments and reduce sulfate to sulfide and produce methane, ammonia, carbon dioxide, and other products. (3) Some elements have been selectively removed from sea water by inorganic processes, such as adsorption on clays and colloidal matter. When this matter is reconstituted by diagenetic and other changes, some components are redissolved. The abundance of lithium and possibly boron and other elements may be controlled to a considerable extent by these inorganic processes. (4) The interstitial water may react chemically with enclosing sediments and produce dolomite, reconstituted clays, and other minerals. The high loss of magnesium relative to calcium in most connate waters is probably caused by such reactions.

Volcanic hot-spring waters of different compositions have been discussed in an accompanying paper (White, 1957). The most significant type is believed to be dominated by sodium chloride, and is best explained as originating from dense gases driven at high temperature and pressure from magma and containing much matter of low volatility that is in solution because of the solvent properties of high-density steam. This dense vapor is condensed in and greatly diluted by deeply circulating meteoric water. Most other types of volcanic water are believed to be derived from the sodium-chloride type.

Volcanic sodium-chloride waters are similar in many respects to connate waters but are believed to be distinguishable by relatively high lithium, fluorine, silica, boron, sulfur, CO_2, arsenic, and antimony; by relatively low calcium and magnesium; and by lack of hydrocarbons, water-soluble organic compounds, and perhaps ammonia and nitrate. Relatively high boron and combined CO_2 are alone not reliable indicators of a volcanic origin.

During compaction, rocks lose most of their interstitial high-chloride water; much additional water may then be lost during progressive metamorphism, and the content changes from about 5 per cent in shale to perhaps 1 per cent in gneiss. This expelled water is here called metamorphic. Because of pressure and permeability gradients, it must normally escape upward and mix with connate and meteoric water. Even though large quantities must exist, no example of metamorphic water has been positively identified.

Some thermal springs in California are high in salinity and relatively low in temperature and apparent associated heat flow. Some are clearly connate in origin. Other springs are characterized by very high combined carbon dioxide and boron, relative to chloride. Their compositions are considerably different from known connate and volcanic waters and are believed to be best explained by a metamorphic origin.

Although some major types of deep water seem to be recognizable, there is much danger of oversimplifying the problems. Many waters are no doubt mixtures of different types, and some of high salinity result from dissolution of salts by meteoric water.

CONTENTS

TEXT

	Page
Introduction	1660
Statement of the problem	1660
Acknowledgments	1661
Definitions	1661
Terms defined elsewhere	1661
Meteoric water	1661
Connate water	1661
Metamorphic water	1662
Juvenile water	1662
Comparison of ocean, connate, and volcanic waters	1662
Method of approach	1662
Temperature and heat flow	1663
Isotopes of the waters	1663
Total dissolved solids	1667
Halogen elements	1668
Alkali elements	1668
Sulfur	1669
Carbon dioxide	1670
Boron	1671
Combined nitrogen	1671
Hydrocarbons and other organic compounds	1672
Silica	1672
Alkaline-earth elements	1673
Other elements	1673
Nonvolcanic hot springs	1674
Discussion	1674
Hot springs with connate water	1674
Hot springs with metamorphic water	1678
Thermal waters of mixed origin	1679
Conclusions and speculations	1679
References cited	1679

ILLUSTRATIONS

Figure	Page
1. Simplified diagram illustrating the relationships of genetic types of water considered in this paper	1662

TABLES

Table	Page
1. Analyses of oil-field brines of the chloride types, compared with ocean water	1664
2. Analyses of volcanic hot springs of the sodium-chloride type, compared with average igneous rocks	1665
3. Tentative criteria for recognition of major types of ground water of different origins	1666
4. Approximate ranges in chemical components of ocean water, oil-field brines dominated by chloride, and volcanic sodium-chloride springs	1666
5. Analyses of thermal waters believed to contain water of connate or metamorphic origin	1676

INTRODUCTION

Statement of the Problem

Many thermal springs occur in areas of recent or active volcanism, are high in temperature and in associated total heat flow, and have chemical characteristics that indicate a close relation to volcanism (White, 1957). Other springs have only low to moderate temperatures but are more highly mineralized than ordinary ground waters. Do some of these springs also contain volcanic water, or is the mineral matter derived from nonvolcanic sources? This is part of a broader problem: How many different genetic types of ground water exist, and what means are available to distinguish each from the others?

When fine-grained marine sediments are deposited, their initial water content is commonly 50 per cent or more of the wet weight (Emery and Rittenberg, 1952, p. 747–755). Most of this water is driven off during compaction and progressive metamorphism. Clarke (1924, p. 631) computed the average water content for the following rocks: Shale, 5.0 per cent; slate, 3.8; and schist, 2.0. No averages are given for gneiss, but individual analyses (p. 630) range from 0.4 to 1.8 per cent. For comparison, average granodiorite, which is at least in part ultrametamorphosed or fused sedimentary rock, contains only 0.65 per cent of water (Nockolds, 1954, p. 1014). The evidence is clear that very large quantities of connate and metamorphic water must be driven off from sediments during compaction, progressive metamorphism, and reconstitution to minerals containing little water. Some of this water no doubt reaches the surface in thermal springs.

Some effort has been made in the past to distinguish ocean water from oil-field brines (Revelle, 1941; Vinogradov, 1948; Piper et al. 1953, p. 90–92). The characteristics of volcanic or magmatic water as compared with the waters driven from sediments are much less well known.

The writer has considered magmatic or volcanic waters of different types in an associated paper (White, 1957) and concluded that the sodium-chloride type is derived directly from high-density vapors at considerable depth and is most representative of magmatic emanations. Other types seem to be for the most part derived from the sodium-chloride type. The volcanic sodium-chloride waters are here compared with ocean water and with other possible types of "deep" water.

This paper reviews the characteristics of different kinds of "deep" water, within the limitations of available data. Analyses containing many of the components now believed to be diagnostic are surprisingly scarce. The most diagnostic components seem to be: all the halogen and alkali elements; all the ionic and molecular species of sulfur, carbon, boron, and nitrogen; silica; organic compounds; and isotopes of the water. The alkaline-earth elements and many other metals are of considerable interest but are of less diagnostic value because quantities in natural solutions are so dependent on pH or other components in the water. The more soluble components that rarely attain saturation are particularly significant.

The writer wishes to emphasize the tentative nature of the conclusions reached here. Additional data will surely change some of these conclusions but others, it is hoped, will become established criteria for distinguishing different types of water.

Acknowledgments

The writer is greatly indebted to many individuals for stimulating conversations and ideas that have led to the writing of this paper. Of initially great value was the published account of Wilbur Springs, California by G. A. Waring (1915, p. 99–103) that brought to the writer's attention the anomalously low temperature, high salinity, and high iodine content of these springs. These characteristics suggested a nonvolcanic origin, and are now interpreted to indicate connate water. Mr. Waring has also assisted in some of the bibliographic study for the present work.

Mr. Cole R. McClure of the California Division of Water Resources has been very helpful in discussing the characteristics of oil-field brines and in making available a number of brine analyses, of which the Rio Vista sample of Table 1 is typical. Data on the stable isotopes of some of the waters have been provided by Dr. Harmon Craig of the University of Chicago and Scripps Institution of Oceanography, and on tritium by Dr. Fred Begemann of the University of Chicago. Some of the problems have been discussed most helpfully with Dr. E. G. Zies of the Geophysical Laboratory. The writer is greatly indebted to his associates on the U. S. Geological Survey, especially to H. L. James, V. E. McKelvey, Edgar H. Bailey, P. F. Fix, S. Muessig, and C. H. Sandberg, who have read the manuscript. A. H. Lachenbruch of the Geological Survey has been most helpful with considerations of heat flow of the different types of thermal springs. Much benefit has also been derived from discussions and communications with the following members of the U. S. Geological Survey: C. S. Howard, V. T. Stringfield, H. E. LeGrand, S. K. Love, W. F. White, C. L. McGuinness, J. H. Feth, J. D. Hem, and F. H. Rainwater. None of these men, however, should be held responsible for interpretations and conclusions.

Definitions

Terms defined elsewhere.—The following terms were defined in an accompanying paper (White, 1957): thermal spring, hydrothermal, magmatic water, plutonic water, volcanic water, and mixed water. Other definitions that apply particularly to the present paper are:

Meteoric water.—Meteoric water, according to Lane (1908) includes connate water as well as "rain, vadose, or pluvial waters". As used here, the term meteoric applies to water that was recently involved in atmospheric circulation. The age of meteoric ground water is slight when compared with the age of the enclosing rocks and is not more than a small part of a geologic period. The writer's usage is very similar to that of Meinzer (1923, p. 31) except that age is specifically considered.

Connate or fossil water.—Connate water, as the term is used in the present paper, is "fossil" water that has been out of contact with the atmosphere for at least an appreciable part of a geologic period. It consists of the fossil interstitial water of unmetamorphosed sediments and extrusive volcanic rocks and water that has been driven from the rocks. The term has been defined and redefined a number of times since it was first proposed by Lane (1908, p. 502–507; see Rogers, 1917, p. 20–21; Meinzer, 1923 p. 31; Case, 1955). Lane clearly recognized that some brines consist of mixtures of connate and meteoric water, and that some connate water may have been nonmarine. Much of the confusion stems from Lane's implication, followed by Meinzer (1923, p. 31) that connate water has remained since burial with the rocks in which it occurs. This is an unworkable restriction that probably never holds in detail (*see* for example, Emery and Rittenberg, 1952, p. 751–763) and should be abandoned. For those who insist that the term connate should be retained solely for waters "born with" the immediately enclosing rocks, the term fossil water is suggested.

Another major source of confusion stems from the fact that no oil-field brine is identical in composition to sea water. The drastic changes that occur in sea water after burial have been

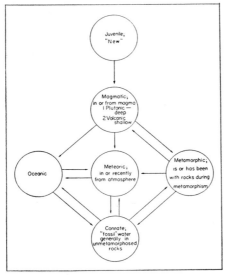

FIGURE 1.—SIMPLIFIED DIAGRAM ILLUSTRATING THE RELATIONSHIPS OF GENETIC TYPES OF WATER CONSIDERED IN THIS PAPER

reviewed by Emery and Rittenberg (1952), and by Schoeller (1956) and are a major part of the present study; in general, the changes are believed to result from normal reactions of water with sediments and associated organic matter, induced in considerable part by bacterial action and also influenced by elevated temperature and time. "Formation water", defined as the water present in rocks immediately before drilling (Case, 1955), may be a useful nongenetic term for waters of unknown age and origin but should not replace the genetic concept of connate or fossil water.

Chebotarev (1955) and others have held that all oil-field brines are entirely meteoric water that has circulated deeply and dissolved mineral matter. A little isotope evidence is already available to indicate that the D/H ratios of oil-field brines is significantly higher than in most surface waters (Rankama, 1954, p. 155, 157–159). Additional unpublished measurements on the isotopes of oxygen as well as hydrogen (Harmon Craig, personal communication) indicate that oil-field brines are derived from ocean water and are unlike most meteoric waters.

Metamorphic water.—Metamorphic water as here defined is water that is or that has been associated with rocks during their metamorphism. Meinzer (1923, p. 31) had a somewhat similar concept termed water of dehydration. Metamorphic water is probably derived largely from hydrous minerals during recrystallization to anhydrous minerals. In its broadest sense, connate or fossil water could also include metamorphic water, because water of hydration of minerals of sedimentary rocks was at one time involved in atmospheric circulation. In the process of recrystallization, however, some, if not most, oxygen atoms were exchanged for the oxygen atoms of silicate materials; on an atomic scale, identity is not maintained. As used here, connate water is considered to consist largely of water that was interstitial and was driven off before metamorphism. It presumably grades into metamorphic water, just as sedimentary rocks grade into metamorphic rocks.

Juvenile water.—"New" water that is in, or is derived from, primary magma or other matter and has not previously been a part of the hydrosphere. Volcanic or magmatic waters (for definitions, *see* White 1957) are not assumed to be juvenile unless indicated by significant evidence.

COMPARISON OF OCEAN, CONNATE, AND VOLCANIC WATERS

Method of Approach

Relationships of the types of water defined here or in the previous paper (White, 1957) are illustrated diagrammatically in Figure 1. Only the oceanic, meteoric, and one type of magmatic water can be sampled and identified with certainty.

The existence of connate water has been questioned by Chebotarev (1955) and others, but these skeptics have not explained satisfactorily how all the original interstitial water of marine sediments was flushed from deeply buried rocks or why so much highly soluble matter was retained; they have also failed to explain the detailed chemical and isotopic composition of the interstitial water.

Most geologists will grant that oil-field brines probably contain some "fossil" connate water of an age that is not greatly different from that of the enclosing rocks. Many oil-field waters, on the other hand, contain little dissolved matter (Crawford, 1940; 1942; 1949); they are probably for the most part meteoric, most of the original connate water having been flushed out.

Other oil-field waters have 5–10 times the salinity of present ocean water and are also characterized by very high calcium as well as sodium chlorides. Their origin has long been debated (Mills and Wells, 1919; Russell, 1933; De Sitter, 1947; Chebotarev, 1955) and cannot be discussed in detail here. Most geologists have favored some process of subsurface concentration to account for their compositions. These high-density brines are relatively abundant and are generally in pre-Tertiary rocks; they are generally not associated with salt beds or unconformities; and they show remarkable similarities in chemical composition over large areas (Hoskins, 1947; Jeffords, 1948; Meents et al., 1952; and De Sitter, 1947, p. 2033–2034). These characteristics point strongly to a connate origin; salts were concentrated by processes that are not yet well understood but which involve selective removal of water.

For comparison with other waters, oil-field brines dominated by chloride have been selected as the most likely representatives of connate waters or simple dilutions of such waters (Table 1). Chebotarev (1955, p. 153) has stated that oil-field waters dominated by sulfate occur at a mean depth of 1800 feet, by bicarbonate, at a depth of 2300 feet, and by chloride, at 3500 feet. The chloride waters are therefore more likely to be connate water or simple dilutions of such waters.

Selected analyses of volcanic hot springs of the sodium-chloride type are shown in Table 2. Of the different types of water in volcanic association, the sodium chloride type is held to be most representative of deep volcanic water (White, 1957).

The writer's approach in comparing selected waters that are most likely to contain the desired deep types, and his objectives, are different from Schoeller's, who has recently completed an outstanding monograph (1956) comparing oil-field waters with ordinary ground and mineral waters. Schoeller's main purpose was to bring out chemical contrasts between oil-field waters and other waters not associated with petroleum. In Schoeller's statistical treatment, all oil-field waters of varied chemical type, complexity of history and likely content of connate water, are considered together and are compared with the nonpetroliferous waters. Some of the latter are probably entirely meteoric, but others, judging from the present study, contain connate and perhaps metamorphic water. Schoeller's study and this one overlap in part; where conclusions are in apparent disagreement, the cause is likely to be found in the differences in scope and treatment of the data.

No metamorphic water has yet been identified with certainty, but in a later section of this paper evidence is presented to show that some thermal springs are probably of connate and metamorphic origin.

No way is yet known to distinguish between juvenile and recycled magmatic water. This paper also does not treat in detail criteria to distinguish the different "deep" waters from meteoric water that has come in contact with salt beds. It is hoped that this can be done adequately when more chemical and isotopic data are available.

To orient the reader, the tentative criteria that are concluded to distinguish major types of ground water of different origins are summarized in Table 3. The known ranges in chemical components of ocean, oil-field, and volcanic waters of the chloride types are shown in Table 4.

Each characteristic or chemical component is discussed in turn for the oceanic, connate, and volcanic waters. The physical and geochemical reasons for the differences are thus more apparent than if summarized separately for each type of water.

Temperature and Heat Flow

The temperatures of oil-field brines in situ are largely dependent upon depth below the surface. Temperature gradients in sedimentary basins are generally in the order of 1°C. per 150 feet of depth, and the rate of flow of heat to the surface is approximately 1.2×10^{-6} cal./sq. cm/sec. (Birch, 1954, p. 645). When connate brines from depth escape rapidly along faults or other permeable aquifers, they should have temperatures that are considerably above "normal" for their new environment. Some warm springs that are believed to have such an origin are discussed later.

Temperatures and heat flow associated with volcanic hot springs are commonly very high. (Table 2; White, 1957).

Isotopes of the Waters

The isotopic compositions of hot spring waters will be discussed in detail in papers by

TABLE 1.—ANALYSES OF OIL-FIELD BRINES OF THE CHLORIDE TYPES, COMPARED WITH OCEAN WATER
(In parts per million)

	Ocean[1]	Huntington Beach, Orange Co., California[2]	Rio Vista Field, Solano Co., California[3]	Seaboard Field, Fresno Co., California[4]	Elliott Co., Kentucky[5]	Oil test, Colusa Co., California[6]
Temp. °C.		>50				21.3
pH			7.3	6.4		7.2
SiO_2	7	90	30	36		13
Fe	0.02	0.79			0.52	
Al	1.9					
Ca	400	177	98	2190	12,300	590
Mg	1270	127	3.4	832	3350	90
Ba	0.05	24		42		
Sr	13	2.6			628	130
Na	10,600	9400	3770	14,800	40,200	6100
K	380	124	50	251	191	61
Li	0.1		2.0			2.7
NH_4	0.05	(160)	25	31		40
As	0.017					
CO_3		0	0	0		0
HCO_3	140	1500	1760	193	28	60
SO_4	2650	1.0	5.9	0.0	66	
Cl	19,000	14,400	4960	29,000	93,900	11,000
F	1.4	0.6	1.8	0.6		
Br	65	99	16	108	817	
I	0.05	35	17	21	10	
NO_3	0.7		3.3	44		
NO_2	0.05			0.0		
B	4.6		108	8.3		20
Sum	34,500	26,100	10,900	47,600	151,000	18,100
Ratios by wt.						
HCO_3/Cl	0.0074	0.10	0.35	0.0067	0.00030	0.0055
SO_4/Cl	0.14	0.00007	0.0012	0.00000	0.0007	
F/Cl	0.00007	0.00004	0.00036	0.00002		
Br/Cl	0.0034	0.0069	0.0032	0.0037	0.0087	
I/Cl	0.000003	0.0024	0.0034	0.00072	0.00011	
B/Cl	0.00024		0.022	0.00029		0.0018
K/Na	0.036	0.013	0.013	0.017	0.0048	0.010
Li/Na	0.00001		0.00053			0.00044
Ca + Mg/Na + K	0.15	0.032	0.027	0.20	0.39	0.11

[1] Ocean water (Rankama and Sahama, 1950, p. 287, 290). Some components range greatly in concentration; upper limits shown here.

[2] Well 6/11-11A2, The Texas Co. (Piper et al., 1953, p. 207, 210). NH_4 not determined—average of two other analyses from Huntington Beach (p. 210). Jenson's (1934) average of 7 from Huntington Beach is nearly 200 ppm.

[3] Amerada Petroleum Corp. well Drouin No. 4, sec. 25, T. 4 N., R. 2 E., producing gas and brine from 4400 to 4420 ft. Collected by C. R. McClure, Calif. Division of Water Resources, May 13, 1955. Analyst, R. O. Hansen, U. S. Geological Survey.

[4] Seaboard well No. S. T. U. 305–13, sec. 13, T. 15 S., R. 17 E., from 4700-foot depth. Collected by Calif. Division of Water Resources, April 20, 1954.

[5] Kentucky and Ohio Gas Co. No. 1 Hutchenson well producing from Corniferous Formation, 1506 feet (McGrain, 1953, p. 12, 14).

[6] Old oil-test well on Bear Creek, 1 mi. SE of Wilbur Springs, sec. 27, T. 14 N., R. 5 W.; seeping overflow. From Knoxville Formation (W. B. Myers, personal communication). Analysts, H. Almond and S. Berman, U. S. Geological Survey.

TABLE 2.—ANALYSES OF VOLCANIC HOT SPRINGS OF THE SODIUM-CHLORIDE TYPE, COMPARED WITH AVERAGE IGNEOUS ROCKS

(In parts per million)

	Steamboat Springs, Washoe Co., Nevada[1]	Morgan, Tehama Co., California[2]	Norris Basin, Yellowstone Park, Wyoming[3]	Upper Basin, Yellowstone Park, Wyoming[4]	Wairakei, New Zealand[5]	Average igneous rocks[6]
Temp. °C.	89.2	95.4	84	94.5	>100	
pH	7.9	7.83	7.45	8.69	8.6	
SiO_2	293	233	529	321	386	593,000
Fe				Tr.		50,000
Al				0		81,300
Ca	5.0	79	5.8	4	26	36,300
Mg	0.8	0.8	0.2	Tr.	<0.1	20,900
Sr	1	10[7]				220
Na	653	1400	439	453	1130	28,300
K	71	196	74	17	146	25,900
Li	7.6	9.2	8.4		12.2	22
NH_4	<1	<1	0.1	0	0.9	
As	2.7	2.2	3.1			5
Sb	0.4	0.0	0.1			1(?)
CO_3	0	0	0	66		
HCO_3	305	52	27	466	35	1620
SO_4	100	79	38	15	35	1560
Cl	865	2430	744	307	1930	314
F	1.8	1.5	4.9	21.5	6.2	700
Br	0.2	0.8	0.1			1.6
I	0.1	<0.1	<0.1			0.3
B	49	88	11.5	3.7	26.2	3
S_2O_3				2		
H_2S	4.7	0.7	0	0	1.1	
CO_2					11	
Sum	2360	4580	1890	1680	3740	841,000
Ratio by wt.						
$HCO_3^{[8]}/Cl$	0.35	0.021	0.036	1.9	0.018	5.2
SO_4/Cl	0.12	0.033	0.051	0.049	0.018	5.0
F/Cl	0.0021	0.0006	0.0066	0.070	0.0032	2.2
Br/Cl	0.0002	0.0003	0.0001			0.005
I/Cl	0.0001	<0.00004	<0.0001			0.001
B/Cl	0.057	0.036	0.015	0.012	0.013	0.01
K/Na	0.11	0.14	0.17	0.038	0.13	0.91
Li/Na	0.012	0.0066	0.019		0.011	0.00078
Ca + Mg/Na + K	0.0080	0.050	0.012	0.008	0.020	1.1

[1] Spring 8, Main Terrace, collected by White, August 9, 1949, analyzed by W. W. Brannock, U. S. Geological Survey.

[2] Growler spring. The group is located just south of Lassen Volcanic National Park, sec. 11, T. 29 N., R. 4 E. Collected by White, July 29, 1949, analyzed by W. W. Brannock, U. S. Geological Survey.

[3] Unnamed spring with periodic discharge and precipitating abundant silica, 200 feet southwest of Pearl geyser. (Allen and Day, 1935, p. 482.) Collected by White, August 3, 1951, analyzed by P. W. Scott and W. W. Brannock, U. S. Geological Survey.

[4] Sapphire Pool (Allen and Day, 1935, p. 103, 249).

[5] Water erupted from drill hole 4, eastern part of thermal area (Wilson, 1955, p. 37). Carbonic acid recalculated as CO_2, HBO_2 as B, H_2SiO_3 as SiO_2; 0.15 ppm HS included in H_2S.

[6] Rankama, 1954, p. 135. All C and S assumed to be oxidizable to carbonate and sulfate so that ratios of all analyses may be compared.

[7] Spectrographic determination.

[8] All carbonate and bicarbonate calculated as HCO_3.

TABLE 3.—TENTATIVE CRITERIA FOR RECOGNITION OF MAJOR TYPES OF GROUND WATER OF DIFFERENT ORIGINS

	Chemical composition	Isotopic composition	Heat relations
Meteoric water	Controlled by surface waters, bedrocks	Same as or close to surface waters	Near "normal"
Sea water (direct from ocean)	Same as or very similar to ocean (very low in I and and other nutrients)	Same as or close to ocean	Near "normal" or sub-subnormal
Connate water (chloride types)	Enriched in I, B, SiO_2, combined N, Ca, soluble organic components and low in SO_4 and Mg relative to ocean	$D/H \lessgtr$ ocean, O^{18}/O^{16} > ocean	"Normal" to moderately thermal
Metamorphic water	Little known. High in combined CO_2 and B, low in Cl relative to ocean?	Little known. $D/H \lessgtr$ ocean? O^{18}/O^{16} > ocean?	"Normal" to moderately thermal
Magmatic water (volcanic sodium-chloride type)	Relatively high in Li, F, SiO_2, B, S, CO_2, low in Ca, Mg, combined N(?)	Little known. $D/H <$ ocean? O^{18}/O^{16} > ocean	Strongly thermal in active systems

TABLE 4.—APPROXIMATE RANGES IN CHEMICAL COMPONENTS OF OCEAN WATER, OIL-FIELD BRINES DOMINATED BY CHLORIDE, AND VOLCANIC SODIUM-CHLORIDE SPRINGS

Ratios, by weight	Ocean*	Oil-field brines†	Volcanic hot springs**
HCO_3/Cl††	0.0074	0.0001–1	0.01–3
SO_4/Cl	0.14	0.00000–1	0.01–0.5
F/Cl	0.00007	0.00001–0.001	0.0005–0.1
Br/Cl	0.0034	0.0001–0.01	0.0001–0.001
I/Cl	0.000003	0.00003–0.02	0.00001–0.0005
B/Cl	0.00024	0.00001–0.02	0.01–0.1
K/Na	0.036	0.001–0.03	0.03–0.3
Li/Na	0.00001	0.0001–0.003	0.003–0.03
Ca + Mg/Na + K	0.153	0.01–5	0.001–0.2
Concentrations, ppm			
SiO_2	0.04–8	10–100	100–700
Ca	400	10–20,000	1–100
Mg	1272	1–10,000	0.1–10
Sr	13	1–3000	0.1–10
Ba	0.05	nil–5000	0–1
NH_4	<0.005–0.05	0–500	nil?
NO_3	0.001–0.7	nil–100	nil?
As	0.002–0.017	?	0.1–10
Sb	Trace	?	0.0–1
H_2S	0–60	0–3000	0–10
CH_4	?	large	nil
Water-soluble organic compounds	slight	large?	nil

* Rankama and Sahama, 1950, p. 290.
† From Table 1 and other data (*see* text).
** From Table 2 and other data (*see* text).
†† All combined carbonate and bicarbonate calculated as HCO_3.

Harmon Craig, G. Boato, and the present writer (*see* initial paper, Craig, Boato and White, 1956), and will therefore be mentioned only briefly here.

The isotopic analyses of only a few connate waters have been published (Rankama, 1954. p. 159, 252). These few indicate lower deuterium and higher O^{18} contents than for ocean water. Deuterium content of the Wilbur oil test well (Table 1) is also a little lower (2 per cent) relative to the standard than ocean water, and the O^{18} is higher (0.3 per cent). According to Harmon Craig (personal communication), Texas brines are isotopically very similar to water from the oil test well.

The isotopes of the volcanic sodium-chloride waters are very similar to those of the surface waters of each specific area in which a spring emerges (Craig *et al.*, 1956). The hydrogen of the hot springs is almost identical to that of the surface waters, but the oxygen is commonly several tenths of a per cent heavier. If the differences were caused by volcanic water mixing with meteoric water, the hydrogen of each volcanic water must be nearly the same as that of the particular meteoric water with which it mixes. This demand on coincidence is highly improbable; a much more reasonable explanation, supported by some evidence, is that part of the oxygen of meteoric water has exchanged with the heavier oxygen of silicates. This "oxygen shift" may well conceal a small difference in oxygen and hydrogen that is actually caused by a contribution of volcanic water. The upper limit of the volcanic contribution appears to be in the order of a few per cent for the analyzed spirngs. There is no compelling isotopic evidence for the presence of any magmatic water in these springs. On the other hand, the discharge of heat and mineral matter are of such magnitude and nature that a volcanic source seems essential (White, 1957). Except for the unlikely possibility that the mineral matter is evolved anhydrously, water is also an essential volcanic component. It is believed to be present in small amount in the near-surface mixtures (probably 5 per cent or less), but is diluted with meteoric water almost beyond recognition by isotopic methods.

Tritium, the radioactive isotope of hydrogen, with a half life of about 12.5 years, is produced in the atmosphere by cosmic-ray bombardment of nitrogen (Libby, 1954). Tritium is a minor component of relatively young meteoric water but is not detectable in waters out of contact with the atmosphere for 50 years or more.

Tritium has been found in volcanic hot springs in quantities that are similar to those of associated meteoric waters (Von Buttlar and Libby, 1955). The tritium data not only confirm the conclusion that the water of the hot springs is dominantly of surface origin but also prove that subsurface travel times of a least part of the meteoric water are slight, probably not exceeding a few years.

No tritium analyses of oil-field brines are known to the writer. The T content of Wilbur Springs, which is shown below to have the characteristics of connate water, is very low (Fred Begemann, personal communication). The small amount of tritium in the water is believed to result from slight dilution with meteoric water near the surface.

Total Dissolved Solids

Oil-field waters contain from a little more than 1 per cent of the dissolved matter of sea water to about 10 times as much. Table 1 contains a range in salinity that includes nearly all oil-field brines that are dominated by chlorides. Crawford (1942) and others believe the dilute waters result from the flushing action of circulating meteoric waters. Apparently few oil-field brines in Tertiary rocks contain more than 15,000 ppm of chloride (*see* for example, Jenson, 1934), as compared to about 19,000 for average sea water. The lower salinity of the connate waters is commonly believed to result from some dilution by meteoric water. De Sitter, however (1947, p. 2039), has suggested that the brines with more dissolved matter than sea water have been concentrated by "sieve" action of fine-grained sediments, which behave as semipermeable membranes allowing water molecules to pass but retarding salt ions. If this or some other mechanism involving separation of water from ions has formed the high-concentration brines, the connate water that had escaped should contain less dissolved matter than the original brine. This could explain, at least in part, the scarcity of California brines of Tertiary age with chlorinities above that of sea water (Jensen, 1934). Timm and Maricelli (1953), on the other hand, believe that high-concentration brines escape first and leave dilute connate water behind. In spite of the fact that the mechanism is not clear, it appears that connate waters evolve into types that are somewhat more dilute, and other types that are more concentrated than sea water, without involving dilution with meteoric water or contact with salt beds. Many connate waters, however, are clearly mixed with meteoric water.

Volcanic hot spring waters generally range from about 1000 to 5000 ppm of dissolved matter. They are therefore more dilute than most oil-field brines. High-density volcanic waters before mixture with meteoric water (White, 1957) probably contain at least as much dissolved matter as ocean water and perhaps much more.

Halogen Elements

Fluorine is relatively low in ocean water and seems to be slightly higher in most oil-field waters in spite of the fact that calcium is also generally enriched (Schoeller, 1956, p. 94). F is relatively high in volcanic hot springs of the sodium-chloride type and attains concentrations as high as 20 ppm (Table 2) in alkaline waters with little calcium. Even though the volcanic sodium-chloride waters are high in F some volcanic gases have far higher F/Cl ratios (Zies, 1929, p. 4); this suggests that much has been fixed at depth as fluorite or in silicates such as the micas.

The Br/Cl ratio of ocean water is 0.003 (Table 1), which is very similar to that of average igneous rocks, 0.005. (The reliability of the bromine content of igneous rocks, as for many trace elements, is uncertain; see Correns, 1956, p. 184–187). Oil-field brines commonly have about the same ratio (Schoeller, 1956, p. 97). No major difference is to be expected, in view of the scarcity of bromine-concentrating organisms (Vinogradov, 1953, p. 106). This contrasts strongly with differences in iodine ratios as will be seen.

The apparent Br/Cl ratio of the volcanic hot springs (about 0.0002) may be an order of magnitude lower than that of average connate water. More study is necessary to determine whether the difference is real or is a result of lack of precise analyses of dilute volcanic waters.

Differences in the iodine-chlorine ratios are marked. The ratio for ocean water is 0.000003 (Table 1); the mean for all oil-field brines is about 0.001 (Schoeller, 1956, p. 101, 106). Iodine is apparently essential in the growth processes of all plants and animals, and is particularly enriched in marine algae (Vinogradov, 1953, p. 69–73, 91–93, 153–154; Goldschmidt, 1954, p. 602–612). According to Vinogradov (1953, p. 70–72), and Dzens-Litovsky, (1944), the fine-grained carbonaceous ocean sediments are very rich in iodine, with a concentration in the order of 0.01 per cent or more than a thousand times that of sea water. Iodine is apparently deposited with the remains of marine organisms; when the organic compounds decompose a considerable part of the iodine dissolves in the interstitial water. Much iodine remains in the sediments, however, and the I/Cl ratio is more than a thousand times higher than in ocean water (from calculations based on Correns, 1956, p. 226).

The average I/Cl ratio of the calcium-chloride brines of high salinity (Cl > 100,000 ppm) is about 0.0001, or in the order of 30 times that of sea water (Schoeller, 1956, p. 101, 109–111; Chajec, 1949; Lamborn, 1952; McGrain, 1953; Schoewe, 1943). The sodium-chloride brines, for the most part Tertiary in age, tend to have much higher I/Cl ratios, in the order of 0.003 (Schoeller, 1956, p. 101; Bemmelen, 1949, p. 185–190; Piper et al., 1953, p. 207–216; and Sawyer, Ohman, and Lusk, 1949).

The iodine-chlorine ratios of the volcanic sodium-chloride springs (Table 1) are in the order of 0.0001 and are lower than in all connate brines, except for some of the highly saline calcium-chloride waters. As might be expected, the I/Cl ratios of igneous rocks and volcanic hot springs tend to be intermediate between sea water and most connate brines.

Alkali Elements

The K/Na ratio is relatively high in rivers and in dilute ground waters (Schoeller, 1956, p. 136) and commonly ranges from 0.3 to 0.6. Waters from granitic and rhyolitic rocks tend to be particularly high in potassium. The only surface waters that are consistently higher are acid-sulfate hot springs in rhyolitic rocks, where alkalies have been derived directly from chemical attack on the adjacent rocks (White, 1957). Schoeller has shown that, on the average, the ratio decreases with increasing salinity. The K/Na ratio of ocean water is only 0.036 and that of most oil-field brines is still lower; the mean is 0.02 for the waters studied by Schoeller (p. 137). These also indicate decreasing ratios with increasing salinity. These changes reflect progressive base exchange of potassium in the waters for sodium from silicate minerals, and the conversion of low-potassium to high-potassium clays.

The K/Na ratio of volcanic sodium-chloride springs is commonly about 0.1 (Table 2). It seems to be highest (up to about 0.25) in nearly neutral springs high in free CO_2, but is as low as about 0.03 in some of the most alkaline dilute waters with pH's of 9 or more. The limited data also suggest, in contrast to other waters, that the ratio tends to increase with increasing salinity. These relationships are interpreted to mean that volcanic waters near their source are relatively high in potassium; perhaps the K/Na ratio is as high as 0.5 in waters from silicic volcanic rocks. As the waters react with adjacent rocks to form sericite, potassium clays, and in some places adularia, they are enriched in sodium and combined CO_2 and are impoverished in potassium. The K/Na ratios of the most impoverished volcanic hot springs waters are about equal to the average of oil-field brines.

Lithium has seldom been determined in oil-field waters. The lithium content of the Rio Vista analysis (Table 1) is typical of a dozen analyses of central California gas fields and is also very similar to that of the Wilbur oil-test well, and oil-field waters analyzed by Vinogradov (1948, p. 26), and Howard (1906, p. 284). One Hungarian oil-field water, according to an old analysis quoted by Telegdi-Roth (1950, p. 81), is reported to be high in Li, but others (p. 83, 84) contain little. Many more reliable analyses are needed, but the meager data suggest that the very low Li/Na ratio of sea water (0.00001) is increased an order of magnitude or more in many oil-field waters (0.0001–0.003). The geochemistry of lithium is not sufficiently well known to explain the suggested enrichment. Lithium is concentrated somewhat in plants relative to soils, but it has no known physiological function (Goldschmidt, 1954, p. 134–135). It is concentrated in marine shales (p. 135), and some may be released when clay minerals are reconstituted. More data are clearly needed.

In contrast to connate brines, the waters of volcanic hot springs have commonly been analyzed for lithium. Of all natural waters that occur at the surface, those of the volcanic sodium-chloride type are the most enriched in lithium. The Li/Na ratios of the analyses of Table 2 range from 0.006 to about 0.02, and the average of all analyses of this type of water that are known to the writer is about 0.01. Many volcanic hot springs of the sodium-chloride type have approximate ratios of alkalies, by weight, of 100 Na:10 K:1 Li. The corresponding atomic ratios are 100 Na:6 K: 3.3 Li. The atoms of lithium, a relatively rare atomic species, are nearly as abundant as those of potassium. Stated in another way, lithium constitutes about 0.3 per cent or 3000 ppm of the dissolved solids of Steamboat Springs water but only about 22 ppm of average igneous rock (Rankama, 1954, p. 135) and 46 ppm of the average glassy rhyolites and dacites of western United States (Coats, 1956, p. 76).

Lithium in the waters has been enriched about 100 times over its content in igneous rocks, but sodium has been enriched only 10 times in spite of similar solubilities and base-exchange properties. The high content of lithium in volcanic waters, with Li/Na ratios that are generally one or two orders of magnitude higher than probably connate waters, may prove to be a very significant criterion in distinguishing these waters.

The high lithium content of the volcanic sodium chloride springs is a weighty argument favoring a volcanic origin for the alkalies, which were probably transported as soluble alkali chlorides in water or very dense vapor (White, 1957). Allen and Day (1935, p. 114–118) reached no certain conclusion but favored the transport of halogens as volatile alkali halides rather than as halogen acids in spite of the known low volatilities of the halides. The possibilities for transport of nonvolatile matter in solution in dense vapors were indicated by Schröer (1927) and Ingerson (1934). The evidence, considering all the halogens and alkalies, favors a dominantly volcanic source for both groups of elements. The original ratios of the alkali metals, however, have clearly been modified by later exchange of cations with wall rocks. Sodium is generally enriched in the waters by exchange, potassium is lost, and lithium may decrease slightly, judging from relations at Steamboat Springs, where the lithium content of hydrothermally altered granodiorite appears to be somewhat greater than that of fresh rocks.

Very little is known of the rubidium and cesium contents of various waters. These elements should be given more attention (Borovik-Rommanova, 1946, p. 145–180).

Sulfur

Nearly all sulfur of sea water is present as sulfate; the average concentration is about 2600 ppm and the SO_4/Cl ratio is 0.14 (Table 1), but in basins of restricted circulation some of the sulfur is present as H_2S in quantities as high as 60 ppm (Goldschmidt, 1954, p. 529). Most oil-field brines, on the other hand, are relatively low in sulfate as compared to sea water (De Sitter, 1947; Schoeller, 1956, p. 115; Chebotarev, 1955, p. 159). Rogers (1917, p.

93–94) believed that low sulfate was the most characteristic feature of all oil-field waters. The sulfate-chloride ratios of the oil-field brines of Table 1 range from essentially zero to about 0.001.

Other oil-field waters are considerably higher in sulfate. Of the 667 analyses considered by Chebotarev (1955, p. 153), anions of 28 are dominated by sulfate. The mean depth of the sulfate waters was 1800 feet, of bicarbonate waters, 2300, and of chloride waters, 3500 feet. The relative shallowness and low chloride content of the sulfate waters is an indication that they are related to oxidation and flushing action of meteoric water and that, of all types of oil-field waters, they probably contain the lowest proportion of connate water. The sulfate of sea water buried with sediments has generally been reduced entirely or in large part to sulfide by the action of sulfur-reducing bacteria (Bastin and Greer, 1930; Ginter 1934; Emery and Rittenberg, 1952, p. 789–791; Beerstecher, 1954, p. 58–60). The bacteria reduce sulfate, utilizing energy available from oxidation of organic material to CO_2 and other substances (Schoeller, 1956, p. 115–119). Emery and Rittenberg have shown that most of the sulfate in basin sediments off the coast of Southern California has been reduced to H_2S and precipitated as sulfide when the sediments have been buried to depths of only a few feet. Much of the sulfide that is formed reacts with ferric oxides to form iron sulfides (Emery and Rittenberg, 1952, p. 791–793), but some may remain in solution as H_2S or as the HS^-ion. H_2S has not commonly been determined in oil-field brines; the available data are summarized by Schoeller (1956, p. 115). According to Rogers (1917, p. 100–101) H_2S does not exceed 350 ppm in the waters of the East Coalinga field, and is generally less than 50 ppm. H_2S is generally absent in Texas brines, but a content as high as 1000 ppm was noted by Berger and Fash (1934) and 2400 ppm by Ginter (1934, p. 922); 12 per cent of the accompanying gas was H_2S.

The sulfur of volcanic hot springs occurs dominantly as sulfate, but the waters commonly contain a few ppm of H_2S, and a few per cent is in the accompanying gases. Some but perhaps not all volcanic waters at depth contain considerable sulfide, some of which is apparently oxidized by mixing meteoric waters containing dissolved oxygen.

The sulfate of volcanic sodium-chloride springs (Tables 2, 4) commonly ranges from 50 to more than 100 ppm, and the sulfate-chloride ratio from about 0.01 to 0.3 or more; it is believed to be much higher in volcanic waters found at the surface than in most connate waters.

Carbon Dioxide

Ocean water contains an average of 140 ppm of combined CO_2, calculated as HCO_3. The HCO_3/Cl ratio is about 0.007. The corresponding ratios and CO_2 concentrations of most oil-field brines of low to moderate salinity are higher than ocean water (Table 1). Of the 667 oil-field waters considered by Chebotarev (1955, p. 153, 159), the anions of only 34 were dominated by bicarbonate. As mentioned above, the mean depth of these samples was 2300 feet, intermediate between the mean depth of sulfate waters (1800 feet) and chloride waters (3500 feet). The bicarbonate waters are less likely, because of their greater depth, to be as strongly influenced as the sulfate type by flushing action of meteoric water, but they are also not likely to contain as high a proportion of connate water as the chloride type.

Schoeller (1956, p. 116–120) has summarized evidence for a general decrease in sulfate with depth, accompanied by an increase in combined CO_2. The latter in turn is usually associated with a decrease in calcium because of mass-action effect and the low solubility of calcium carbonate.

Bicarbonate and free CO_2 are products of the decomposition of organic matter and the activity of sulfate-reducing bacteria. The energy is supplied by oxidation of organic matter, with the oxygen derived from sulfate. CO_2 is then available for chemical reaction. This may explain the relatively high bicarbonate of many oil-field brines of moderate salinity.

Oil-field gases nearly always contain some CO_2 and a few consist largely of this gas (Anderson and Hinson, 1951; Miller, 1936, 1937; Dobbin, 1935; Schoeller, 1956, p. 87). This CO_2 was probably derived largely from organic carbon.

The highly saline calcium-chloride brines generally contain little combined CO_2 in spite of pH's that are generally in the range of 5–7. These brines seem to be enriched in Ca relative to Na to such an extent that, because of mass action, the solubilities of carbonate and bicarbonate are very low. There is no evidence, however, for a corresponding increase of CO_2 in the associated gases. The scarcity of CO_2 in or near the high-density calcium-chloride brines is a puzzling fact.

Volcanic hot springs of the sodium-chloride type commonly contain much combined CO_2; HCO_3/Cl ratios range from less than 0.1 to at least 2 (Table 2). All these waters also contain free CO_2 not reported in ordinary water analyses. The free CO_2 content was not determined in the Steamboat analysis shown in Table 2, but a somewhat similar spring with a pH of 6.5 contained about 470 ppm of free CO_2 (computed from analysis of total CO_2 by Harmon Craig).

Near their source, the dense volcanic emanations probably contain much free CO_2 but little combined CO_2 (White, 1957); after condensation of steam, the waters react with wall rocks and dissolve alkali and alkaline-earth elements as soluble bicarbonates. The quantity of bicarbonates and carbonates in solution is a rough measure of the rock alteration produced by the waters, with the exception of any alteration by free halogen acids. The pH gives a rough indication of the quantity of free CO_2 available for further alteration. High pH's indicate that most of the original CO_2 has already reacted, or that the water was so hot upon arrival at the surface that most of the free CO_2 was boiled off before collection of the sample (White, Sandberg, and Brannock, 1953, p. 119–120).

Many volcanic hot springs have higher ratios of combined CO_2 to chloride than most oil-field brines but there is much overlap. A high content of CO_2 is probably less diagnostic of volcanic waters than many geologists have believed.

Boron

The average boron content of sea water is about 4.6 ppm (Table 1), and the boron-chloride ratio is roughly 0.0002. The data for oil-field brines are summarized by Schoeller (1956, p. 156). Few brines in the United States have been analyzed for boron until very recently but California brines apparently contain from 10 to 100 times as much, relative to chloride, as sea water (Scofield, 1933, p. 127–128; Table 1).

A relatively high content of boron characterizes most Russian oil-field brines (Tzeitlin, 1936). The boron-chloride ratios show a considerable scatter, but Kazmina's graph (1951, p. 302) indicates a strong tendency for inverse relationships, and boron is relatively low where chloride is high and vice versa. The sodium-chloride brines, with salinities near or below sea water, have an average boron content of 106 ppm, or 0.68 per cent of the mineral content according to Tageeva (1942), but brines of very high salinity, also high in calcium, may have little more boron than normal sea water. The relatively low solubility of the calcium borates may explain the low boron content of brines that are high in calcium (Schoeller 1956, p. 157).

Boron is greatly enriched in some marine sediments (Landergren, 1945; Gulyaeva, 1948, p. 833–835; Goldschmidt, 1954, p. 280–288), but little is known of its chemical association. The boron content of argillaceous marine sediments generally ranges from 30 to 1000 ppm, in contrast to 4.6 ppm in average ocean water and 3 ppm in average igneous rock. It is clear that some of the boron of sediments goes back into solution in connate brines.

The boron content of volcanic sodium-chloride springs is usually in the order of 10 to 100 ppm; the boron-chloride ratios are generally from about 0.01 to 0.1, as compared to 0.00001 to 0.02 for most oil-field brines. A high boron-chloride ratio is usually an indicator of volcanic water, as compared to ordinary oil-field brine. Waters of possible metamorphic origin with very high B/Cl ratios are considered below.

Combined Nitrogen

Combined nitrogen is one of the chief nutrients for organisms in sea water, thereby explaining the low concentration of ammonium and nitrate in the sea. Emery, Rittenberg, and Orr (Emery and Rittenberg, 1952, p. 776–780; Emery, Orr, and Rittenberg, 1955, p. 300–303; Rittenberg, Emery, and Orr, 1955, p. 29–38) have shown that much nitrogen-bearing organic matter in the ocean and in bottom sediments is decomposed soon after burial, presumably by bacteria. Some amino acids are more stable and decompose only after long intervals of time or at rather high temperatures (Abelson, 1954). Nitrate is produced in oxidizing environments, but ammonia forms in the reducing environment provided by sediments with appreciable organic content.

Ammonium ion, although generally not determined, seems to be one of the characteristic components of oil-field brines. Jensen's (1934) analyses of California oil-field brines indicate that nearly all contain ammonium with a mean of about 90 ppm. For California waters with salinities close to sea water, the NH_4 content is in the order of 150 ppm. It is present in nearly all Illinois oil-field brines (Meents et al, 1952) and ranges up to 309 ppm and averages about 40 ppm. The data for several other areas are summarized by Schoeller (1956, p. 145). The mean concentrations, in mg/l are: Roumania, 155; Polasna-Krasnokamsk, Russia, 122; Hungary, 10.6.

Nitrate has been searched for in only a few brines and seems to be enriched in some, relative to ocean and surface waters. Nitrate is formed in preference to, or from, ammonia where the oxidation potential is sufficiently high (Rittenberg et al., 1955, p. 29–38).

Ammonia has commonly been considered a volcanic component because it occurs in many volcanic and hot-spring gases and in fumarole deposits. If it is volcanic, however, its very erratic distribution has never been explained satisfactorily. It is absent in most Yellowstone hot springs (Allen and Day, 1935, p. 249, 373, 469), and it shows no systematic relation to hydrogen or other components (p. 86) that are likely to be volcanic in origin. Shepherd (1938, p. 322) suspected that ammonia is nonvolcanic and did not find it in any of his volcanic gases, even though elemental hydrogen and nitrogen are abundant in some of the samples. Rubey (1955, p. 642–645) concluded that there is no strong evidence for a magmatic source of ammonia. Rayleigh (1939) on the other hand, believed that combined nitrogen, presumably as ammonia, occurs in granite. Urey (1952, 1956) maintains that ammonia and methane were both important components of an early reducing atmosphere; presumably he also believes they should be present in magmatic gases.

Both ammonium and nitrate have been searched for but not found at Steamboat Springs, whose waters emerge from granitic rocks. It is possible that ammonia is not volcanic and that the ammonia of volcanic fumaroles and hot springs is derived entirely from sediments. A systematic search should be made for combined nitrogen in volcanic gases and waters that emanate from granitic and high-grade metamorphic rocks. If none is found, an organic origin is indicated for all combined nitrogen.

Combined nitrogen is characteristic of connate brines but may be lacking in waters of direct volcanic origin.

Hydrocarbons and other Organic Compounds

Hydrocarbons are, of course, abundant in gases and oils associated with oil-field brines. The lower hydrocarbons of the methane group are generally the most abundant. Little is known about the water-soluble organic acids and phenols that occur in many, if not most, connate brines (Shapiro, 1951; Sukharev, 1951). Smith (1931, p. 899) and Sukharev (1951) have been particularly interested in the napthenic acids in waters and their possible relation to petroleum deposits.

Methane and occasionally ethane have been found in the gases of some volcanic hot springs, but their distribution is highly erratic. Methane is absent, for example, in 5 of the 40 analyses of Yellowstone spring gases reported by Allen and Day (1935, p. 86) but is as much as 13 per cent of the total in one of the analyses. The proportion of methane seems unrelated to other components, including free hydrogen. It was searched for but not found in the gases emerging from granitic rocks at Steamboat Springs. Shepherd (1931, p. 81–82; 1938, p. 321, 326) was unable to find methane in Kilauea gases or in any of the gases he obtained by heating volcanic rocks, in spite of rather abundant hydrogen in some of the samples. He concluded that methane indicates contamination from sedimentary rocks or, presumably, from organic matter covered by lava flows, and the writer concurs. Water-soluble organic acids, phenols, and hydrocarbons, when present in a water, are indicative of a connate contribution from sedimentary rocks.

Silica

The silica of ocean water ordinarily ranges from less than 0.1 to about 8 ppm in spite of the fact that it is being contributed continuously in higher concentration by rivers. Its solubility with respect to amorphous silica in cold sea water is about 75 ppm according to Krauskopf (1956). The activity of silica-secreting organisms, the very slow rates of dissolution at low temperature, and perhaps differences in solubility of opaline and amorphous silica account for the low silica content of the ocean. The silica of oil-field brines is somewhat higher; it commonly ranges from about 10 to 50 ppm but rarely reaches as much as 90 ppm (Table 1). In view of the fact that the solubility with respect to amorphous silica at temperatures found in oil fields is from about 150 to more than 400 ppm, White, Brannock, and Murata (1956) suggest that diatoms are slowly dissolved, but seldom at a rate that attains saturation. As the diatoms dissolve, quartz and chalcedony, which are low-solubility forms of silica, are presumably deposited.

The solubility of amorphous silica in hot-spring waters (White, Brannock, and Murata 1956) ranges from about 110 ppm at 25°C. to about 315 ppm at 90°C. "Soluble" silica seems to be nonionized monomeric silicic acid (probably H_4SiO_4). It may polymerize very slowly from superstaturated waters to form colloidal molecules. Because of the sluggishness of the reactions, supersaturated hot spring waters are not uncommon, and many volcanic sodium-chloride waters are saturated at vent temperatures. The silica content of the volcanic springs of Table 2 ranges from 150 to more than 500 ppm.

The silica content of volcanic hot springs is nearly always significantly higher than that of oil-field brines. Volcanic springs of the calcium-bicarbonate type (White, 1957; White, Brannock, and Murata, 1956, p. 51), however, are commonly as low as 50 or 60 ppm, and heated connate brines that are rapidly altering or dissolving silicate rocks such as serpentinite may contain amounts within the normal range for volcanic hot springs.

Alkaline-Earth Elements

Concentrations of the alkaline-earth elements are strongly dependent on pH, temperature, and other substances in the waters.

Magnesium is strongly dominant over calcium in sea water; the ratio is about 3 to 1 (Table 1). Nearly all oil-field brines, however, show a reversal of relationships with the Mg/Ca ratio as low as 0.03. The change probably has some bearing on the origin of diagenetic dolomite.

Probably because carbonate and bicarbonate are relatively high, the more dilute brines commonly have a lower ratio of $\frac{Ca + Mg}{Na + K}$ than ocean water. Calcium is greatly enriched however, in most, if not all, of the highly saline oil-field brines and in a few waters with salinities similar to sea water (Hudson and Taliaferro, 1925). Many of the very saline brines are relatively low in pH (Meents et al., 1952) and range from 5 to 7 in spite of the fact that combined CO_2 is very low. The low content of combined CO_2 seems to be related to the very high calcium content and to calcium-carbonate equilibria.

The calcium content of most volcanic hot springs in igneous rocks is very low relative to the alkalies because of the low solubility of calcium and magnesium carbonates at high temperature (Garrels and Dreyer, 1952, p. 339, 340; Miller, 1952; and White, 1957). In the waters, calcium is generally greatly dominant over magnesium, probably because the latter is selectively fixed in argillic minerals during rock alteration. Ca and Mg are commonly very low even in neutral and slightly acid waters with abundant free CO_2, probably because these waters have been at much higher temperatures at depth. Volcanic spring waters of the calcium-bicarbonate type (White, 1957) are believed to attain their character at relatively low temperature through the mixing in limestone beds of a little volcanic water with very abundant meteoric water. This is consistent with the fact that springs of this type seldom if ever have boiling temperatures at the surface.

The ratios of calcium and magnesium are likely to be similar in oil-field brines and volcanic sodium-chloride springs, but in the latter the concentrations relative to the alkalies are generally low.

Strontium is probably present in all connate and volcanic waters but has seldom been determined. It seems to be most abundant in oil-field brines that are low in combined CO_2 and attains concentrations as high as 2600 ppm in West Virginia brines (Price et al., 1937, p. 67, 98), and 3550 ppm in Pennsylvania brines (Mills and Wells, 1919, p. 39). The solubilities of strontium carbonate and sulfate are high enough to permit detectable concentrations in most natural waters.

Barium is relatively abundant in oil-field brines that have little or no sulfate (Heck, 1940). Some West Virginia brines, for example, contain little sulfate but as much as 5500 ppm of barium (Price et al., 1937, p. 50–51, 91). The high barium content is clearly related to bacterial reduction of sulfate in the water and in sediments (Heck, 1940). Barium goes into solution when all or nearly all sulfate in the water has been reduced. Barite is readily precipitated if these waters are later oxidized or if they mix with normal meteoric water.

The barium content of volcanic hot springs is seldom appreciable, perhaps because of the relatively high sulfate content that characterizes these waters. If volcanic water at depth is low in sulfate, it may, of course, contain barium.

Other Elements

More than 1,000 ppm of iron has been reported in a few oil-field brines (Schoeller, 1956

p. 151–153) but most contain about 10 ppm or less. The content differs greatly in waters that are otherwise very similar (Meents et al., 1952). Only traces of iron are found in volcanic hot springs of the sodium-chloride type.

As much as 330 ppm of AL_2O_3 has been found by Meents et al. (1952, p. 36) in Illinois brines, most of which are very high in salinity. The average alumina content is 20–30 ppm. Alumina, generally in smaller quantities, has been reported in many other oil-field brines. Volcanic hot springs of the sodium-chloride type are notably low in alumina; they seldom if ever contain more than 1 ppm.

Manganese is a minor component of many oil-field brines and is rather commonly in quantities as much as 1 ppm and exceptionally more than 10 ppm (Meents et al., 1952, p. 37; Schoeller, 1956, p. 151). The quantity in volcanic sodium-chloride springs has seldom been determined but appears generally to be less than in oil-field brines.

A trace of arsenic was found by Howard (1906, p. 284) in a West Virginia brine and the surprising amount of 5.8 ppm was reported by Telegdi-Roth (1950, p. 81) from the same Hungarian brine that purportedly contains 84 ppm of Li. Mercury has been reported from wells in the Cymric field of California (Stockman, 1947). Oil-field brines should be searched systematically for arsenic, antimony, and mercury to determine whether appreciable quantities are present. Volcanic hot springs of the sodium-chlorice type commonly contain at least 1 ppm of arsenic and detectable amounts of antimony. Their mercury content is not yet well known because of analytical problems.

NONVOLCANIC HOT SPRINGS

Discussion

Many springs of low to moderate temperature have compositions differing markedly from springs that most clearly have volcanic contributions. The dilute waters are best explained as deeply circulating meteoric waters that are heated by rock conduction and have picked up some additional mineral matter. Some moderately to highly mineralized meteoric waters have no doubt come in contact with salt beds. The possibility must now be recognized, however, that some springs may be fed by connate water from sediments that are being compacted, or by water from rocks that are being progressively metamorphosed. According to Clarke (1924, p. 631), as mentioned previously, average shale has 5.0 per cent water, slate has 3.8 per cent, and schist 2.0 per cent. No averages are given for gneiss, but individual analyses range from 0.4 to 1.8 per cent (p. 630) and have a probable average that is not much greater than the 0.65 per cent reported for average granodiorite (Nockolds, 1954, p. 1014). It is evident that very large quantities of water, not represented by normal oil-field brines or by volcanic hot springs, are driven off from sediments during their progressive compaction and metamorphism. Water-saturated shale that decreases in porosity from 20 to 10 per cent during compaction must lose 10^{11} liters of water per km^3 of sediments. Shale with a water content of 5 per cent also must lose approximately 10^{11} liters of water per km^3 if metamorphosed to gneiss with a water content of 1 per cent. In both situations the quantity of water driven off from a cubic kilometer of rock happens to be about the same, and each could supply a discharge of 20 liters or 5 gallons per minute for 10,000 years.

Hot Springs with Connate Water

The analyses of two hot spring waters believed to be dominantly connate in origin are shown in Table 5. Wilbur Springs, California, has a maximum surface temperature of about 60°C. and Tuscan Springs, California, a maximum of about 30°C. Mercy Springs, California, is also believed to be connate in part, but rather extensively diluted by meteoric water. The combination of high salinity and low temperature are the characteristics that first interested the writer in Wilbur Springs and that led to the present conclusions.

In spite of their relatively low temperatures, these saline springs could conceivably be volcanic if: (1) they are young enough, and flow through cool rocks that have not yet been heated extensively, or (2) they are cooled by mixture with enough cold meteoric water, or (3) their discharge is so small that most of the heat is carried off by normal rock conduction. Relative youth is an unlikely explanation, at least for Wilbur, because the springs are closely associated with mercury deposits (Waring, 1915, p. 99–106; White 1955, p. 130–132) that apparently are not forming now but probably were formed in Quaternary time. The high salinity and great excess of sulfide over sulfate argue strongly against cooling by extensive mixture with meteoric water.

The rate of discharge of saline water at Wilbur Springs and within ⅛ mile of the resort is relatively small (about 40 gpm, computed to chlorinity of main Wilbur Spring) as compared to Steamboat Springs (about 700 gpm including subsurface discharge from the system). Because of great differences in salinity, however, the discharge of chloride at Wilbur Springs is about 1.7 kg per minute or 70 per cent of the 2.5 kg that is discharged at Steamboat Springs. The drainage area of Sulphur Creek above Wilbur Springs discharges additional water that is at least twice the total quantity of Wilbur, calculated to the same chlorinity. At least 6 kg per minute of chloride from water similar to that of Wilbur is discharged from a total drainage area of 25 km^2.

The flow of heat accompanying the discharge of water at Steamboat Springs is about 7 x 10^6 cal./sec., which is equivalent to the normal flow of heat from 600 km^2 of the earth's surface (White, 1957). This heat flow, from an area of about 5 km^2, is 120 times the "normal" flow (Birch, 1954). Only the heat contained in water above the mean annual temperature is included in these calculations.

The only obviously abnormal flow of heat in the Sulphur Creek area is in the water of Wilbur and other similar springs, which flow from rocks of the Knoxville Formation of Upper Jurassic age, adjacent to or near serpentine intrusions. The springs are not particularly gaseous and, because of the low temperatures, relatively little heat is lost in steam or water vapor, or by rock conduction. The flow of heat above the mean annual temperature in the discharging water is about 4 x 10^5 cal./sec. from 25 km^2, which is nearly equivalent to what the "normal" flow of heat by conduction should be from the same area (3 x 10^5 cal./sec.). The geothermal gradient is no doubt somewhat above the normal gradient; the total heat flow, including that of flowing water and rock conduction, is probably 3–5 times the "normal" for an area of this size but falls far short of the 120 multiples of "normal" flow from the 5 km^2 area of Steamboat Springs. The very abnormal heat flow at Steamboat must be explained as volcanic heat (White, 1957), but the slightly abnormal flow at Wilbur Springs appears explainable by the upflow of moderately heated water of non-volcanic origin.

The analyses of Wilbur and Tuscan Springs are similar to oil-field brines in a number of respects (Tables 1, 4, 5). The high iodine content of Wilbur and near-by springs has been recognized for many years (Watts, 1893; Waring, 1915, p. 102, 105), and Tuscan Springs is now known to be enriched in iodine. Both spring groups are also high in ammonium, in the ratio of sulfide to sulfate, and low in potassium and lithium relative to sodium. These chemical characteristics are believed to point strongly to a connate rather than a volcanic origin, supporting the evidence of heat flow. The similarity in composition between the waters of Wilbur Springs and the oil-test well 1 mile southeast of Wilbur is apparent (Tables 1, 5). Inflammable gases, presumably largely methane, accompanies the Tuscan waters (Waring, 1915, p. 289) and hot springs in the Wilbur Springs area (Watts, 1893, p. 182–183). In addition, isotopes of the waters of Wilbur and Tuscan are similar to those of Texas oil-field brines, according to Harmon Craig (personal communication). The quantities of some other components, such as the relatively high silica and low calcium, suggest a relation to volcanic hot springs. The boron content of Wilbur and Tuscan waters is higher than for any oil-field brine known to the writer, but the boron-chloride ratios are very similar to that of the Rio Vista gas-field brine. The bicarbonate content of Wilbur is exceptionally high, and some other components fall within the known ranges of both connate waters and volcanic hot springs.

The characteristics of Wilbur and Tuscan Springs are best explained by a dominantly connate origin. The differences between these spring waters and known oil-field brines may be explained in three ways: (1) Enough volcanic water is also present to account for the differences in characteristics, but not enough to affect the heat flow drastically. (2) Connate water, in changing from ocean water to oil-field brine, is approaching the composition of volcanic water in many respects as organic compounds decompose and as minerals, such as clays, are reconstituted and release adsorbed ions. Some of these changes have progressed a little further in the Wilbur and Tuscan waters in most oil-field brines. The high silica content of Wilbur may be due to the relatively high temperature of the source environment or to contact with rocks such as serpentine that may react readily to release monomeric silica. (3) As indicated below the exceptionally high bicarbonate and boron content suggests the presence of some metamorphic water. The writer favors the latter possibility, although the presence of a small percentage of volcanic water mixed with dominately connate water cannot be disproved by present evidence.

Many other springs are believed to be connate, at least in part. Bad Hall in Austria

TABLE 5.—ANALYSES OF THERMAL WATERS BELIEVED TO CONTAIN WATER OF CONNATE OR METAMORPHIC ORIGIN
(In parts per million)

	Wilbur Colusa Co. California[1]	Tuscan Tehama Co. California[2]	Mercy Fresno Co. California[3]	Bad Hall Austria[4]	Skaggs Sonoma Co. California[5]	Aetna Napa Co. California[6]	Sulphur Bank Lake Co. California[7]
Temp. °C.	57	30.1	45.8		54½	22	57
pH	7.24	8.3	8.6		7.2	8.1	
SiO_2	190	99	75	8.9	74	97	86
Fe		0.7[8]		6.1			0
Al	1[8]	1[8]		Tr.			
Ca	1.4	22	43	245	14	24	11
Mg	58	17	nil	156	5.4	72	0
Ba	1[8]	<4[8]					
Sr	10[8]	14[8]		13.0			
Na	9140	8080	830	6460	912	328	1550
K	460	51	7.1	10.4	33	6.0	50
Li	14	2.0	0.13	2.3	0.46	0.3	
NH_4	303	59	4.9	52		0.2	507
As	0.0						0(?)
Sb	0.0						0(?)
CO_3	0	103	31	425	0	0	0
HCO_3	7390	1150	13	0	2480	1060	3830
SO_4	23	67	5		6.5	0.5	119
Cl	11,000	11,800	1300	10,700	58	143	741
F	1.1	4.8	0.4		8.0	0.6	
Br	15	nil(?)	nil(?)	78		0(?)	Tr.
I	16	5.0	20	39		0.7	1.4
NO_3		0	0.5	0.03	0.5	2.1	
NO_2							
B	293	201	10	41	92	36	720
H_2S	178	185			0		Present
CO_2				26			
Sum	29,100	21,900	2350	18,200	3680	1770	7620

Ratios by wt.						
HCO₃⁹/Cl	0.67	0.11	0.059	0.040	43	7.4	5.2
SO₄/Cl	0.0021	0.0057	0.004	0.00000	0.11	0.003	0.16
F/Cl	0.0001	0.00041	0.0003		0.14	0.004	
Br/Cl	0.0014	<0.0001(?)	<0.0001(?)	0.0073		(<0.01)	<0.001
I/Cl	0.0015	0.00042	0.015	0.0037		(0.005)	0.0019
B/Cl	0.027	0.017	0.0077	0.0038		0.25	0.97
K/Na	0.050	0.0063	0.0085	0.0016	0.036	0.018	0.032
Li/Na	0.0015	0.00025	0.00016	0.00036	0.0005	0.0009	
Ca + Mg/Na + K	0.0062	0.0048	0.051	0.062	0.020	0.29	0.0069

[1] Mainspring (No. 22 of Waring, 1915, p. 102). Collected by White, August 3, 1949, analyzed by W. W. Brannock, U. S. Geological Survey. Density, 25°C., 1.016. Br and I determined in sample collected June 30, 1951.

[2] Largest flowing spring, collected by White, June 9, 1954; rises in large concrete tank (Waring, 1915, p. 290–291). Analyzed by H. Kramer and H. Almond, U. S. Geological Survey; contains 0.3 ppm Cu by spectrographic determination.

[3] Main Spring (Waring, 1915, p. 78–79). Collected by White, June 13, 1955, analyzed by H. Almond and S. Berman, U. S. Geological Survey.

[4] Guntherhöhen spring. Schmölzer, 1955. Also contains traces of Mn and Cu. May be nonthermal.

[5] Middle of 3 principal springs, collected by D. F. Hewett, August 20, 1953, analyzed by R. N. Martin of U. S. Geological Survey. Li from sample collected by White, from upper spring January 15, 1954 and analyzed by H. Kramer; latter also contained 0.2 ppm Fe, 0.1 Al, 0.01 Mn, 0.02 Cu, 1.0 Sr, 0.007 Cr, and only 0.04 Li by spectrographic determination.

[6] Soda spring, sec. 1, T. 9 N., R. 6 W. Collected by Calif. Division of Water Resources, May 3, 1955, analyzed by J. J. Popper, U. S. Geological Survey. Hotter (35° C.) and more highly mineralized water (1850 ppm HCO₃, 292 Cl, and 78 B) occurs in the shaft of an old quicksilver mine at the Aetna resort, but analyses are less complete (Waring, 1915, p. 156–159).

[7] From diamond drill hole supplying water to bathhouse. Collected by C. P. Ross, October 1938, analyzed by M. Fleischer, U. S. Geological Survey. Br and I from 1863 analysis (Whitney, 1865, p. 99–100).

[8] Spectrographic analysis.

[9] All carbonate and bicarbonate calculated as HCO₃.

(Schmölzer, 1955) is an example (Table 5). It is not clear whether these springs are cold or slightly thermal but Schmölzer believes their high iodine content indicates a connate origin.

Other analyzed springs (Waring, 1915, unless noted) with many or most of the chemical aspects of connate water are: Byron, California (p. 110–111); Fouts, California (p. 207); Neys, California (p. 265); Richardson, California (p. 292); Alhambra, California (p. 294); Complexion, California (p. 298); Tolenas, California (p. 163); Clifton, Arizona (Hem, 1950, p. 82); Saratoga, New York (Strock, 1941, p. 857); Kuan-Tsu-Ling, Formosa (Pan, 1952); Kwanshirei, Formosa (Ishizu, 1915, p. 162); Hamman Lif, Tunisia (Berthon, 1927); Bad Hamm, Germany (Himstedt, 1907, p. 163); Arima, Japan (Ikeda, 1949, p. 363); Debrecen and Hajdu-szoboszlo, Hungary (Papp, 1949, p. 300); and springs in Turkmenian Russia (Smolko, 1932).

Hot Springs with Metamorphic Water

No natural water has been proved to originate from sediments as a result of progressive metamorphism, although the almost universal decrease in water content with increase in grade of metamorphism demands the existence of migrating water of this origin. Little attention has been given in the past to metamorphic water, or to criteria for its recognition. It presumably escapes very slowly from rocks that are being progressively metamorphosed, and no doubt ordinarily mixes with large proportions of meteoric and connate water in overlying rocks. All geologists are familiar with old metamorphic rocks that are now revealed after extensive erosion of the original cover, but very little thought has been given to the probability that metamorphism is now taking place at depth in some parts of the world.

Many springs described by Waring (1915, p. 155–249) have compositions that are similar in part to connate water but are very high in combined CO_2 and commonly in boron, with relatively low chloride. Their temperatures, with few exceptions, are less than 60°C., and are always less than boiling. Many are in the California Coast Ranges north of San Francisco, and issue from or near rocks of the Franciscan group. The springs show a considerable range in composition but also have some features in common. As examples, analyses of Skaggs, Sulphur Bank, and Aetna springs are included in Table 5. All three springs are closely associated with mercury deposits (White, 1955, p. 117–120, 125, 130).

The isotopes of the water of Sulphur Bank are almost identical to those of Wilbur and Tuscan Springs (Harmon Craig, personal communication). The Sulphur Bank water is so different from the near-by Clear Lake water that it can have little admixed meteoric water.

The dissolved solids of these springs consist for the most part of sodium bicarbonate; HCO_3/Cl ratios are exceptionally high for known connate or volcanic waters. The boron-chloride ratios are among the highest of all analyzed waters. The few springs of this type that have been analyzed for ammonium and nitrate are relatively high in combined nitrogen.

Can these waters be at least in part of metamorphic origin? It is reasonable to suppose that, as sedimentary rocks are compacted and progressively metamorphosed, connate water high in chloride is first driven off. As the ratio of interstitial water to originally combined water decreases, the content of chloride in the escaping water may also decrease, because chloride is not abundantly adsorbed on clays and is rare in other hydrous minerals. When these minerals are recrystallized to minerals with lower water content or to anhydrous minerals, water with relatively little chloride is presumably freed.

With sufficient metamorphism, quartz combines with carbonates to form silicates and releases CO_2 (*See*, for example, Urey, 1956 p. 1126–1127). It is reasonable to suppose that some metamorphic waters, therefore, are high in CO_2 and in HCO_3/Cl ratios. James (1955, p. 1462, 1484–1485) indicates that the reaction between dolomite and quartz has produced some tremolite in the upper part of the biotite zone of metamorphism at a temperature that he concludes is close to 200°C. Where silicates are formed in the presence of only a little water, a CO_2-rich vapor phase or two fluid phases, one very high in CO_2, may form. At low temperatures where the associated water is below its boiling point, the reaction of carbonate and silica may take place to a limited extent until the water is saturated with CO_2. Conceivably, this could even occur in the low-grade metamorphic zones, but limited so much by the solubility of CO_2 that it has not been recognized. Another possibility is the reaction of carbonate and silica to form minerals such as prehnite and laumontite, and CO_2 in very low-grade metamorphic zones (Fyfe, 1955).

Little is known of the geochemistry of boron

in metamorphic rocks, or of the possibilities for enrichment of boron in metamorphic waters. Boron is greatly enriched in argillaceous marine sediments and ranges from about 15 to 300 ppm (Goldschmidt, 1954, p. 285), as compared with average igneous rocks, which contain only about 3 ppm (Rankama and Sahama, 1950, p. 226; Goldschmidt, 1954, p. 281). If many igneous rocks are formed by melting or partial fusion of sediments, as numerous geologists believe, most of the boron must be driven off before final crystallization. Although much boron may be evolved in the recycled magmatic water, it is reasonable to suspect that a large part could be driven off at an earlier stage in metamorphic water. This hypothesis remains to be tested by future geochemical studies.

Data on the boron content of glassy volcanic rocks (Coats, 1956, p. 76) and of thermal waters (some data supplied by D. F. Hewett of the U. S. Geological Survey) suggest that California, western Nevada, and southern Oregon constitute a high-boron province. If this is a fact, metamorphic waters in other parts of the world may not be characterized by boron-chloride ratios as high as in the California waters that have been described.

H. E. LeGrand of the U. S. Geological Survey (personal communication) has wisely suggested that a word of caution be added here. Care should be taken to guard against classifying many puzzling waters as metamorphic because no more obvious explanation is apparent.

Thermal Waters of Mixed Origin

It is assumed in the previous discussion that any of the deep waters may contain from negligible to high proportions of meteoric water but are still characterized chemically and identified by the deep components. Many other combinations of mixture are possible—deep waters with each other, with water immediately from the ocean by subsurface circulation, or with meteoric water that has come in contact with salt-bearing sediments.

A large number of thermal springs are relatively low in dissolved solids (1500 ppm or less) and differ appreciably from near-by meteoric waters. The waters may be almost entirely meteoric in origin; they perhaps contain traces of highly mineralized water of deep origin that have modified the composition only enough to make it unrecognizable, or additional mineral matter may have been leached from rocks.

There is, therefore, much danger of oversimplifying the problems of identification. This paper is a preliminary attempt to recognize some of the outstanding genetic types of ground water.

Conclusions and Speculations

Tentative criteria are proposed to distinguish waters of different origins; the results may be of significance for many fields of geology. For example, the ground water supply of many areas is threatened by salt-water encroachment. In coastal areas, the two principal possibilities are the direct invasion of sea water, and connate water from marine sediments. These two types should now be clearly distinguishable in many circumstances.

Probably all changes that occur in rocks— from diagenetic to metamorphic changes—are reflected in the compositions of the interstitial waters. The ability to correlate and to interpret these changes is still in its infancy. Direct economic applications of this principle are likely to appear first in the search for petroleum. The major components of oil-field brines have been of some interest in the past, but the minor components have been largely overlooked. Local differences in chemical and isotopic compositions of oil-field brines must be significant, but ability must be developed to interpret these differences in terms of the history and associations of the waters.

Many economic geologists assume that ore deposits are formed from hydrothermal solutions, and that these solutions are magmatic in origin. The present work indicates that hydrothermal solutions of many different origins may be transporters of ore. An important field for future research is to analyze chemically and isotopically the fluid inclusions of ore and gangue minerals for components that are likely to indicate the origin of the water and probably also of the ore components.

References Cited

Abelson, P. H., 1954, Paleobiochemistry, p. 97–101 *in* Ann. Rep. of Director of Geophys. Lab. for year 1953-54: Carnegie Inst. Washington Pub. 1235, p. 95–145

Allen, E. T., and Day, A. L., 1935, Hot springs of the Yellowstone National Park: Carnegie Inst. of Wash. Pub. 466, 525 p.

Anderson, C. C., and Hinson, H. H., 1951, Helium-bearing natural gases of the United States: U. S. Bur. Mines Bull. 486, p. 1–141

Bastin, E. S., and Greer, F. E., 1930, Additional data on sulfate-reducing bacteria in soils and waters of Illinois oilfields: Am. Assoc. Petroleum Geologists Bull., v. 14, p. 153–159

Beerstecher, E., Jr., 1954, Petroleum microbiology: London, Elsevier Press, v. 1, p. 1–375

Bemmelen, R. W. van, 1949, The geology of Indonesia: The Hague, Govt. Printing Office, v. 2, Economic geology, p. 185–190

Berger, W. R., and Fash, R. H., 1934, Relation of water analyses to structure and porosity in the West Texas Permian Basin, p. 869–899 *in* Wrather, W. E., and Lahee, F. H., *Editors*, Problems of petroleum geology, (Sidney Powers memorial volume): Am. Assoc. Petroleum Geologists, 1073 p.

Berthon, L., 1927, Étude sur les sources thermo-minérales de la Tunisie: Serv. des mines et de la Carte Géol., premier facs., régiones de Gabes et de Tunis, p. 1–177

Birch, F., 1954, The present state of geothermal investigations: Geophysics, v. 19, p. 645–659

Borovik-Romanova, F., 1946, Rubidium in the biosphere: Trudy Biogeokhim. Lab., Akad. Nauk SSSR., v. 8, p. 145–180

Case, L. C., 1955, Origin and current usage of the term "connate water": Am. Assoc. Petroleum Geologists Bull., v. 39, p. 1879–1882

Chajec, W., 1949, Iodine and bromine in brines from petroleum boreholes: Nafta, v. 5, p. 366–372

Chebotarev, I. I., 1955, Metamorphism of natural waters in the crust of weathering: Geochim. et Cosmochim. Acta, v. 8, p. 22–48, 137–170, 198–212

Clarke, F. W., 1924, The data of geochemistry: U. S. Geol. Survey Bull. 770, p. 630–631

Coats, R. R., 1956, Uranium and certain other trace elements in felsic volcanic rocks of Cenozoic age in Western United States: U. S. Geol. Survey Prof. Paper 300, p. 75–78

Correns, C. W., 1956, The geochemistry of the halogens, p. 181–233 *in* Ahrens, L. H., Rankama, K., and Runcorn, S. K., *Editors*, Physics and chemistry of the earth, v. 1, 317 p.: McGraw-Hill Book Co., Inc., N. Y.

Craig, H., Boato, G., and White, D. E., 1956, The isotopic geochemistry of thermal waters: Nat. Research Council Nuclear Sci. Ser., Rept. no. 19, Nuclear processes in geologic settings, p. 29-44

Crawford, J. G., 1940, Oil-field waters of Wyoming and their relation to geologic formations: Am. Assoc. Petroleum Geologists Bull., v. 24, p. 1214–1329

Crawford, J. G., 1942, Oil-field waters of the Montana Plains: Am. Assoc. Petroleum Geologists Bull., v. 26, p. 1317–1374

—— 1949, Water analysis characteristics of oil-field waters of Rocky mountain region: Colo. School Mines Quart., v. 44, no. 3, p. 188–210

De Sitter, L. U., 1947, Diagenesis of oil-field brines: Am. Assoc. Petroleum Geologists Bull,, v. 31, p. 2030–2040

Dobbin, C. E., 1935, Geology of natural gas rich in helium, nitrogen, carbon dioxide, and hydrogen sulfide, p. 1053–1072 *in* Wrather, W. E. and Lahee, F. H., *Editors*, Geology of natural gas: Am. Assoc. Petroleum Geologists, 1073 p.

Dzens-Litovsky, A. I., 1944, Iodine and bromine in the brine of mineral lakes, in natural brines, and in oil waters: Acad. Sci. U. R. S. S., Compt. Rend. (Doklady), v. 45, p. 65–66

Emery, K. O., and Rittenberg, S. C., 1952, Early diagenesis of California basin sediments in relation to origin of oil: Am. Assoc. Petroleum Geologists Bull., v. 36, p. 735–806

Emery, K. O., Orr, W. L., and Rittenberg, S. C., 1955, Nutrient budgets in the oceans, p. 299–309 *in* Essays in the Natural Sciences in honor of Captain Allan Hancock: Los Angeles, Univ. Southern Calif. Press, 345 p.

Fyfe, W. S., 1955, Lower limit of the green-schist facies (Abstract): Geol. Soc. America Bull., v. 66, p. 1649–1650

Garrels, R. M., and Dreyer, R. M., 1952, Mechanism of limestone replacement at low temperatures and pressures: Geol. Soc. America Bull., v. 63, p. 325–379

Ginter, R. L., 1934, Sulphate reduction in deep subsurface waters, p. 907–925 *in* Wrather, W. E., and Lahee, F. H., *Editors*, Problems of petroleum geology (Sidney Powers memorial volume): Am. Assoc. Petroleum Geologists, 1073 p.

Goldschmidt, V. M., 1954, Geochemistry: Oxford, Clarendon Press, p. 1–730

Gulyaeva, L. A. 1948, Boron content in recent sea silts: Akad. nauk SSSR Doklady, v. 60, p. 833–835

Heck, E. T., 1940, Barium in Appalachian salt brines: Am. Assoc. Petroleum Geologists Bull., v. 24, p. 486–493

Hem, J. D., 1950, Quality of water of the Gila River Basin above Coolidge Dam, Arizona: U. S. Geol. Survey Water-Supply Paper 1104, p. 82

Himstedt, F., 1907, Deutsches Baderbuch: bearbeitet unter Mitwirkung des K. Gesundheitsamtes, p. 1–535

Hoskins, H. A., 1947, Analyses of West Virginia brines: W. Va. Geol. Survey, Rept. Inv. 1, p. 1–22

Howard, C. D., 1906, Occurrence of barium in the Ohio Valley brines and its relation to stock poisoning: West Virginia Univ. Agr. Expt. Sta. Bull. 103, p. 281–295

Hudson, F. S., and Taliaferro, N. L., 1925, Calcium chloride waters from certain oil-fields in Ventura Co., Calif.: Am. Assoc. Petroleum Geologists Bull., v. 9, p. 1071–1088; 1926, v. 10, p. 775–778

Ikeda, N., 1949, Geochemical studies on the hot springs of Arima, pt. 2: Chem. Soc. Japan Jour., Pure Chemistry Sec., v. 70, p. 363–366

Ingerson, E., 1934, Relation of critical and super critical phenomena of solutions to geologic processes: Econ. Geology, v. 29, p. 454–470

Ishizu, R., 1915, The mineral springs of Japan, pt. 2: Tokyo Hygienic Lab., p. 1–203

James, H. L., 1955, Zones of regional metamorphism in the Precambrian of northern Michigan: Geol. Soc. America Bull., v. 66, p. 1455–1488

Jeffords, R. M., 1948, Graphic representation of oilfield brines in Kansas: Kansas State Geol. Survey Bull., v. 76, pt. 1, p. 1–12

Jensen, J., 1934, California oilfield waters, p. 953–985, *in* Wrather, W. E., and Lahee, F. H., *Editors*, Problems of petroleum geology (Sidney Powers memorial volume); Am. Assoc. Petroleum Geologists, p. 1073.

Kazmina, T. I., 1951, Boron-chloride coefficient in waters of petroleum deposits: Akad. nauk SSSR Doklady, v. 77, p. 301–303

Krauskopf, K. B., 1956, Dissolution and precipitation of silica at low temperatures: Geochim. et Cosmochim. Acta, v. 10, p. 1–26

Lamborn, R. E., 1952, Additional analyses of brines from Ohio: Ohio Dept. Nat. Res., Div. Geol. Survey, Rept. Inv. no. 11, p. 1–56

Landergren, S., 1945, Contribution to the geochemistry of boron; II, The distribution of boron in some Swedish sediments, rocks, and iron ores; the boron cycle in the upper lithosphere: Arkiv Kemi, Miner. Geol. (K. Svenska Vetensk.) Bd. 19, H. 5, A no. 26, p. 1–31

Lane, A. C., 1908, Mine waters and their field assay: Geol. Soc. America Bull., v. 19, p. 501–512

Libby, W. F., 1954, Radiocarbon and tritium dating. Possibility of using these isotopes in industry: Z. Elektrochem., v. 58, p. 574–585

McGrain, P., 1953, Miscellaneous analyses of Kentucky brines: Ken. Geol. Survey, Rept. Inv. no. 7, p. 1–16

Meents, W. F., Bell, A. H., Rees, O. W., and Tilbury, W. G., 1952, Illinois oilfield brines, their geologic occurrence and chemical compositions: Ill. State Geol. Survey, Petroleum Bull. 66, p. 1–38

Meinzer, O. E., 1923, Outline of ground-water hydrology, with definitions: U. S. Geol. Survey Water-Supply Paper 494, p. 1–71

Miller, J. C., 1936, Occurrence, preparation, and utilization of natural carbon dioxide: Am. Inst. of Mining Engineers Tech. Pub. No. 736, p. 3–22

—— 1937, Carbon dioxide accumulations in geologic structures: Am. Inst. of Mining Engineers Tech. Pub. No. 841, p. 1–28

Miller, J. P., 1952, A portion of the system calcium carbonate-carbon dioxide water, with geologic implications: Am. Jour. Sci. v. 250, p. 161-203

Mills, R. van A., and Wells, R. C., 1919, The evaporation and concentration of water associated with petroleum and natural gas: U. S. Geol. Survey Bull. 693, p. 1–104

Nockolds, S. R., 1954, Average chemical compositions of some igneous rocks: Geol. Soc. America Bull., v. 65, p. 1007–1032

Pan, K., 1952, Chemical composition of the hot spring in Kuan-Tsu-Ling (Taiwan): Bull. Agr. Chem. Dept., Nat'l. Taiwan Univ., v. 1, p. 22–26 (analysis in Chem. Abstracts, 1954, v. 48, p. 11,998)

Papp, F., 1949, Les eaux medicinales de la Hongrie: Hidrologiai Kozlony, v. 29, p. 295–300

Piper, A. M., et al., 1953, Native and contaminated ground waters in the Long Beach-Santa Ana area, Calif.: U. S. Geol. Survey Water-Supply Paper 1136, p. 1–320

Price, P. H., Hare, C. E., McCue, J. B., and Hoskins, H. A., 1937, Salt brines of West Virginia: W. Va. Geol. Survey, v. 8, p. 1–203

Rankama, K., 1954, Isotope geology: London, Pergamon Press, Ltd., 535 p.

Rankama, K., and Sahama, T. S. 1950, Geochemistry: Univ. of Chicago Press, 912 p.

Rayleigh, Lord, 1939, Nitrogen, argon, and neon in the earth's crust, with applications to cosmology: Royal Soc. London Proc., v. A170, p. 451–464

Revelle, R., 1941, Criteria for recognition of sea water in ground waters: Am. Geophys. Union Trans., v. 22, p. 593–597

Rittenberg, S. C., Emery, K. O., and Orr, W. L., 1955, Regeneration of nutrients in sediments of marine basins: Deep-sea research, v. 3, p. 23–45.

Rogers, G. S., 1917, Chemical relations of the oil-field waters in San Joaquin Valley, California: U. S. Geol. Survey Bull. 653, p. 1–119

Rubey, W. W., 1955, Development of the hydrosphere and atmosphere with special reference to probable composition of the early atmosphere, p. 631–650 in Poldervaart, Arie, Fditor, Crust of the earth: Geol. Soc. America Special Paper 62, 762 p.

Russell, W. L., 1933, Subsurface concentration of chloride brines: Am. Assoc. Petrol. Geologists Bull., v. 17, p. 1213–1228

Sawyer, F. G., Ohman, M. F., and Lusk, F. E., 1949, Iodine from oil-well brines: Indus. and Eng. Chemistry, v. 41, no. 8, p. 1547–1552

Schmölzer, A., 1955, The geochemistry of iodine-brine springs: Chem. Erde, v. 17, p. 192–210

Schoeller, H., 1956, Geochimie des eaux souterraines. Application aux eaux des gisements de pétrole: Société des Editions, Paris, p. 1–213; 1955, Rev. Int. Franc. Petrole, v. 10, p. 181–213, 219–246, 507–552, 671–719, 823–874

Schoewe, W. H., 1943, Kansas oil-field brines and their magnesium content: Kansas Geol. Survey, Bull. 47, pt. 2, p. 37–76

Schröer, E., 1927, Untersuchungen über den Kritischen Zustand, I.: Zeitshr. physikal. chemie, v. 129, p. 79–110

Scofield, C. S., 1933, South coastal basin investigations, quality of waters: Calif. Dept. Public Works, Div. Water Resources Bull. 40-A, p. 127–128

Shapiro, S. A., 1951, The content of phenols in water of "Naftusya" No. 1 and No. 2 springs at the Truskavets health resort: Ukrain. Khim. Zhur., v. 17, p. 447–484

Shepherd, E. S., 1931, Gases in rocks and volcano gases: Carnegie Inst. Washington Yearbook. v. 30, 1930–31, p. 78–82

—— 1938, The gases in rocks and some related problems: Am. Jour. Sci., 5th ser., v. 35-A, p. 311–351

Smith, J. E., 1931, Venezuelan oil-field waters: Am. Assoc. Petroleum Geologists Bull., v. 15, p. 895–909

Smolko, G. I., 1932, The iodic springs in the western part of the Turkmenian S. S. R.: Glavnoe geologo-razvedochnoe upravlenie, Trudy, v. 175, p. 1–72

Stockman, L. P., 1947, Mercury in three wells at Cymric: Petroleum world, Feb. 1947, p. 37

Strock, L. W., 1941, Geochemical data on Saratoga mineral waters, applied in deducing a new theory of origin: Am. Jour. Sci., v. 239, p. 857–898

Sukharev, G. M., 1951, Estimation of the problem of the presence of petroleum deposits from the hydrochemical and temperature characteristics: Akad. nauk SSSR Doklady, v. 77, no. 4, p. 645–647

Tageeva, N. V., 1942, Fluorine and boron in natural waters and their bearing on the occurrence of petroleum: Compt. Rend., Acad. Sci., U. R. S. S., v. 34, p. 117–120

Telegdi-Roth, K., 1950, Chemical composition of the waters of exploratory and producing oil and gas wells in Hungry: Földtani Közlony, v. 80, p. 17–98

Timm, B. C., and Maricelli, J. J., 1953, Formation waters in southwest Louisiana: Am. Assoc. Petroleum Geologists Bull., v. 37, p. 394–409

Tzeitlin, S. G., 1936, Der Borgehalt in Ölfeldwassern: Compt. Rend. (Doklady) Acad. Sci. U. R. S. S., v. 1 (X), no. 3, p. 123–126

Urey, H. C., 1952. On the early chemical history of the earth and the origin of life: Nat. Acad. Sci., Proc., v. 38, p. 351–363

—— 1956, Regarding the early history of the earth's atmosphere: Geol. Soc. America Bull., v. 67, p. 1125–1128

Vinogradov, A. P., 1948, Distribution of chemical elements in subterranean waters of various origins: Akad. Nauk S. S. S. R., Trudy Lab. Gidrogeol. Problem im. F. P. Saverenskogo, v. 1, p. 25–35

—— 1953, The elementary chemical composition of marine organisms: Mem. No. 2, Sears Foundation for Marine Research, Yale University, p. 1–647

Von Buttlar, H., and Libby, W. F., 1955, Natural distribution of cosmic-ray produced tritium. II: Jour. Inorganic Nuclear Chemistry, v. 1, no. 1, p. 75–91

Waring, G. A., 1915, Springs of California: U. S. Geol. Survey Water-Supply Paper 338, p. 1–410

Watts, W. L., 1893, Colusa County, California; Mines of Sulphur Creek and vicinity: Calif. State Min. Bur., 11th Rept., 1891–1892, p. 181–187

White, D. E., 1955, Thermal springs and epithermal ore deposits: Econ. Geology, 50th Annual Volume, p. 99–154

—— 1957, Thermal waters of volcanic origin: Geol. Soc. America Bull., v. 68, p. 1637–1658

White, D. E., Brannock, W. W., and Murata, K. J., 1956, Silica in hot spring waters: Geochim. et Cosmochim. Acta, v. 10, p. 27–59

White, D. E., Sandberg, C. H., and Brannock, W. W., 1953, Geochemical and geophysical approaches to the problem of utilization of hot spring water and heat: 7th Pacific Sci. Cong. (Wellington, New Zealand) Proc., v. 2, Geology, p. 490–499

Whitney, J. D., 1865, Geology: Calif. Geol. Survey, v. 1, p. 99

Wilson, S. H., 1955, Geothermal steam for power in New Zealand, Chap. 4; Chemical investigations: New Zealand Dept. Sci. Indus. Research Bull. 117, p. 27–42.

Zies, E. G., 1929, The Valley of Ten Thousand Smokes: Nat. Geog. Soc. Contr. Tech. Papers, v. 1, no. 4, p. 1–79

U. S. GEOLOGICAL SURVEY, MENLO PARK, CALIFORNIA

MANUSCRIPT RECEIVED BY THE SECRETARY OF THE SOCIETY, MARCH 26, 1957

PUBLICATION AUTHORIZED BY THE DIRECTOR. U. S. GEOLOGICAL SURVEY

Erratum

On page 289, second column, line 16 from the bottom: the sentence "Some of these changes have progressed a little further in the Wilbur and Tuscan waters in most oil-field brines." *should read* "Some of these changes have progressed a little further in the Wilbur and Tuscan waters than in most oil-field brines."

EVIDENCE ON HISTORY OF SEA WATER FROM CHEMISTRY OF DEEPER SUBSURFACE WATERS OF ANCIENT BASINS[1]

KEITH E. CHAVE[2]

La Habra, California

ABSTRACT

Subsurface waters of ancient basins are thought to be remnants of sea water entrapped with the sediments at the time of their deposition. The post-depositional alteration of these waters is investigated in order to evaluate the possibility of using their chemical composition as an indicator of ancient sea water chemistry.

A review of literature on subsurface water chemistry indicates that the reliability of the data is, in general, poor. Using the best analyses available, the concentrations of Cl, K, Ca, Mg, Sr, Br, and I in waters from rocks ranging in age from Pliocene to Ordovician are compared. There appear to be no significant compositional trends with time. The post-depositional processes altering the water chemistry are discussed. It is concluded that the magnitude of the modifying processes are so great that it is unlikely that evidence on ancient sea-water chemistry can be obtained from the study of subsurface waters.

INTRODUCTION

The waters enclosed in the sedimentary rocks of ancient basins are generally considered to be remnants of sea water entrapped with the sediments at the time of their deposition. As Rubey (1951) pointed out in his classical paper on the geologic history of sea water, "... connate waters have been so altered in composition that they do not represent at all accurately the original water of deposition." The chemical changes, which modify the interstitial waters, begin shortly after the water becomes separated from free circulation with the open ocean, and often continue to occur long after sedimentation. The processes which modify the water chemistry are of two types: mixing with foreign waters, which are usually fresh meteoric waters, and reactions with enclosing mineral grains and organic materials. If the nature and magnitude of these post-depositional changes can be evaluated, speculation on the chemistry of ancient sea water will be possible. At the present state of knowledge this can not be done.

The purpose of this paper is to bring together data on subsurface water chemistry, examine its reliability, discuss some hypotheses about the processes which cause the chemical changes, and perhaps point out a few directions of work which could prove fruitful in solving the problems.

---

[1] Read before the International Oceanographic Congress at New York, September, 1959. Manuscript received, October 13, 1959.

[2] California Research Corporation. Present address: Department of Geology, Lehigh University, Bethlehem, Pennsylvania.

SAMPLING AND ANALYSIS OF SUBSURFACE WATERS

Several thousand chemical analyses of subsurface water, from many ages of rocks and many parts of the world, appear in the literature. Several writers, notably, Rogers (1917), Mills and Wells (1919), Russell (1933), de Sitter (1947), and Chebotarev (1955), have attempted to explain the differences between various subsurface waters and between subsurface waters and modern sea water. Many of the conclusions reached by these writers have been clear and valid. Other conclusions have been seriously affected by the inclusion of unreliable data derived either from questionable sampling or from inaccurate analytical techniques.

Most samples of subsurface water are taken from wells drilled in search of petroleum. Samples are normally taken during or immediately after the completion of the drilling operation. Contamination of water samples by drilling and completion fluids is a serious problem. Samples which are taken long after the completion of the well are often contaminated by foreign waters leaking into the well through breaks in the casing. Truly representative samples of subsurface water are rarely obtained. The validity of a sample can be evaluated only by careful study of well history.

Most modern oil wells are drilled with rotary rigs which circulate water-base mud in the bore hole during the drilling operation. The muds are composed of water, clay, and additives. The mud waters range from fresh surface water to sea water to saturated sodium chloride brine. The clays are normally swelling montmorillonites. The additives can be almost anything—barite, phosphate, tannins, lime, gypsum, caustic soda, salt, petroleum products, chicken feathers, walnut shells, ground-up rubber tires, and other odd materials. The specific gravity of the mud is normally considerably higher than that of water, so that filtrate—water with various dissolved materials—is forced into porous zones, commonly for distances of tens of feet. This filtrate water, mixed with subsurface water, is commonly produced from the well and sampled and analyzed as true subsurface water.

When a well is prepared for production, steel casing is cemented into the hole. The water in the cement may penetrate the permeable zones and contaminate the subsurface water. Samples of water are commonly taken for analysis immediately after cementing.

Oil and gas production from carbonate reservoirs is stimulated by acid treatment. This involves forcing up to several thousand gallons of hydrochloric acid into the producing zone. Chemical evidence of the acid can be found in waters produced from these wells months after treatment (W. F. Meents, personal communication).

In the past, many wells were drilled with cable-tool rigs. Some are today. Cable-tool holes are drilled without mud and with a minimum of casing. Samples of water taken from the bottom of a cable-tool hole, with a bailer, are commonly contaminated with waters entering the well bore from zones above the bottom of the hole.

The problems of sampling oil and gas wells to obtain representative subsurface waters are complex. A careful study of the history of the well is necessary to evaluate fully the possibilities of contamination. Certainly, published analyses of subsurface waters must be treated as suspect, if a discussion of sampling techniques is not included. One possible index of contamination by drilling and completion fluids is pH. Many of the fluids involved have either high or low pH (e.g., limestone, gypsum, caustic muds, cement waters, and acid stimulants). Waters with pH values around neutral can, in the absence of other well data, perhaps be considered less likely to be contaminated by drilling and completion fluids than those with high and low pH.

The analysis of a sample of subsurface water is usually a complicated task. Standard procedures, such as those outlined by Meents *et al.* (1952), are designed for simple solutions with limited numbers of significant cations and anions.

Samples of subsurface water, even when not seriously contaminated, contain materials not normally included in the analysis. These can be co-analyzed with other ions or can make certain techniques impossible by precipitating indicators or by modification of color in colorimetric analysis. Some of the more difficult materials to cope with in these respects are the tannins in drilling muds and organic and sulfur compounds occurring naturally in the waters.

One final consideration with respect to the analysis of subsurface water: sodium and potassium are rarely analyzed. In a standard analysis the major cations and anions are determined, and sodium and potassium are calculated as the difference and expressed as sodium. Thus, reported sodium values, unless otherwise indicated, not only are not sodium or sodium and potassium alone, but are a record of the analytical errors in a normally complex procedure.

Most published reports on subsurface water chemistry do not contain any discussion of sampling or analytical problems or techniques and, therefore, the reliability of the data presented is difficult to evaluate. The lack of awareness of these problems on the part of some writers is clearly and perhaps amusingly illustrated by the common practice of reporting calculated sodium values to five or six significant figures.

In order to reach valid conclusions from analyses of subsurface water samples, one must consider not only the reliability of samples and analyses, but how representative the samples are of the volume of water they are intended to typify. The significance placed on an individual analysis must be in accord with the natural variability of the samples. Data on samples taken from the same well at different times are given by Elliott (1953) and Plummer and Sargent (1931). Analyses of these re-samples are shown in Table I. Apparently each analysis was intended by the writer to be typical of the specific zone in the well. The reasons for the differences between re-samples are not clear. Perhaps they represent natural chemical differences of waters in a specific stratigraphic zone in a local area; perhaps they represent contamination.

In summary, uncontaminated samples of subsurface waters are difficult to obtain and analyze. In addition, waters produced from wells vary significantly in their chemical composition from time to time. Interpretations based on published

TABLE I. ANALYSES OF WATERS SAMPLED FROM SAME WELL AT DIFFERENT TIMES

Time Difference	Ca	Mg	Parts Per Million HCO$_3$	SO$_4$	Cl	Reference
7 Mos.	(5460 (5290	1140 1850	215 469	1520 2500	81,800 78,400	Elliott
% Difference*	3%	47%	74%	49%	4%	
1 Mo.	(7790 (2750	848 709	191 445	1600 2630	73,900 41,100	Elliott
% Difference	95%	18%	80%	49%	57%	
8 Mos.	(5670 (6210	1580 1600	191 149	691 719	86,300 85,100	Elliott
% Difference	9%	1%	25%	4%	1%	
24 Mos.	(2330 (2350	637 696	288 431	2560 2450	38,300 37,900	Elliott
% Difference	1%	9%	40%	4%	1%	
18 Mos.	(11,900 (2,660	2490 741	34 385	538 2210	111,000 45,100	Elliott
% Difference	127%	108%	168%	122%	84%	
48 Mos.	(2190 (2280	555 590	703 725	2560 2770	30,800 31,800	Elliott
% Difference	4%	6%	3%	8%	3%	
1 Mo.	(2440 (2370	592 603	722 778	2780 2780	31,800 21,700	Elliott
% Difference	3%	2%	7%	0%	1%	
2 Mos.	(222 (241	83 83	476 549	- -	13,000 13,600	Plummer and Sargent
% Difference	8%	0%	14%	-	4%	
6 Mos.	(52 (80	48 364	671 1270	0 14	8720 8650	Plummer and Sargent
% Difference	42%	153%	60%	>100%	1%	
2 Mos.	(- (54	45 58	1130 1340	29 0	4960 5030	Plummer and Sargent
% Difference	>100%	25%	17%	>100%	1%	

* % Difference = $\dfrac{2 \times \text{difference}}{\text{sum}}$

figures must be carefully examined in light of these considerations. There is a great need for new, carefully taken data in the field of subsurface water chemistry.

WATER ANALYSES

In the subsequent discussion, published water analyses are used extensively. The reliability of many of these analyses is certainly open to question. In order to make the interpretations as meaningful as possible, several criteria for selection were applied. Many analyses were rejected as unreliable. Several analyses of unknown reliability are included in order to present some data from as much of the post-Cambrian time spectrum as possible. The sources of the data used in this paper are here discussed.

Rogers, G. S. (1917).—Rogers presents 60 analyses of Tertiary waters from California. Some of the samples are from flowing wells, some are bailed from cable-tool holes. Some were taken by the author, others by various oil companies. There are no specific data on the sampling or analytical techniques used. The pH of the samples, a value which commonly provides an index of contamination by drilling or completion fluids, is not included. Only the potassium and chloride values for four saline samples were taken from Rogers. Data for other elements in California waters are included in the present paper (Tables II and III).

Plummer, F. B., and Sargent, E. C. (1931).—Two hundred nineteen analyses of waters from the Cretaceous Woodbine sandstone of East Texas are given by Plummer and Sargent. The writers discuss sampling procedures in general but do not give sampling data on specific samples. The analytical techniques used are discussed in detail. No pH value is given; thus there is no index of drilling or completion fluid contamination. Twenty-one analyses with chlorinities higher than 19,000 ppm are taken from Plummer and Sargent.

Price, P. H., et al. (1937).—One hundred eighty-four analyses of upper Paleozoic waters of West Virginia are presented by Price *et al.* Most of the samples were bailed from cable-tool holes. There are no statements on casing in the holes. No analytical techniques are discussed. The average concentrations for Ca, Mg, Sr, K, Br, I, and Cl presented by Price *et al.* are used in order to extend the time range for data on these elements back to the Paleozoic. The analyses presented must be treated as highly suspect.

Schoewe, W. H. (1943).—Schoewe presents 166 analyses of Paleozoic subsurface waters from Kansas. There is some general discussion of

TABLE II. ANALYSES OF PLIOCENE WATERS FROM MIDWAY-SUNSET AREA, KERN COUNTY, CALIFORNIA

Depth	Na	Ca	Mg	SO_4	Cl	CO_3	HCO_3	T.S.	pH	ppm I	How Sampled
4061'-4168	11,500	1,370	468	49	21,300	0	451	35,200	7.19	52.5	Tank
3143'-3395'	14,000	910	710	30	24,700	0	1020	41,400	6.78	-	Well Head
3013'-3337'	11,400	1,180	492	7	21,000	0	164	34,200	6.82	-	Tank
2851'-3000'	12,300	870	264	142	20,700	0	655	35,000	7.3	74.5	Well Head
3610'-3680'	13,800	990	447	4	24,100	0	293	39,600	7.15	-	Lead Line
3364'-3500'	13,600	817	300	22	23,000	0	451	38,300	7.45	-	Well Head
3298'-3567'	12,800	1,140	429	25	22,800	0	353	37,600	7.21	-	Lead Line
4493'-4550'	11,800	744	456	4	20,500	0	499	34,000	7.50	-	Lead Line
3539'-3585'	11,400	1,190	433	113	20,700	0	198	34,100	7.25	-	Lead Line
3215'-3490'	12,400	961	358	25	21,300	0	927	36,000	7.10	85.9	Formation Test
3192'-3530	14,300	1,860	886	21	27,800	0	134	45,000	7.35	-	Tank
3175'-3310'	12,000	1,220	520	1	22,200	0	140	36,400	6.76	-	Tank
2630'-2683'	11,900	1,040	587	4	21,800	0	313	35,700	6.72	69.8	Tank
4398'-4509'	12,800	1,420	778	1	24,400	0	384	39,800	6.87	-	Well Head
2195'-2232'	12,200	405	497	4	20,200	0	1330	34,600	7.50	-	Lead Line
3440'-3630'	13,700	1,040	356	112	23,800	0	253	39,200	7.39	13.6	Lead Line
3672'-3742'	14,400	947	921	3	26,200	0	470	42,900	6.87	-	Lead Line
3375'-3600'	11,900	1,240	492	32	21,900	0	217	35,900	7.23	-	Well Head
2822'-2898'	11,200	1,130	453	19	20,400	0	375	33,700	6.65	-	Well Head
3413'-3465'	12,900	1,160	442	16	23,200	0	315	38,200	7.3	-	Formation Test
3552'-3630'	12,000	1,040	363	1	21,100	0	463	35,000	7.00	-	Well Head
4398'-4509'	12,300	1,250	574	7	22,700	0	378	37,300	7.00	24.1	Tank
3507'-3580'	13,200	1,630	688	74	24,900	0	308	41,000	6.81	207	Formation Test
2632'-2710'	11,300	850	520	4	20,100	0	697	33,500	7.06	50.1	Lead Line
4531'-4643'	12,000	1,110	391	5	21,300	0	458	35,300	6.72	61.5	Lead Line
3353'-3452'	13,600	527	270	3	22,300	0	585	37,300	7.71	-	Tank
3539'-3585	11,400	1,170	434	114	20,700	0	198	34,100	7.25	-	Lead Line
3565'-3678'	13,800	990	447	4	24,100	0	294	39,600	7.15	-	Lead Line
3249'-3343'	12,300	636	314	6	20,800	0	332	34,400	7.41	-	Tank

sampling. No analytical techniques are mentioned. pH values are not listed. Twenty-seven analyses of Ordovician Arbuckle waters are discussed in the present paper. They include all of Schoewe's Arbuckle limestone analyses with chlorinities greater than 19,000 ppm.

Jessen, F. W., and Rolshausen, F. W. (1944).—These authors present 116 analyses of Oligocene Frio waters from the Texas Gulf Coast. They do not mention sampling or analytical techniques. No pH values are given. Eighty-nine analyses with chlorinities greater than 19,000 ppm are used in the present paper. There is no measure of their reliability.

Meents, W. F., et al. (1952).—Meents presents several hundred carefully made analyses of Paleozoic subsurface waters from Illinois. He presents a good discussion of sampling problems and techniques. The analytical techniques used are described in detail. pH values are given for many samples. The present paper utilizes Meents' data for Pennsylvanian, Chester series and Ste. Genevieve waters. Those samples with chlorinity greater than 19,000 ppm and pH between 6.5 and 7.5 were selected.

Elliott, W. C. (1953).—Elliott presents 70-odd analyses of Pennsylvanian and Permian waters from West Texas. Sampling procedures are discussed. Elliott tabulates the dates of acid treatment and sampling of wells. No analytical techniques are mentioned. No pH values are given. Data on 42 Canyon reef samples taken more than 4 months after acid treatment are used in the present paper.

McGrain, P. 1953).—Fifty-seven partial and complete analyses of Paleozoic waters from Kentucky are given in this Kentucky Geological Survey Report. There is no mention of sampling or analysis. Four analyses with chlorinities greater than 19,000 ppm are used in the subsequent discussion because of the trace constituents determined.

Province of Alberta (1955).—The Alberta Oil and Gas Conservation Board publishes a looseleaf compilation of Paleozoic and Mesozoic water analyses from Alberta. Sampling information and pH values are usually given. Some analyses are made in the Conservation Board laboratories; others come from various oil company laboratories. No analytical techniques are discussed. Cretaceous and Leduc D-3 analyses in the 1955 compilation with chlorinities greater than 19,000 ppm and pH values between 6.5 and 7.5 are used in the subsequent discussion.

Additional analyses.—Two groups of analyses of waters from the Tertiary of California are included in the present paper in order to extend the age range of the data. Table II contains analyses of 29 Pliocene water samples from the Midway-Sunset area, Kern County, California. All of the samples have chlorinities greater than 19,000 ppm and pH between 6.5 and 7.5. The sampling source is indicated. The analyses were made at the Standard Oil Company of California, Producing Department Laboratory, Taft, California. D. Silcox was the analyst. Standard techniques described by Meents *et al.* (1952) were used in the analyses.

Nineteen strontium and chloride analyses of California subsurface waters are listed in Table III. The samples were taken by the writer at the well-heads of producing wells. All of the wells sampled had produced fluids for more than 6 months and had not been treated by contaminating chemicals during that time. The strontium analyses were made on an emission spectrograph at the La Habra Laboratory of the California Research Corporation. R. G. Smalley was the analyst.

A summary of the sources and apparent re-

TABLE III. STRONTIUM CONTENT OF CALIFORNIA WATERS

Age	County	ppm Cl	ppm Sr	pH
Pliocene	Los Angeles	18,400	52	7.9
Pliocene	Los Angeles	18,800	53	7.8
Pliocene	Los Angeles	13,800	68	7.4
Pliocene	Los Angeles	11,400	33	7.7
Pliocene	Los Angeles	12,600	40	7.5
Pliocene	Los Angeles	10,800	27	7.8
Pliocene	Los Angeles	10,300	29	7.8
Pliocene	Los Angeles	19,500	61	7.6
Pliocene	Los Angeles	12,400	34	7.7
Pliocene	Los Angeles	19,100	55	7.3
Pliocene	Los Angeles	19,300	70	7.6
Miocene	Los Angeles	12,600	130	7.4
Miocene	Kern	11,300	110	7.8
Oligocene	Ventura	16,700	120	7.5
Oligocene	Ventura	12,900	59	7.0
Oligocene	Ventura	18,300	59	7.1
Oligocene	Ventura	12,200	74	6.8
Oligocene	Ventura	16,100	28	7.1
Oligocene	Ventura	17,200	45	7.7

TABLE IV. SOURCES OF WATER ANALYSES USED IN DISCUSSION

Reference	Year	Analyzed Ions		pH	Discussion of Procedures for		Apparent Reliability
		Major	Trace		Sampling	Analysis	
Rogers, G. S.	1917	Yes	Some	No	No	No	Poor
Plummer, F. B., and Sargent, E. C.	1931	Yes	No	No	Some	Yes	Fair
Price, P. H., et al.	1937	Yes	Yes	No	Some	No	Fair
Schoewe, W. H.	1943	Yes	Some	No	Yes	No	Fair
Jessen, F. W., and Rolshausen, F. W.	1944	Yes	No	No	No	No	Poor
Meents, W. F.	1952	Yes	Some	Some	Some	Yes	Good
Elliott, W. C.	1953	Yes	No	No	Yes	No	Fair
McGrain, P.	1953	Yes	Yes	No	No	No	Poor
Province of Alberta	1955	Yes	Some	Some	Some	No	Fair
This paper, Standard Oil of Calif.	1959	Yes	Some	Yes	Yes	Yes	Good
This paper, Calif. Research Corp.	1959	No	Yes	Yes	Yes	Yes	Good

liability of the data used in this paper is presented in Table IV.

In the subsequent discussion most of the data are presented as the arithmetic means of the concentrations or ratios of elements in groups of waters from limited stratigraphic units in local areas. It is hoped that by handling the data in this way, rather than as individual water analyses, errors due to sampling, analysis, or natural variability will be minimized.

Composition of Subsurface Waters

Subsurface waters contain dissolved solids in a range from essentially zero to greater than 30 per cent (300,000 ppm). In basins with dominantly marine sediments, waters having salinities less than that of sea water are indicated by geologic evidence to be formed by dilution with fresh meteoric waters. Waters that are hypersaline (i.e., $>$ 19,000 ppm Cl) can usually be assumed to be undiluted. Movement of meteoric water into the deeper parts of a basin can normally be recognized by areal salinity mapping. The freshness of the waters increases by several orders of magnitude toward outcrops of continuous aquifers. In places where aquifers are discontinuous, due either to stratigraphic change or structure, strong salinity gradients do not exist. Examples of salinity maps illustrating entry of meteoric waters are presented by Meents (1952), Plummer and Sargent (1931), Dott and Ginter (1930), and others. Figure 9 of Meents (1952) for example, showing the concentration of Ste. Genevieve brines, illustrates a relatively static condition. The salinity change from the margin to the center of the basin is less than a factor of ten. Outcrops of the Ste. Genevieve limestone in the Illinois basin are rare, since over most of the basin the Ste. Genevieve is overlapped by later Mississippian sediments. Plate VIII of Plummer and Sargent (1931), showing the salinity of Woodbine brines in East Texas, illustrates a typical example of entry of meteoric water into a sedimentary basin. The Woodbine crops out extensively on the west and north flanks of the basin. The sand pinches out south and eastward in the basin. Salinity changes of four orders of magnitude are shown. Waters with salinities greater than 35,000 ppm are probably little modified by mixing with meteoric water.

In order to simplify the problem of investigating sea-water history, subsurface waters which have obviously been mixed with meteoric waters should be avoided. The following discussion is limited to waters which apparently have not been mixed with meteoric waters, and therefore in most cases have chlorinities greater than 19,000 parts per million.

Post-Depositional Modification of Sea Water

Reactions between mineral grains and water take place both in the circulating ocean system and within the deposited sediments. Reactions in the open water modify ocean-water chemistry. Only the reactions taking place in interstitial waters, after separation from the open sea by sedimentation and compaction, influence the interstitial and eventually the subsurface-water chemistry.

Many of the reactions which take place within the sediments have been studied from the point of view of the mineral or rock composition. Very little work has been done on the water chemistry, which, because the water is the source of much of the material involved in the mineral changes, logically should reflect the complement of the mineralogical changes.

The chemical differences between sea and subsurface water are discussed in the following paragraphs. Possible mechanisms of the changes are suggested in each case. These should be considered as working hypotheses, subject to modification as further field and experimental data become available. The data used in the figures are listed in Table V.

TOTAL DISSOLVED SOLIDS AND CHLORINITY

Subsurface waters which have not been diluted by fresh meteoric waters are higher in dissolved solids and chlorinity than modern sea water. In terms of chlorinity, the range of concentration is from 19,000 to about 200,000 ppm. There is no clear-cut relation between chlorinity and age of the water within the limits of available samples (Fig. 1).

The process which produces hypersalinity in

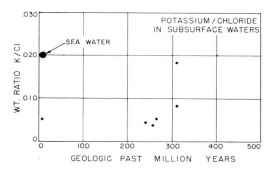

FIG. 2.—Mean potassium to chloride ratios for 6 groups of subsurface waters ranging in age from Pliocene to Devonian.

subsurface water apparently does not seriously affect the relative proportions of dissolved ions. Figures 2–4 and 6–8 illustrate the average ratios of the concentration of 6 ions to chloride concentration in many groups of waters. These ratios in each case lie in a relatively narrow range even though the average chloride concentrations range from 12,000 to 153,000 ppm.

Possible mechanisms for concentrating dissolved salts in subsurface water have been proposed by several writers. Summaries of these are presented by Russell (1933), de Sitter (1947), and Levorsen (1954). Evaporation and entrainment in expanding subsurface gas (Mills and Wells, 1919) and density stratification due to molecular settling are unsatisfactory mechanisms in terms of geologic and chemical consideration (Russell, 1933). Derivation of hypersaline waters from normal evaporite sequences is possible in a few cases, but in many basins, such as the Illinois basin and the California basins, it is not possible. Mechanisms which involve ionic absorption on clay or ionic filtering through clay membranes appear to be the most satisfactory, although no direct field or laboratory observations are available to support the idea.

If clays are involved in this concentration process, an interesting question is raised. Waters such as those in the lower Paleozoic of Illinois and Kansas, where shales are exceedingly rare, are similar in chlorinity and over-all composition to waters such as those from the Tertiary of the Gulf Coast and the Cretaceous of Alberta, where shales are very common. If a large volume of shale is necessary to concentrate the solids in these large volumes of water, then the waters such as those in the lower Paleozoic of Illinois and

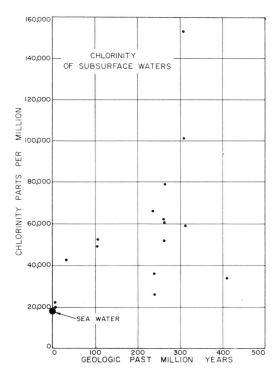

FIG. 1.—Mean chlorinity of 16 groups of subsurface waters ranging in age from Pliocene to Ordovician. Only waters with chlorinities higher than modern sea water are shown.

TABLE V. SOURCES AND TABULATION OF DATA USED IN FIGURES 1–7

No.	Reference	Date	Location	Formation	Age	Lithology	Number of Samples	K/Cl	Ca/Cl	Mg/Cl	Sr/Cl	I/Cl	Br/Cl	Mean Chlorinity ppm	Chlorinity Range ppm
1	Rogers, G. S.	1917	California	-	Plio.	ss	4	.0055	-	-	-	-	-	19,900	18,700–20,700
2	Plummer, F. B. and Sargent, E. C.	1931	E. Texas	Woodbine	Cret.	ss	21	-	.046	.0068	-	-	-	52,000	19,000–112,000
3	Price, P. H. et al	1937	W. Virginia	Salt	Penn.	ss-ls	106	.0045	.118	.024	.0035	.00005	.0057	36,400	100–104,000
4	"	"	"	Maxton	Miss.	ss	19	.0038	.119	.027	.0031	.00008	.0060	52,900	5830–103,000
5	"	"	"	Big Injun	Miss	ss	33	.0052	.127	.028	.0033	.00011	.0075	60,700	70–113,000
6	"	"	"	Oriskany	Dev.	ss-ls	6	.0181	.115	.017	.0066	.00007	.0074	153,000	135,000–169,000
7	Schoewe, W. H.	1934	Kansas	Arbuckle	Ord.	ls	27	-	.082	.021	-	-	-	34,100	19,100–113,000
8	"	"	"	"	"	"	17	-	-	-	-	-	.00067	39,500	19,300–113,000
9	Jessen, F. W. and Rolshausen, F. W.	1944	Texas	Frio	Olig.	ss	89	-	.046	.0057	-	-	-	42,400	19,000–73,000
10	Meents, W. F. et al	1952	Illinois	-	Penn.	ss	7	-	.029	.015	-	-	-	26,200	21,100–32,300
11	"	"	"	Chester	Miss.	ss	64	-	.052	.019	-	-	-	62,200	31,800–87,600
12	"	"	"	Ste Genevieve	Miss.	ls	52	-	.069	.026	-	-	-	79,200	49,400–95,400
13	Elliott, W. C.	1953	W. Texas	Canyon	Penn.	ls	42	-	.072	.017	-	-	-	66,400	31,800–179,000
14	McCrain, P.	1953	Kentucky	Corniferous	Dev.	ls	4	.0084	.094	.029	.0035	.00012	.0070	58,700	13,600–93,900
15	Alberta O.&GCB	1955	Alberta	-	Cret.	ss	52	-	.057	.018	-	-	-	49,900	21,800–101,000
16	"	"	"	-	"	"	15	-	-	-	-	-	-	46,100	22,500–101,000
17	"	"	"	Leduc D-3	Dev.	dol.	13	-	.132	.034	-	.00041	.0038	101,000	44,700–167,000
18	"	"	"	"	"	"	7	-	-	-	-	.00015	.0052	110,000	71,100–167,000
19	This Paper	1959	California	-	Plio	ss	29	-	.047	.022	-	-	-	22,400	20,100–27,800
20	"	"	"	-	"	"	9	-	-	-	-	.0032	-	22,000	20,100–24,900
21	"	"	"	-	"	"	11	-	-	-	.0031	-	-	15,100	10,300–19,500
22	"	"	"	-	Mio.	ss	2	-	-	-	.010	-	-	12,000	11,300–12,600
23	"	"	"	-	Olig	ss	6	-	-	-	.0042	-	-	15,600	12,200–18,300

304

Kansas must have originated elsewhere, in shaly environments, perhaps in the geosynclines hundreds of miles from the present location of the waters. Other reactions involving clay minerals are here discussed.

SODIUM AND POTASSIUM

Sodium is the principal cation in subsurface waters. Potassium concentrations are low. As no published data indicate actual analysis for sodium all the reported values must be assumed to be calculations. In light of the possibilities for error in calculating sodium values, reported differences in concentration can not be considered significant. For this reason sodium is not considered in this paper.

Figure 2 shows the average K/Cl ratios of groups of subsurface waters ranging in age from Pliocene to Devonian. All of the ratios are lower than those in modern sea water. No significant trend with age is indicated.

Potassium is removed from solutions of seawater composition by clay minerals. Some removal is by ion exchange (Kelley and Liebig, 1934 and others). Apparently this process is rapid, probably reaching equilibrium in sea water prior to isolation of water in the sediment. Absorption and fixation of potassium with associated clay mineral growth or modification have been described by Whitehouse and McCarter (1956) and Weaver (1958). This process proceeds more slowly than the ion exchange process and, therefore, may take place in sediments in which interstitial waters are to some extent isolated from sea water. Such a mechanism would explain K/Cl ratios in subsurface waters lower than those in modern sea water. Again it should be noted that clay minerals are involved.

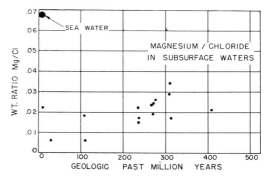

Fig. 4.—Mean magnesium to chloride ratios for 15 groups of subsurface waters ranging in age from Pliocene to Devonian.

CALCIUM AND MAGNESIUM

The average Ca/Cl and Mg/Cl values for suites of waters ranging in age from Pliocene to Ordovician are shown in Figure 3 and 4. In all cases Ca/Cl is higher and Mg/Cl is lower than in modern sea water. The apparent trend of increasing Ca/Cl and Mg/Cl with increasing age can perhaps best be attributed to a lithologic difference of the producing formations, the higher values of the two ratios usually occurring in waters from carbonate reservoirs.

Calcium and magnesium concentrations in the various groups of waters are illustrated in Figure 5. Concentrations are given in moles per million parts of the total solution (parts per million/mol. wt.) with the concentrations normalized so that the chlorinity of each sample equals the chlorinity of modern sea water. The data are presented as trend lines for each group of waters. The trends are the maximum likelihood line[3] calculated from the individual water analyses of the group.

[3] The maximum likelihood line is a straight line which maximizes the probability that the measured values are the true values as defined by the line. It is calculated as follows:

$$\text{slope}, a = \pm \sqrt{\frac{\bar{y^2} - (\bar{y})^2}{\bar{x^2} - (\bar{x})^2}}$$

$$\text{intercept } b = \bar{y} - a\bar{x}$$

where x and y are the measured values of the parameters involved, and $\bar{x}$ and $\bar{y}$ are mean values.

The positive value of the square root term for the slope was chosen in each case by examination of the data. The length of the trend line indicates the principal range of the data.

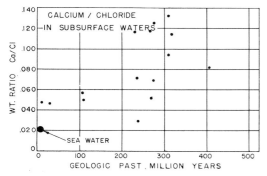

Fig. 3.—Mean calcium to chloride ratios for 15 groups of subsurface waters ranging in age from Pliocene to Ordovician.

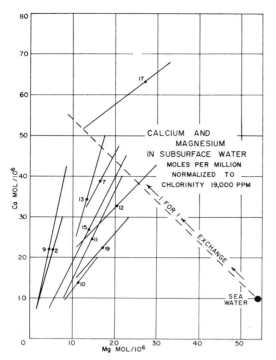

TABLE VI. CORRELATION COEFFICIENTS FOR CA AND MG DATA IN FIGURE 5

No. from Table 5	Source	No. of Samples	Correlation Coefficient
2	Woodbine ss.	20	.387
7	Arbuckle ls.	27	.541
9	Frio ss.	89	.090
10	Pennsylvanian ss.	7	.483
11	Chester ss.	64	.210
12	Ste. Genevieve ls.	52	−.084
13	Canyon ls.	42	.222
15	Cretaceous ss.	52	.575
17	Leduc D-3 dol.	13	.396
19	Pliocene ss.	29	.293

FIG. 5.—Relation of calcium to magnesium in 10 groups of subsurface waters ranging in age from Pliocene to Ordovician. Values are in moles per million (parts per million/mol.wt.), normalized to sea-water chlorinity. Trend lines are maximum likelihood lines for groups of individual water analyses from a single rock sequence in a local area. Numbers on trend lines refer to data sources in Table V. A theoretical mole-for-mole exchange curve for modification of modern sea water is indicated.

The maximum likelihood lines shown in Figure 5 represent the trend of, in general, poorly correlated variables. The correlation coefficients, r, between the calcium and magnesium values are listed in Table VI.

Because of the small number of samples in most groups of water samples, the statistical significance of the correlation coefficients is rather low. It is perhaps significant that, with the exception of the Ste. Genevieve suite, all of the coefficients are positive.

Two features of the relation between calcium and magnesium in subsurface waters relative to modern sea water, as illustrated in Figure 5, require explanation. These are (1) a loss of more moles of magnesium than are added as calcium, and (2) the positive correlation between calcium and magnesium concentrations. It appears that more complicated processes than simply mole-for-mole exchange as in dolomitization as suggested by White (1957) or in base exchange in clays as suggested by several writers, are required.

Clarke (1924) presents analyses which suggest a post-depositional uptake of magnesium in argillaceous sediments. A composite analysis of 52 green and blue modern marine muds, with $CaCO_3$ removed from the values, shows CaO 2.04%, MgO 2.17%, or a Mg/Ca ratio of 0.90. An average of 51 Paleozoic shales, low in calcium carbonate, shows CaO 1.41, MgO 2.32, and a Mg/Ca ratio of 1.40. The mechanism of the magnesium uptake may be growth of chlorites as described from laboratory experiments by Whitehouse and McCarter (1956) or diagenetic growth of muscovite or illite as described by Weaver (1958). In either case magnesium could be removed from interstitial waters without a compensating addition of calcium.

A second mechanism is necessary to produce a positive correlation between the calcium and magnesium concentrations. A process which might explain this relation involves the dissolution of a small amount of $CaCO_3$ from shell material in the enclosing sediment—the amount of solution being controlled, perhaps, by a limited lowering of pH, produced in the bacterial reduction of dissolved sulfate ion in the presence of organic matter. This process is discussed later. The extent of the lowering of the pH and therefore the amount of $CaCO_3$ solution would be limited by the amount of sulfate present in the original water—2,600 ppm in modern sea water. The correlating increase in magnesium is greater than would be expected from the solution of average shell material or limestone. Perhaps the best explanation of this increase is, that with addition of calcium to the interstitial water, exchangeable magnesium is removed from clay minerals, being replaced by calcium. This process would produce a positive correlation between the two ions.

The higher Ca/Cl values, such as are indicated for waters from the Devonian dolomites of Alberta and the Arbuckle limestone of Kansas, could be the result of further addition of sulfate ion from associated evaporites and further lowering of pH by bacterial reduction of the sulfate ion in the presence of organic material.

The suggested reactions explaining changes in calcium and magnesium as well as potassium in subsurface water involve argillaceous sediments, probably in large quantities. As discussed previously, this suggests migration of some waters for distances of several hundred miles.

STRONTIUM AND BARIUM

The average Sr/Cl ratios for suites of subsurface waters ranging in age from Pliocene to Devonian are shown in Figure 6. All of the values of this ratio are higher than that of modern sea water. No trend with time is indicated. Several published analyses indicate Ba/Cl ratios higher than those of modern sea water. The barium data are not included in this paper because the mineral barite is a common constituent of drilling mud and the possibilities of contamination are great.

The source of the strontium which is added to the subsurface water is a problem. Clays apparently do not carry significant amounts of strontium into the sediments. The increase in strontium concentration in subsurface water is considerably greater than could be accounted for by solution of carbonates, including the strontium-rich skeletal aragonites, unless the calcium were reprecipitated later, free of strontium.

If strontium were present in sediments as celestite ($SrSO_4$), it would dissolve into the interstitial water as the sulfate concentration therein decreased by bacterial activity, and raise the Sr/Cl ratio in the water. Celestite has not been reported as a mineral constituent of modern sediments, although celestite radiolaria have been reported in the ocean. A similar mechanism would account for higher Ba/Cl ratios in subsurface water, if barite could be shown to be a common constituent of modern sediments.[4]

SULFATE

The sulfate concentration in modern sea water is approximately 2,600 ppm. Most subsurface waters which are not mixed with meteoric waters are very low in sulfate. The SO_4/Cl weight ratio of modern sea water is 0.13. Values of this ratio for subsurface waters rarely exceed 0.10 and commonly are less than 0.05.

The mechanism of sulfate reduction is discussed by Ginter (1934), Zobell (1946), and others. It involves the action of anaerobic bacteria in the presence of oxidizable organic material. H_2S and CO_2 or organic acids are normally produced in the process. Most of the H_2S probably precipitates as metal sulfides. The reduction process can produce either higher or lower pH, depending on conditions and materials involved (ZoBell, 1946).

Sulfate reduction is the one process in the complex of changes involved in modification of entrapped sea water which can be placed in time. Interstitial waters from four modern sediment cores from the Gulf of Mexico were analyzed for sulfate and chloride by George Bien of API Project 51 at Scripps Institution of Oceanography. The data were given to this writer for publication. The analytical procedures are described by Bien (1954). The analyses are given in Table VII.

Cores PL-557 and 558 were obtained approximately 7 miles east of the easternmost distribu-

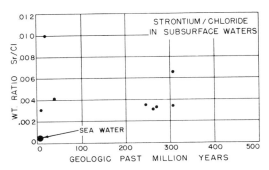

FIG. 6.—Mean strontium to chloride ratios of 8 groups of subsurface waters ranging in age from Pliocene to Devonian.

---

[4] Since preparation of the manuscript, Professor U. G. Whitehouse has suggested a mechanism which would explain the strontium content of these waters. His experiments have shown that montmorillonite will scavenge Sr from sea-water type solutions in non-exchange sites. The Sr is later lost from these sites and replaced by water or large ions. Whitehouse suggested that, in the case of interstitial waters of marine sediments, removal of Sr from the clay might be influenced by the lowering of SO_4 concentration by bacteria.

TABLE VII. SULFATE:CHLORIDE RATIOS IN INTERSTITIAL WATERS OF GULF OF MEXICO CORES

Depth cm	PL-557				PL-558				PL-559				PL-560			
	SO_4/Cl	pH	% Sd	% Clay	SO_4/Cl	pH	% Sd	% Clay	SO_4/Cl	pH	% Sd	% Clay	SO_4/Cl	pH	% Sd	% Clay
	Water		Sediment		Water		Sediment		Water		Sediment		Water		Sediment	
Bottom Water	.17	8.0			.16	6.7			.14	7.9			.20	-		
0	.12	6.9	2.	33.	.16	7.4	.5	74.	.14	7.1	.5	66	.15	7.5	18.	58
15	.10	6.9	.5	52.	.17	7.4	.5	76.	.10	7.1	.5	78	.19	7.5	3.5	74
30	.04	6.9	1.	31.	.16	7.5	2.5	60.	.08	7.1	0	74	.19	7.4	2.	75
60	.04	7.1	4.	34.	.15	7.5	.5	77.		7.1	.5	73	.17	7.5	6.5	72
120	.03	7.1	.5	43.	.14	7.3	.5	80.	.04	7.3	.5	70	.14	8.0	20.5	59
130									.03	7.2	.5	73				
160	.03	7.1	0	36.												
195													.14	7.7	75.	9.
240					.03	7.6	.5	72.								

tary of the Mississippi delta. Cores PL-559 and 560 are from about 12 miles east of the delta. The water depth in all places is about 30 feet.

The interstitial waters in cores PL-557 and 559 show a marked reduction in SO_4/Cl ratio, 15 cm. below the sediment surface. The waters at this depth must be sufficiently separated from the overlying sea water to prevent complete remixing. In core PL-558 evidence for SO_4 reduction is not good shallower than 240 cm. In core PL-560 there is no clear-cut evidence for sulfate reduction in the interstitial water. Perhaps the sandiness of core PL-560 allows circulation of ocean water to a depth of 195 cm. and prevents development of anaerobic conditions necessary for SO_4 reduction. The chloride concentration in all the waters analyzed is approximately that of normal sea water.

The sulfate content of the interstitial and eventually subsurface water should remain low unless further sulfate is added by a process such as solution of anhydrites or gypsum in the geologic section. If sulfate-reducing bacteria and organic material are present when sulfate is added, the sulfate concentration in the water should not increase greatly. If the bacteria or the organic matter are absent, the sulfate concentration in the waters can become high. Crawford (1940) reports exceptionally high sulfate concentrations in three waters from the Chugwater redbeds of Wyoming—10,500, 12,400, and 16,500 parts per million sulfate. Organic matter does not normally occur in these beds; gypsum and anhydrite do occur.

BICARBONATE

The bicarbonate ion concentration in subsurface waters which have not been mixed with meteoric waters is apparently low. The significance of the published bicarbonate and carbonate values is open to question. Dissolved carbon dioxide is often present in subsurface waters. It is also a common minor constituent of natural gas. If dissolved gas is present, it is lost during and after the normal sampling of the water. Calcium carbonate occasionally precipitates out of water samples in visible amounts in the time between sampling and analysis of subsurface waters. For these reasons the $CO_2 - CO_3^= - HCO_3^=$ relationship in subsurface waters is not considered in detail in this paper.

BROMINE AND IODINE

Average Br/Cl and I/Cl values for groups of subsurface waters are shown in Figures 7 and 8. With the exception of the Br/Cl ratio from Ordovician water of Kansas, all values are higher than in modern sea water.

Two mechanisms appear to control the bromine and iodine concentration in subsurface water. In the process of complete evaporation of sea water these two elements are the last to be precipitated. Thus subsurface waters associated with evaporite

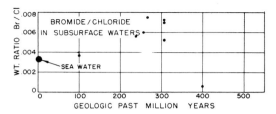

FIG. 7.—Mean bromide to chloride ratios of 8 groups of subsurface waters ranging in age from Cretaceous to Ordovician.

deposits could be enriched in bromine and iodine through mixing with the final liquor of the evaporation sequence.

Subsurface waters can probably take up bromine and iodine from the tissues of marine organisms deposited with sediments, as suggested by Mrazec (1926). Vinogradov (1953) presents a compilation of analyses of the chemical composition of marine organisms in which he shows that certain organisms, particularly algae, concentrate large amounts of bromine and iodine. The Phaeophyceae with the largest concentrations contain as much as 0.2 per cent bromine and 0.6 per cent iodine, dry weight. Other plants and some animals concentrate large amounts of these elements. High concentrations of bromine and iodine in subsurface waters are probably caused by either or both of these processes.

Evidence on History of Sea-Water Chemistry

The subsurface waters of ancient basins probably represent remnants of sea water entrapped with the sediments at the time of their deposition. Chemical evidence suggests that some of these waters may have moved considerable distances since their deposition. The most probable driving force for the migration of these waters is sediment compaction, in which case the movements must have taken place shortly after deposition when compaction rates are greatest. Under these conditions water movements should be in the direction of least resistance, largely laterally, and in the absence of major unconformities the waters should exist today in sediments of approximately the same age as those in which they were deposited.

The foregoing discussion has illustrated the fact that many of the chemical characteristics of subsurface waters, ranging in age from Pliocene to

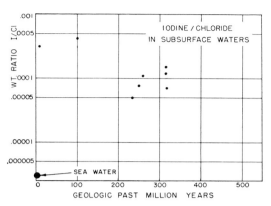

FIG. 8.—Mean iodine to chloride ratios of 8 groups of subsurface waters ranging in age from Pliocene to Devonian. Note logarithmic scale of I/Cl ratio.

Ordovician, are similar. In addition, it has been shown that these waters are all different from modern sea water.

These observations could be explained, in general, in one of two ways: either the subsurface waters have compositions similar to the water which was trapped during sedimentation—in which case modern sea water would be unique in many of its properties relative to ancient sea waters—or all waters which are entrapped with sediments are post-depositionally modified. Much chemical evidence has been presented indicating the latter to be most reasonable. On this basis the following conclusions can be drawn.

1. There is no evidence from subsurface water chemistry for changes in the relative proportions of the major dissolved ions in sea water since the beginning of the Cambrian. The observed differences in subsurface water chemistry appear to be best explained in terms of reactions with the enclosing rocks.

2. No evidence on absolute concentration of dissolved ions in ancient sea water can be obtained, at the present state of knowledge, from subsurface waters. Mechanisms controlling the salinity of these waters are not understood.

3. Minor differences in the composition of the sea water entrapped with sediments at the time of their deposition could not be recognized through the study of subsurface waters because of the magnitude of the post-depositional processes affecting these waters.

4. The present available data on subsurface

water chemistry are not sufficiently reliable to use for speculation on possible gross changes in seawater chemistry through time. With future careful collection and analysis of samples, perhaps this point can be examined.

REFERENCES

ALBERTA PETROLEUM AND NATURAL GAS CONSERVATION BOARD, 1955, *Pool and Field Water Analyses.* Looseleaf.

BIEN, G. S., 1954, "High Frequency Titration of Micro Quantities of Chloride and Sulfate," *Analytical Chem.*, Vol. 26, pp. 909–11.

CHEBOTAREV, I. I., 1955, "Metamorphism of Natural Waters in the Crust of Weathering, Pts. I–III," *Geoch. Cosmoch. Acta*, Vol. 8, Nos. 1–4, pp. 22–48, 137–70, and 198–212.

CLARKE, F. W., 1924, "The Data of Geochemistry," 5th ed., *U. S. Geol. Survey Bull. 770.*

CRAWFORD, J. G., 1940, "Oil-Field Waters of Wyoming and Their Relation to Geological Formations," *Bull. Amer. Assoc. Petrol. Geol.*, Vol. 24, pp. 1214–1329.

DE SITTER, L. U., 1947, "Diagenesis of Oil-Field Brines," *ibid.*, Vol. 31, pp. 2030–40.

DOTT, R. H., AND GINTER, R. L., 1930, "An Iso-Con Map for Ordovician Waters," *ibid.*, Vol. 14, pp. 1215–19.

ELLIOTT, W. C., 1953, "Chemical Characteristics of Waters from the Canyon, Strawn, and Wolfcamp Formations in Scurry, Kent, Borden, and Howard Counties, Texas," *Petrol. Eng.*, Vol. 25, pp. B77–B89.

GINTER, R. L., 1934, "Sulphate Reduction in Deep Subsurface Waters," *Problems of Petroleum Geology*, Amer. Assoc. Petrol. Geol., pp. 907–925.

JESSEN, F. W., AND ROLSHAUSEN, F. W., 1944, "Waters from the Frio Formation, Texas Gulf Coast," *Trans. Amer. Inst. Min. Metal. Eng.*, Vol. 155, pp. 23–38.

KELLEY, W. P., AND LIEBIG, G. F., 1934, "Base Exchange in Relation to Composition of Clay with Special Reference to Effect of Sea Water," *Bull. Amer. Assoc. Petrol. Geol.*, Vol. 18, pp. 358–67.

LEVORSEN, A. I., 1954, *Geology of Petroleum.* 703 pp. W. H. Freeman and Company, San Francisco.

MCGRAIN, P., 1953, "Miscellaneous Analysis of Kentucky Brines," *Kentucky Geol. Survey R17.* 16 pp.

MEENTS, W. F., BELL, A. H., REESE, O. W., AND TILBURY, W. G., 1952, "Illinois Oil-Field Brines," *Illinois Geol. Survey, Illinois Petrol. No. 66.* 38 pp.

MILLS, R. VAN, AND WELLS, R. C., 1919, "The Evaporation and Concentration of Water Associated with Petroleum and Natural Gas," *U. S. Geol. Survey Bull. 693.*

MRAZEC, L., 1926, "Vorlesungen über die Lagerstalten des Erdols II," *Tiel Petroleum Z.*, Vol. 22, p. 901, quoted in RANKAMA, K., AND SAHAMA, T. G., 1950, *Geochemistry*, University of Chicago Press, p. 358.

PLUMMER, F. B., AND SARGENT, E. C., 1931, "Underground Waters and Subsurface Temperatures of the Woodbine Formation in Northeast Texas," *Univ. Texas Bull. 3138.* 178 pp.

PRICE, P. H., HARE, C. E., MCCUE, J. B., AND HOSKINS, H. A., 1937, "Salt Brines of West Virginia," *West Virginia Geol. Survey Vol. III.* 203 pp.

ROGERS, G. S., 1917, "Chemical Relations of the Oil Field Waters in San Joaquin Valley, California," *U. S. Geol. Survey Bull. 653.* 119 pp.

RUBEY, W. W., 1951, "Geologic History of Sea Water," *Bull. Geol. Soc. America*, Vol. 62, pp. 1111–48.

RUSSELL, W. L., 1933, "Subsurface Concentration of Chloride Brines," *Bull. Amer. Assoc. Petrol. Geol.*, Vol. 17, pp. 1213–28.

SCHOEWE, W. H., 1943, "Kansas Oil Field Brines and Their Magnesium Content," *Geol. Survey Kansas Bull. 47*, Pt. 2, pp. 41–76.

VINOGRADOV, A. P., 1953, *The Elementary Chemical Composition of Marine Organisms*, translated by Sears Foundation for Marine Research, Yale University.

WEAVER, C. E., 1958, "The Effects and Geologic Significance of Potassium 'Fixation' by Expandable Clay Minerals Derived from Muscovite, Biotite, Chlorite and Volcanic Material," *Amer. Mineral.*, Vol. 43, pp. 839–61.

WHITE, D. E., 1957, "Magmatic, Connate and Metamorphic Waters," *Bull. Geol. Soc. America*, Vol. 68, pp. 1659–82.

WHITEHOUSE, U. G., AND MCCARTER, R. S., 1956, "Diagenetic Modification of Clay Mineral Types in Artificial Sea Water, in Clays and Clay Minerals," *5th National Conf. on Clays and Clay Minerals*, pp. 81–119.

ZOBELL, C. E., 1946, *Marine Microbiology.* 240 pp. Chronica Botanica Company, Waltham, Mass.

Copyright © 1962 by the Economic Geology Publishing Co.

Reprinted from *Econ. Geol.*, **57**(7), 1045–1061 (1962)

STUDIES OF FLUID INCLUSIONS I: LOW TEMPERATURE APPLICATION OF A DUAL-PURPOSE FREEZING AND HEATING STAGE [1]

EDWIN ROEDDER

CONTENTS

	PAGE
Abstract	1045
Introduction	1046
Design of the apparatus	1048
The cell	1048
Auxiliary equipment	1050
Operation	1051
Heat-exchange medium	1051
Sample preparation	1051
Freezing the inclusions and the problem of metastability	1052
Determination of the first melting temperature	1054
Determination of the freezing temperature	1055
Lighting	1056
Calibration, precision, and accuracy	1057
References	1060

ABSTRACT

The design and operation of a microscope freezing stage developed for use at magnifications up to 500× are described. It makes possible studies of low-temperature phase changes such as the freezing of a saline water phase, and hence an estimate of the total salt concentration, in fluid inclusions as small as 10 microns (10^{-6} milligram in weight). The crystal or polished mineral plate containing the inclusions is viewed while immersed in a thermostated heat exchange medium (acetone) circulating rapidly in order to minimize thermal gradients. The stage permits easy operation at temperatures down to $-35°$ C, with electrical control to $\pm 0.05°$ C, and to much lower temperatures with manual control. With substitution of silicone oil for acetone, the same stage can be used for heating experiments up to $+250°$ C. Calibration points in the low range indicate the accuracy of freezing temperature determinations on optimum material to be better than $\pm 0.1°$ C. The low relief of ice crystals in water solution places considerable importance on sample selection, preparation, and lighting.

As a result of almost ubiquitous and drastic supercooling (metastability), $-35°$ C is inadequate to freeze most inclusions. Holding at $-78.5°$ C (acetone + solid CO_2) for thirty minutes is generally adequate, although a few samples require extended immersion in liquid nitrogen at $-196°$ C to cause freezing of even a part of their inclusions. Such extensive supercooling is not possible with most surface waters owing to the presence of abundant extraneous solid nuclei for the crystallization of ice.

[1] Publication authorized by the Director, U. S. Geological Survey.

It is believed that the exceedingly slow flow rates for most deep-seated waters trapped in inclusions permitted settling of any such nuclei.

Most inclusions are nearly opaque when solidly frozen; with gradual warming they become translucent rather suddenly when the amount of liquid formed is adequate to fill in between the minute solid grains of ice and salts. This is called the *first melting temperature* and, although inexact, does provide some useful information as to the gross composition of these multicomponent systems.

More significance can be attached to the *freezing temperature,* defined as that temperature at which the last crystal, usually ice, melts in the inclusion, under *reversible* equilibrium conditions. Reversibility is verified by causing renewed growth of the last small crystal with a slight drop in temperature. As a result of supercooling, this test can be used at any temperature up to, but not including, the actual equilibrium freezing temperature. As long as the last crystal phase to melt is ice, the depression of the freezing temperature provides an approximate measure of the amount of salts present in solution. The combination of chemical analytical data on leachates, giving the *ratios* of the various salts present in inclusions, with freezing data, giving the *concentration* of salts, makes possible much more complete characterization of the included fluids than would be possible from either type of data alone.

INTRODUCTION

THERE have been many studies of the composition of the fluids trapped as fluid inclusions in minerals. Although these studies have shown that most such inclusions contain a water solution of salts in various ratios, with or without other phases, very few of them have provided data on the concentration of these salts in the water solutions. The concentrations are of considerable importance to studies of the origin, nature, possible mixing, and wall-rock alteration effects of the ore-forming fluids, to an understanding of the nature of reactions between minerals and pore liquids during diagenesis and metamorphism, and to the pressure corrections to be applied to inclusion filling-temperature determinations. It is thus highly desirable to investigate the methods by which such concentrations may be readily measured on a wide variety of samples.

Most of the methods for determining the concentrations—such as determination of the index of refraction of the liquid from total reflection at the liquid-mineral interface (1, 2, 16); actual drilling into single inclusions followed by analysis of the fluid extracted (5); decrepitation in a gas analysis train followed by leaching and analysis of the salts (13; 23, p. 266); and crushing in vacuo for analysis for water, followed by leaching and analysis of the salts (a technique to be described in a paper now in preparation)—require special samples or relatively large amounts of inclusion fluid and hence impose serious sample limitations. As the abundance of inclusions in minerals increases very rapidly with decrease in inclusion size in the range from 100 μ to 10 μ diameter, it is obvious that a procedure whereby one could obtain a measure of the approximate concentration of salts in single fluid inclusions of small size would be most desirable.

It appears that the only parameter that can be measured on a single, small, aqueous fluid inclusion in the range of 10–20 microns that would be a function of the approximate concentration of salts in solution in the inclusion is the depression of the freezing point of that solution. This is based on the assumption of the predominance of sodium chloride among the saline components in the fluids, as has been verified by analysis of inclusions. Obviously, optical methods must be used for recognizing the presence of solid phases such as ice in determining the freezing point, as the total amount of liquid is far too small (approximately 10^{-6} milligram or 1 mγ of liquid) for the use of such physical methods as cooling curves.

There have been comparatively few attempts to use cooling devices in the study of fluid inclusions. Davy found in 1822 (5) that some organic fluids in quartz crystals from France froze at 13.3°,[2] and Brewster in 1823 (1, p. 102) used a drop of ether to cool down what were later shown to be liquid CO_2 inclusions. In 1858 Sorby (29) attempted to determine the freezing point for fluid in some fluid inclusions but supercooling phenomena interfered (30). Spezia, in 1904 (31), determined the point of formation and the expansion of a gaseous phase in certain inclusions upon cooling below room temperature. Deicha (6, 7, 8, 8a, p. 57), and Deicha and Taugourdeau (9), reported on similar studies. Cameron, Rowe, and Weis (4, p. 224) used a wooden cell as a cooling stage, with a current of cooled air, to obtain the depression of the freezing point for liquid inclusions in beryl from pegmatites in Connecticut ($-5°$ to $-12°$), and Weis (33, p. 677) found that freezing temperatures ranged from $-3°$ to $-19°$ in some inclusions in beryl from pegmatites in South Dakota. The technique used involved complete freezing, followed by gradual warming in the cell to the melting point; frosting was avoided in part by permitting the temperature of the entire equipment, including the microscope, to drop to about $-29°$ (Weis, personal communication, 1961). Yermakov (34, p. 154) and Grechny (12) briefly mention the use of cooling stages, and Little (18) briefly describes the use of liquid air, ethyl chloride spray, or a dish of $CaCl_2$ solution cooled with dry ice on the microscope stage, but reports that frosting precluded operation at temperatures less than $0°$.

In order to measure with precision the depression of freezing points for fluid inclusions, it was decided that total immersion of the sample in a refrigerated thermostated heat exchange medium would be most desirable. This would minimize thermal gradients and hence permit a more precise measurement. As most of the phenomena to be observed with the cooling stage occur in the range from about $0°$ to $-25°$, considerably more precision in measurement is needed than in heating experiments, where the data may be spread over many hundreds of degrees. Thermal gradients within heating stages seriously limit the accuracy of many filling-temperature determinations, so it was decided that the same microscope cell used for cooling should be designed to permit heating as well.

I would like to express my appreciation for the help of J. F. Abell, E. Curtis, and R. Fones, all of the U. S. Geological Survey, in the design, construction, and maintenance of the equipment. The idea for the study came from Sorby's classic paper of 1858 on inclusions (29).

[2] All temperatures are given in degrees Centigrade (i.e. Celsius).

DESIGN OF THE APPARATUS

The apparatus was designed to permit precisely controlled and measured heating or cooling of samples, in the range $-35°$ to $+250°$, while under observation with the microscope in transmitted light at magnifications up to 500 diameters. Although it was specifically designed to avoid several of the experimental problems of control and measurement that formerly plagued determination of filling temperatures and freezing temperatures of fluid-filled inclusions in minerals, it is suitable for a variety of phase studies on inclusions and other materials.

The Cell.—The essential feature of the cell design is the total immersion of the sample in circulating thermostated fluid, as shown in Figure 1. The unit consists of an Invar or steel cell (a) having an inlet and outlet (b and c) for the thermostated fluid. The ends of the cell are Invar or steel caps (d), notched for a spanner wrench, which apply pressure to Teflon gaskets (e) above and below fused silica windows (f).[3] Within the cell, a loosely fitting slotted brass ring (g) supports a glass plate (h) with slotted periphery, on which the sample (x) is placed. If there are several samples to be run simultaneously, they are made into a mosaic on a glass plate, which is then placed on (h). A light spiral spring (i) acts to hold down the sample plate, preventing flutter from the circulating fluid moving by the samples. (In lieu of this spring, threaded notched retaining rings may be used above and below the sample glass plate.) Surrounding the cell is a cup (j) made of Marinite (asbestos-diatomaceous earth mixture), thermally insulating the cell from the metal cup (k) into which it is inserted; thumb screw (l) is used to locate and lock the cell in this unit. Brass plate (m) is mounted with clamps and centering pins (n) on the stage of a petrographic or stereoscopic microscope, with glass plate (o) centered. The cell is supported on plate (p) which is connected to plate (m) through east-west and north-south translation screws (adapted from a microscope mechanical stage—see insert on Figure 1), represented diagrammatically by (q). These permit the operator to examine a series of inclusions, as vernier scales on the adjustments allow rapid relocation of specific areas in succession during the run.

A loose fitting split Marinite insulating shield (r) of appropriate inside diameter slides on top of the cell and permits the microscope objective (s) to move freely up and down, and sideways. Tubes (t) are for the introduction of a stream of air into this shield, and below the cell. For low-temperature operation only, a sleeve of transparent plastic film has been found to be more suitable for the upper shield. It is fastened to the cell and to the microscope tube above the turret nose piece, permitting introduction of dry air without loss of visibility or flexibility. Dry air is used to prevent fogging of the objective and frosting of the windows (f) that would otherwise preclude observation at temperatures below ambient. The desiccant used must of course be adequate to lower the dew point of the air to less than the temperature of

---

[3] An earlier design had the upper window recessed below the exit bore to avoid trapping bubbles, but this precaution was found to be unnecessary. Epoxy resin was also used at first to cement in the fused silica plates but the circulating acetone gradually softened the cement and eventually leaked.

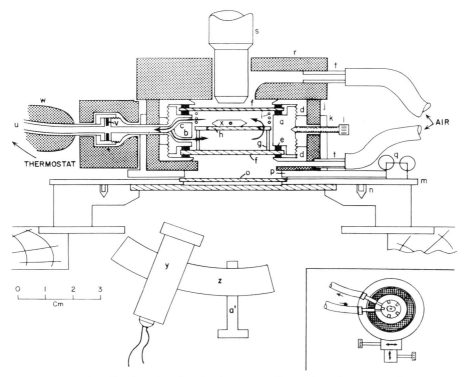

Fig. 1. Cooling (or heating) stage. Somewhat diagrammatic.

the cell. A layer of liquid acetone or alcohol on the top window is a much less satisfactory alternative. Glass plate (o) helps to act as a lower frost shield, and protects the microscope substage from possible leakage. A stream of air into the shield helps keep the microscope objective cool at elevated temperatures.

For high-power observations on thick plates during heating runs, long working distance universal stage objectives must be used occasionally, but special thin (1/32 inch) silica upper cell windows permit the use of regular objectives on most runs. The practical limit is 500× magnification, which is adequate for inclusions as small as 10 microns in diameter. Higher magnification, obtained using a 25× ocular, has little to recommend it as it gives little additional resolving power. The height of the metal ring (g in Fig. 1) is chosen such that the samples are brought sufficiently close to the upper cell window to permit focusing on the inclusions in question and yet not impede the flow of thermostated fluid above the sample plate. Where mineral plates of different thickness are to be viewed at high magnification, small glass plates are used as shims for the thinner sample plates.

The cell design permits observation over a circle of 35 mm diameter, so that a number of different mineral plates can be cemented together and studied simultaneously. Rubber cement is convenient for cementing together mosaics of samples for low-temperature runs, as it is readily removable and is insoluble in acetone, but it cannot be tolerated in the optical path as it becomes birefringent and opaque when very cold; if it is permitted to thicken before use it can be placed between the sample plates of the mosaic, and at the edge of the mosaic, without flowing under the plates. For both high and low temperatures sodium silicate is best, dried overnight at 90°; it can be removed without harm to the sample by immersion in water at 90° overnight. Thin wires of soft lead (electrician's fuse wire) can be bent around sample plates as a quick temporary mount.

Auxiliary Equipment.—The auxiliary equipment consists of a thermostated bath and pump, ½ ton refrigeration unit, light, microscope, and two variable-voltage transformers. The heat exchange fluid for circulation through the cell is pumped from a seven-liter thermostated bath [4] maintained about two feet from the cell. Transfer of this medium is through flexible continuous Sylphon bellows-type copper tubing of 7 mm bore, with braided metallic cover (u), Figure 1, with standard flat fittings using flat Teflon washers (v) to avoid soldered joints, which would melt, and to minimize the burn hazard inherent in the pumping of oil at temperatures up to 250°.[5] A precision mercury thermometer of appropriate range, graduated in tenths of a degree, one of a set of 5 to cover the entire range, is placed in the return line from the cell, at a point just before it enters the main bath. Immersed in the bath are the pump, mercury sensing element for the thermostat, two electric heater coils, and a cooling water coil for more rapid cooling from elevated temperatures. For low temperatures an expansion coil of fifty feet of ¼ inch OD refrigeration tubing was placed against the inside wall of the bath and connected to the electric refrigeration unit, mounted immediately below the thermostat. To obtain the needed cooling capacity at low temperatures, it was found necessary to install a bypass valve around the back pressure regulator of this unit.

The thermostated bath and delivery tubes must be carefully insulated throughout to minimize corrections and to cool the apparatus to $-35°$, as the refrigeration unit is rated for only $-40°$ at the expansion valve. Combinations of cork, glass wool, asbestos, and aluminum foil were used because most of the more efficient thermal insulators for refrigeration use will not stand the higher temperatures involved in filling temperature determinations, or acetone fumes, or are not flexible. The delivery tubes in particular must be well insulated to minimize corrections; several layers of fine glass wool separated with aluminum foil wrapping (w) proved flexible and thermally adequate.

As manufactured the 500- and 1,000-watt heaters can only be turned "on" or "off" by the relays. This was unsatisfactory, as the thermal lag of the sensing element was large enough to cause major fluctuations (up to 1°) in the temperature of the bath before the controls could be actuated, even though the element was sensitive to 0.01°. This was found to be true both at low

---

[4] A Zeiss Hoeppler Colura "Ultrathermostat" was used. The mercury-filled control thermometers on this unit are adjusted by a continuous screw (one full turn equals 0.3°), permitting precise adjustment.

[5] Teflon should not be heated over 250° without adequate ventilation, as its gaseous decomposition products are poisonous.

temperatures, using refrigeration, and at elevated temperatures. To eliminate this problem, the unit was rewired so that the power relay of the electronic thermostat controls only the 500-watt heater, using power from a variable voltage transformer. The 1,000-watt heater was operated independently of the thermostat control through another variable voltage transformer. By the adjustment of these two transformers the total power input to the heaters could be controlled to an amount just less than that needed with the thermostat off, and just a little more than that needed with the thermostat on. The thermostatic control thus makes only a small change in the total heat input in the system and hence causes only a very small amount of temperature hysteresis. For fine control at reduced temperatures it is sometimes necessary to reduce the amount of heat pumped out of the bath by controlling the refrigerant pressure with a back pressure regulator and gauge. In this manner it is possible to maintain any given temperature, with total cycling from thermostat control action well under 0.1°, or to raise the temperature at rates from 0.1 to 8 degrees per minute.

Cells of similar design, but with a single 10-inch copper pipe silver soldered into the side were used for temperatures below the limit of the refrigeration unit, for long duration cooling runs, and for quick reconnaissance runs. The pipe provides a handle for immersion of the cell into cold baths, and is bent up slightly so that the fluid will not run out when the cell is periodically placed on the microscope. Cell temperature is maintained by immersion in an acetone bath whose temperature is controlled manually with lumps of CO_2 and a low-temperature thermometer calibrated at the sublimation point for CO_2 at 1 atmosphere ($-78.5°$), the triple point of CO_2 ($-56.6°$), the melting point of mercury ($-38.9°$), and the ice point ($0.0°$). The thermal mass of the cell is adequate to permit fairly detailed observations without significant temperature change.

OPERATION

Heat-exchange Medium.—Acetone has been found to be the most suitable heat-exchange medium for low-temperature operation, and Dow-Corning Silicone Oil 550 for high-temperature operation. Viscosity increase at low temperatures precludes the use of silicone fluid for both ranges, and the vapor pressure of cold acetone is too low to constitute much of a fire hazard. At room temperature the pump will circulate 6 liters per minute of acetone, or 2 liters per minute of silicone oil. This oil is rather expensive, but is fairly stable even at 250° and is easily miscible with acetone, from which it can be readily separated by a water extraction of the acetone. Thus a simple acetone flush is used on changing from silicone to acetone; on the reverse change the residue of acetone in the thermostat, piping, and cell must be evaporated completely with a flow of air, or serious foaming occurs.

Sample Preparation.—Very careful sample preparation is essential. Sample plates must have either highly polished surfaces or very smooth cleavage or crystal planes, and should be as thin and small in mass as possible, for minimum thermal lag. Although readily recognizable fluid inclusions can be found in most mineral plates after only a little experience on the part of the observer, the optical requirements for freezing studies limit the method essentially to those inclusions that show a completely clear view of the entire inclusion when placed in the acetone-filled cell. Ice crystals in liquid present far less contrast than do gas bubbles, so inclusions that are suitable for filling temperature determinations are not necessarily suitable for freezing studies. Dark borders or areas due to total reflection, poor images due to superimposed inclusions or fractures, and confusing detail from irregular walls are objectionable and eliminate most inclusions from study. Such features must be avoided in the selection process or they must be minimized either by cutting and polishing new planes appropriately located in the sample plate, or by adjustments in the lighting (see below). The larger inclusions are generally easier to study, but seldom have good optical characteristics; in addition, as will be shown in the next article in this series, they frequently have leaked. Most of the determinations were made on inclusions between 0.01 and 0.10 mm diameter.

Freezing the Inclusions and the Problem of Metastability.—Although the refrigeration unit will maintain a temperature low enough ($-35°$) to keep most aqueous inclusions frozen, it was found necessary to use temperatures well below this to overcome an amazingly stubborn metastability in most inclusions, first described by Sorby (29, p. 471). For example, in the present work it was found that many inclusions in sphalerite, with actual freezing temperatures (disappearance of last ice crystal at equilibrium) about $-3°$, would not freeze even when held for an hour at $-35°$. This metastable supercooling, although very annoying in the experimental work, actually provides us with some indirect but useful information concerning the cleanliness and rates of flow of the fluids that have been trapped in fluid inclusions.

Supercooling of pure water and various salt solutions has been extensively studied (11, 15, 17, 19, 21, 22, 24, 28, 30). Apparently pure water cannot be made to freeze at $0°$ except when in contact with ice (28). *Homogeneous* nucleation of ice is a random process, controlled by the probability of spontaneous formation of a suitably sized cluster of molecules in ice configuration, and hence is time and volume dependent. It occurs at $\sim -36°$ for droplets 75–95 μ diameter (15), and 1 μ drops freeze in 0.6 seconds at $-41°$ (22); the nature of the surrounding medium, if uncontaminated with dust, has no measurable effect. *Heterogeneous* nucleation wherein an extraneous phase such as a dust particle acts as the nucleus is a function of the size and nature of the nuclei, and as even momentary exposure to ordinary air provides large numbers of nuclei of varying efficiency (22), it will also be volume dependent.

Most salts in solution lower the temperature at which nucleation (homogeneous or heterogeneous) occurs, apparently by binding the water in hydrated ions (15, 21), but Hoffer (15) quotes work of Kiryukhin and Pevsner showing that actual crystals of NaCl in saturated droplets did not aid the nucleation of ice. Airborne mineral dusts are effective in raising the nucleation temperature of ice in water droplets as much as $12°$, and Dorsey (11) has found that the nuclei in most natural (near surface) waters preclude supercooling below about $-10°$, although a considerable part of these nuclei may be meteoric in origin (11a).

The actual cause of the extensive supercooling found possible in fluid inclusions is not known. Nucleation in them is obviously volume dependent, as the larger inclusions freeze first, but the large amount of supercooling found for most inclusions in the present work indicates that they are quite clean and free from suitable nuclei for ice formation.[6] Presumably this requires very slow flow rates for the fluids at the time of trapping, thus avoiding the transport of fine debris in suspension. Apparently the structures of many minerals do not favor nucleation of ice, as one structure is present in the form of the walls of the inclusions, and, unless supersaturation (30) occurs generally, others should be present, even in small inclusions, as precipitated grains of 0.01 μ radius or larger (24). This radius is theoretically adequate to preclude much supercooling if they are structurally suitable as nuclei.

Other methods known to minimize supercooling include the use of jarring (28, 35) and an electrostatic field (25). Neither 120 cycle per sec jarring, nor a 20,000 cycle per sec ultrasonic field seemed to influence nucleation in inclusions. Although an electrostatic field has been shown to reduce drastically the amount of supercooling possible in insect larvae (25), it could not be conveniently applied here.

Recent studies by R. G. Layton at the University of Utah (personal communication, 1962; ms in preparation) indicate certain minute crystal imperfections may act as nucleation centers, but as many inclusions are the result of crystal imperfections, they would be expected to have some structural irregularities in their walls.

To overcome the metastable supercooling of inclusions, the tubing and cell, with samples in place, are first cooled by pumping refrigerated acetone. The pump is then turned off, and the cell, still connected to the thermostat by its flexible delivery tubes, is immersed in a large acetone-dry ice bath at $-78.5°$. This has been found adequate to freeze practically all aqueous inclusions so far examined, but many of the smaller ones will not freeze in the first 15 minutes in the bath at $-78.5°$; very few samples have resisted 30 minutes. Mere thermal lag is not involved here, as large inclusions in the same plates usually freeze within the first few minutes. The only real exceptions found were in two Iceland spar samples, ER61-42 and USNM 104292, in a calcite crystal from the native copper deposits in Michigan (USNM C1866), in halite samples ER61-48 and USNM R1202, and in an emerald from Colombia, ER61-6, to be discussed in the next article in this series. Many of the inclusions in these samples do not freeze after a week or more at $-78.5°$, or even $-196°$.[7] The freezing time may be saved in part by keeping a series of sample plates in the acetone-dry ice mixture, and inserting one quickly into the cold cell as soon as the thermostated bath temperature is sufficiently low.

---

[6] Simple statistical probabilities of spontaneous nucleation may be involved in the nucleation of the very small inclusions, as one that is 3 μ in diameter contains only 10^{11} molecules of water.

[7] It is possible that at least some of the very small inclusions cooled to very low temperatures form vitreous or amorphous ice (3; 10, pp. 396 and 643; 27; 32) and recrystallize later at higher temperatures. The larger supercooled inclusions are definitely liquid, however, as movement of the gas bubbles can be seen in some. The cubic modification of ice (20, 27) is probably not formed.

A metal jig on a sturdy arm hinged nearby, holding the cell high to prevent leakage on opening, helps to make sample changing fast (when cold), and safe (when hot).

Freezing is not without hazard to the samples, as occasionally they break open on freezing, even when the gas bubble is far larger than the expansion to be expected on freezing. The freezing of these strongly supercooled fluids appears to be practically instantaneous (data on the rate of crystallization of ice vs. degree of supercooling (14) would indicate that they freeze in approximately 10^{-5} seconds), so it can only be surmized that nucleation may occur at the surface of the gas bubble, and the growth of ice thus shields it from being effective as an expansion chamber. Such breakage has been observed only in samples having large inclusions (>1 mm), particularly in those inclusions having very small gas bubbles. Strangely enough, the mineral plates generally do not crack from thermal shock, even when liquid nitrogen is used, but the notched glass sample plates frequently break in liquid nitrogen. The cell itself has been cycled from $-78.5°$ to $+250°$ many times, and occasionally immersed in liquid nitrogen at $-196°$, but no problems arising from differential expansion and contraction have occurred in this design.

Determination of the First Melting Temperature.—Once the inclusions are frozen, the cell is placed in the Marinite cup on the microscope stage, examined briefly at $-78.5°$, and then the pump is turned on, rapidly bringing the cell up to the temperature of the bath. The temperature is then raised, either with the transformers or the thermostat control, to the temperature of first melting. This is the approximate temperature during warming at which liquid *in appreciable amounts* is formed in the inclusion. Except for very small inclusions and those of very low salinity, even thin frozen inclusions are almost completely opaque in transmitted light at $-78.5°$, particularly if the cooling was sufficiently rapid to result in freezing under conditions of strong supercooling. On slow warming these multicomponent systems become translucent rather suddenly, the change frequently taking place within a fraction of a degree.[8] This signals the formation of appreciable amounts of liquid in the inclusion, and is illustrated in Figure 2 A and B. After this first liquid is formed, gradual recrystallization of the very fine ice crystals occurs at constant or rising temperature, yielding a coarse granular mass (C, Figure 2). In small inclusions this eventually yields a very few (or even one) ice crystals with edges marked by fine, dark lines. Although these lines are exceedingly thin, they presumably are the fluid phase, as they widen to form a visible liquid phase on warming. The translucency upon first melting is due to this small amount of liquid phase wetting the surfaces of the ice (and salt) crystals. Occasionally inclusions with low salinities will freeze to ice so transparent that it cannot be distinguished from liquid except by occasional jagged or irregular boundaries separating supposed "liquid" from gas. With melting, however,

---

[8] In pure NaCl solutions this change occurs over an interval of less than 0.1°, and hence is useful as a calibration point. There is a possibility, however, of formation of metastable phases upon rapid freezing, and hence of obtaining first melting at a temperature below the stable eutectic or solidus temperature. In multicomponent systems there is also a possibility of poor contact between three or more solid phases causing the apparent first melting to occur at above the stable eutectic temperature.

the distinction between liquid and crystals becomes readily visible in all such inclusions. If a two-phase inclusion is not frozen, it may frequently be recognized as such without warming all the way to 0°, by small movements of the gas bubble.

Only part of the samples examined gave recognizable first melting temperatures; a number froze to clear ice and others showed evidence that they had melted between the temperature at which they were frozen ($-78.5°$) and the temperature at which they were first observed.

Determination of the Freezing Temperature.—The freezing temperature is defined as that temperature at which the last crystal, usually ice, melts in the inclusion, under reversible equilibrium conditions. Actually the reversi-

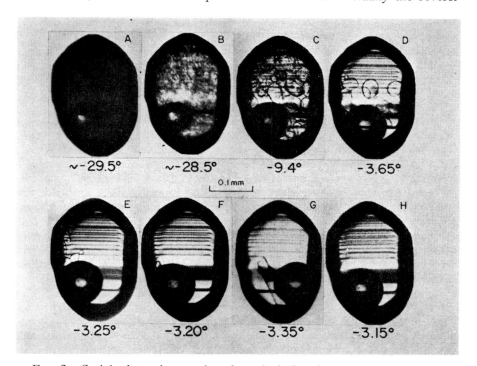

Fig. 2. Serial photomicrographs of an inclusion in a transparent plate of sphalerite from Creede, Colorado ("ER57-34"), taken after equilibration at the temperatures indicated; A and B show that the first melting temperature is at $\sim -29°$; C, D, and E show a decreasing group of rounded ice crystals in a solution; in F only one tiny crystal remains; after F was taken, the temperature was dropped 0.15°, causing typical growth of the ice crystal as two parallel plates, and the movement of the gas bubble, as shown in G; this crystal melted completely on raising the temperature (H) to 0.05° above that of F; the freezing temperature is thus taken to be $-3.15°$. The expansion of the water on freezing results in a reduced bubble diameter in A and B; another photograph taken at $+20°$ also shows a slight contraction of the bubble, from that in H, caused by the thermal expansion of the liquid solution; this expansion makes the gas bubble disappear at slightly over 200°. The horizontal bars are from oscillatory stria on the cavity walls.

bility is between *growth* and *melting* of the last remaining crystal slightly below the temperature at which it melts completely; metastable supercooling precludes attaining reversibility between *nucleation* and melting. Just before the last crystal of ice melts the temperature is dropped slightly. If the last crystal grows at one setting (or remains unchanged) and melts at a higher setting, the freezing temperature must lie between these two; where the two temperatures differ only slightly, the freezing temperature is taken as the higher one. Sharp blades and platelets of ice will usually grow out from the sides of the remaining rounded ice crystal on dropping the temperature only 0.1°. This is illustrated by a series of photomicrographs F, G, and H in Figure 2.

If the ice crystal disappears into a dark corner of the inclusion at a critical time, and cannot be revealed by adjustments of the light below, it is sometimes possible to detect its presence by dropping the temperature slightly. If the ice crystal has melted, this will cause no change, owing to the practically inevitable supercooling, but if it has not completely melted, dropping the temperature will cause it to grow and become visible again. It is thus possible, though time consuming, to get freezing points for inclusions with large dark margins.

One fortunate aspect of the apparatus is that it permits useful reconnaissance results to be obtained on a very rapid heating schedule of one to three degrees per minute. This rapid heating schedule gave results within a few tenths of a degree of those obtained with more careful work; this is probably the result of a cancellation of errors that can be expected whenever the size of the sample is such that the thermal lag in heating it is approximately equal to the thermal lag involved in heating the mass of mercury in the bulb of the measuring thermometer.

Lighting.—One of the most severe limitations in the use of the cooling cell lies simply in the problem of getting light through the inclusion where desired. Only a very small percentage of the fluid inclusions in a given sample will be suitable for freezing determinations owing to the common occurrence of large, dark borders from total reflection at the inclusion-mineral interface. This is particularly true for sphalerite which has a very high index of refraction. The cell design precludes effective use of dark field condensers. During moderately fast melting, the bubble, to which the last ice crystal is usually attached, moves about the inclusion. This is presumably due to the Marangoni effects from changes in the surface tension of the liquid (26) occurring on the addition of fresh water (melted ice) to the brine solution present. Even when an inclusion has a very large clear area in its center, it is not uncommon for the last crystal to move off into a dark area at the side of the inclusion at a critical point in the run. Trouble from this source can be minimized, and a greater variety of inclusion shapes can be used, if the light source is made adjustable as indicated in Figure 1. The light (y), a moderately high intensity focusing-type microscope light, is mounted on a curved bracket that slides on an arc (z), so located that the center of curvature, about which the axis of the light moves as a radius, is located at the sample point (x). This whole unit, in turn, can be rotated about a vertical post (a^1) arranged coincident with the microscope optical axis extended. In use, the light, which is under the microscope table, is adjusted on its arc or swung about to get light into that portion of the inclusion where it is needed. When using a stereoscopic microscope, tilting of the light may preclude true stereoscopic vision in some inclusions, but the correct setting for the stronger eye is easily found.

In most optically good inclusions the ice crystals are outlined by a clear dark line and obvious Becke line effects from their low index of refraction (Fig. 2), but where the optics are poor, and in working with dark minerals such as sphalerite, it is desirable to block off all excess light coming through the sample mosaic. This is most conveniently done by painting with India ink on the underside of the glass plate holding the mosaic. By this means light transmission is limited to the samples only, and very intense light can be used without having serious glare between the individual samples.

With normally isotropic minerals such as fluorite, the thick sample plates used frequently have too much irregular birefringence to permit identification of the presence of ice crystals from their own very low birefringence, except in large inclusions. The birefringence of crystals of $NaCl \cdot 2H_2O$ is readily seen, however, when they form in strongly saline inclusions. In studying inclusions in birefringent mineral plates it is best to keep one or the other polarizer in and rotated so that it is parallel to one of the vibration directions of the sample plate; this eliminates the bothersome double images obtained in thick plates, as mentioned by Sorby in 1858 (29). With the host mineral at extinction, optical properties of the crystals in the inclusions, such as birefringence and extinction angles, may be estimated by simple statistical procedures, if enough grains at random orientation are visible.

Calibration, Precision, and Accuracy.—Calibration of the apparatus is somewhat of a problem, in that it is not easy to have a true calibration run that duplicates all aspects of the experimental runs themselves. For a variety of reasons, some of which are obvious, the thermometer in the return line from the cell does not read the same temperature as that of the fluid in the inclusion. Even if the thermometer were placed in the cell itself, which was not done, for practical reasons, this would still be true. To minimize ambiguity, tiny droplets of various solutions were placed in a series of very small capillary tubes, sealed off on one end.[9] No more than one millimeter of length of the tubing was filled to avoid bursting on freezing. These were then quickly frozen in dry ice, and the open end sealed off with an oxygas flame, so that the entire sealed "synthetic inclusion" was <1 cm long, and yet had lost only negligible amounts by evaporation. Runs were made on a series of such inclusions containing various concentrations of sodium chloride and other salts in water, and various organic compounds, including National Bureau of Standards sample 559-5s, *n*-dodecane (m.p. $-9.60°$) and sample 2136-5, *o*-xylene (m.p. $-25.16°$). Erratic results were obtained in some of the salt solutions, evidently caused by a visible transference of droplets of water from one end of the tube to the other by evaporation and recondensation before the inclusion was frozen. More consistent results were obtained by setting up a

---

[9] Although ordinary organic melting point capillaries are fairly satisfactory, commercially available microcapillaries down to 0.020 mm I.D. were used in part.

strong thermal gradient down the length of the tube during the freezing operation, so that all water was transferred to the cooler end. As a result of these calibration runs, a deviation curve (Fig. 3) showing the difference between the observed temperature and the actual temperature in the cell was made, and all readings have been corrected by the use of this curve. The applied correction, added algebraically to the thermometer reading, is almost linear from $-0.3°$ at $0°$ to $-0.45°$ at $-25°$; at least a part of this correction is due to heat input from the mechanical energy of the 45-watt circulating pump motor, and from heat flow into the cell and the return tubing, as the thermometer itself was found to read only $+0.11°$ high when checked at the ice point. Frequent recalibration has shown the drift in the curve to be only a few hundredths of a degree.

The actual salt concentration within inclusions can be obtained from the depression of the freezing point only by indirect methods, as inclusion fluids

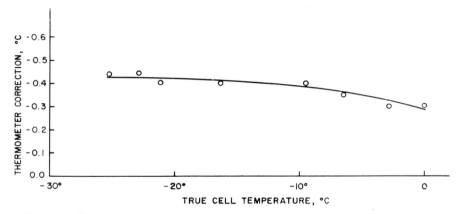

Fig. 3. Calibration curve showing the points used and the temperature corrections obtained, as deviations vs. true cell temperature at thermal equilibrium. The correction is added to the thermometer readings, i.e., the true temperature is colder than indicated by the thermometer. The points indicated, with the assumed correct melting point, liquidus, or eutectic temperatures are as follows: H_2O m.p., $0.0°$; 5 percent NaCl liq., $-2.9°$; 10 percent NaCl liq., $-6.5°$; n-dodecane m.p., $-9.60°$; 20 percent NaCl liq., $-16.3°$; NaCl-H_2O eutec., $-21.1°$; CCl_4 m.p., $-22.8°$; o-xylene m.p, $-25.18°$.

are generally multicomponent systems. If the *ratio* of the various salts can be obtained by one of the various chemical procedures that have been used for the analysis of inclusions, a strong synthetic solution of this particular composition can be made, and synthetic capsule inclusions of this solution, diluted to various concentrations, can then be run to establish a curve of the depression of the freezing point for the particular mixture concerned. As long as individual fluid inclusions do not deviate too far from this ratio of salts, close approximations of the true concentrations can be determined from

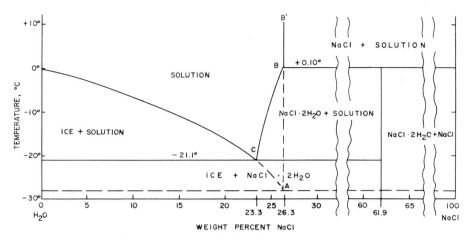

Fig. 4. The system H_2O-NaCl. Data from various sources (Int. Crit. Tables IV, p. 235, 1928).

the freezing temperature. In lieu of this, the depression of the freezing point for pure solutions of NaCl can be used as a crude guide (Fig. 4).[10]

In view of the variations in freezing temperature to be expected from variations in the ratios of the salts present, the precision of temperature measurement possible with the cell ($\pm 0.05°$) has been found to be adequate; when performed with due care repeat determinations on the same inclusion by different observers usually fall within such a range. There are a series of factors, however, that have been found to cause serious errors in the accuracy of the determinations, as much as $-1°$ or even more in rare instances. Most of these are basically nonequilibrium conditions resulting from time-dependent processes such as chemical and thermal diffusion. The following factors are particularly involved, and unfortunately all of them tend to *raise* the apparent freezing temperature over the true temperature:

1. Large volume or density of sample plate.
2. Large mass of inclusion fluid.
3. Low salinity, resulting in a larger amount of melting near the final temperature.
4. Elongate or highly irregular inclusions, requiring considerable diffusion to homogenize the liquid phase after an increment of melt water is added.
5. Interference by the gas bubble, partially or completely isolating the salt solution from the ice crystals.

The first three of these can be eliminated by briefly dropping the cell temperature just before final melting to verify the reversibility of the equilibria, but this will only be misleading in the fourth and fifth cases, as the apparent equilibrium conditions seen will not involve the entire inclusion system. Very

[10] The freezing point is lowered by increased pressure within the inclusion, but the effects are minor. Thus if the inclusion contains liquid CO_2, the most likely source of internal pressure in inclusions, the freezing point of a water phase would be lowered $< 0.3°$ due to the pressure alone.

large inclusions, such as the synthetic ones, permit significant compositional stratification by the floating of ice crystals and the sinking of $NaCl \cdot 2H_2O$ crystals. Considerable time is required to avoid erroneously high results from this cause.

The magnitude of the errors due to thermal lag may be large. Thus, if a determination is made by simply raising the temperature gradually to the temperature of disappearance of the last crystal, the errors due to non-equilibrium associated with the factors above may be additive. A good example of this is found in the behavior of small (i.e., 1 mm^3) "synthetic inclusions" of n-dodecane in sealed glass capillary tubes, used as a calibration point. Dodecane crystals grew at an uncorrected thermometer reading of $-9.25°$ and melted at $-9.20°$ in these tubes, as determined by applying the sensitive tests of changes in birefringence and the position of sharp crystal terminations relative to fixed points. On simple heating of these tubes at $0.05°$ per minute, the last crystal melted at $-8.80°$, i.e., an overshoot of $0.42°$; at $0.13°$ per minute, they melted at $-8.65°$; and at $0.72°$ per minute, they melted at $-7.8°$. Although the walls of these synthetic inclusions are thin, the mass of fluid is relatively large. They are made of a pure compound, so the heat of melting for the entire mass must all be absorbed *after* the temperature has risen to the melting point. The resultant overshoot is thus much larger than that obtained with natural fluid inclusions, in which the bulk of the heat of melting of the ice is absorbed well before the final temperature is attained, and the fluid mass is usually 10^{-3} to 10^{-6} that of the tubes of n-dodecane.

The second paper in this series will give data obtained with the apparatus on a wide variety of samples and their application to the solution of geologic problems.

U. S. GEOLOGICAL SURVEY,
   WASHINGTON, D. C.,
    May 18, 1962

REFERENCES

1. Brewster, David, 1823, On the existence of two new fluids in the cavities of minerals, which are immiscible, and possess remarkable physical properties: Edinburgh Phil. Jour., v. 9, p. 94–107.
2. Brewster, David, 1826, On the refractive power of the two new fluids in minerals, with additional observations on the nature and properties of these substances: Royal Soc. Edinburgh Trans., v. 10, p. 407–427.
3. Burton, E. F., and Oliver, W. F., 1935, The crystal structure of ice at low temperatures: Royal Soc. London Proc., A153, p. 166–172.
4. Cameron, E. N., Rowe, R. B., and Weis, P. L., 1952, Fluid inclusions in beryl and quartz from pegmatites of the Middletown district, Connecticut: Am. Mineralogist, v. 38, p. 218–262.
5. Davy, Sir Humphry, 1822, On the state of water and aeriform matter in cavities found in certain crystals: Royal Soc. London Philos. Trans., p. 367–376.
6. Deicha, Georges, 1950a, Essai de manométrie minéralogique: Bull. Soc. française de Minéral. et de Cristallog., v. 73, p. 55–62.
7. Deicha, Georges, 1950b, Example de recherches de venues pneumatolytiques dans le val Leventina (Tessin): Soc. Geol. de France, Compte rendu sommaire, no. 11, p. 186–188.
8. Deicha, Georges, 1953, L'emploi de la platine refrigérante dans l'appréciation du raport constituants fugaces / eau caractéristique des fluides d'origine profonde: Soc. Geol. de France, Compte rendu sommaire, no. 16, p. 351–353.

8a. Deicha, Georges, 1955, Les lacunes des cristaux et leurs inclusions fluides: Masson and Co., Paris, 126 p.
9. Deicha, Georges, and Taugourdeau, Phillipe, 1952, Étude microcinématographique des inclusions fluides: Soc. Geol. de France, Compte rendu sommaire, no. 11, p. 213–215.
10. Dorsey, N. E., 1940, Properties of ordinary water-substance: Am. Chem. Soc. Monograph 81, Reinhold Publishing Corp., N. Y., 673 p.
11. Dorsey, N. E., 1948, The freezing of supercooled water: Am. Philos. Soc. Trans., v. 38, p. 247–328.
11a. Fletcher, N. H., 1961, Freezing nuclei, meteors, and rainfall: Science, 134, no. 3476, p. 361–367, Aug. 11.
12. Grechny, Ya. V., 1957, [Crystal formation in binary melts]: p. 65–70 in [Growth of crystals; Reports at the First Conference on Crystal Growth 5–10 March, 1956] (in Russian) Acad. Sci. USSR, Moscow.
13. Grushkin, G. G., and Prikhid'ko, P. L., 1952 [Changes in chemical composition, concentration and pH of gaseous-liquid inclusions in successive fluorspar series] (in Russian): Zapiski Vsesoyuznogo Mineralogicheskogo Obshchestva, Ch. LXXXI, no. 2, p. 120–126.
14. Hillig, W. B., and Turnbull, D., 1956, Theory of growth in undercooled pure liquids: Jour. Chem. Physics, v. 24, no. 4, p. 914.
15. Hoffer, T. E., 1960, A laboratory investigation of droplet freezing: Tech. Note No. 22, Cloud Physics Lab., Univ. of Chicago, Ill., 64 p.
16. Kalyuzhnyi, V. A., 1954, [Measurement of the index of refraction of free liquids and mother liquors included in minerals by using a Fedorov stage] (in Russian): Mineral. Sbornik L'vov. Geol. Obshchestva no. 8, p. 315–344.
17. Kamenetskaya, D. S., 1957, [The effect of impurities on the production of crystallization nuclei in supercooled liquids]: p. 33–38 in [Growth of crystals; Reports at the First Conference on Crystal Growth, 5–10 March, 1956] (in Russian), Acad. Sci. USSR, Moscow.
18. Little, W. M., 1955, A study of inclusions in cassiterite and associated minerals: Ph.D. thesis, Univ. of Toronto, Nov., 1955. (Also in shorter form, in ECONOMIC GEOLOGY, v. 55, p. 485–509, 1960.)
19. Lloyd, D. J., and Moran, T., 1934, Pressure and the water relations of proteins No. I—Isoelectric gelatin gels: Royal Soc. London Proc., 147A, p. 382–395.
20. Lonsdale, Dame Kathleen, 1958, The structure of ice: Royal Soc. London Proc., v. 247A, p. 424–434.
21. Mason, B. J., 1957, The Physics of Clouds: Oxford Univ. Press, London, 481 p.
22. Mossop, S. C., 1955, The freezing of supercooled water: Phys. Soc. B. Proc., v. 68, p. 193–208.
23. Roedder, Edwin, 1958, Technique for the extraction and partial chemical analysis of fluid-filled inclusions from minerals: ECON. GEOL., v. 53, p. 235–269.
24. Roedder, Edwin, 1960, Fluid inclusions as samples of the ore-forming fluids: XXI Internat. Geol. Congr., Proc. of Sec. 16—Genetic problems of ores, p. 218–229.
25. Salt, R. W., 1961, Effect of electrostatic field on freezing of supercooled water and insects: Science, Feb. 17, 1961, v. 133, no. 3451, p. 458–459.
26. Scriven, L. E., and Sternling, C. V., 1960, The Marangoni effects: Nature, July 16, v. 187, no. 4733, p. 186–188.
27. Shallcross, F. V., and Carpenter, G. B., 1957, X-ray diffraction study of the cubic phase of ice: Jour. Chem. Physics, v. 26, p. 782–784.
28. Smith-Johannsen, Robert, 1948, Some experiments in the freezing of water: Science, v. 108, p. 652–654.
29. Sorby, H. C., 1858, On the microscopic structure of crystals, indicating the origin of minerals and rocks: Geol. Soc. London Quart. Jour., v. 14, part 1, p. 453–500.
30. Sorby, H. C., 1859, On the freezing-point of water in capillary tubes: Philos. Magazine, 4th ser., v. 18, no. 118, p. 105–108.
31. Spezia, Giorgio, 1904, Sulle inclusioni di anidride carbonica liquida nella anidrite associata al quarzo trovata nel Traforo del Sempione: Atti R. Accademia delle Scienze di Torino, v. 39, p. 407–418 (also v. 42, p. 409–417, 1907).
32. Vegard, L., and Hillesund, Sigurd, 1942, Die Strukturen einiger Deuterium Verbindungen und ihr Vergleich mit denjenigen der entsprechenden Wasserstoffverbindungen: Avhandl. Norske Videnskaps—Akad. Oslo. I. Mat.-Naturv. Kl., No. 8, p. 1–24.
33. Weis, P. L., 1953, Fluid inclusions in minerals from zoned pegmatites of the Black Hills, South Dakota: Am. Mineral., v. 38, p. 671–697.
34. Yermakov, N. P., 1950, [Studies of mineral-forming solutions] (in Russian): Univ. of Kharkov, 460 p.
35. Young, S. W., and Van Sicklen, W. J., 1913, The mechanical stimulus to crystallization: Jour. Am. Chem. Soc., v. 35, p. 1067–1078.

18

Copyright © 1967 by the Israel Program for Scientific Translation Ltd.

Reprinted from *Chemistry of the Earth's Crust*, Vol. 2, A. P. Vinogradov, ed., Israel Program for Scientific Translation Ltd., Jerusalem, 1967, pp. 260–281

PRINCIPAL GEOCHEMICAL ENVIRONMENTS AND PROCESSES OF THE FORMATION OF HYDROTHERMAL WATERS IN REGIONS OF RECENT VOLCANIC ACTIVITY
(Osnovnye geokhimicheskie obstanovki i protsessy formirovaniya gidroterm oblastei sovremennogo vulkanizma)

V. V. Ivanov

"The liquid which a gas usually encounters under natural conditions is water, i.e., the various aqueous solutions of which there are hundreds of kinds and which determine all geological, geochemical, mineralogical, and biochemical phenomena on the Earth."

"Thus, any aqueous solution takes up some of the gases in an unaltered form, producing an aqueous atmosphere, changes the composition of others creating new gases, and converts yet others into compounds, ions."

V. I. Vernadskii, 1931

1. ROLE OF GROUNDWATER IN THE FORMATION AND DIFFERENTIATION OF EXHALATIONS IN VOLCANIC REGIONS

As is known, regions of recent volcanic activity are distinguished by very intensive and variegated gas-hydrothermal activity which is reflected at the surface by discharges of high-temperature volcanic gases, vapor and thermal waters of different compositions. Their compositions and the regularities governing their formation and distribution are basically very similar for a majority of regions of recent volcanic activity, including Kamchatka and the Kurile Is., Japan, Indonesia, New Zealand, Iceland, and many others.

On the other hand, the intensity and specific features of the gas-hydrothermal activity of individual volcanic regions are conditioned by a complex combination of different geological, geothermal and hydrogeological conditions which show significant variations often even within limited areas.

The principal factors conditioning the composition of surficially discharged volcanic gases and the resulting thermal mineralized solutions are as follows:

1. Activity of deep-seated magma chambers and the thermodynamic conditions prevailing at great depths.

2. Processes of oxidation of deep-seated magmatic and thermometamorphic gases in the course of their rise toward the surface.

3. The geothermal and hydrogeological conditions in the upper horizons of the volcanic massifs.

At present, it has become necessary to introduce significant changes in the generally accepted volcanological principle that all so-called fumarole, solfatarra and mofette gases are typical volcanic exhalations originating at depth which are genetically directly related to magma chambers, and that the time-conditioned changes in their composition observed on the surface reflect changes in the composition of exhalations liberated by the magma. The true nature and causes of the observed postvolcanic gasothermal activity and of its variations are very complex and cannot be regarded as an index of the variation in the composition of gases liberated by the magma at great depths.

Variations in the nature of volcanic exhalations at the surface are largely conditioned by secondary processes due to changes in the intensity of discharge of high-temperature gases arriving from great depths and variations in the thermal regime of the gas-discharging channels. The cooling of these channels below a certain temperature limit leads to condensation of volcanic gases, part of which is absorbed by the groundwater at relatively small depths. This is indicated, on the one hand, by the considerable stability in the composition of gases discharged by molten lava over a very wide temperature range, and on the other hand by the simultaneous surface discharges of gases of different compositions from the same volcanic focus.

The available data on the composition of high-temperature volcanic gases and gases liberated by molten rocks /1, 16, 33, 36, 40/ suggest that at great depths, magma, which is a silicate melt with temperatures not less than 1,000-1,200°, liberates gases of a very complex composition, which always include halogens, sulfur, carbon, hydrogen and several other components (Tables 1 and 2).

TABLE 1

Average composition of volcanic gases from the Kilauea lava cauldron, vol. %*

Components	Content	Components	Content
H_2	0.72	CO	0.65
Cl_2	0.38	CO_2	12.55
S_2	0.53	N_2	8.58
SO_2	2.82	Ar	0.23
SO_3	1.90	H_2O	68.15

* According to Jaggar /33/, average of 26 analyses at 1,200° and 760 mm Hg; the average values were calculated by the present author.

As a rule, the ratios of the principal components remain very stable over a wide temperature range (above 200–250°) in all the typical high-temperature volcanic gases, as follows: $Cl + F < H_2S + S + SO_2 + SO_3 < CO + CO_2 < H_2O$.

TABLE 2

Chemical composition of fumarole gases from the Showa volcano (Japan)*

Component	Content with reference to the condensate, g/l			Component	Content with reference to the condensate, g/l		
	No. 9081 (760°)	No. 9051 (525°)	No. 9063 (220°)		No. 9081 (760°)	No. 9051 (525°)	No. 9063 (220°)
Cl	0.728	0.420	0.433	H_2	0.685	0.381	0.020
F	0.238	0.169	0.035	NH_3	0.0013	0.0008	0.017
Br	0.0011	0.0009	0.0012	N_2	0.567	0.676	1.250
H_2S	0.0080	0.042	1.080	A^{40}	0.0006	—	0.0015
S	0.0037	0.0018	—	CH_4	0.0015	0.018	—
SO_2	1.490	0.716	0.716	SiO_2	0.253	0.289	0.048
SO_3	0.021	0.011	0.0027	Al	0.015	0.014	0.0013
CO	0.050	0.034	—	Fe	0.0013	0.0012	0.0059
CO_2	29.200	25.800	13.000	Ca	0.0046	0.0043	0.021
B	0.039	0.021	0.0056	Mg	0.032	0.014	0.0079
PO_4	0.0028	0.003	0.0008	Na	0.022	0.022	0.013
NO_2	0.00001	0.000001	0.000008	K	0.015	0.011	0.0017
O_2	0.051	0.047	0.023				

* According to Nemoto, Haiakawa et al. (1957) cited by us according to White and Waring /40/. Our table excludes the trace elements (Ni, Cu, Zn, Ge, As, Mo, Ag, Sn, Sb, Pb and Bi) which were spectroscopically determined (contents were less than 1 mg/l).

It was emphasized by A.P.Vinogradov /6/, Krauskopf /16/ and others that in this case several features of the composition of different volcanic gases appear to be conditioned mainly by the degree of their oxidation. On the basis of the available analytical data and a series of calculations, Krauskopf proposed the following probable average composition of magmatic gases (in atm):

Water vapor	10^3 (1,000)
Total carbon as CO_2	$10^{1.7}$ (50)
" sulfur as H_2S	$10^{1.5}$ (50)
" chlorine as HCl	10^1 (10)
" nitrogen as N_2	10^1 (10)
" fluorine as HF	10^{-1} (0.1)

However, these ratios of the principal components can exist only in relatively slightly altered, high-temperature volcanic gases and then only in cases of intensive liberation of gases from the magma, these gases heating the gas-conducting channels to very high temperatures thus preventing condensation of gases and their contact with liquid groundwaters. The composition of gases discharged at the surface may resemble that of the primary, deep-seated, magmatic exhalations only in the case when the high-temperature volcanic gases completely "break through" the underground hydrosphere, "pressing" the waters away from the gas-conducting channel.

In the absence of these conditions, the composition of volcanic gases must necessarily undergo significant changes, aqueous solutions removing halogens, boron and CO, followed by SO_2 and partly by H_2S and CO_2. This process causes significant changes in the anionic composition of the groundwaters themselves depending on the composition of the dissolved gases, producing (see below) acid chloride, sulfate-chloride, or purely sulfate thermal solutions, the cationic composition of which is principally determined by the processes of their interaction with the country rocks.

In the past, the extraordinarily important role played by groundwaters in changing the composition of gases flowing through them was especially emphasized by V.I.Vernadskii. This phenomenon is especially pronounced in the regions of strong volcanic activity. In our opinion, groundwaters represent a factor responsible for the differentiation of gases into different types, including their separation into the two fundamental groups, the high-temperature chloride-sulfide-CO_2 ("fumarole") gases and the low-temperature hydrogen sulfide-CO_2 gases ("solfatarra") gases.

We are led to the conclusion that in a majority of cases the separation of volcanic gases into these two principal groups noted at the surface is not due to a change in the composition of deep exhalations, but only to variations in the intensity of process of liberation of volatiles from the magma at great depths, resulting in secondary processes of reaction of gases with the underground hydrosphere and in the marked qualitative change in the composition of gases emerging at the surface.

However, the role played by groundwaters in the gasothermal activity of volcanic regions is not limited to the above processes of alteration of the composition of volcanic gases. Under favorable geological and hydrogeological conditions, groundwaters sometime either directly participate in volcanic eruptions (being entrained and transported to the surface by the volcanic gases) or from independent gas-vapor separations which sometimes assume the nature of hydrothermal eruptions and explosions of considerable force (e.g., in the districts Wuotapu and Tarawera on North Island, New Zealand).

The available data indicate the very wide occurrence in the regions of recent volcanic activity of gas-vapor and vapor exhalations the formation of which is due exclusively to the groundwaters. The rapid conversion of water to vapor (with an attendant abrupt increase in the volume) in regions of recent volcanic activity may be due to a variety of geological causes, including the influx of volcanic gases or lava into lake basins, penetration of groundwaters into the channels through which lava flows, emergence of superheated groundwaters at the surface, etc. Therefore, regions of pronounced volcanic activity produce, in addition to gas exhalations of volcanic origin, exhalations of purely hydrothermal origin (formed from the groundwaters). Quite obviously, the composition of such exhalations reflects the gas composition of the thermal waters themselves; sometimes they contain appreciable quantities of acid gases CO_2 and H_2S.

There is no doubt that the so-called mofette gases, consisting mainly of pure CO_2 at temperatures not usually exceeding 100°, are largely a product of the degassing of carbonated thermal waters and are not volcanic exhalations.

These gasothermal separations, usually containing certain amounts of acid gases, play a far more important role in the hydrothermal alteration of volcanic rocks than was previously assumed, being more important than that played by the volcanic exhalations themselves which are of a more local nature.

The above principles are reflected in the following classification of the gasothermal exhalations in regions of recent volcanic activity, which differs to a certain extent from the common concepts of volcanology, including S.I.Naboko's classification of volcanic exhalations /18/.

It will be seen from Table 3 that our classification recognizes the existence of the following six principal types of the gasothermal exhalations in the regions of recent volcanic activity:

I.	Eruptive	IV.	Mofette
II.	Fumarole	V.	Gas-steam
III.	Solfatarra	VI.	Steam

The types I—III are of volcanic origin, type IV may sometimes be of volcanic origin but is probably of predominantly hydrothermal origin, and types V and VI are of exclusively hydrothermal origin.

Thus, we accept, in a somewhat modified form, the currently existing terminology for gaseous emanations of volcanic origin.

2. PRINCIPAL ENVIRONMENTS OF FORMATION OF THERMAL WATERS IN THE REGIONS OF RECENT VOLCANIC ACTIVITY

As is known, the chemical composition of the natural thermal solutions in regions of recent volcanic activity displays great variance, owing to the diversity of the geochemical environment in the deep-lying zones of such regions.

Conditions for the interaction of gases, vapors, groundwaters and rocks in the regions of volcanic activity are subject to marked variations even within small territories. Therefore, it is quite impossible to refer to any unified "general scheme" of the formation of hydrothermal waters, or to any "single genetic line" for the origin of the thermal waters in volcanic regions, which would be independent of the specific geochemical environments.

The anomalously high temperatures and the wide development of the most recent, hydrogeologically open zones of tectonic fracturing, offering the possibility of deep vertical water exchange, create very favorable conditions in the regions of recent volcanicity for the wide occurrence of underground thermal waters in deep-seated strata and their emergence to the surface, irrespective of whether these waters are or are not genetically related to active volcanic processes.

The available data make possible the identification of the following four principal, fundamentally dissimilar environments for the formation of hydrothermal springs in regions of recent volcanic activity:

I. Near-surface oxidizing environment in areas receiving volcanic gases and thermal fluxes generated by these gases.

II. Deep-seated reducing environment with exceptionally high-temperature conditions (250—300° and higher) owing to proximity of magma chambers, with limited influx of magmatic and thermometamorphic gases.

III. Deep-seated reducing environment in the sphere of influence of the thermometamorphic processes, usually under conditions of somewhat high temperatures.

TABLE 3

Classification of gasothermal exhalations in the regions of recent volcanic activity

Principal group of exhalations			Temperature, °C		Conditions of emergence at the surface	Principal types of exhalations	Principal gas components
I. Volcanic	Primary		High-temperature	Up to 1,200°	From volcanic craters during eruptions	I. Eruptive (of very complex composition)	Gases of the group Cl, F, S, C, B, etc., H_2O
				180–700	a) from crater lava cauldrons b) from incandescent slag cones and lava flows c) from fissures in volcanic rocks	II. Fumarole (chloride–sulfide–CO_2)	HCl, HF, NH_3, H_2S, SO_2, CO, CO_2, B, H_2, CH_4, N_2, H_2O
		Showing secondary alteration	Medium-temperature	100–180	a) from cooling lava masses b) from fissures in volcanic rocks	III. Solfatarra (sulfide–CO_2)	H_2S, SO_2, CO, CO_2, H_2, CH_4, N_2, H_2O
			Low-temperature	Less than 100°	a) from thermal crater and caldera lakes b) from fumarole thermal sources and mud pots		H_2S, CO_2, CH_4, N_2, H_2O
II. Hydrothermal				Less than 100	From carbonated thermal waters	IV. Mofette (CO_2)	CO_2
				About 100	From high-temperature (superheated) water rising under pressure in zones of tectonic faults	V. Gas-vapor (nitrogen–CO_2)	H_2O, CO_2, N_2, H_2S, sometimes CH_4
				Less than 100	From descending cold water which is heated in the rocks	VI. Vapor (oxygen–nitrogen)	H_2O, N_2, and O_2

265

IV. Deep-seated reducing environment with normal or high temperatures (usually up to 100–150°) lying outside the sphere of influence of volcanic or thermometamorphic processes.

Each one of the above environments is distinguished by its own specific geochemical conditions.

3. THERMAL WATERS OF UPPER OXIDIZING ZONES OF ACTIVE VOLCANOES

Thermal waters forming in the upper horizons of volcanic massifs under the direct influence of volcanic gases and geochemical processes generated by the latter are the most typical volcanic waters which we designate as fumarole thermal waters /9, etc./. This is the most specific type of natural thermal solutions and is not formed under any other geological conditions.

These thermal waters are formed at very different hypsometric levels (sometimes more than 1,000 meters above sea level) usually owing to local infiltration atmospheric waters which become saturated with fumarole gases.

In the case of high activity of a magma chamber and an intensive liberation of gas exhalations to the surface, high-temperature volcanic gases of complex composition (chloride-sulfide-CO_2, with fluorine, boron and other components) may reach the surface; on dissolving in underground (or surface) waters, these gases form acid sulfate-chloride or almost purely chloride thermal solutions.

These thermal waters display several specific features, including the presence of free HCl and H_2SO_4, a markedly acid reaction (pH less than 3.0, sometimes less than 1.0), a very complex cationic composition which depends on the conditions of leaching of volcanic rocks, frequent very high concentrations of iron and aluminum partly in combination with chlorides (!), high contents of HBO_2 and, especially, H_2SiO_3, etc.

Characteristic features of these thermal waters are the unusually high value of the K/Na ratio (up to 1), approximately of the same order of magnitude as that in volcanic rocks themselves (according to A. P. Vinogradov /7/, the ratio K/Na = 0.43–0.71 in intermediate and basic rocks), and the relatively (with respect to other alkalis) high content of lithium (Li/Na up to 0.025). These specific features of the thermal waters under consideration can obviously be explained by the type of leaching of the volcanic rocks by the highly acid nature of the thermal waters.

Depending on the conditions of influx of gases, there may arise either undergound fissure-vein waters which sometimes feed large boiling thermal springs, or hot crater lakes, or finally small funnels with natural highly concentrated condensates of volcanic gases. All these thermal waters produced by only slightly altered volcanic gases coming from deep-seated zones are arbitrarily designated by us as thermal waters of "deep formation" (Table 4).

When the activity of magma chambers and the temperature of the gas-conducting channels are low, the volcanic gases (Cl, F, B, SO_2) may be partially dissolved in groundwaters at a certain depth, so that the gases arriving at the surface have been "filtered" through waters, are markedly altered, usually show a $H_2S - CO_2$ composition, and are at a considerably lower temperature. In such cases, sulfate-chloride thermal waters

TABLE 4

Brief characterization of the composition of upwelling thermal waters of the upper oxidizing environment of active volcanoes

Serial No.	Site and year of sampling	Water temperature, °C	M, g/l	Formula of the ionic composition	Cl, mg/l	H_2SiO_3, mg/l	Characteristic ionic ratios			pH
							K/Na	Li/Na	Cl/SO_4	
I. Thermal waters of "surface formation"										
1	USSR Iturup I. Sernyi Zavod (Glavnyi Spring; 1954)	43.0	2.7	$\frac{(SO_4 + HSO_4)\ 97}{Al\ 49\ Ca25H23}$	38.0	153.4	0.42	—	0.02	2.0
2	Kamchatka Uzon volcano (Chernyi Spring; 1951)	85.0	2.5	$\frac{SO_4\ 99}{NH_4\ 48\ Fe25H15}$	13.0	242.0	0.44	—	0.007	2.3
3	Japan Nazudak volcano (Moto-Yu Spring; 1954)	69.9	2.9	$\frac{(SO_4 + HSO_4)\ 85}{H46Al29}$	168.8	190.8	0.32	—	0.08	1.5
II. Thermal springs of "deep formation" a) natural condensates of volcanic gases										
4	USSR Paramushir I. Ebeko volcano (central vent of the Verkhnii crater; 1955)	100.0	66.8	$\frac{Cl\ 99}{H\ 98}$	63041.2	76.8	0.83	0.022	63	−1.7
5	New Zealand White Island volcano (rapidly boiling crater lake; 1939)	—	106.4	$\frac{Cl\ 91}{Mg30\ Fe20Na16H15Al11}$	61840.0	230.0	0.13	—	5.8	0.5

TABLE 4 (continued)

Serial No.	Site and year of sampling	Water temperature, °C	M, g/l	Formula of the ionic composition	Cl, mg/l	H_2SiO_3, mg/l	Characteristic ionic ratios			pH
							K/Na	Li/Na	Cl/SO_4	
				b) waters of crater lakes							
6	USSR Paramushir I. Ebeko volcano (central vent of the Verkhnii crater; 1955)	Up to 60.0	4.8	$\dfrac{Cl61(SO_4 + HSO_4)38}{H38(Na+K)23Al19}$	1619.9	252.0	0.13	--	0.8	1.3
7	Kunashir I. (Golovninskoe hot lake; 1953)	30.0	1.7	$\dfrac{Cl83(SO_4 + HSO_4)17}{(Na+K)51Ca22H11}$	844.2	118.3	0.06	--	3.3	2.5
				c) (waters of thermal springs)							
8	USSR Paramushir I. (Verkhne-Yur'evskii Spring; 1955)	95.0	17.6	$\dfrac{SO_457Cl43}{H45Al21}$	4928.0	317.0	0.63	0.024	0.56	0.9
9	Kunashir I. (Nizhne-Mendeleevskii Spring; 1954)	91.3	4.3	$\dfrac{Cl45(SO_4+HSO_4)55}{H33Al17Fe15Ca12}$	982.5	351.5	0.91	--	0.4	1.7

Note. Water samples Nos. 1, 2, 4, 6, 7, 8, and 9 were collected by the author, and the analyses were performed by S. S. Krapivina and E. F. Prokof'eva; the analysis of sample No. 3 is cited according to Kimura et al. /41/, and that of sample No. 5 according to Hamilton and Baumgart /32/.

268

analogous to those mentioned above are likewise formed in lower horizons of active volcanoes, while only purely sulfate solutions, resulting mainly from the oxidation of H_2S, are formed in the near-surface horizons. These thermal waters are distinguished by the almost total absence of chlorides, presence of free H_2SO_4, acid reaction, complex composition of cations (including Fe, Al, sometimes NH_4, etc.) and the absence of boron. They are called the thermal waters of "surface formation."

On the whole, the formation of fumarole thermal springs, which are present-day thermal solutions in the full sense of the word, is conditioned by a complex combination of several processes which will be briefly described below (see Table 8).

4. THERMAL WATERS OF A REDUCING ENVIRONMENT AT DEPTH IN THE REGIONS OF ACTIVE VOLCANIC FOCI

The formation of natural thermal solutions in deep (1 — 2 km and more) strata of the region of active volcanoes occurs under exceptionally high-temperature conditions (250 — 300° and more) created by the flow of thermal waters from magmatic masses in a markedly reducing environment due to the low total oxygen potential under a high hydrostatic pressure reaching 100 — 200 kg/cm² (100 — 200 atm) at a depth of 1 — 2 km.

Geologically, these thermal waters are related to thick formations of Quaternary volcanic rocks (Iceland, New Zealand) or to volcanogenic-sedimentary Tertiary deposits (Kamchatka, Kurile Is., Japan).

Under these conditions, the thermal waters are typical high-pressure waters obeying general laws of the dynamics of artesian groundwaters, but with a markedly reduced viscosity because of the high temperatures.

Depending on the geostructural conditions and the collector properties of rocks, these thermal waters are characterized by the fissure or fissure-porous circulation conditions in the deep zone, but their arrival at the surface (often with the formation of geyser areas, etc.) is always related to large zones of tectonic fracturing (Wairakei and Wuotapu in New Zealand, Geizery and Pauzhetskie thermal springs of Kamchatka, etc.).

The composition of thermal waters forming under these conditions is strikingly uniform in all regions of volcanic activity. Usually they show a nitrogen-CO_2 gas composition (with the prevalence of CO_2 at great depths, some N_2 and insignificant amounts of H_2S), a low total gas content (commonly below 200 — 300 ml/l), relatively low mineralization (1.5 — 5.0 g/l), uniform chloride-sodium ionic composition (sometimes with high contents of HCO_3), frequently insignificant content of sulfates (usually below 100 mg/l), commonly high content of HBO_2, and the highest concentrations of H_2SiO_3 — up to 500 — 600 mg/l (Table 5).

A characteristic feature is a value of the ratio Li/Na higher than that usual for waters (relative enrichment of the waters in lithium).

At present, there are various views on the origin of the thermal waters under consideration. Several workers /39, 41, and others/ regard these thermal waters as typical volcanic waters the composition of which is entirely conditioned by the magmatic exhalations, including the appearance of chloride. The principal argument in favor of this view is that there is a relatively high concentration of lithium which is considered to be the product of

TABLE 5

Brief characterization of the composition of certain thermal waters of a reducing environment at depth in the regions of active volcanic foci

Serial No.	Site and year of sampling	Water temperature, °C	M, g/l	Formula of the ionic composition	SO_4, mg/l	H_2SiO_3, mg/l	HBO_2, mg/l	Characteristic ionic ratios				pH
								K/Na	Li/Na	Cl/SO_4	Cl/Br	
	USSR Kamchatka											
1	Pauzhetskie (Pul'siruyushchii Spring; 1958)	>100	3.1	$\dfrac{Cl94}{(Na+K)94}$	72.7	211.8	traces	0.07	—	21	366	8.2
2	Geizernye (Fontan Geyser; 1951)	>100	2.0	$\dfrac{Cl87}{(Na+K)93}$	102.8	308.3	"	0.10	—	8	428	8.6
	Kunashir I.											
3	Goryachii Plyazh (Glavnyi Spring; 1954)	>100	4.6	$\dfrac{Cl96}{(Na+K)93}$	54.2	309.4	"	0.18	—	43	690	8.1
	New Zealand											
4	Wairakei (borehole No. 4; 1955)	>100	3.9	$\dfrac{Cl97}{(Na+K)95}$	35.0	502.0	106.0	0.13	0.011	55	—	8.6
	Iceland											
5	Krizuvik (main borehole; 1957)	>100	2.0	$\dfrac{Cl87}{(Na+K)94}$	74.0	658.0	—	0.10	—	11	424	9.2
	USA											
6	"Morgan" Spring, California	95.4	4.6	$\dfrac{Cl96}{(Na+K)97}$	79.0	295.9	35.2	0.14	0.007	30	3030	7.8
7	"Norris Basin," Yellowstone Park	84.0	1.9	$\dfrac{Cl93}{(Na+K)93}$	38.0	671.8	46.0	0.17	0.018	20	7440	7.4

Note. Water samples Nos., 1, 2, 3, and 5 were collected by the author, and analyses were performed by S. S. Krapivina; the composition of sample No. 4 is cited according to Wilson /41/, and that of samples Nos. 6 and 7 according to White /40/.

magmatic exhalation in view of the alleged impossibility of its formation by the leaching of rocks.

We do not regard this argument to be sufficiently convincing since lithium possesses a great migration capacity under supergene conditions (A. I. Ginzburg /8/), and apparently this capacity is especially pronounced in acid waters. Therefore, we are of the opinion that the chemical composition of these thermal waters may also be conditioned by several other processes.

The presence of CO_2 in these thermal waters is irrefutable proof of the influx of certain amounts of gases of "deep" origin, which may be either magmatic and/or thermometamorphic.

However, the above statement does not yet prove that the main ionic composition of waters is exclusively determined by the deep exhalations. It is impossible to assume a complete absence of leaching of rocks during the slow and prolonged circulation of high-temperature waters at great depths in volcanic and especially volcanogenic-sedimentary rocks which were only slightly flushed.

The presence of chlorine in all thermal waters circulating in deep-seated volcanic rocks, including all nitrogenous thermal waters genetically unrelated to any magmatic processes, indicates that it was probably introduced into the thermal waters in question as the result of leaching of volcanic rocks. Chlorides in these rocks may always be present as products of ancient volcanic sublimates in sufficient amounts to ensure Cl accumulation in water in the low concentrations (maximum up to 2 g/l) actually noted. This is even more true of volcanogenic-marine deposits which may retain remnants of the marine saline complex.

In our opinion, the characteristic feature of the thermal waters in question, i.e., the vanishingly small content and sometimes almost total absence (in the most high-temperature waters) of sulfates, is explained by the instability of sulfates under high-temperature, reducing conditions at great depths, this being in good agreement with the absence of sulfate minerals in the high-temperature zones of the Earth's crust /3, 14/. The resulting hydrogen sulfide cannot likewise be retained by the thermal waters at great depths because it should combine with iron which is invariably present in volcanic rocks. Obviously, this also holds for hydrogen sulfide of magmatic origin in the case of its penetration into thermal waters at great depths. This provides an adequate explanation for the insignificant H_2S content in all the thermal waters under consideration. Very characteristically, H_2S waters are totally unknown at large depths in regions of volcanic rocks; this appears to be a characteristic regional hydrogeological regularity.

Thus, we are of the opinion that under the mentioned high-temperature reducing conditions in volcanic regions, all thermal waters, which possibly are of different initial compositions and origins, will ultimately be converted to chloride-sodium, sulfate-free waters containing CO_2 (and small amounts of N_2 and CH_4) and a certain, usually low content of calcium bicarbonates which depends on the presence of calcite formations in the volcanic rocks themselves.

In contrast to the above-enumerated processes of formation of the thermal waters at depth, their upwelling into the near-surface horizons (of much lower pressure) is accompanied by totally different processes which lead to certain variable changes in the composition. Changes of this kind include liberation of vapor and the attendant concentration and cooling of the thermal waters, their degassing (liberation of CO_2, H_2S and other gases), precipitation of calcium carbonates, and increased alkalinity of the thermal waters (see Table 8).

5. THERMAL WATERS IN A REDUCING ENVIRONMENT AT DEPTH WITHIN THE SPHERE OF INFLUENCE OF THERMOMETAMORPHIC PROCESSES

Considerable territories within regions of recent volcanic activity are not subject to the direct influence of magmatic gas exhalations which are usually concentrated in the form of small aureoles only in the vicinity of active volcanic centers. However, the thermal fluxes generated by the magma chambers undoubtedly spread to considerably larger areas at great depths, causing thermal metamorphism of thick strata of rocks of different compositions and origins. At temperatures of 350 – 400°, these processes cause the liberation by rocks of primary CO_2 /2, 30, 36, etc./ which rises to the Earth's surface, saturating groundwaters of deep-lying horizons. Obviously, these processes are of greater intensity and occur on a larger scale in the presence of carbonate deposits or calcite vein formations in the rock section of the volcanic regions. Under these circumstances, very high CO_2 concentrations can arise in the waters of deep-lying horizons or fissure systems, even at very high temperatures under conditions of restricted water exchange.

Experiments have shown /17, 25/ that under pressure of 200 kg/cm^2 (corresponding to a hydrostatic pressure at depths of the order of 2 km) and at temperatures of 200 – 300° the water may contain up to 60 – 70 g/l dissolved CO_2.

These phenomena quite naturally lead to the very wide occurrence of carbonated thermal waters in many regions of recent as well as Early Quaternary volcanic activity (Kamchatka, Minor Caucasus, the volcanic Massif Central, etc.). The formation of carbonated thermal waters in deep-lying, geologically fairly closed structures may occur in a variety of rocks (sedimentary, volcanic, intrusive) by way of carbonation of groundwaters of different initial compositions and origins (marine, atmospheric, or mixed).

Thus, the formation over large areas of carbonated thermal waters not directly related to active volcanic exhalations is conditioned by the thermal conditions in deep-lying strata and may occur both in regions of recent and of Early Quaternary volcanic activity as a result of the same processes.

The chemical composition of carbonated thermal waters may vary broadly, but usually it is dominated by chlorides and sodium, as in the case of other waters which circulate at depth. The other characteristic components of the waters in question are the bicarbonates which accumulate in carbonated waters owing to the leaching of rocks, but may also be present in certain quantities in the initial waters themselves (bicarbonate-chloride waters of marine genesis).

In regions of recent volcanic activity consisting mainly of volcanic and volcanogenic-sedimentary rocks which do not contain large concentrations of readily soluble salts, the carbonated thermal waters are distinguished generally by a low overall mineralization (less than 8.0 g/l) and relatively low content of HCO_3 ions. The carbonated thermal brines which are encountered in a borehole in Tenmango-No-Yu in Japan, with a total mineralization of 76 g/l, constitute a rare exception.

All carbonated thermal waters are distinguished by their high contents of H_2SiO_3, frequently high boron concentrations, and sometimes high arsenic concentrations (for instance, the Nalachevskie thermal waters of Kamchatka). The most probable source of these solutes is the water-containing rocks.

Table 6 shows the chemical composition of certain carbonated thermal waters in regions of recent and Early Quaternary volcanic activity in the USSR and other countries. Characteristic features of a majority of carbonated thermal waters are the intermediate values of the K/Na ratio (mostly 0.03 - 0.2), low values of the Li/Na ratio (0.002 - 0.004) and varying Cl/Br ratio which often approaches values typical of marine waters.

The temperature of carbonated thermal waters is dependent on their depth of circulation, often under high-temperature conditions. However, the temperature of these thermal waters on emergence at the surface under natural conditions does not exceed 60 - 80°. This is explained by the intensive, almost complete degassing of carbonated thermal waters of very high temperatures under the near-surface conditions, the other likely cause being the conditions of slow circulation (for small discharge) leading to their considerable cooling on the journey toward the surface.

The liberation of CO_2 by carbonated thermal waters in the near-surface rock horizons may lead to an abrupt disturbance in the carbonate equilibrium of the thermal waters, precipitation of $CaCO_3$, and a considerable increase in the pH value (see Table 8). The degassing of these thermal waters may generate gas jets of CO_2 which formerly were usually regarded as exhalations of purely volcanic origin.

6. THERMAL WATERS UNDER REDUCING CONDITION AT DEPTH, BEYOND THE SPHERE OF THE INFLUENCE OF MAGMATIC AND THERMOMETAMORPHIC PROCESSES

Notwithstanding the exceptional role played by magmatic and thermometamorphic processes in the creation of very specific geochemical environments in the regions of recent volcanic activity, these processes do not occur continuously throughout these regions. The latter contain considerable areas where recent volcanic activity does not manifest itself, thermometamorphism develops only at very large depths, and the geothermal regime within the upper rock formation (1 - 2 km) is normal or only slightly higher than normal.

In such regions, in the presence of considerable, deep, open tectonic faults, conditions are provided for deep infiltration of atmospheric waters and for the formation of thermal nitrogenous fissure waters, the chemical composition of which is not related to active magmatic and thermometamorphic phenomena, but is exclusively determined by processes of leaching of rocks (mainly volcanic and volcanogenic-sedimentary).

On the whole, the conditions governing the formation of these thermal waters are analogous to those conditioning the formation of thermal waters in ancient platforms and folded regions which were involved in the

TABLE 6

Brief description of the composition of thermal waters under reducing conditions at depth in the sphere of thermometamorphic processes

Serial No.	Site and year of sampling	Water temperature, °C	Dissolved CO_2, g/l	M, g/l	Formula of the ionic composition	H_2SiO_3, mg/l	Characteristic ionic ratios			pH
							K/Na	Li/Na	Cl/Br	
	USSR									
	Kamchatka									
1	Nalachevskie (Novyi Spring; 1951)	74.8	0.457	4.4	$\frac{Cl72SO_415}{(Na+K)75Ca21}$	191.1	0.17	—	370	6.4
2	Oksinskie (Bol'shoi Spring; 1951)	59.5	0.200	3.1	$\frac{HCO_453SO_433}{(Na+K)89}$	133.5	0.03	—	638	6.5
	Iturup I.									
3	Dachnye (Osnovnoi Spring; 1953)	34.0	0.750	5.9	$\frac{Cl60HCO_338}{(Na+K)70Ca15}$	174.2	0.04	—	900	6.0
	North Caucasus									
4	Pyatigorsk (borehole No. 19; 1955)	60.0	0.601	6.8	$\frac{Cl42HCO_336SO_422}{(Na+K)65Ca27}$	30.0	0.07	0.002	332	6.8
	Lesser Caucasus									
5	Dzhermuk (borehole No. 4/51; 1961)	61.5	0.400	4.8	$\frac{HCO_460SO_426Cl14}{(Na+K)82Ca11}$	90.0	0.07	0.004	379	6.8
	Czechoslovakia									
6	Karlovy Vary [Carlsbad] (Sprudel Spring; 1955)	73.0	—	6.5	$\frac{HCO_340SO_440Cl20}{(Na+K)89}$	88.4	0.06	0.002	440	—
	France									
7	Massif Central, La Bourboule (Chaussy Spring; 1950)	54.7	0.651	5.6	$\frac{Cl61HCO_335}{(Na+K)96}$	152.0	0.05	0.003	409	6.7
	Japan									
8	Arima District (borehole Tenmango-No-Yu; 1949)	94.0	0.370	76.7	$\frac{Cl99}{(Na+K)81Ca17}$	0.199	0.23	0.003	796	5.8

Note. Samples Nos. 1, 2, 3 were collected by the author and analyzed by S. S. Krapivina, M. S. Suetina and E. F. Prokof'eva; the analysis of sample No. 4 is cited according to I. Ya. Panteleev /22/; sample No. 5 was analyzed by O. A. Bozoyan, 1961; sample No. 6 is described according to A. M. Ovchinnikov /21/; sample No. 7, according to Urbain /38/; sample No. 8, according to Kimura et al. /34/.

TABLE 7

Brief description of the composition of certain thermal waters in a reducing environment at depth outside the sphere of influence of magmatic and thermometamorphic processes

Serial No.	Site and year of sampling	Water temperature, °C	M, g/l	Formula of the ionic composition	H_2SiO_3, mg/l	Characteristic ionic ratios K/Na	Characteristic ionic ratios Cl/Br	pH
	USSR							
	Kamchatka							
1	Nichikinskie (1950)	79.4	1.2	$\frac{SO_459Cl34}{(Na+K)\,92}$	126.0	0.04	1850	8.4
2	Malkinskie (1950)	82.7	0.6	$\frac{SO_438Cl36CO_323}{(Na+K)\,95}$	130.0	0.04	970	9.4
3	Nizhne-Paratunskie (1950)	61.0	1.5	$\frac{SO_469Cl26}{(Na+K)\,61Ca37}$	81.0	0.03	997	8.2
	Kunashir I.							
4	Dobryi Spring (1954)	68.2	1.4	$\frac{Cl63SO_434}{(Na+K)\,72Ca23}$	73.6	0.03	795	8.0
	Continental regions							
5	Kul'durskie (1958)	72.0	0.3	$\frac{(HCO_3+CO_3)\,45Cl31HSiO_311}{(Na+K)\,94}$	110.9	0.05	—	9.3
6	Garginskii (1939)	74.5	1.0	$\frac{SO_472HCO_311}{(Na+K)\,90}$	42.8	0.03	—	8.2
	France							
	The Pyrenees							
7	Cauterets (Casares Spring; 1924)	45.6	0.2	$\frac{Cl38SO_430HSiO_329}{(Na+K)\,99}$	88.0	0.04	205	9.4

Note. Water samples Nos. 1, 2, 3, 4 were collected by the author and analyzed by E. F. Prokof'eva and S. S. Krapivina; data for sample No. 5 are cited according to V. V. Ivanov /12/, for sample No. 6 according to V. V. Krasintseva /15/, and for sample No. 7 according to Urbain /37/.

Principal Geochemical Environments and Processes of the Formation of Hydrothermal Waters

Principal processes of formation of thermal

Characteristic environments for the formation of thermal waters	Principal types of thermal waters	Principal components of the gas and the ionic-salt composition	Origin of initial waters
I. Upper oxidizing, within the sphere of fumarole gases (of active volcanoes)	H_2S-CO_2 (fumarole), sulfate-chloride and sulfate, with a complex composition of cations, strongly acid	$\dfrac{CO_2,\ H_2S}{Cl,\ SO_4,\ H,\ Fe,\ Al,\ Na,\ Ca;\ H_2SiO_3,\ HCl,\ H_2SO_4}$	Present-day infiltration atmospheric waters, with the participation of waters of magmatic and thermometamorphic origin in the case of thermal waters of "deep formation"
II. Deep reducing, in regions of active volcanic foci	Nitrogen-CO_2, chloride, sodium, strongly superheated, alkali (when emerging at the surface)	$\dfrac{CO_2,\ N_2,\ \text{traces of }H_2S}{Cl,\ Na;\ H_2SiO_3,\ HBO_2}$	Infiltration atmospheric waters, with a possible insignificant participation of waters of marine, magmatic and thermometamorphic origin
III. Deep reducing within the sphere of thermometamorphic processes	Carbonated, of different anionic compositions, usually calcium-sodium, slightly acid	$\dfrac{CO_2}{HCO_3,\ SO_4,\ Cl,\ Na,\ Ca;\ H_2SiO_3,\ HBO_2}$	Ancient infiltration atmospheric waters, possibly with an insignificant participation of ancient waters of marine origin
IV. Deep reducing, outside the sphere of magmatic and thermometamorphic processes	Nitrogenous, usually chloride-sulfate sodium, alkali	$\dfrac{N_2}{Cl,\ SO_4,\ Na,\text{sometimes}\ Ca;\ H_2SiO_3}$	Infiltration atmospheric waters

344

waters in regions of recent volcanic activity

Principal formation processes	Principal metamorphic processes (under near-surface conditions)	Secondary phenomena and processes
1. Influx of volcanic gases and vapors into groundwaters, intensive heating of the waters 2. Passage of readily soluble gases (HCl, HF, B and certain others) into the ionic-salt composition of water with the formation of acid solutions 3. Oxidation of sulfurous gases (H_2S, SO_2) by free oxygen and by oxides of metals in the volcanic rocks, enriching the waters in H_2SO_4 4. Leaching of volcanic rocks by acid thermal solutions containing HCl, and H_2SO_4 (or H_2SO_4 alone), with enrichment of waters in Fe, Al, Ca, Na, H_2SiO_3, etc.	1. Increase in the overall mineralization of the thermal waters on interaction with rocks 2. Degassing of the thermal waters (removal of CO_2, H_2S, etc.) 3. Decrease in acidity (increase in the pH of the thermal waters)	1. Formation of H_2S-CO_2 gas jets 2. Decomposition and leaching of rocks (which lose Fe, Al, SiO_2, etc.) 3. Influx of sulfur into the rocks and its deposition at the surface
1. Heating of groundwaters under anomalously high-temperature conditions at depths of 250 — 300 m and more 2. Influx into the waters of insignificant amounts of gases of thermometamorphic (CO_2) and possibly magmatic (HCl, B, H_2S, CO_2, etc.) origin 3. Leaching of volcanic or volcanogenic-sedimentary rocks by weakly acid waters which become enriched in Cl, SO_4, Na, B and H_2SiO_3 4. Reduction of dissolved SO_4 to H_2S, which combines with the iron in rocks	1. Separation of vapor and corresponding increase in concentration of hydrothermal solution 2. Degassing of thermal waters (removal of CO_2, N_2, CH_4, traces of H_2S) 3. Precipitation of $CaCO_3$ 4. Change of weakly acid reaction of waters to alkaline 5. Formation of SiO_2 geyserite deposits	1. Formation of gas-vapor jets (CO_2 and H_2S) 2. Penetration of gas-vapor jets into the ground and surface waters with the formation of weakly mineralized sulfate or bicarbonate-sulfate surface thermal solutions ("pseudofumarole") 3. Leaching of volcanic rocks by acid solutions with the formation of variegated clays
1. Heating of groundwaters, usually under somewhat high-temperature conditions 2. Saturation with CO_2 of groundwaters of different initial compositions and origins 3. Leaching of different rocks by weakly acid thermal waters which become enriched primarily in carbonates of Ca and Mg and in H_2SiO_3	1. Degassing of the thermal waters (removal of CO_2) 2. Precipitation of $CaCO_3$ 3. Lowering of acidity (increase in pH)	1. Formation of CO_2 gas jets 2. Saturation with CO_2 of the upper horizons of ground (also surface) waters, with the formation of secondary carbonated bicarbonate magnesium-calcium waters
1. Heating of groundwaters under normal or somewhat high-temperature conditions 2. Leaching of rocks (mostly volcanic) by alkaline thermal waters which become enriched in H_2SiO_3 and other components		

most recent intensive orogenesis. This is the "province of nitrogenous alkali thermal waters of the regions of most recent tectonic movements," which forms an extensive belt within the continental part of the USSR /13/.

These thermal waters are commonly of very uniform composition. They are nitrogenous (N_2 > 95%, often constituting nearly 100% of total gases) with very low gas content, low total mineralization (< 1.5 g/l), commonly of chloride-sulfate sodium composition, siliceous (H_2SiO_3 up to 130—150 mg/l), and showing a distinct alkaline reaction (pH > 8.0) (Table 7).

A comparison between the nitrogenous poorly mineralized alkali thermal waters of Kamchatka forming in volcanogenic rocks and the nitrogenous thermal waters of the granite massifs of the continental regions of the USSR shows the former to have a somewhat higher total mineralization (averages of 1.1 and 0.4 g/l, respectively), and higher contents of chlorides (averages of 150 and 30 mg/l, respectively) and sulfates (400 and 120 mg/l, respectively) and sodium (250 and 100 mg/l, respectively). Nevertheless, the overall nature of these thermal waters is identical, sulfates and sodium being the predominant anions and cations in both types. The K/Na ratio is of approximately the same order of magnitude in both groups of nitrogeneous thermal waters (0.03 – 0.05), and the Cl/Br ratio is usually very high (up to 1,000 and more), being characteristic of infiltrating leaching waters.

Table 8 provides a general description of the fundamental processes of formation of thermal waters.

Conclusions

1. In the regions of recent volcanic activity, groundwaters play an important role in the differentiation of gasothermal exhalations from deep-seated magmas.

2. These regions contain, in addition to exhalations of volcanic origin, widely occurring, purely hydrothermal exhalations formed owing to vaporization of groundwaters and causing intensive hydrothermal alteration of rocks.

3. The presence of high-temperature volcanic gases of complex composition, intensive deep-seated thermometamorphic processes, high-temperature regime and deep infiltration of atmospheric waters along the zones of tectonic faulting — all these lead to the formation of four principal types of geochemical conditions, giving rise to different types of thermal waters in regions of recent volcanic activity.

4. The saturation of waters with volcanic gases in the upper oxidizing zone of active volcanoes leads to the formation of highly acid H_2S-CO_2 (fumarole) thermal waters, sulfate or chloride-sulfate, showing a complex cationic composition, with probable participation of waters of magmatic origin in their deep-seated varieties.

5. The thermal waters formed in the deep reducing high-temperature conditions in regions of active volcanoes are nitrogen-CO_2 (weakly acid at great depths, alkaline at the surface) chloride-sodium, highly siliceous, strongly superheated, and of a very uniform composition. Gases of thermometamorphic and possibly magmatic origin participate in the formation of these thermal waters.

6. CO_2 thermal waters of different ionic compositions are formed in a reducing environment at depth, in the sphere of influence of the thermal effect of magma chambers, from ancient infiltration waters and possibily also some residual marine waters.

7. Nitrogenous, alkaline weakly mineralized thermal waters are the only ones formed in a reducing environment at depth outside the sphere of influence of magmatic and thermometamorphic gases, mainly as a result of the leaching of different volcanic rocks by atmospheric waters.

8. On their way toward the surface, as a result of prolonged interaction with rocks, the H_2S-CO_2, nitrogenous-CO_2 and carbonated thermal waters undergo significant changes induced by secondary geochemical processes. However, these processes do not convert the thermal waters from one genetic type to another.

Laboratory of Volcanology
of the Academy of
Sciences of the USSR

REFERENCES

1. BASHARINA, L. F. Vulkanicheskie gazy na razlichnykh stadiyakh aktivnosti vulkanov (Volcanic Gases at Different Stages of Volcanic Activity). — Trudy Laboratorii vulkanologii AN SSSR, No. 19. 1961.
2. BELOUSOV, V. V. Ocherki geokhimii prirodnykh gazov (Essays on the Geochemistry of Natural Gases). — ONTI. 1937.
3. BETEKHTIN, A. G. Gidrotermal'nye rastvory, ikh priroda i protsessy rudoobrazovaniya (Hydrothermal Solutions, Their Nature and the Processes of Ore Formation). — In sbornik: "Osnovnye problemy v uchenii o magmatogennykh rudnykh mestorozhdeniyakh", Izd. AN SSSR, 1955.
4. VERNADSKII, V. I. O klassifikatsii prirodnykh gazov (Classification of Natural Gases). — In sbornik: "Prirodnye gazy", No. 2, Soyuzgeolrazvedka. 1931.
5. VERNADSKII, V. I. Vodnoe ravnovesie zemnoi kory i khimicheskie elementy (Water Equilibrium of the Earth's Crust and the Chemical Elements). — Priroda, No. 8—9. 1933.
6. VINOGRADOV, A. P. Khimicheskaya evolyutsiya Zemli (The Chemical Evolution of the Earth). — Izd. AN SSSR. 1959.
7. VINOGRADOV, A. P. Srednee soderzhanie khimicheskikh elementov v glavnykh tipakh izverzhennykh gornykh porod zemnoi kory (Average Contents of Chemical Elements in the Principal Types of Igneous Rocks in the Earth's Crust). — Geokhimiya, No. 7. 1962.
8. GINZBURG, A. I. Nekotorye osobennosti geokhimii litiya (Some Features of the Geochemistry of Lithium). — Trudy Mineralogicheskogo muzeya AN SSSR, No. 8. 1957.
9. IVANOV, V. V. Gidrotermy ochagov sovremennogo vulkanizma Kamchatki i Kuril'skikh ostrovov (Thermal Springs of Present-Day Volcanic Foci in Kamchatka and the Kuriles). — Trudy Laboratorii vulkanologii, No. 12. 1956.
10. IVANOV, V. V. Osnovnye stadii gidrotermal'noi deyatel'nosti vulkanov Kamchatki i Kuril'skikh ostrovov i svyazannye s nimi tipy termal'nykh vod (Principal Stages of Hydrothermal Activity of Volcanoes in Kamchatka and the Kurile Islands and the Related Types of Thermal Waters). — Geokhimiya, No. 5. 1958.
11. IVANOV, V. V. O proiskhozhdenii i klassifikatsii sovremennykh gidroterm (Origin and Classification of Present-Day Thermal Springs). — Geokhimiya, No. 5. 1960.
12. IVANOV, V. V. Osnovnye zakonomernosti rasprostraneniya i formirovaniya termal'nykh vod Dal'nego Vostoka SSSR (Principal Distribution Patterns and the Formation of Thermal Waters in the Soviet Far East). — In sbornik: "Voprosy formirovaniya i rasprostraneniya mineral'nykh vod SSSR", Izd. Tsentral'nogo Instituta Kurortologii i Fizioterapii Ministerstva zdravookhraneniya SSSR. 1960.
13. IVANOV, V. V., A. M. OVCHINNIKOV, and L. A. YAROTSKII. Karta podzemnykh mineral'nykh vod SSSR (masshtab 1 : 7,500,000) (Map of Underground Mineral Waters in the USSR, to a Scale of 1 : 7,500,000). — Izd. Tsentral'nogo Instituta Kurortologii i Fizioterapii Ministerstva zdravookhraneniya SSSR. 1960.

14. KORZHINSKII, D. S. Faktory mineral'nykh ravnovesii i mineralogicheskie fatsii glubinnosti (Factors of Mineral Equilibria and Mineralogical Depth Facies). — Trudy Instituta Geologicheskikh Nauk AN SSSR, seriya petrograficheskaya, issue 12, No. 5. 1940.
15. KRASINTSEVA, V. V. Khimicheskii sostav glavneishikh termal'nykh istochnikov Buryat-Mongolii (Chemical Composition of the Principal Thermal Springs in Buryat-Mongolia). — In sbornik: "Voprosy izucheniya kurortnykh resursov SSSR", Izd. Tsentral'nogo Instituta Kurortologii i Fizioterapii Ministerstva zdravookhraneniya SSSR. 1955.
16. KRAUSKOPF, K. B. The Use of Equilibrium Calculations in Finding the Composition of a Magmatic Gas Phase. — In Abelson, P. H., ed., Researches in Geochemistry, pp. 260–278. 1959.
17. MALININ, S. D. Sistema H_2O-CO_2 pri vysokikh temperaturakh i' davleniyakh (The System H_2O-CO_2 at High Temperatures and Pressures). — Geokhimiya, No. 3. 1959.
18. NABOKO, S. I. Vulkanicheskie eksgalyatsii i produkty ikh reaktsii (Volcanic Exhalations and Their Reaction Products). — Trudy Laboratorii vulkanologii AN SSSR, No. 16. 1959.
19. OVCHINNIKOV, A. M. Gidrogeologicheskie usloviya gidrotermal'nykh protsessov (Hydrogeological Conditions of Hydrothermal Processes). — Byull. MOIP, otd. geol., Vol. 32 (5). 1957.
20. OVCHINNIKOV, A. M. Usloviya formirovaniya mestorozhdenii uglekislykh vod (Conditions for the Formation of Carbonated Waters). — In sbornik: "Voprosy formirovaniya i rasprostraneniya mineral'nykh vod SSSR", Izd. Tsentral'nogo Instituta Kurortologii i Fizioterapii Ministerstva zdravookhraneniya SSSR. 1960.
21. OVCHINNIKOV, A. M. Zakonomernosti rasprostraneniya i formirovaniya uglekislykh gidroterm (Distribution Patterns and Formation of Carbonated Thermal Springs). — In sbornik: "Problemy geotermii i prakticheskogo ispol'zovaniya tepla Zemli", Trudy Pervogo Vsesoyuznogo soveshchaniya po geotermicheskim issledovaniyam v SSSR v 1956, AN SSSR, Vol. 2. Moskva. 1961.
22. PANTELEEV, I. Ya. Gidrogeologiya i genezis Kavkazskikh mineral'nykh vod (Hydrogeology and Genesis of the Caucasian Mineral Waters). — In sbornik: "Voprosy formirovaniya i rasprostraneniya mineral'nykh vod SSSR", Izd. Tsentral'nogo Instituta Kurortologii i Fizioterapii Ministerstva zdravookhraneniya SSSR. 1960.
23. SAVCHENKO, V. P. Zakony upravlyayushchie sistemoi zhidkost' + gazy, i ikh prilozhenie dlya vyyasneniya geneziza prirodnykh gazov (Laws Governing the System Liquid + Gases, and Their Application to Studies of the Genesis of Natural Gases). — In sbornik: "Prirodnye gazy", No. 11, Leningrad, ONTI. 1936.
24. SAUKOV, A. A. Neskol'ko zamechanii o gidrotermal'nykh rastvorakh i gidrotermal'nykh mestorozhdeniyakh (A Few Remarks on Hydrothermal Solutions and Hydrothermal Deposits). — Voprosy geokhimii, II, Trudy IGEM AN SSSR, No. 46. 1960.
25. KHITAROV, N. I. and S. D. MALININ. O ravnovesnykh fazovykh otnosheniyakh v sisteme H_2O-CO_2 (Equilibrium Phase Relationships in the System H_2O-CO_2). — Geokhimiya, No. 7. 1958.
26. KHITAROV, N. I. Voprosy formirovaniya gidrotermal'nykh rastvorov (Formation of Hydrothermal Solutions). — Trudy Laboratorii vulkanologii AN SSSR, No. 19. 1961.
27. SHCHERBINA, V. V. Okislitel'nye i vosstanovitel'nye reaktsii v magme (Oxidation and Reduction Reactions in Magma). — In book: "Akademiku D. S. Belyankinu k 70-letiyu so dnya rozhdeniya", Izd. AN SSSR. 1946.
28. ELLIS, A. Chemical Equilibrium in Magmatic Gases. — Am. J. Sci. 246, pp. 464–502. 1954.
29. BODVARSSON, G. Physical Characteristics of Natural Heat Resources in Iceland. U. N. Conf. New Sources Energy. 1961.
30. CHAMBERLIN, R. T. The Gases in Rocks. — Publs. Carnegie Inst. Washington, No. 106. 1908.
31. ELLIS, A. J. Geothermal Drillholes — Chemical Investigations. U. N. Conf. New Sources Energy. 1961.
32. HAMILTON, W. M. and J. L. BAUMGART. White Island. N. Z. Dept. Scient. and Industr. Res. Bull., p. 127. 1959.
33. JAGGAR, T. A. Magmatic Gases. — Amer. J. Sci., Vol. 238, No. 5. 1940.
34. KIMURA, K., Y. YOKOGAMA, and N. Ikeda. Geochemical Studies on the Minor Constituents in Mineral Springs of Japan. — Union geol. et geophys. internat. Assoc. Int. hydrol. sci. Assamblee Gen. de Rome, t. II. 1954.
35. LLOYD, E. T. The Hot Springs and Hydrothermal Eruptions of Waiotapu. — N. Z. J. Geol. and Geophys., Vol. 2, No. 1. 1959.
36. SHEPHERD, E. S. The Gases in Rocks and Some Related Problems. — Amer. J. Sci., ser. 5, Vol. 35 A. 1938.
37. URBAIN, M. P. Analyse des eaux minérales de Cauterets. — Institute d'Hydrologie et de Climatologie, Paris, Vol. 5. 1927.

38. URBAIN, M. P. Analyse des eaux minérales de la Baurboule et du Mont-Dore. — Institute d'Hydrologie et de Climatologie, Paris, Vol. 27. 1956.
39. WHITE, D. E. Thermal Waters of Volcanic Origin. — Bull. Geol. Soc. Amer., Vol. 68. 1957.
40. WHITE, D. E. and G. A. WARING. A Review of the Chemical Composition of Gases from Volcanic Fumaroles and Igneous Rocks. — Geological Survey Research. 1961.
41. WILSON, S. H. Chemical Investigations Geothermal Steam for Power in New Zealand. — N. Z. Dept., Scient. and Industr. Res. Bull., Vol. 117. 1955.

Erratum

On page 345, Table 8, the sixth line from the bottom under heading "Principal formation processes" should read: "in HCO_3, Ca and Mg, and in H_2SiO_3."

Copyright © by Microforms International Marketing Corporation

Reprinted from *Geochim. Cosmochim. Acta,* **28**, 1323–1357 (1964) with permission of Microforms International Marketing Corporation as exclusive copyright licensee of Pergamon Press journal back files.

Natural hydrothermal systems and experimental hot-water/rock interactions

A. J. ELLIS and W. A. J. MAHON

Chemistry Division, D.S.I.R., Petone, New Zealand

(*Received* 16 *August* 1963)

Abstract—Analyses of waters from many New Zealand hydrothermal areas (both volcanic and non-volcanic) are presented for discussion. The natural hot-water compositions are compared with those resulting from the experimental interaction of water at 150–350°C with volcanic rocks and a greywacke from the central part of the North Island of New Zealand. Appreciable quantities of the minor components Cl, B, F, As and NH_3 were liberated into solution from the rocks along with silica and alkalis. The ease of solution of the former group of elements, the kinetics of solution, and the slight degree of rock alteration, showed that they existed to a large extent on surfaces in the rocks (particularly for crystalline rocks) rather than in silicate structures.

It appears that volcanic thermal water compositions could be approached in nature by the reaction of high-temperature water with rock, without requiring a contribution from a "magmatic" fluid rich in the typical hydrothermal phase elements. It would be difficult to define the so-called "magmatic" solutions supposedly involved in the genesis of volcanic hydrothermal systems, as key elements such as lithium, caesium, chloride and boron would at equilibrium be concentrated into a hot-water phase interacting with either crystalline silicates or a rock melt. Preliminary experimental work at 500–600°C adds support to this suggestion.

The composition of solutions obtained from the interaction of greywacke and water was of similar type to that of warm springs occurring in this rock.

INTRODUCTION

HOT-SPRING areas occur in many regions of the world and several have been described in detail, e.g. Yellowstone Park U.S.A. by ALLEN and DAY (1935), Iceland by BARTH (1950), Kamchatka U.S.S.R. by PIIP (1937), and the North Island of New Zealand by WILSON (1961). WHITE (1957a, b) presented a general discussion on thermal waters and of the mechanisms that could lead to their formation. He reviewed the range of chemical compositions of thermal waters found in various parts of the world, and suggested tentative chemical criteria that could be used to distinguish waters of volcanic, metamorphic or connate origin.

Both the temperature and composition of thermal waters vary widely. The temperature of surface springs ranges from a few degrees above mean atmospheric temperatures up to the boiling point of water for the local atmospheric pressure. Drill-holes in many volcanic thermal regions reveal temperatures higher than are found in the surface springs, and the maximum temperature found to date was over 300°C in a hole drilled to 5230 ft in Imperial Valley, California. In the hot waters, the concentration of dissolved material ranges from a few hundred parts per million, up to the concentrated 30% brine obtained from the Imperial Valley drill-hole (WHITE, ANDERSON and GRUBBS, 1963).

The source of the waters, the origin of the heat, the mechanism of heat transfer to the water and the source of the dissolved chemicals have been the objects of discussion in many papers over the last 100 years. Agreement has been reached on many points, particularly on the predominance of local meteoric water in volcanic hydrothermal systems. A major point of controversy is the amount, if any, of water from a "magmatic" source present in these systems, and whether the dissolved material in the thermal waters and some of their heat content is derived from such a source. Early viewpoints of workers such as SUESS (1903), that volcanic spring waters were almost entirely magmatic fluids are untenable when evidence from deuterium and O^{18} determinations (CRAIG, BOATO and WHITE, 1956) are taken into consideration. These investigators showed that the volcanic thermal waters in general had isotopic compositions similar to the local surface waters. The possibility that 5–10 per cent of another type of water was present could not however be disregarded.

For high-temperature volcanic hydrothermal systems, difficulties are experienced in proposing simple convective models in which all the heat is derived by conduction from hot rock. It is difficult to maintain sufficient heat transfer over the life of a hydrothermal system because of the low thermal conductivity of rock. To overcome this difficulty, additional postulates must be made, such as a short period of flow following the heating of a stored body of water or convection within a molten magma and heat transfer to water across a thin solidified rock zone. The addition of "magmatic" water is often suggested as an extra supply of heat, which would at the same time explain the chemical composition of the hot waters. The concept that meteoric waters are heated by both conduction and magmatic steam from a molten rock intrusion to produce thermal waters (often at 200–300°C) has been used by several recent authors including WHITE (1957a), BODVARSSON (1961) and one of the present writers (ELLIS and WILSON, 1960).

The relatively low-temperature (about 100°C) hydrothermal systems in the Tertiary Basalts of Iceland have been explained satisfactorily by conductive heating alone, and the low concentrations of dissolved chemicals in the waters were assumed to have been derived by solution from the country rock (EINARSSON, 1942).

In this paper, attention is focused on the composition of waters found in hydrothermal systems of recent volcanic areas, such as the Taupo Volcanic Zone of the North Island of New Zealand. Three main types of thermal water compositions are found in these areas. The waters described are all from New Zealand, but the types are common to other volcanic zones of the world.

A. *Neutral chloride waters* with a predominance of sodium and potassium chlorides, close to saturation for their temperature with calcite and silica, and containing also arsenic, boric acid, fluoride, bromide, sulphate, bicarbonate, ammonia, lithium, rubidium and caesium. The chloride/sulphate ratio is high. These waters are found in extensive reservoir systems which extend to deep levels of at least a mile. In the underground systems the hot waters (often 200–300°) contain dissolved gases (mainly carbon dioxide and hydrogen sulphide), the total gas to water proportion being in the range 0·01–0·1 mole per cent. The pH of the deep hot underground water is about neutral compared with that for pure water at the same temperature (pH 5·7 at 260°C), but when the waters reach the surface and lose steam and carbon dioxide they become slightly alkaline due to the buffering action of the bicarbonate, silicate and borate ions.

B. *Acid–sulphate–chloride waters* which contain comparable concentrations of chloride and sulphate and are acid (pH 2–5), can originate in two ways.

(1) In a variation of Type-A waters, a high proportion of the associated sulphide has been oxidized at depth to bisulphate ions. Because of the change in acid dissociation constant of the bisulphate ion with temperature, the waters change from hot neutral pH waters underground to cooler acid waters at the surface (ELLIS and WILSON, 1962).

(2) In active volcanic areas, high-temperature, low-pressure steam rises from hot rock at a shallow level to condense in surface waters. The thermal waters often contain high fluoride concentrations derived from the hydrogen fluoride in volcanic steam. With a decrease in steam temperature the fluoride, chloride and sulphur gases, in order, decrease in abundance, so that acid–sulphate–chloride–fluoride waters merge into acid–sulphate–chloride waters, then into acid–sulphate waters of class C(2). Many of the constituents in the waters are derived by surface leaching of rocks by the acid solutions containing hydrochloric acid and sulphuric acid from sulphide and sulphur dioxide oxidation.

C. *Acid sulphate waters.* These are waters low in chloride content, formed, and made acid by the condensation of a low temperature (up to about 250–300°) steam phase in surface waters, and the oxidation of the hydrogen sulphide contained in the steam. They may be found *either* in (1) solfatara areas where steam rises from underlying hot water of Type A or B(1), *or* in (2) areas of surface volcanic activity where in the cooler stages, gases containing mainly carbon dioxide and hydrogen sulphide remain in the vapour phase.

Mixtures of the principal types of water sometimes occur. For example, some acid–sulphate–chloride waters are formed by mixing of Types A- and C(1)-waters in areas where chloride water springs exist close to the outcrop of the water table at the surface and perched acid–sulphate pools occur at higher levels (e.g. in the Waiora Valley near Wairakei). These "mixed" waters are not to be confused with the Type-B waters formed by a different mechanism.

Opinion as to the origin of the chemicals present in the higher-temperature hydrothermal systems (particularly Types A and B(1) which are of greatest importance because of their extent and volume) has alternated between derivation from a magmatic fluid, and from country rock. As elements such as boron, fluorine, lithium, potassium, rubidium and caesium are present in unusually high concentrations in volcanic thermal waters and are known to be concentrated in the residual fluid of a crystallizing rock melt, the theory of a magmatic origin has been supported by most writers since the time of the major researches of ALLEN and DAY (1935) at Yellowstone Park. Little was known of the concentration of many of the elements of interest in the volcanic and associated rocks of the various thermal regions, so that the alternative view of derivation by interaction of hot water with country rock has been largely neglected. The concentration in waters of major rock constituents such as sodium, potassium, calcium and silica have been recognized as being influenced by rock/water interaction (e.g. FENNER, 1936), but elements in minor concentration in average igneous rocks such as chlorine, fluorine, boron and nitrogen have been referred to a deep magmatic fluid without first testing adequately whether they could be derived by simple rock/water interaction at shallow levels. HAGUE (1911) in early studies at Yellowstone Park presented some analyses to show that constituents in the spring waters were derived from local rocks at shallow levels, but the results were too limited in scope. In this work Hague followed the earlier writings of BUNSEN (1847) on investigations of Iceland springs, but these waters are not really comparable to Yellowstone waters because of their lower temperatures and mineral contents. ALLEN and DAY (1935) later wrote that the concentration of elements such as chlorine, fluorine, sulphur and arsenic from rocks into the spring waters was "unthinkable".

It was decided to test by experiment the quantities of various elements which were liberated from rocks typical of the Taupo Volcanic Zone of New Zealand when they were exposed to hot water. Accordingly, eight rocks from this area were reacted with water at temperatures ranging from 150 to 350° and at a pressure of about 500 bar. (A few preliminary runs have also been made at 500–600° and 1000–1500 bar pressure.) The resulting solutions were analysed at intervals of time to see the rate at which the various elements were liberated from the rocks, and to see for each element whether an equilibrium between water and solid was set up, or whether continuous leaching from the rock occurred. The composition of the experimental solutions is compared with that of hot waters of New Zealand hydrothermal areas. Typical analyses of waters from both high-temperature hydrothermal systems and from warm springs are now given for this purpose.

COMPOSITION OF NEW ZEALAND THERMAL WATERS

Areas of active and recent volcanism

Table 1 gives some representative analyses of thermal waters from the Taupo Volcanic Zone of the North Island of New Zealand. Deep drill-holes of 1500–4000 ft depth exist at Wairakei, Waiotapu and Kawerau; Rotorua has shallow holes down to about 800 ft; the remaining representative areas have no existing holes. Figure 1 is a map showing the hydrothermal areas of New Zealand.

(a) *Active volcanic areas.* The sample from the 1933 Crater on White Island is an example of a water from an active andesite volcanic crater. It is acid, high in mineral content and contains many constituents leached from the rocks. Steam temperatures up to 570°C have been measured in fumaroles on White Island (unpublished report, Dominion Physical Laboratory, N.Z., D.S.I.R., 1956), and the steam has a high content of sulphur gases (H_2S, SO_2, SO_3) and hydrogen chloride, together with ammonia, hydrogen fluoride and boric acid. WILSON (1959) discussed in detail the chemistry of the fumaroles and pools on the Island. Fumarole condensates up to molar in hydrochloric acid were collected. The low value for the ratio Cl/F, and high ratios of Cl/B and Cl/As are notable features of the waters, although recent analyses show that Cl/B ratios for pools in the 1933 Crater range from 50 to 70 at similar chloride concentrations.

(b) *Recent volcanic areas.* The remaining analyses in Table 1 are of neutral sodium chloride waters, except for Rotokaua which is given as an example of a water from a deep-seated acid–chloride–sulphate area. All these occur in a volcanic

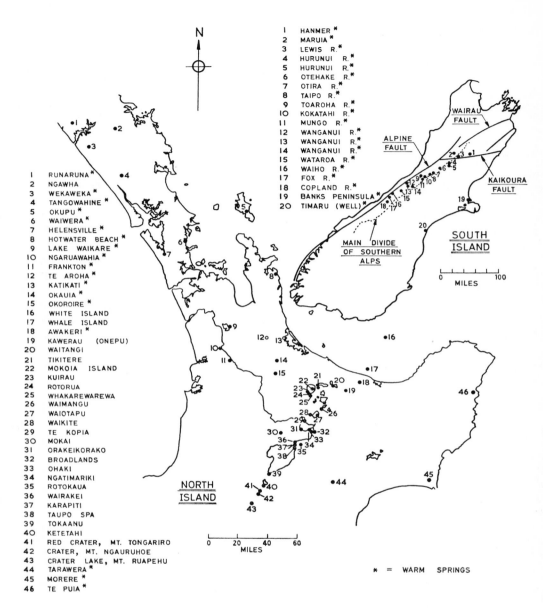

Fig. 1. Thermal spring areas of New Zealand.

Table 1. New Zealand volcanic thermal waters

Thermal area	pH							Concentrations in waters in ppm						
		Li	Na	K	Rb	Cs	Ca	Mg	F	Cl	Br	I	SO$_4$	As
Hole 4/1 Wairakei	7·9	12·6	1300	192	2·9	2·2	23	0·02	6·9	2140	5·7	<0·2	33	4·3
Hole 20 Wairakei	8·4	13·8	1300	220	3·1	2·6	18	0·04	8·2	2215	5·5	0·4	35	4·7
Hole 44 Wairakei	8·6	14·2	1320	225	2·8	2·5	17	0·03	8·3	2260	6·0	0·3	36	4·8
Spring 97 Wairakei	6·8	6·8	665	68	0·7	1·7	45	4·2	4·4	1110	2·5	0·4	72	1·8
Spring 190 Wairakei	7·5	10·0	950	62	n.d.	n.d.	20	0·05	5·8	1596	n.d.	n.d.	56	2·8
Hole 6 Waiotapu	8·9	6·6	860	155	2·4	0·8	10	0·06	7·5	1450	4·7	0·2	52	3·1
Hole 7 Waiotapu	8·8	6·4	790	90	0·7	0·9	10	n.d.	5·3	1310	3·7	1·0	86	n.d.
Spring 64 Waiotapu	5·7	9·0	1220	160	2·3	1·7	35	n.d.	5·5	2000	7·2	0·4	145	4·9
Spring 20 Waiotapu	8·6	4·0	450	22	0·3	0·6	9	n.d.	5·2	688	2·0	0·8	93	1·1
Hole 7a Kawerau	6·9	7·6	915	152	0·85	0·85	3·5	0·16	1·2	1473	n.d.	n.d.	60	n.d.
Hole 8 Kawerau	7·5	7·1	833	149	0·70	0·50	1·0	n.d.	1·4	1314	n.d.	n.d.	60	1·9
Spring 2 Onepu/Kawerau	6·2	2·7	330	49	n.d.	n.d.	13	n.d.	1·4	445	n.d.	n.d.	158	n.d.
Spring 4 Onepu/Kawerau	7·6	3·3	398	53	0·27	0·25	13	n.d.	1·9	544	1·6	0·8	96	n.d.
Hole 219 Rotorua	9·4	2·5	375	35	0·27	0·23	<1	0·06	n.d.	355	0·1	0·2	12	0·08
Hole 137 Rotorua	9·4	1·4	565	31	0·26	0·31	<1	0·22	4·0	632	2·1	0·7	30	0·30
Spring 98 Orakeikorako	8·3	4·0	280	42	0·15	0·22	2·5	0·5	8·5	284	0·6	0·2	220	0·30
Spring 14 Tokaanu	7·2	14·8	1170	116	1·5	2·6	25	n.d.	1·5	1956	5·5	0·6	42	5·5
Spring 6 Rotokaua	2·5	7·8	990	102	1·7	2·0	12	10	<1	1433	4·0	0·5	520	—
Taupo, Terraces Hotel	7·4	4·6	405	47	0·23	0·13	11	2·3	1·1	537	n.d.	n.d.	101	0·4
White Island*	Acid	n.d.	7670	1000	n.d.	n.d.	2560	7310	870	61840	40	6	10500	6

Concentrations in waters open to atmospheric pressure.
n.d. = not determined.
Total CO$_2$, H$_2$S, etc, includes both the molecular form and the derived ions.
Hole and spring numbers as recognized by N.Z. Department of Scientific and Industrial Research.

Table 1 cont.

Thermal area	Concentrations in waters in ppm							Atomic ratios						
	pH	Total SiO$_2$	Total HBO$_2$	Total NH$_3$	Total CO$_2$	Total H$_2$S	Cl/B	Cl/F	Cl/Br	Cl/As	Na/K	Na/Li	Na/Ca
Hole 4/1 Wairakei	7.9	590	112	0.25	25	—	23.6	165	850	1050	11.5	31	100
Hole 20 Wairakei	8.4	590	110	0.20	17	—	24.9	145	910	1000	10.0	28	125
Hole 44 Wairakei	8.6	640	117	0.15	19	—	23.9	145	850	990	10.0	28	135
Spring 97 Wairakei	6.8	235	57	0.22	88	1.9	24.1	135	1000	1300	16.6	30	26
Spring 190 Wairakei	7.5	245	82	0.37	38	2.0	24.0	145	—	1200	26	29	83
Hole 6 Waiotapu	8.9	470	56	0.9	65	—	32	105	690	990	9.4	39	150
Hole 7 Waiotapu	8.8	n.d.	63	n.d.	90	—	25.7	130	800	—	14.9	37	140
Spring 64 Waiotapu	5.7	490	117	11.5	235	6	21.1	195	620	860	13.0	41	60
Spring 20 Waiotapu	8.6	380	27	0.4	58	5	31	71	770	1300	35	34	87
Hole 7a Kawerau	6.9	760	273	n.d.	115	—	6.7	660	—	—	10.2	36	450
Hole 8 Kawerau	7.5	770	255	1.5	135	—	6.4	500	—	1450	9.5	35	1500
Spring 2 Onepu/Kawerau	6.2	245	85	4	85	n.d.	6.5	170	—	—	11.5	37	44
Spring 4 Onepu/Kawerau	7.6	240	102	4	110	n.d.	6.6	150	750	—	12.8	36	53
Hole 219 Rotorua	9.4	405	25.4	<0.05	206	36	17.2	—	8000	9000	18.2	45	>700
Hole 137 Rotorua	9.4	314	32.3	<0.05	143	74	24.2	85	680	4500	31	120	>1000
Spring 98 Orakeikorako	8.3	280	13.6	0.55	190	1.3	25.8	17.9	1100	2000	11.3	21	190
Spring 14 Tokaanu	7.2	220	246	2.6	170	0.05	9.8	700	800	750	17.2	24	81
Spring 6 Rotokaua	2.5	340	183	1.6	144	0.2	9.7	>800	810	—	16.5	38	140
Taupo, Terraces Hotel	7.4	235	38	0.1	180	0.4	17.5	260	—	3000	14.7	27	64
White Island*	Acid	180	26	17	n.d.	30	2900	38	3500	22000	13	n.d.	5.2

* From WILSON (1959); water from the centre of 1933 Crater, which also contained Al^{3+}, 2030 ppm; Fe^{3+}, 140 ppm; Fe^{2+}, 11340 ppm; Mn^{3+} 260 ppm; Sr^{2+}, 10 ppm; S$_4$O$_6^{2-}$, 170 ppm; and H$^+$, 196 ppm.

zone of predominantly rhyolitic rocks (pumice, rhyolite and ignimbrite flows), and the waters reach the surface at boiling point. With the exception of Rotokaua, the waters from this recent volcanic area have rather similar chemical characteristics. The pH values range from about 6 to 9, rarely being higher or lower.

There is about a ten-fold variation in the total ion concentrations of the various spring waters, those from Orakeikorako being the lowest, and from Tokaanu the highest. The concentrations in the underground reservoir systems feeding the springs, and tapped by drill-holes are not the same as in the surface discharges. For example, at Wairakei the underground reservoir temperature is about 260°, and in allowing this water to separate steam in flashing to atmospheric pressure, a concentration factor of about 1·46 is introduced. The factor is rather higher at Waiotapu and Kawerau, where underground temperatures of up to 295 and 285° respectively have been measured by N.Z. Ministry of Works engineers. In the areas tested by deep drilling, the concentrations of chloride in the underground waters range from about 800–1600 ppm.

All the waters contain several parts per million of lithium (Na/Li atomic ratios range from 20–40 with one exception). Rubidium and caesium are also present at the level of 0·01–0·001 of the potassium molar concentration (ELLIS and WILSON, 1960; GOLDING and SPEER, 1961).

Calcium contents do not exceed 50 ppm, and are lowest in waters from Kawerau drill-holes where underground temperatures are highest. This suggests that the ion is limited in concentration by a temperature-dependent solubility equilibrium. The increase in the calcium content of waters at Wairakei as they travel through the field and cool has been used as a means of indicating water movement (ELLIS and WILSON, 1960). The magnesium concentrations are so low that in the past they have not been determined satisfactorily by titration in the presence of calcium. By atomic absorption flame photometry, concentrations in the deep underground waters at 250–280° prove to be of the order of 0·01–0·1 ppm, and are probably dependent on a solubility equilibrium involving magnesium carbonate or magnesium hydroxide. At the surface, at lower temperatures, magnesium concentrations in the waters increase to several parts per million, and are still higher in acid pools.

As discussed by MAHON (1963) the fluoride contents are dependent on underground temperatures and solution compositions, but the usual concentrations of this ion are in the range 1–12 ppm. Spring waters in an area often have lower Cl/F ratios than water tapped from the deep reservoir, due to leaching of fluoride from the country rock by the water on its upward journey.

With one exception, a Rotorua drill-hole, the chloride-to-bromide atomic ratios are in the range 600–1100 and are rather higher than the value of 660 for sea water (ELLIS and ANDERSON, 1961). The atomic ratios of chloride to arsenic are usually about 1000, being slightly higher for the more dilute waters. They are particularly high at Rotorua (RITCHIE, 1961).

The sulphate contents are erratic, but it is of interest that the minimum value of 12 ppm SO_4^{2-} corresponds to about the content that would be formed by quantitative oxidation of sulphide by oxygen dissolved in water at 15°. Higher contents presumably come from oxidation underground by ferric iron, or from hydrolysis of old buried sulphur deposits ($4S + 4H_2O \rightleftharpoons 3H_2S + H_2SO_4$). The springs selected are of good flow, so that surface oxidation of hydrogen sulphide would not be important.

The regular analysis of the discharges from about 70 drill-holes at Wairakei shows that silica contents of the waters correspond closely to the solubility of quartz at the measured underground temperatures. As water cools from 260° in rising through the country, the silica in the waters decreases by depositing quartz in the rock. However, below about 150–200° the quartz/solution equilibria is too slow to maintain equilibrium silica concentrations and further deposition in the rock takes place only when the solubility of amorphous silica is exceeded, close to, or at the surface (ELLIS, 1961). The solubility of amorphous silica at 100° is about 400 ppm according to KENNEDY (1950), and the experiments below show that silica solubilities of similar magnitude can be expected from glassy volcanic rocks.

With respect to boron, there are two groups of water systems; those with atomic ratios of Cl/B of about 20–30, and the other group with Cl/B ratios of 6–10. Kawerau, the highest temperature area is in the latter group, as is Tokaanu and Rotokaua, but the deep underground temperatures in the latter two areas are not known. The tendency for the group with the low Cl/B ratios to also have low fluoride concentrations hints at high-temperature water reservoirs. As noted, the low pH of the Rotokaua waters is only a surface phenomenon.

Other minor constituents have been determined on the neutral chloride type waters. As typical examples the concentrations of various elements in the Wairakei drill-hole waters are as follows; Sr, 0·1 ppm; Ba, 0·01 ppm; Sb, 0·1 ppm; P, 0·05 ppm; Se, 0·01 ppm; Mo, 0·02 ppm; Pb, 0·002 ppm; Ni, 0·001 ppm.

As discussed by WILSON (1961), the general concept of a deep and major water storage system beneath each hydrothermal area is supported by the constancy of ratios of elements within each area (e.g. Wairakei and Kawerau). Rotorua and Waiotapu appear to be areas where more than one reservoir system exists in the top few thousand feet.

The concentration of gaseous constituents (CO_2, H_2S, NH_3) is really only of significance when given as contents in solution in the underground hot water before steam separation. In springs and drill-hole discharges, the concentration of gases in the waters depends on the way in which the waters pass from the underground storage to the surface, i.e. the amount of steam and gas separation and loss before a water is sampled. The content of total CO_2, H_2S and NH_3 appearing in the water analyses in Table 1 is the summation of the molecular and ion forms of the compounds, expressed as the molecule, found by analysis. They are not the concentrations in the underground waters, as steam separation and temperature changes influence many interacting acid/base equilibria involving silicic, hydrofluoric, boric and carbonic acids, hydrogen sulphide, the bisulphate ion and ammonia.

For Wairakei drill-hole discharges it is possible to add together the separate gas contents of the steam and water phases collected at a particular pressure from a discharge to obtain an adequate idea of the gas contents of the underground 260° water. At Kawerau and Waiotapu this knowledge of underground conditions is not so certain, as the drill-holes often tap at their bases a water/steam mixture derived from the parent liquid phase. Average compositions of gases from the three areas of deep drilling are given in Table 2, together with approximate gas contents

Table 2

Area	Approx. average drilling depth (ft)	Average gas content of total discharge (moles/100 moles H_2O)	% gas composition				
			CO_2	H_2S	HCS	H_2	N_2
Wairakei	2000	0·020	92·8	4·2	0·9	1·8	0·3
Waiotapu	3000	0·10	88·0	10·3	0·2	1·0	0·5
Kawerau	3000	0·25	94·0	2·6	2·1	0·3	1·0

HCS = total hydrocarbons, mainly methane (compositions, as volume percentages).

in the parent hot water systems. Carbon dioxide and hydrogen sulphide are the predominant gases.

Warm spring areas (Table 3)

(a) *Springs in non-volcanic areas.* In the South Island of New Zealand at Hanmer, Maruia and the Wanganui River, hydrothermal systems rise in Mesozoic or Paleozoic greywacke and pass through Recent sands and gravels. Major faulting and uparching of the old rocks has occurred in the zone about the flanks of the Southern Alps, and the hot waters are thought to be associated with these rock movements and a high geothermal gradient (HEALY, 1948). Springs at Tarawera and Awakeri in the North Island also rise in greywacke close to the margin of rhyolitic volcanic rocks of the Taupo Volcanic Zone.

On the East Coast of the North Island at Morere and Te Puia, springs occur in quite a different environment in an area of Tertiary sediments. There are no signs of volcanic activity that could be considered as a heat source, and the origin of the heat is unknown.

The spring waters from both greywacke and sediments are characterized by high Na/K, Na/Li, Na/Rb, Na/Cs ratios and low silica concentrations. The springs associated with greywacke (except Wanganui River) have low Cl/B ratios, but in Morere and Te Puia waters, although the boric acid concentrations are considerable, the Cl/B ratios are high because of high sodium chloride concentrations which approach the level found in sea water. The fluoride concentrations are similar to those for springs of volcanic origin. Calcium and magnesium concentrations are high only at Morere and Te Puia. These ions are not accompanied by an equivalent concentration of bicarbonate, but correspond with low fluoride contents in the waters. Morere and Te Puia springs also have particularly high bromide and iodide concentrations, as might be expected from their sedimentary environment (ELLIS and ANDERSON, 1961), but their ammonia contents are low.

Iodide is relatively high in concentration in the Tarawera spring compared with that in South Island greywacke springs, and in volcanic spring waters. All the waters are low in sulphate, and have negligible amounts of arsenic.

(b) *Springs in old volcanic areas.* Te Aroha springs are situated along a fault scarp in an area of Tertiary and Pleistocene andesites. The country has extensive sulphide mineralization and a high thermal gradient. Helensville and Waiwera

Table 3. Warm spring waters

Area	Temp (°C)	pH	Li⁺	Na⁺	K⁺	Rb⁺	Cs⁺	Ca²⁺	Concentrations in waters in ppm Mg²⁺	F⁻	Cl⁻	Br⁻
North Island												
Ngawha: Jubilee Bath	42–83*	6·4	8·0	830	63	0·3	0·55	7·8	2·5	0·3	1250	2·6
Waiwera: Hotel Baths	40	7·2	1·7	720	8	<0·2	0·2	38	2·2	1·6	1110	3·2
Helensville: Hotel Supply	65	6·5	2·3	600	22	<0·1	<0·1	63	3·0	2·2	1030	n.d.
Te Aroha: CO₂ Geyser	85	7·5	2·0	3500	108	<0·1	<0·1	8·2	3·9	0·3	518	1·6
Awakeri: Pukaahu Spring	54	8·6	0·24	120	8	<0·02	<0·02	1·9	0·3	3·9	42	<0·2
Spring, Tarawera, Napier-Taupo Rd.	49	8·4	1·9	500	9	<0·1	<0·1	12	0·1	11·5	660	1·9
Morere: Baths 1 and 2	62	6·7	5·4	6100	100	<0·1	<0·1	3900	137	0·4	16000	8
Te Puia: Bath	65	6·8	n.d.	4550	22	n.d.	n.d.	815	8	1·5	8300	n.d.
South Island												
Hanmer: Bath	54	8·0	1·7	360	7	<0·1	<0·1	6·5	0·2	4·4	451	1·3
Maruia: Pool	58	7·9	1·6	130	5	<0·02	<0·02	6·5	0·3	2·5	99	0·7
Wanganui River Spring	39	6·2	0·9	170	14	0·08	0·05	15	1·5	2·0	188	n.d.
Lyttelton Tunnel Spring	21·5	7·1	n.d.	174	18	<0·02	<0·02	152	163	0·7	513	1·9

* Temperatures at surface and base of pool.

Table 3 cont.

Area	I⁻	SO₄²⁻	As	Total CO₂	Concentration in waters in ppm Total SiO₂	Total HBO₂	Total NH₃	Cl/B	Cl/F	Atomic ratios Cl/Br	Na/K	Na/Li	Na/Ca
North Island													
Ngawha: Jubilee Bath	1·0	347	0·2	490	178	3690	140	0·42	2200	1100	22	31	185
Waiwera: Hotel Baths	1·2	1	<0·02	77	40	39	1·9	35	370	780	150	130	33
Helensville: Hotel Supply	n.d.	1	0·06	87	72	54	0·05	23·5	250	—	46	78	16·5
Te Aroha: CO₂ Geyser	0·6	321	0·4	8050	120	651	3·4	0·98	920	730	55	530	740
Awakeri: Pukaahu Spring	0·5	29	n.d.	106	70	15	<0·1	3·5	5·8	>500	25	150	110
Spring, Tarawera, Napier-Tapuo Rd.	2·5	82	n.d.	111	42	322	2·4	2·5	31	780	94	80	73
Morere: Baths 1 and 2	25	21	n.d.	25	28	198	1·5	100	20000	450	104	340	2·7
Te Puia: Bath	18	110	<0·01	60	52	290	2·1	35	2900	—	350	—	9·7
South Island													
Hanmer: Bath	0·7	43	n.d.	140	46	216	3·4	2·6	55	780	87	64	96
Maruia: Pool	0·1	25	n.d.	—	70	12·5	1·7	9·8	21	320	44	24	35
Wanganui River Spring	n.d.	7	n.d.	298	70	6·9	0·13	34	50	—	21	57	20
Lyttelton Tunnel Spring	<0·2	110	<0·01	—	93	4·7	0·01	135	400	610	16	—	2·0

thermal systems are in an area of basalts and andesites of similar age, and the hot waters emerge through Miocene sandstone (HEALY, 1948). Ngawha in North Auckland occurs in an area of deep Cretaceous shales and sandstones, and in which there have been extensive Quaternary basaltic eruptions. The heat is assumed to be derived from a basaltic intrusion within or beneath the sediments, but isolated rhyolite intrusions are also known in North Auckland. As drilling has revealed temperatures over 100° within a few hundred feet of the surface, its classification as a warm spring area is arbitrary, as is its assessment as an old volcanic area. In the South Island, the Lyttelton thermal springs are probably related to the basaltic volcanic activity of late Tertiary age about Banks Peninsula. Water temperatures are low and range from 21–27°C.

Notable features of the Ngawha waters are the very high boric acid, bicarbonate and ammonia concentrations. Boric acid is also high at Te Aroha, as are the bicarbonate ion concentrations. The silica contents for these two areas are the highest of the warm springs.

All the waters are of neutral pH, and have chloride concentrations of about 500–1000 ppm. The ratios of sodium to rare alkalis are similar to those for non-volcanic springs in greywacke country, but calcium and magnesium contents are in general higher, particularly for the Lyttelton tunnel water.

Sulphate and ammonia contents are extremely variable and arsenic concentrations uniformly low. The ratios of chloride to bromide are similar to those for springs in recent volcanic areas.

Previous Work on Rock Leaching

BASHARINA (1958) extracted many water-soluble constituents from a pyroxene andesite ash from Bezymiany volcano. The following amounts, expressed on an air-dried rock basis, were extracted into distilled water at room temperature: chloride, 760–5300 ppm; fluoride, 30–67 ppm; bromide traces to 21 ppm; sulphate, 2370–4690 ppm; bicarbonate, 120–1040 ppm; sodium, 100–1240 ppm; potassium, 24–345 ppm; calcium, 840–4890 ppm; magnesium, 173–388 ppm; silica, 25–200 ppm; boric acid, 16–42 ppm. The pH values of the solutions were 3·2–6·8. Rather similar results were reported by TOVAROVA (1958), but in addition several tens of ppm of ammonia were extracted. These ashes probably contained hydrogen chloride and sulphur gases etc., adsorbed during the eruption, and their very acid characteristics would only be temporary.

The interaction of acid sulphate and chloride solutions at 100°C with finely ground Kamchatka basalts, andesites and dacites was examined by NABOKO and SILNICHENKO (1960). The concentrations of major rock constituents leached into, and held in solution, depended on the pH rather than on the rock compositions. The experiments provided a reasonable model for explaining the compositions of the acid surface waters of active volcanic areas, which may contain high concentrations of iron, aluminium, calcium and magnesium.

KHITAROV (1957) studied the interaction of water with feldspars, micas and granite at higher temperatures and much higher pressures than in the present work, but his interests were in the concentrations of the major rock constituents passing into solution.

Experimental

The following technique and apparatus was used in the laboratory investigation of solution compositions resulting from the interactions of various rock types with water.

(a) Procedure

Grains (1–5 mm) of the rock under investigation were packed into 105 cm^3 stainless-steel pressure vessels and after sealing, the vessels were heated dry in 3-ft electric tube furnaces to the required temperatures. The pressure vessels were connected at their bases, by capillary stainless-steel tubing, to a Sprague air-driven liquid pump unit, capable of delivering and maintaining water pressures up to 30,000 lb/in^2. After the vessels reached thermal equilibrium, water was pumped in until the pressure was 5000 lb/in^2, giving a rock-to-water weight ratio of about unity for the rhyolite pumice experiments, and about two for the other rocks. The water pressure was sufficient to maintain a single liquid–water phase in the vessels for all the runs. Temperatures at the top and bottom of the vessels were measured by chromel–alumel thermocouples.

At intervals during the runs, samples of water were taken from the top of the vessel through a capillary line and a valve, the pump maintaining the pressure, and replacing the water sampled with distilled water. Sampling times depended on the reactivity of the rock and the temperature at which the experiment was being conducted. For example, with rhyolitic pumice, a reactive rock, samples were taken after intervals of a few hours but for most other rocks the time intervals were increased to at least 24 hr. By taking samples of 10 cm^3, the resultant dilution of the reaction solutions by fresh water was approximately 15 per cent. It was confirmed that mixing did occur in the times between sampling. At the completion of each experiment the rock was removed from the vessel, air dried and stored for subsequent examination.

(b) Methods of analysis

In the reaction solutions, ten to thirteen constituents were determined on 10 cm^3 of reaction solution by micro-analytical techniques. Sodium, potassium, lithium, rubidium and caesium were determined by flame photometry. Calcium and magnesium were determined by both emission flame photometry and atomic absorption flame photometry (DAVID, 1960).

Chloride was determined by differential potentiometry using silver/silver chloride electrodes (BLAEDEL et al., 1952). For fluoride, a distillation from perchloric acid and silica was followed by a thorium nitrate–chromazurol titration (MILTON et al., 1947). Silica was determined photometrically after the formation of silicomolybdic acid. The water samples were first heated with alkali to get the silica into the monomeric reactive form.

Sulphate was reacted with barium chromate to form insoluble barium sulphate and the chromate ions liberated in the reaction were determined photometrically (IWASAKI et al., 1958). Boron was determined photometrically after reaction with carminic acid in concentrated sulphuric acid (SMITH et al., 1955). The estimation of ammonia consisted of a micro-distillation, followed by the reaction of the distillate with Nessler's reagent. Arsenic was determined by the Gutzeit method. The pH values of a number of samples were determined by mixing with indicator solutions on spot plates and comparing with standard pH buffer solutions.

Minor rock constituents were determined as follows: chloride by the procedure of KURODA and SANDELL (1950), fluoride according to CHU and SCHAFER (1955), ammonia by WLOTZKA's (1961) method and boron by a spectrographic technique developed in this laboratory (SEWELL, 1963).

Results

(a) Rock petrology and compositions

The eight rocks investigated cover the range of compositions and types typical of the Taupo Volcanic Zone of the North Island of New Zealand. Four of the eight were of rhyolitic composition, and this reflects the preponderance of silica-rich volcanics in this region. Typical major-constituent analyses made in this laboratory for some fifty rocks from the central volcanic zone were brought together by STEINER

(1958a), while REED (1957) reviewed the compositions of many New Zealand greywackes. Brief petrological descriptions of the rocks used in the present experiments were provided by Dr. A. Ewart of the N.Z. Geological Survey, who also assisted in choosing the specimens from surface outcrops or quarrys so that only fresh and unaltered material was used. They were selected from areas away from natural hydrothermal activity. After the reaction with water a further examination of the rocks was made in thin section by Dr. Ewart to check on hydrothermal alteration. The numbers following the rock types are New Zealand grid references of the localities from which the rocks are taken.

Pumice (N33–5528). This was a rhyolitic pumice; 98% glass together with andesine plagioclase, minor hypersthene and magnetite phenocrysts.

Obsidian (N94/462490). This rock was of rhyolitic composition and consisted of practically 100% fresh glass.

Ignimbrite (N53/274923). This rhyolitic ignimbrite consisted of 60% glass which was fresh and unaltered. Crystalline material included plagioclase, quartz, minor biotite, hornblende and magnetite. Some pumice inclusions were present.

Rhyolite (N94/605462). This was a flow rhyolite showing a well-developed spherulitic structure. Between the spherulites, patches of glass showed crystallization to a microcrystalline mosaic, probably of feldspar and tridymite. About 5% plagioclase phenocrysts and minor magnetite and hypersthene constituted the remainder of the rock.

Dacite (N94/635376). A matrix consisting of andesine plagioclase, glass and cristobalite, constituted 78 per cent of this rock. Also present were plagioclase and quartz phenocrysts together with hypersthene, hornblende, augite and magnetite.

Andesite (N111/970870). This rock consisted of a largely devitrified glass matrix (65 per cent), together with plagioclase phenocrysts (25 per cent), hypersthene (10 per cent) and minor augite and magnetite.

Basalt (N84/418745). Fresh crystals of plagioclase feldspar, pigeonite and a matrix largely consisting of oxidized interstitial glass made up the bulk of this rock.

Greywacke (N33–0013). The rock consisted of poorly sorted angular fragments of quartz, sodic plagioclase, potash feldspar, some rhyolitic volcanic fragments, together with biotite, rare chlorite, laumontite and muscovite.

(b) Minor constituents of rocks

Surveys of the chloride, boron, fluoride and ammonia contents of New Zealand volcanic rocks are at present being made, and will be reported elsewhere. The concentrations of the elements shown in Table 4 for the eight rocks used in the present work are typical for the rock types.

The nitrate nitrogen of the rocks was also determined by the method of WLOTZKA (1961), but the contents were all under 5 ppm and are of little significance.

Table 4

Constituent rock	Cl	F	HBO_2	NH_3
Pumice	990	440	70	31
Obsidian	900	400	100	—
Ignimbrite	690	410	200	—
Rhyolite	600	300	80	19
Dacite	120	290	45	30
Andesite	190	190	90	50
Basalt	360	180	30	20
Greywacke	12	280	140	280

(c) *Compositions of solutions from rock/water interaction*

Table 5 gives the compositions of the solutions after contact with various rocks for periods up to 300 hr, at temperatures in the range 150–350°. Samples were taken in each run at 24, 48, either 72 or 120, and 300 hr. In runs with obsidian and greywacke, samples were also taken at 480 hr. All the solutions were analysed, but little of importance is lost by reporting only the analyses for 24 and 300 hr, or in the case of 300° runs at 24, 120 and 300 hours. Trends which are more complicated than a simple increase or decrease in concentration-with-time are discussed in greater detail later.

In many analyses it will be apparent that the equivalent concentrations of anions and cations reported do not balance. The concentration of bicarbonate was not determined, and it is likely that this common ion makes up the apparent deficiency of anions in solution compositions.

For the ignimbrite and rhyolite, single runs were made at 350°, the results appearing in Table 6. The sampling schedule was as for the other rocks.

Pumice was very reactive, and solution compositions were obtained at short intervals of time. Figure 2 shows the change in concentration with time for the constituents in solution at 180°, while Fig. 3 gives this information for the 270° experiment. Figure 4 summarizes the concentrations of constituents found in six runs with pumice at various temperatures at the longest common time of 72 hr allowed for reaction. Lithium was determined only on the 72 hr, 310° experiment solution, the concentration being 0·5 ppm.

On a small number of solution samples estimations of arsenic, rubidium, caesium and magnesium were made. The results are summarized in Table 7. In the discussion rubidium and caesium determinations on solutions from some higher temperature reactions are given.

(d) *Hydrothermal alteration of rocks*

Following the reactions, a selection of the rocks were re-examined in thin section under the microscope, and by X-ray diffraction. Determinations were also made of water contents. This work was concentrated on the highest temperature experiments where the maximum alteration could be expected.

With the exception of the pumice and obsidian, the rocks showed little change, even after reaction with water at 350°C. No alteration was apparent for basalt and dacite in the 350° experiments, and for greywacke only slight oxidation of matrix occurred. With andesite there was a slight devitrification of the margins of the groundmass glass. The ignimbrite under these conditions showed intense oxidation and the earliest stage of devitrification in its glassy matrix. After reaction with water at 350°, the obsidian grains showed three zones of alteration. On the outside there was a micro-crystalline zone, followed by an opaque zone of crypto-crystalline material (possibly cristobalite and feldspar) and an inner zone of hydrated glass. The pumice from the 310° experiment showed strong strain polarization, presumably due to the increased hydration of the glass during the run. No definite devitrification could be identified under the microscope.

Each rock was examined by X-ray diffraction before and after reaction with water at 350° (pumice at 310°). No changes were noted in the X-ray patterns, except

Table 5. Constituent concentrations (ppm) in reaction solutions

Rock:		Obsidian	Dacite	Andesite	Basalt	Greywacke
Temperature (°C)	Time of reaction (hr)					
(a) Chloride and (fluoride)						
150	24	4·4(<1)	12(4·4)	8(1·5)	21(<1)	15(<1)
	300	3(<1)	10(4·4)	12(5·0)	28(3·5)	4(<1)
200	24	7(<2)	16(4·5)	17(4·0)	58(6·5)	9(<1)
	300	4(<2)	12(11)	20(5·0)	61(9·0)	4(<1)
250	24	5(9·5)	22(16)	23(10)	71(17)	8(<2)
	300	9(9·5)	40(15)	52(10)	250(6)	3(<2)
300	24	9(5)	28(20)	40(10)	140(17)	12(<2)
	300	54(43)	28(18)	48(20)	250(3·0)	3(2)
350	24	39(21)	60(23)	115(11)	460(7·5)	10(<2)
	120	45(20)	37(16)	150(10)	n.d.(n.d.)	6(2)
	300	320(8·5)	37(12)	270(7·5)	390(3·0)	5(2)
(b) Boric acid as HBO_2 and (ammonia)						
150	24	n.d.(1·1)	0·3(0·9)	n.d.(1·8)	2·0(3·4)	4·9(1·2)
	300	0·8(0·6)	0·6(0·4)	1·3(1·8)	2·0(n.d.)	4·9(0·7)
200	24	n.d.(2·1)	1·7(0·8)	1·5(4·2)	2·8(1·3)	12(1·2)
	300	n.d.(1·9)	2·6(1·9)	3·5(6·1)	2·6(n.d.)	3·3(2·2)
250	24	1·2(1·7)	3·0 n.d.	5·7(7·0)	3·0(3·1)	12(4·2)
	300	1·5(1·3)	7·5(3·3)	10(3·2)	5·4(3·7)	7·0(n.d.)
300	24	0·6(1·9)	5·3(4·6)	4·7(3·2)	18(4·0)	17(4·4)
	300	0·7(0·1)	3·1(2·2)	5·8(3·9)	17(5·6)	7(9·0)
350	24	0·1(1·0)	12(15)	10(12)	n.d.(12)	36(34)
	120	1·0(0·6)	3·8(7·6)	18(7·0)	n.d.(n.d.)	11(18)
	300	35(0·6)	3·5(4·3)	33(6·7)	10(4·2)	8·0(18)
(c) Silica and (sulphate)						
150	24	150(n.d.)	160(n.d.)	235(n.d.)	230(n.d.)	200(n.d.)
	300	155(15)	390(32)	240(36)	235(45)	220(75)
200	24	290(n.d.)	790(n.d.)	560(n.d.)	830(n.d.)	520(n.d.)
	300	460(30)	890(24)	400(32)	750(39)	415(135)
250	24	770(n.d.)	1320(n.d.)	890(n.d.)	720(n.d.)	760(n.d.)
	300	860(14)	990(42)	1150(n.d.)	1170(35)	550(135)
300	24	910(n.d.)	1540(n.d.)	1060(n.d.)	1540(n.d.)	700(n.d.)
	300	800(4)	1360(28)	1240(24)	1360(28)	530(145)
350	24	1150(n.d.)	1600(n.d.)	1800(n.d.)	950(n.d.)	1060(n.d.)
	120	1000(n.d.)	1340(n.d.)	1760(n.d.)	n.d.(n.d.)	910(n.d.)
	300	900(10)	1400(10)	1800(n.d.)	2600(10)	890(40)

Table 5 cont.

Rock:		Obsidian	Dacite	Andesite	Basalt	Greywacke
Temperature (°C)	Time of reaction (hr)					
(d) *Sodium and (potassium)*						
150	24	30(3)	36(10)	55(21)	70(22)	65(25)
	300	30(2)	42(9)	75(24)	85(24)	60(11)
200	24	55(5)	55(12)	60(21)	100(31)	80(11)
	300	65(5)	55(11)	85(27)	95(29)	70(5)
250	24	80(7)	58(13)	55(15)	125(31)	55(6)
	300	90(9)	65(10)	65(18)	150(47)	70(4)
300	24	75(9)	65(14)	70(23)	125(46)	40(4)
	300	110(21)	70(7)	70(22)	340(150)	50(5)
350	24	70(12)	52(10)	68(27)	230(130)	25(5)
	120	70(13)	36(6)	92(30)	n.d.(n.d.)	25(5)
	300	440(110)	38(6)	160(33)	175(87)	25(4)
(e) *Lithium and (calcium)*						
150	24	0·15(20)	0·10(4)	0·05(7)	n.d.(9)	0·1(13)
	300	0·10(7)	0·10(4)	0·10(3)	n.d.(5)	0·1(4)
200	24	0·25(6)	0·20(5)	0·05(<2)	0·5(n.d.)	0·35(5)
	300	0·25(1)	0·20(2)	0·10(3)	n.d.(4)	0·25(3)
250	24	0·35(0·4)	0·35(6)	0·1(4)	0·4(5)	0·7(4)
	300	0·45(0·3)	0·65(9)	0·5(4)	0·3(3)	0·5(5)
300	24	0·35(1)	0·5(7)	0·35(18)	0·1(4)	0·5(2)
	300	0·5(1)	0·6(0)	0·6(n.d.)	1·5(5)	0·9(5)
350	24	0·2(1)	0·4(3)	0·7(12)	0·4(5)	0·6(3)
	120	0·2(n.d.)	0·4(5)	0·9(6)	n.d.(n.d.)	0·8(n.d.)
	300	1·3(1)	0·4(0)	1·5(12)	0·5(3)	0·6(4)

n.d. = not determined.

Table 6. Solution compositions at 350°

Constituent (ppm)	Reaction Time (hr)	Cl	F	HBO_2	NH_3	SiO_2	SO_4	Na	K	Li	Ca
Ignimbrite	24	530	8·5	21	3·3	870	n.d.	280	136	0·9	5
	120	420	5·0	30	3·2	840	17	205	98	0·7	2
Rhyolite	24	160	16	5·0	4·1	825	n.d.	95	22	1·3	4
	120	110	19	1·5	3·9	850	8	75	14	1·2	4

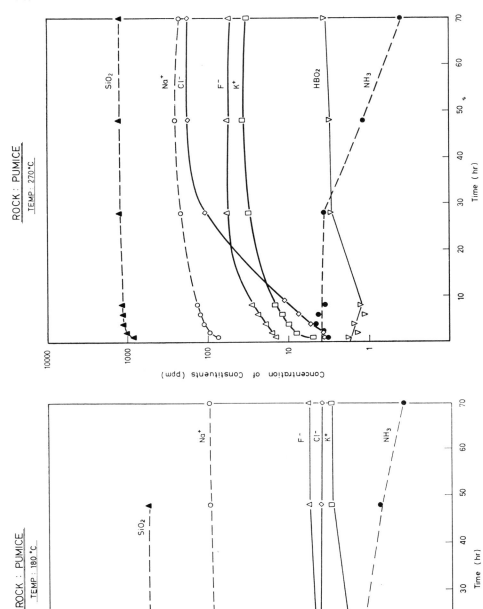

Fig. 3. Change in concentration with time of constituents dissolved from rhyolitic pumice at 270°C.

Fig. 2. Change in concentration with time of constituents dissolved from rhyolitic pumice at 180°C.

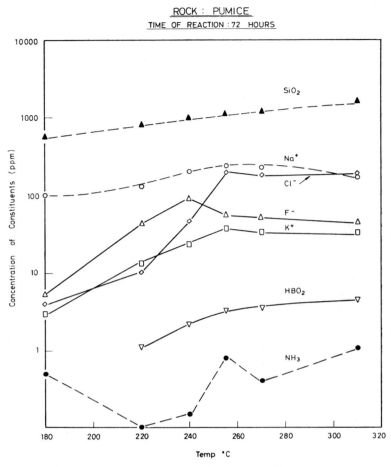

Fig. 4. Variation with temperature in the concentration of constituents dissolved from rhyolitic pumice.

Table 7

Rock type	Temperature (°C)	Time (hr)	Concentrations (ppm)			
			Mg	As	Rb	Cs
Basalt	350	300	0·20	n.d.	<0·1	<0·1
Obsidian	350	480	n.d.	n.d.	0·2	<0·1
Ignimbrite	350	480	0·03	n.d.	<0·1	<0·1
Pumice	310	72	0·17	0·85	n.d.	<0·1
Dacite	350	300	n.d.	<0·1	n.d.	n.d.
Andesite	350	300	n.d.	1·3	n.d.	n.d.
Greywacke	350	480	0·75	n.d.	n.d.	n.d.

for pumice for which slight crystallization to feldspar and quartz was apparent. The proportion of crystalline alteration product in the obsidian must have been too small to be apparent on an X-ray pattern.

The total water content of the air-dried rock particles was determined before and after exposure to water at the highest temperatures of the experiments (pumice over a range of temperatures)—see Table 8.

Only pumice and obsidian gained water, the remainder, except rhyolite, losing small amounts.

EXAMINATION OF RESULTS

(a) General

The tables show that the interaction of water with various rock types can liberate appreciable quantities of many of the elements characteristic of natural thermal waters. The exact concentrations in the reaction solutions are, with some exceptions, not of great significance as the effective ratio of rock to water in a natural hydrothermal system could vary widely. Also, it must be considered that a selective extraction of constituents from rock could occur during the lifetime of a natural system which consisted of a deep convective cycle of meteoric water. The results cannot be related directly to the composition of natural hot waters, but must be interpreted with reference to reaction rates, chemical equilibria and knowledge as it becomes available of the deep hydrology of thermal areas.

As the times of interaction were very short compared with the periods available for reaction in natural hydrothermal systems, the results must be interpreted by judging as far as possible what the element concentrations in the experimental solutions would be after a much longer reaction period. This can be done in some cases only by reference to known equilibria, e.g. the high silica concentrations in the reaction solutions would not persist in nature, but would decrease until equilibrium with quartz was established. In most experiments only slight alteration of the rock occurred, and it must be considered whether or not from a chemical equilibrium viewpoint an increase in the extent of alteration would increase the amount of an element in solution.

Widely different reactivities were observed for the various rock types. The relative rates at which constituents were liberated are of interest as these are a reflection of the way in which the elements are held in the rocks.

The results for each element are now reviewed, and the concentrations in the reaction solutions are compared with the total amount of the element available in the rocks to see the effectiveness of the extraction processes. As similar results for an element were often given by several rocks, it is not necessary to discuss each rock type in detail. The dilution factor of 15 per cent each time a sample was taken during a run should be remembered.

A distinction must be drawn between major and minor rock constituents. Sodium, potassium, calcium, magnesium, silica, as major constituents, are present in the rock in excess of the amounts that could be dissolved in the water under the experimental conditions, and a saturated equilibrium state defined by the rock composition, temperature and pressure should eventually be set up. Constituents such as chloride, fluoride, ammonia, arsenic, sulphur, lithium, rubidium, caesium, which are present in the rocks to the extent of less than $0 \cdot 1$ per cent, could, unless limited by sparingly-soluble compound formation or by inclusion in a rock alteration mineral structure, be transferred largely to the water phase. For all of the latter group of elements the equilibrium rock/water distribution coefficient is likely to favour the water phase at the temperatures of interest (e.g. (Cs/K feldspar)/(Cs/K solution) at 400° is 0·25 according to EUGSTER, 1955).

(b) Individual constituents

(i) Chloride. The pattern of reaction was often a fast build-up in chloride concentration which was then followed by a further, more gradual, increase in chloride concentration. The proportion of chloride in the rocks which was easily removed during the initial rise in concentration differed considerably for the various rock samples. Figure 5 shows approximately the relative proportions of "easily-removed" chloride in some of the rock types, along with similar results for boron.

For basalt at 300–350° about 75 per cent of its total chloride was easily removed by water. For andesite a rapid increase in reaction rate occurred after about 120 hr at 350°, after about 45 per cent of its total chloride content was removed into solution. The chloride content of the rock dropped from 190 ppm to 50 ppm during the 350° experiment, giving a good mass balance with the chloride found in solution.

The amount of chloride removed from the pumice in the 310° experiment after 1–3 days remained at about 25–30 per cent of the chloride in the rock, the analysis of the rock before and after 3 days exposure giving chloride concentrations of 990 ppm and 680 ppm.

The lower chloride content of the dacite was reflected in the chloride content of its solutions, but in addition, the percentage of easily-removed chloride in the rock was low (about 25 per cent). The rhyolite contained the lowest percentage (about 13 per cent) of total chloride readily available for extraction; this rock did not lose water during the reaction. Almost 40 per cent of the chloride in the ignimbrite was readily removed by 350° water, and it is of significance that this rock also lost about 2 per cent of water from its structure during the reaction and had started to devitrify. Only small amounts of chloride were extracted from greywacke but the rate of reaction was rapid even at 150°. About 50 per cent of the total chloride in the rock was readily available for solution.

A large proportion of the chloride in many of the rocks was readily available for extraction (particularly for basalt, andesite and greywacke), although the degree of reaction of the rocks with water as shown by microscopic and X-ray examination was slight in all cases except for obsidian and pumice. The water content figures in Table 8 show that the glassy rhyolitic rocks absorbed several per cent of water before crystallization occurred.

The initial rate ($dc/dt = k$) of increase in chloride concentration in solution from rhyolite pumice was examined as a function of temperature between 220–270°, and the Arrhenius activation energy, E, obtained from the equation $k = A \exp(-E/RT)$, was found to be equal to $2·7 \pm 1$ kcal/mole. This is of the order expected for an absorption/desorption process, but is also in the vicinity of the activation energy of about 4–5 kcal/mole associated with rate control by ion diffusion in solution (MOELWYN–HUGHES, 1948). The kinetics suggest that a high proportion of the easily-liberated chloride in this rock was held simply on surfaces in the structure,

Table 8. Total water content (%) of rocks (air-dried)

Rock	Before exposure	After exposure
Pumice	3·8	6·0 (220°)
Pumice	3·8	9·4 (260°)
Pumice	3·8	10·1 (310°)
Pumice	3·8	13·2 (3 days, 340°)
Obsidian	0·24	4·5
Rhyolite	0·14	0·14
Ignimbrite	4·3	2·5
Dacite	0·44	0·34
Andesite	0·70	0·62
Basalt	1·12	0·80
Greywacke	2·76	2·30

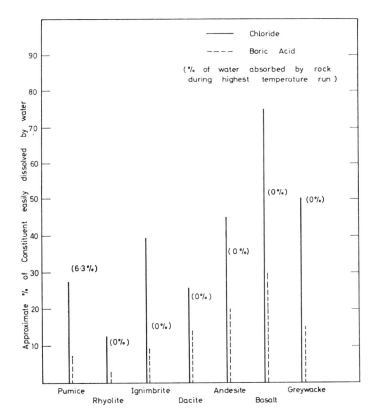

Fig. 5. Proportions of the total chloride and boron in a rock easily dissolved by hot water.

although during the highest-temperature and largest-time experiments the slight crystallization of the pumice surface would have liberated some chloride from solid solution in the glass. However, the initial rate of liberation of chloride into solution was not controlled by either the rate of devitrification or the rate of diffusion of chloride in the pumice glass otherwise the activation energy would have been higher. The solution of such a high proportion of the total chloride from rocks such as basalt and andesite without there being any visible alteration of the rock minerals or structure is a strong indication that much of the chloride in these rocks is not held in solid solution in rock silicates, but on surfaces in the rocks. The proportion of easily-liberated chloride to total chloride varies from rock to rock and could be expected to be lowest for the massive glassy rocks. The dacite sample proved to be rather anomalous, but it is possible that the easily-leached chloride has already been washed out in nature.

The second increase in chloride concentration with time after the initial solution could be due to the permeation of water into the silicate glasses with liberation of the chloride dissolved in their structure.

(*ii*) *Fluoride.* The results for this element were included by MAHON (1963) in a paper on the chemistry of fluorine in hydrothermal systems. It was apparent that the stable concentrations of fluoride that could be retained in the reaction solutions were limited by the solubility of calcium fluoride in the silica-bearing solutions. However, when a glassy rock, such as pumice, reacted with water, high metastable concentrations of fluoride were formed in solution (see Figs. 3 and 4 compared with Fig. 6). These concentrations, given time, would decrease to the stable level, but for some rocks (particularly the pumice), and at lower temperatures, the experiment times were not sufficient for equilibrium fluoride concentrations to be achieved. For the more crystalline and calcium-rich basic volcanic rocks, high metastable fluoride concentrations were not found, the concentrations of fluoride and calcium being of the order expected from the solubility of fluorite in neutral solutions containing silica. The low concentration of fluoride in the greywacke solutions shows that the element is held within a stable mineral in the rock.

Figure 6 shows the results from recent experiments on the solubility of fluorite in water, and in water saturated with respect to amorphous silica. Earlier results on calcium fluoride solubility in water from BOOTH and BIDWELL (1950) seem too high by a factor of about two. The earlier method involved a filtration within a pressure vessel at the temperature of interest. For a substance more soluble in water at lower temperatures this procedure could lead to high values, as small particles could pass through the filter with the solution, and later dissolve when the bomb was cooled.

The present results were obtained by sampling the hot liquid phase in equilibrium with fluorite crystal chips and with silica-gel within a stainless-steel pressure vessel. The liquid was sampled from its upper volume so that solid particles were not expelled into the sampling line, and equilibrium was checked by daily sampling until constant values were obtained for the temperature. The results of ELLIS (1963) showed that it was not necessary to consider the presence of HF in the vapour phase. Iron, as evidence of corrosion, was not present in the reaction solutions.

It appears that for the fluorite/water system the solution process is essentially an ionic reaction $CaF_2 \rightleftharpoons Ca^{2+} + 2F^-$ up to about 230°. Above this temperature the quantities of fluorine liberated into solution become higher than the stoichiometric proportions required to balance the calcium. The reaction tends to become $CaF_2 + 2H_2O \rightleftharpoons Ca(OH)_2 + 2HF$ and the calcium concentrations are limited by the insoluble nature of calcium hydroxide. The addition of silica allows a more favourable equilibrium of type $CaF_2 + H_2O + SiO_2 \rightleftharpoons CaSiO_3 + 2HF$ to occur, and in this case the fluoride and calcium concentrations liberated into solution are

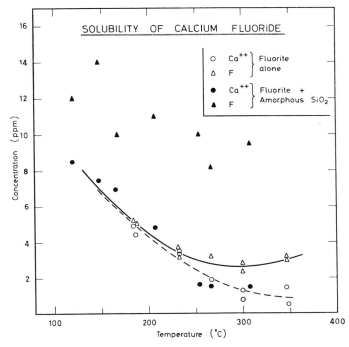

Fig. 6. The solubility of calcium fluoride in pure water and in water saturated with amorphous silica.

non-stoichiometric down to lower temperatures. The ratio HF/F^- in solution depends on the solubility of $Ca(OH)_2$ or of calcium silicates at any particular temperature, and on the acid dissociation constant of hydrofluoric acid. The pH of the calcium fluoride solutions would be within about a pH unit of neutral whether or not silica was present, due to the small ionization constants of all substances present. From ELLIS (1963) the acid dissociation constant of hydrofluoric acid at 250° is approximately 3×10^{-6}.

If the rock/water reaction solutions were appreciably acid or alkaline, their calcium and fluoride concentrations could not be expected to agree with those in Fig. 6. However, since pH measurements made on the reaction solutions showed close to neutral conditions (pH 6·5–8·0), approximate comparisons are valid.

(iii) *Boron*. As for chloride, certain proportions of the total boron in the rock structures were easily removed into solution. However, the fraction of total boron in this category was lower than for chloride, a higher proportion of total boron in rocks being apparently bound within the silicate structures. Figure 5 shows the fraction of boron in the rocks readily available for solution, in comparison with the chloride. Again the proportion was highest for basalt and andesite.

The high concentrations of boron liberated into solution at 350° from obsidian are significant, and suggest that, given time for appreciable alteration of their structure, other rocks would liberate increased amounts of boron into solution. This was confirmed by analysis of rhyolitic rock from Wairakei drill-cores showing extensive hydrothermal alteration, which had an average HBO_2 content of only 10–30 ppm, compared with 80–200 ppm in similar, unaltered rock.

The rhyolite gave boron concentrations which decreased rapidly with time during the 350° experiment. A similar behaviour was noted for greywacke. An alteration product formed on the surface of the rocks may have concentrated the boron into its structure. Some clay minerals are known to have a high capacity for boron (HARDER, 1961).

The rate of boron liberation from the rocks was usually less than that for chloride, considering the proportions of the elements in the rocks. For greywacke at temperatures below about 250°, boron was liberated at a rate faster than that for the igneous rocks.

The atomic ratios of Cl/B in the solutions from the igneous rocks after the readily-dissolved constituents had been liberated ranged from about 10 for andesite up to about 45 for the pumice. Greywacke gave ratios as low as 0·35.

(iv) *Ammonia.* As shown in Table 4, the ammonia content of the igneous rocks ranged from 20–50 ppm, but greywacke contained 280 ppm NH_3. Only a minor proportion of the rock ammonia appeared in solution. For rocks of rhyolitic composition, the ammonia in solution ranged from 1–4 ppm, corresponding to less than 20 per cent of the content in the rocks. For basalt and andesite the concentrations ranged up to 12 ppm, or equal to up to about 30 per cent of the total in the original rock.

A characteristic pattern occurred in the ammonia concentrations in solution for runs with rhyolitic rocks and dacite. An initial period of rapid build-up of ammonia occurred until a maximum was reached, then the concentration declined at a rate greater than that due to dilution during sampling. Alteration products such as clays, zeolites or micas on the surface of the rock particles apparently absorbed ammonia into their structures. This is in agreement with the results of WLOTZKA (1961) which showed high ammonia contents in these types of mineral, and with our observations of up to 1000 ppm NH_3 in hydrothermally altered rhyolitic rocks. An equilibrium between solution and alteration products is set up, and the experiments show that the equilibrium solution concentration is of the order of 1 ppm NH_3.

For basalt and andesite the concentrations of ammonia in solution did not decrease appreciably with time. This may have been due to the slower rate of formation of alteration products from these rocks, or to the alteration products having little capacity for ammonia. The basalt results show a minimum during the 300 and 350° runs which suggests that ammonia was first taken up on the rock surface by an unstable alteration product which later changed to a second phase which had a lower capacity for ammonia.

With greywacke only 12 per cent of the ammonia was dissolved from the rock at a rate comparable to that for the igneous rocks, after which the concentrations decreased slowly.

(v) *Sulphur.* The total sulphur content of the rocks was not determined, but earlier analyses of similar types have shown the contents to be usually in the range 100–500 ppm. In contact with hot water containing oxygen any sulphides in the rock would hydrolyse to hydrogen sulphide, then the latter would partly oxidize to sulphate. For all the rocks, the sulphate found in the reaction solutions was in the range 10–40 ppm, the highest values being at the lowest temperatures. The oxygen in the distilled water would be sufficient to create about 15 ppm SO_4^{2-} by sulphide oxidation and some additional oxygen would be held around the rock particles. Oxygen availability probably limited the sulphate concentrations in the low-temperature experiments, but at the highest temperatures a solubility equilibrium appears to be the limiting factor, probably that for calcium sulphate, as the lowest sulphate concentrations were obtained from basalt. REZNIKOV and ALEINIKOV (1953) gave the solubility of $CaSO_4$ in water at 301, 331 and 360° as 20, 9·5 and 6·3 ppm respectively, while DICKSON et al. (1963) showed for 100 bar pressure a solubility decreasing from 826 ppm at 101° to 56 ppm at 217°.

The pressure vessels, when opened after a run, smelled strongly of hydrogen sulphide which evidently existed to the extent of at least several ppm in the solutions. No tests for carbon dioxide were made although this gas could be formed by decomposition of carbonates and organic material in the rocks.

The higher amounts of sulphate liberated into solution from the greywacke point to the presence of sulphate in the rock. The highest temperature solutions appear to be supersaturated with respect to calcium sulphate and there is a trend with time towards lower values.

(vi) *Arsenic.* From the work of ONISHI and SANDELL (1955) the arsenic content of the igneous rocks used in the experiments is likely to be between 2 and 10 ppm. Rhyolitic rocks usually contain more arsenic than do dacites, andesites and basalts. The solution formed by the pumice/water reaction after 72 hr at 310° contained 0·8 ppm arsenic and for andesite after 300 hr at 350° the solution contained 1·3 ppm. An appreciable amount of this element in the rock is therefore available for solution by water.

(vii) *Sodium and potassium.* The concentrations of these elements in the reaction solutions depended on the temperature and the rock type. Concentrations of 200–400 ppm of sodium were taken into solution from pumice, ignimbrite, obsidian and basalt at temperatures over 200° during the reaction period. The remaining rocks generally liberated between 50–150 ppm of sodium into solution. Potassium concentrations were always lower than those of sodium.

For pumice at temperatures over 180°, sodium and potassium concentrations in the waters were about constant for the first 8 hr of exposure. With longer reaction times alkali concentrations increased and the ratios of Na/K decreased, which indicated that potassium was at this stage being liberated more rapidly than sodium. Above about 260–270° stable concentrations and ratios were attained for this rock within the experimental period. For obsidian the trends with temperature were similar to those for pumice, but the solution concentrations were lower. At 350°

the results for rhyolite were very similar to those for obsidian. For ignimbrite the concentrations were much higher, and the Na/K ratios were as low as 3·5.

The alkali concentrations in solutions in contact with andesite and dacite were low, and showed little variation with time after 24 hr exposure, or with temperature. Even at 350° after 300 hr the dacite showed little reaction, but andesite reacted to give about 200 ppm of alkalis in solution. Ratios of Na/K were mainly constant and lower than for the rhyolitic rocks. The amounts of sodium and potassium in solutions in contact with basalt increased with rising temperature, although at higher temperatures (250–350°) little further increase occurred. Na/K ratios were low (4–5) and similar to those obtained from andesite. There was a tendency for the ratios to decrease with increasing temperature.

Examples of stable Na/K ratios attained in solution for the various rocks at different temperatures are given in Table 9. For comparison, ORVILLE (1963) obtained for solutions in equilibrium with both sodium and potassium feldspars, ratios ranging from 3·3 at 600° to greater than 5·3 at 400°.

The ratios for the rhyolitic rocks and basalt will have most significance due to the appreciable solution of constituents from these rocks by water during the time of the experiments.

Table 9. Stable values of atomic ratios Na/K attained in experiments

Rock Temperature (°C)	Rhyolitic* rocks	Dacite	Andesite	Basalt	Greywacke
200	17–30	8·5	5	5·5	12–25
250	10–15	7·5–11	6	5·5	15–30
300	8·5–10	8–17	5·5	4·0	14–18
350	3·5–9	9–12	4–8	3·5	8–11

* Includes results for pumice, rhyolite, obsidian, ignimbrite.

(viii) *Lithium.* The concentration of lithium in the solutions from the various rocks was of similar magnitude at a given temperature. Previous analyses of several rhyolitic pumices, flow rhyolites and ignimbrites showed their lithium contents to range from 20–80 ppm. The results for greywackes were similar, but for andesites and basalts, lower contents of about 15 ppm and 5 ppm respectively, were obtained. The maximum concentration of this element found in the reaction solutions was about 1·5 ppm; the amounts were usually greatest at the highest temperature.

(ix) *Rubidium and caesium.* Analysis of the rocks used in the experiments showed that the caesium contents were 2 ppm or less. The rubidium contents of the four rocks of rhyolitic composition ranged from 40 to 80 ppm and in the dacite, andesite, basalt, and greywacke, the contents were 19, 16, less than 2, and less than 2, respectively. At temperatures up to 350° in the times allowed for reaction, little rubidium or caesium passed into solution. This is not surprising as these ions are presumably held in silicate lattices throughout the rock, and basalt and ignimbrite showed only superficial alteration by the solutions. An exception was obsidian at 350° where appreciable alteration occurred, and a small amount of rubidium was found in the resultant solution.

Some preliminary work was done in extending the reaction temperatures to 500–600°C. In pressure vessels of the type used by KENNEDY (1950b), rocks were reacted with water at 1500 bar pressure for a week at a rock/water ratio of unity (for pumice 0·5), and the resultant solutions were analysed after rapid cooling. The caesium concentration in the solution from the pumice after reaction at 600° was 0·6 ppm, and from the rhyolite 1·4 ppm (this solution also contained 2·0 ppm of rubidium). These additional experiments confirm that with major alteration of the rock, most of the caesium is liberated into solution, as its ion size is too large for it to be favoured as a replacement for other alkalis in hydrothermal alteration silicates.

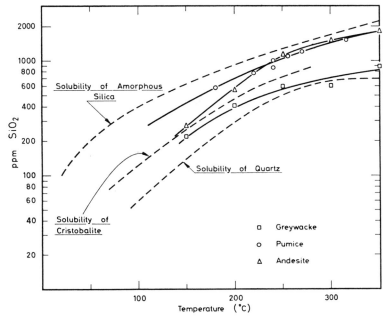

Fig. 7. The concentrations of silica in the solutions derived from the reaction of various rocks with water.

(x) *Magnesium.* In all the solutions examined, the magnesium concentrations were less than 1 ppm, and the concentrations showed little relationship to the magnesium contents of the rock, e.g. basalt and pumice at similar temperatures gave approximately equal solution concentrations. Magnesium is strongly sorbed into rock alteration products.

(xi) *Silica.* As might be expected for the major rock constituents, the concentration of silica in solution rapidly attained an initial equilibrium concentration, which for all rocks lay between the concentrations for saturation with respect to quartz and to amorphous silica. Figure 7 shows the changes in the initial equilibrium silica concentrations with temperature, and includes lines for the solubility of amorphous silica and quartz (KENNEDY, 1950a) and of cristobalite (FOURNIER and ROWE, 1962).

The silica results for pumice and dacite followed the trend of the line for amorphous silica, but the solubilities were about 20 per cent lower. The concentrations in solutions from andesite and basalt were little different from those for pumice at temperatures above 225°, but at lower temperatures the concentrations were much less. The general approach to the amorphous silica solubility line reflects the essentially glassy nature of the volcanic rocks. From the silica concentrations, the activity of silica in the glasses of dacite and pumice was about 0·75 to 0·85 of that for pure amorphous silica. As the molecular fraction of SiO_2 in these glasses is about 0·77 to 0·80, a close to ideal solution behaviour is indicated for silica in the rock glasses. At temperatures over about 200°, the results for andesite and basalt tend towards the amorphous silica solubility curve.

The silica concentrations for obsidian are not shown in Fig. 7 but fell closer to the cristobalite solubility line than to that for amorphous silica. This agrees with the petrological observation of cristobalite as the first product of crystallization of obsidian, but the reason for the different behaviour of the two glasses, pumice and obsidian, is not clear.

Greywacke gave the lowest concentrations of silica in solution, the values above 250° approaching the solubility of quartz. Higher relative silica concentrations at lower temperatures suggest the presence of a small proportion of amorphous silica or cristobalite in the greywacke, which recrystallized readily to quartz at higher temperatures. However, this was not apparent from the petrological examination.

Relevance of Experiments to Natural Systems

(a) Rock-to-water ratios

The experimental results for water-soluble constituents such as chloride, fluoride, boric acid and ammonia showed an important fact about their occurrence in rock structures. Some of the more crystalline rocks readily liberated a high proportion of their content of these constituents before showing evidence of hydrothermal alteration, whereas the glass-rich rocks liberated smaller proportions of the elements even though some alteration of the glass occurred. The water-soluble constituents are held not only in solid solution in the silicate structures, but to a large extent exist on surfaces of grains, crystals and microfissures in the rock. A breakdown of crystal structures is not required to liberate appreciable amounts of chloride etc., into aqueous solution, and therefore pick-up of these constituents by water can occur more readily than has usually been considered.

The reaction solutions were all of about neutral pH, and much of the readily-soluble material appears to have been extracted by simple solution rather than by extensive reaction of rock minerals with water. Changing the rock-to-water ratio in a static system would change the concentrations of the solution constituents accordingly. In natural circulation systems the ratios of rock to water would be much greater than in the present experiments. From porosity measurements, ratios of ten to twelve appear to be reasonable values for recent volcanic areas, and probably still higher ratios in areas of sedimentary or metamorphic rocks. Chloride concentrations of about 50–400 ppm found in the experiments with volcanic rocks would become of the order of 200–2000 ppm if in a static system the rock-to-water ratio was ten. Similarly, boric acid concentrations would become of the order of 20–150 ppm HBO_2. These examples would be minimum figures as the rock and water reactions had certainly not gone to completion. Maximum concentrations for complete reaction are obviously limited by the element content in the rock (Table 4).

For greywacke rock, the minimum concentrations of chloride and boric acid calculated for a rock/water ratio of ten would be 50 ppm and 180 ppm, respectively.

The results from closed-chemical-reaction systems cannot as yet be applied quantitatively to natural thermal waters involving heating and convection of meteoric water. Any relevant discussion would imply knowledge that does not exist at present on the rate of turnover of natural hydrothermal systems, their age and the types of heating mechanism. As all the trace elements in the rocks are not removed with the same ease, natural circulation of hot water could cause differential extraction of the elements, e.g. chloride removed before boron. Variations in a ratio such as Cl/B between thermal areas in similar rocks could, therefore, give some idea of the relative stage of development of the natural systems. The major difficulty is that the ratio of reservoir-storage volume to the total-circulation volume of water over the lifetime of a natural thermal system is not known. It is quite possible that after the initial production of hot water, a Wairakei-type system is in a process of decay rather than of a continuing thermal-equilibrium convection cycle.

The discussion below is not taken further than to show the concentrations of the most soluble elements such as chloride and boron that could be produced in closed reaction systems at reasonable natural rock-porosity values (about a rock/water ratio of ten in recent volcanic hydrothermal areas).

(b) The reaction time factor

The results suggest that two processes are of importance in the liberation of trace elements from rocks. First, a porportion of chloride, boric acid, ammonia and possibly other constituents is held adsorbed on surfaces in the rock and is readily available for solution by water. This material is likely to be removed by hot water in a relatively short time in the geological sense. Secondly, the trace constituents which are held within silicate lattices, but are otherwise of a type favouring concentration in the water phase, would be released only when the silicates are decomposed or recrystallized by hydrothermal alteration. It is impossible to wait a sufficiently long time for complete rock alteration to occur in the laboratory at reaction temperatures much below 250–300°C, but it seems likely that where temperature-dependent solubility equilibria are not involved, the high-temperature results can be used to suggest the solution concentrations that would occur at lower temperatures if the reaction times were extended. To this extent the highest-temperature experiments were the most significant of the series. The mineralogy of drill-cores from the N.Z. hydrothermal areas (STEINER, 1958b), shows that the end product of hydrothermal alteration of rhyolitic rocks at 200–260° is an assemblage consisting of calcium zeolites, secondary potash-rich feldspars, and hydromica. Analysis of a selection of cores showed that over 80–90 per cent of the original chloride and boron in the rock are lost during natural alteration.

(c) Individual elements

The amounts of chloride and boron which can be released from the volcanic rocks during hydrothermal alteration are more than sufficient to give element concentrations in a coexisting water phase which are comparable with the levels found in the recent New Zealand volcanic hydrothermal systems. Any hot water that has been in contact with fresh volcanic rock must inevitably contain constituents such as chloride and boric acid. The lower availability of chloride and higher amount of boric acid obtainable from greywacke is in good agreement with the compositions of the New Zealand warm spring waters issuing from this rock.

Fluoride, sulphate, calcium and magnesium are available from the rocks in excess of the amounts required to explain their concentrations in natural waters. They are limited in concentration by temperature and pH dependent mineral/solution equilibria, particularly at high temperatures where sulphates, carbonates and fluorides are very insoluble.

The ammonia concentrations also appear to be dependent on a distribution between the solution and clay, zeolite or mica alteration minerals. With the exception of those at Ngawha, the ammonia contents of the natural waters are similar to the concentrations found in the experiments. The levels of concentration in the recent volcanic springs are little different from those in the warm springs. At Ngawha the ammonia is probably derived from the surface sediments high in organic material, through which the waters pass.

ONISHI and SANDELL (1955) concluded from their survey of arsenic in igneous rocks that the arsenic in thermal waters could not have been derived from rock/water reaction. However, the present results suggest that at a rock/water weight ratio of ten, concentrations of at least 6–8 ppm arsenic could be expected in solutions in contact with rhyolite pumice or andesite, providing that the element is not held at lower levels by solubility-product relationships.

The silica concentrations in natural waters are temperature dependent, and give little information about the origin of the waters. The high metastable concentrations of silica in the reaction solutions suggest that natural hot water migrating into fresh volcanic rock could cause extensive solution of silica and later redeposition as quartz. Considerable transport of silica could occur, depending on the relative rates of silica solution and deposition. An outward migration of a hydrothermal reservoir into fresh country may be a very slow process due to the flow channels into the cooler country becoming choked with silica or quartz deposits. The sealing of a system by silica may be important in localizing high-temperature volcanic thermal water reservoirs.

Comparison of Table 1 with Table 9 shows that the atomic ratios of Na/K in the waters at Wairakei, Waiotapu, Kawerau and Rotorua, where waters are drawn from reservoir systems in rhyolitic rock at known temperatures, are in good agreement with values obtained in the experiments. For Kawerau, Waiotapu and Wairakei the underground temperatures are 250–295°C, and for Rotorua about 150–200°C. Both experimental and field results show that an equilibrium Na/K ratio in solution exists at each temperature, and this fact was used by ELLIS and WILSON (1960) to indicate water movement within the Wairakei hydrothermal system. The concept of a reversible temperature-dependent Na/K ratio in the natural solutions probably fails below about 150–200° because of slow reaction rates, and for the same reason the experimental ratios for dacite, andesite and greywacke are probably of little practical significance.

A notable feature of the high-temperature waters from the recent volcanic areas is their high content of the rare alkalis, lithium, rubidium and caesium. The solution experiments show that these elements are dispersed throughout the rock structures, and are not concentrated to any great extent on surfaces as are chloride and boron. There is adequate lithium and rubidium in the rhyolitic rocks to account for the observed concentration of these elements in waters from recent volcanic areas, assuming that rock/water weight ratios of the order of ten can be applied, and that at least 10 per cent of the elements are liberated into solution during a hydrothermal reconstitution of the rock. To obtain caesium concentrations of up to 4·7 ppm which are found in these waters (GOLDING and SPEER, 1961), it would be necessary to have 0·5 ppm caesium in the rock at a rock/water ratio of ten, if complete extraction of caesium occurred. Analysis by an ion-exchange/flame-photometry technique indicated that the caesium content of some New Zealand rhyolitic rocks is within the range 0·75–2 ppm. The preliminary 600° experiments noted above show that with extensive alteration of the rocks by water a high proportion of the caesium can be transferred into solution.

Rare-alkali concentrations in the warm springs of New Zealand are mainly very low (Rb and Cs are negligible except at Ngawha) and there is little distinction between springs in old volcanic areas and those in greywacke or sedimentary areas. The Na/Li ratio for Maruia Spring is of a value common for recent volcanic springs, but the other results for this ratio in Table 3 are generally much higher.

Carbon dioxide and carbonate species have been neglected in this study but it was found in preliminary additional experiments that the main effect of adding carbon dioxide to the reaction vessels was to increase the rate of reaction between rock and water. Bicarbonate formed by reaction of carbon dioxide with rock until approximately neutral pH conditions for the temperature were achieved. It is intended to do further work on rock extraction with carbon dioxide solutions.

Conclusions

The laboratory experiments were designed to show to what extent the chemical composition of natural hot waters could be profitably used as evidence in working out the processes which create natural hydrothermal systems. In discussions of these systems the presence of elements such as ammonia, boron, fluoride, lithium and chloride in high concentrations in thermal waters has often been taken as evidence for magmatic water in the hot discharges. Magmatic water is defined as the water leaving a molten rock system, and some writers (including in the past the present authors) have quoted pegmatites as an analogy where elements such as lithium and boron are concentrated in a residual fluid as the parent molten rock body crystallizes.

From the present study it is apparent that the compositions of natural thermal waters can also be approached by the interaction of hot water with rock at moderate temperatures. This does not necessarily exclude the concept of magmatic water, but does show that the composition of magmatic water (for such water must at times exist in nature whether or not it has any bearing on the hydrothermal systems under discussion) is not unique. It should also be noted that magmatic water is not necessarily water that has risen from greater depths with the rock melt. KENNEDY (1955) showed that a rock melt entering a zone of high water pressure could absorb several per cent of water which would be expelled again later when the rock crystallized. Magmatic water, therefore, may be local meteoric water which has passed through a silicate melt.

The approach which appears most profitable is to assume that where rock and hot water co-exist in nature at temperatures of say, over 150°, the times of contact are sufficient for an equilibrium distribution of certain elements between phases to be established. The times available for reaction are very great compared with those in the laboratory experiments. The water phase would concentrate constituents such as lithium, caesium, chloride and boron that are not readily accommodated in secondary silicate structures. The concentration effect may be greater at moderate temperatures than at high temperatures (e.g. EUGSTER's (1955) results on the distribution of caesium between feldspar and water at temperatures of 400–800°C).

There are actually few elements present in thermal waters at temperatures of up to 300°, whose concentration is not governed by equilibria with solid phases in the confining rocks. The principal exceptions, at least at these temperatures, appear to be chloride, bromide, boron, caesium and possibly arsenic and lithium. The distribution of these elements is likely to be so much in favour of the water phase that it is doubtful if it would be possible to distinguish whether the elements in solution resulted from a process of "extraction" from solid rock at, say 200–300°C, *or* from a water-rich fluid separated from a crystallizing molten rock at perhaps 600–800°C, and which was subsequently diluted with meteoric water.

In the light of the experiments the chemical composition of thermal waters could in many cases be deduced from their temperatures for a given underground rock environment, but it is doubtful whether a unique mechanism of origin of high temperature thermal waters could be derived from their chemical composition. In rocks of a particular composition, bodies of thermal waters of similar type but different origin could exist, some containing water from a high-temperature magmatic source, yet others containing no water whose temperature had exceeded the moderate temperatures of the storage system.

A decision on which is the more likely alternative for a particular area will have to be made largely on the basis of whether or not sufficient rock/water interaction could occur to account for the transfer of chemicals. The study of rock/water stable-isotope equilibria, and of the tritium and C^{14} contents of waters as indicators of circulation times, will also be of great assistance. Unfortunately, judging from New Zealand experience, the rock environment of interest is likely to be at a depth of several miles, and can only be inferred. For hydrothermal systems in old areas of dense rock such as those at Steamboat Springs, Nevada, the magmatic-water approach adopted by WHITE (1957) appears reasonable, but for young areas of permeable volcanic rocks such as in the North Island of New Zealand the simple rock/water interaction mechanism is a possibility that cannot be ruled out on the basis of water compositions.

Concentration of elements in an aqueous residual magmatic fluid is only a special case of a distribution equilibrium which, given time, should establish itself between rock minerals and water at all temperatures in situations where complete reconstitution of the original rock occurs. In view of the current arguments over the genesis of ore-bearing fluids this should be checked by further quantitative experimentation. To this end the rock/water distribution studies at higher temperatures and sub-melting water pressures are being extended to cover a wider range of rocks and elements. The results may be of interest both in the study of hydrothermal systems and of the generation of ore-bearing solutions.

Acknowledgements—The authors wish to express their thanks to Mr. S. H. WILSON for some analyses of spring waters; Mrs M. G. RUNDLE, Mr. W. KITT and Mr. J. R. SEWELL for many analyses of trace constituents in the rocks; and Dr. A. EWART, New Zealand Geological Survey, for his assistance in selecting the rocks for the study and providing petrographic information on their structure and extent of hydrothermal alteration.

REFERENCES

ALLEN E. T. and DAY A. L. (1935) *Hot Springs of the Yellowstone National Park*. Carnegie Institute Washington, Publ. 466.

BARTH T. F. W. (1950) *Volcanic Geology, Hot Springs and Geysers of Iceland*. Carnegie Institute Washington, Publ. 587.

BASHARINA L. A. (1958) Water extracts and gases from Bezymyannyi volcano ash cloud. *Bulletin of Volcanological Studies*, No. 27, pp. 38–42. U.S.S.R. Academy of Sciences, Moscow.

BLAEDEL W. J., LEWIS W. B. and THOMAS J. W. (1952) Rapid potentiometric determination of chloride at low concentrations. *Analyt. Chem.* **24**, 509–512.

BODVARSSON G. (1961) *Physical Characteristics of Natural Heat Resources in Iceland*. Paper E/Conf. 35/G/6, U.N. Conference on New Sources of Energy, Rome.

BOOTH H. S. and BIDWELL R. M. (1950) Solubilities of salts in water at high temperatures. *J. Amer. Chem. Soc.* **72**, 2570–2575.

BUNSEN R. (1847) Ueber den inneren Zusammenhang der pseudovulkanischen Erscheinungen Islands. *Liebigs Ann.* **62**, 1–59.

CHU C.-C. and SCHAFER J. L. (1955) Determination of fluorine in catalysts containing alumina and silica. *Analyt. Chem.* **27**, 1429–1431.

CRAIG H., BOATO G. and WHITE D. E. (1956) The isotopic geochemistry of thermal waters. *Natl. Research Council Nuclear Sci. Ser.* Rept. No. 19, Nuclear processes in geological settings, 29–44.

DAVID D. J. (1960) The determination of exchangeable sodium, potassium, calcium and magnesium in soils by atomic-absorption spectrophotometry. *Analyst* **85**, 495–503.

DICKSON F. W. et al. (1963) The solubility of anhydrite in water from 100–275°. *Amer. J. Sci.* **261**, 61–78.

EINARSSON T. (1942) Uber das Wesen Der Heissen Quellen Islands. *Visindafelag Islendinga* **26**, 1–89.

ELLIS A. J. (1961) *Geothermal Drill-holes—Chemical Investigations*. Paper E/Conf. 35/G/42, U.N. Conf. on New Sources of Energy, Rome.

ELLIS A. J. (1963) The effect of temperature on the ionization of hydrofluoric acid. *J. Chem. Soc.* 4300–4304.

ELLIS A. J. and ANDERSON D. W. (1961) The geochemistry of bromine and iodine in N.Z. thermal waters. *N.Z. J. Sci. Tech.* **4**, 415–430.

ELLIS A. J. and WILSON S. H. (1960) The geochemistry of alkali metal ions in the Wairakei hydrothermal system. *N.Z. J. Geol. Geophys.* **3**, 593–617.

ELLIS A. J. and WILSON S. H. (1961) Hot spring areas with acid–sulphate–chloride waters *Nature, Lond.* **191**, 696–697.

EUGSTER H. P. (1955) Distribution of trace elements. *Carnegie Inst. of Washington Year Book, Geophysical Lab. Repts.* **54**, 112–114.

FENNER C. N. (1936) Bore-hole investigations in Yellowstone Park. *J. Geol.* **44**, 225–315.

FOURNIER R. O. and ROWE J. J. (1962) The solubility of cristobalite along the three-phase curve gas plus liquid plus cristobalite. *Amer. Min.* **47**, 897–902.

GOLDING R. M. and SPEER M. G. (1961) Alkali ion analysis of thermal waters in New Zealand. *N.Z. J. Sci.* **4**, 203–213.

HAGUE A. (1911) Origin of the thermal waters of the Yellowstone Park. *Bull. Geol. Soc. Amer.* **22**, 103.

Harder H. (1961) Einbau von bor in detritische Tonminerale. *Geochim. et Cosmochim. Acta* **21**, 284–294.

Healy J. (1948) *The Thermal Springs of New Zealand*. No. 12, Extrait des procès verbaux des séances de l'Assemblée Générale d'Oslo (19–28 août 1948) de l'Union Géodesique et Géophysique Internationale.

Iwasaki I., Utsumi S., Hagino K., Tarutani T. and Ozawa T. (1958) Spectrophotometric determination of a small amount of sulphate ion. *J. Chem. Soc. Japan (Pure Chem. Sect.)* **79**, 38–44.

Kennedy G. C. (1950a) A portion of the system silica–water. *Econ. Geol.* **45**, 629–653.

Kennedy G. C. (1950b) Pressure–volume–temperature relations in water at elevated temperatures and pressures. *Amer. J. Sci.* **248**, 540–564.

Kennedy G. C. (1955) Some aspects of the role of water in rock melt in "Crust of the Earth", *Geol. Soc. Amer. Special Paper* 62, 489–504.

Khitarov N. I. (1957) Chemical properties of solutions arising as the result of the interaction of water with rocks at elevated temperatures and pressures. *Geochemistry* No. 6, 566–578.

Kuroda P. K. and Sandell E. B. (1950) Determination of chloride in silicate rocks. *Analyt. Chem.* **22**, 1144–1145.

Mahon W. A. J. (1963) Some aspects of the geochemistry of fluorine in the thermal waters of New Zealand. *N.Z. J. Sci.*, in press.

Moelwyn-Hughes E. A. (1947) *The Kinetics of Reactions in Solution*. Oxford University Press, London.

Milton R. F., Liddel H. F. and Chivers E. (1947) A new titrimetric method for the estimation of fluorine. *Analyst* **72**, 43–47.

Naboko S. I. and Silnichenko V. G. (1960) The problem of metamorphism on the interaction of hydrothermal solutions and volcanic rocks. *Novejshij vulkanism i gidrotermy*, Volcanological Lab., U.S.S.R. Academy of Sciences, No. 18, 123–133.

Orville P. M. (1963) Alkali ion exchange between vapor and feldspar phases. *Amer. J. Sci.* **261**, 201–237.

Piip B. I. (1937) *The Thermal Springs of Kamchatka*. U.S.S.R. Academy of Sciences, Moscow.

Reed J. (1957) *Petrology of the lower Mesozoic rocks of the Wellington district*. Geological Survey Bulletin No. 57, New Zealand Department of Scientific and Industrial Research.

Reznikov M. I. and Aleinikov G. I. (1953) Solubility of calcium sulphate in water at high temperatures. *Trudy Moskov Energet. Inst.*, No. 11, 198–203 (C.A. 1959, **53**, 9787).

Ritchie J. A. (1961) Arsenic and antimony in some New Zealand thermal waters. *N.Z. J. Sci.* **4**, 218–229.

Sewell J. R. (1963) SiO-band suppression in the spectrographic analysis of silicate rocks for boron. *Appl. Spectrosc.*, in press.

Smith W. C., Goudie A. J. and Sivertson J. N. (1955) Colorimetric determination of trace quantities of boric acid in biological materials. *Analyt. Chem.* **27**, 295–297.

Steiner A. (1958a) Petrogenetic implications of the 1954 Ngauruhoe lava and its xenoliths. *N.Z. J. Geol. Geophys.* **1**, 325–365.

Steiner A. (1958b) Hydrothermal rock alteration in *Geothermal Steam for Power*. N.Z. Department of Scientific and Industrial Research, Bull. 117.

Suess E. (1903) Ueber heisse Quellen. *Verh. Gen. dtschr Naturf. Ärzte* **74**, Vers. I, 133–151.

Tovarova I. I. (1958) Removal of water soluble substances from pyroclastic rocks of the Bezymyannyi Volcano. *Geochemistry* No. 7, 856–860.

White D. E. (1957a) Thermal waters of volcanic origin. *Geol. Soc. Amer. Bull.* **68**, 1637–1658.

White D. E. (1957b) Magmatic, connate, and metamorphic waters. *Geol. Soc. Amer. Bull.* **68**, 1659–1682.

White D. E., Anderson E. T. and Grubbs D. K. (1963) Deepest geothermal well in the world. *Science* **139**, 919–922.

Wilson S. H. (1959) Physical and Chemical Investigations in *White Island*. N.Z. Department of Scientific and Industrial Research, Bull. 127.

Wilson S. H. (1961) *Chemical Prospecting of Hot Spring Areas for Utilization of Geothermal Steam*. Paper E/Conf. 35/G/35, U.N. Conf. on New Sources of Energy, Rome.

Wlotzka F. (1961) Untersuchungen zur Geochemie des Stickstoffs. *Geochim. et Cosmochim. Acta* **24**, 106–154.

V
Chemistry of the Red Sea and the Dead Sea

Editor's Comments on Papers 20 and 21

20   Craig: *Geochemistry and Origin of the Red Sea Brines*

21   Lerman: *Chemical Equilibria and Evolution of Chloride Brines*

The Red Sea and Dead Sea brines, typical of waters that have a high saline content, contain approximately 300 grams of salt per liter. These two bodies of water have very different chemical compositions. Y. K. Bentor, in his paper of 1961, discussed the age and chemical composition of the Dead Sea, and the origin of salts in the water. In 1967 A. Lerman devised a model of the chemical evolution of the Dead Sea by using contemporary techniques in the complex chemistry of chloride solutions. His results imply that the chemical composition of Dead Sea water changes continually with evaporation and dilution. Lerman's paper of 1970, dealing with the evolution of chloride brine such as in the Dead Sea, is presented as Paper 21.

Hot brines and recent heavy metal deposits found at great depths on the Red Sea floor have attracted the attention of geochemists (Brewer et al., 1965; Miller et al., 1966; Degens and Ross, 1969). The discovery of these deposits was made by the Research Vessel Atlantis II of the Woods Hole Oceanographic Institution. A book on this subject that may be of interest to the reader is *Hot Brines and Recent Heavy Metal Deposits in the Red Sea*, edited by Degens and Ross in 1969. H. Craig's paper of 1969 is presented as Paper 20; he investigates the origin, age, and history of the Red Sea by integrating isotopic and major and minor element distributions in the water.

Abraham Lerman received his M.S. in geology from The Hebrew University of Jerusalem in 1960, and his Ph.D. from Harvard University in 1964. He served as a faculty member at Johns Hopkins University, at the University of Illinois in Chicago, and at the Weizmann Institute of Science, and he was a research scientist at the Canada Center for Inland Waters. Since 1971 he has been with the Department of Geological Sciences in Northwestern University.

The biographical sketch of H. Craig is included in "Editor's Comments on Papers 7 Through 11," Part II.

References

Bentor, Y. K. (1961). Some geochemical aspects of the Dead Sea and the question of its age. *Geochim. Cosmochim. Acta*, **25**, 239–260.

Brewer, P. G., Riley, J. P., and Culkin, F. (1965). The chemical composition of the hot salty water from the bottom of the Red Sea. *Deep-Sea Res.*, **12**, 497–503.

Degens, E. T., and Ross, D. A. (eds.) (1969). *Hot Brines and Recent Heavy Metal Deposits in the Red Sea*. Springer Verlag, Inc., New York, 600 pp.

Lerman, A. (1967). Model of chemical evolution of a chloride lake — The Dead Sea. *Geochim. Cosmochim. Acta*, **31**, 2309–2330.

Miller, A. R., Densmore, C. D., Degens, E. T., Hathaway, J. C., Manheim, F. T., McFarlin, P. F., Pocklington, R., and Jokela, A. (1966). Hot brines and recent iron deposits in deeps of the Red Sea. *Geochim. Cosmochim. Acta.* **30**, 341–359.

20

Copyright © 1969 by Springer-Verlag, Inc., New York

Reprinted from *Hot Brines and Recent Heavy Metal Deposits in the Red Sea*, E. T. Degens and D. A. Ross, eds., Springer-Verlag New York, Inc., 1969, pp. 208–242

Geochemistry and Origin of the Red Sea Brines

H. CRAIG

Scripps Institution of Oceanography
University of California at San Diego
La Jolla, California

Abstract

A steady-state model of the brine waters in Atlantis II Deep is presented. The deuterium and oxygen-18 concentrations in the water, and the dissolved argon content suggest a relative warm near surface Red Sea water as the source of the brine. In evaluating the overall environmental situation in the Red Sea in terms of temperature and salinity, the probable source lies about 800km to the south near the Strait of Bab el Mandeb. By integrating isotope data, and the trace and major element spectra of the brine, the origin, age and history of the brine waters become apparent.

Introduction

The deuterium and oxygen-18 concentrations in the water of the Red Sea geothermal brine show that the water is derived from the normal sea water of the Red Sea (Craig, 1966). Sea water of this isotopic composition is found in the south end of the Red Sea in the vicinity of the southern sill which controls the exchange of water with the Gulf of Aden. This area is underlain by a thick evaporite sequence of several thousand meters lying on the crystalline basement and emerging locally as salt domes. The brine is therefore believed to develop from downward circulating sea water in the region of the southern sill; addition of salt from the evaporite deposits increases its salinity and its temperature rises in accordance with the regional geothermal gradient. Driven by its increased density relative to normal sea water, the brine sinks and flows northward, emerging 2,000m lower and perhaps 1,000km farther north in the brine-filled deeps west of Mecca in the central rift of the Red Sea.

This picture of the origin and history of the brine was developed several years ago (Craig, *op. cit.*); since then new geochemical and chemical data have become available and a more accurate assessment of the chemical changes involved in the formation of the brine can be made. This paper summarizes the isotopic geochemical data, including recent radiocarbon measurements on the brine and overlying sea water (Craig and Lal, 1968), and presents a revised and more detailed discussion of the chemistry of the added salts.

Chemistry of the Brine

Eight reliable chemical analyses of the brines and associated waters are available. The Atlantis II Deep 44°C and 56°C brine layers, and the Chain Deep brine have been analyzed by Brewer and Spencer (1969). The Discovery brine (44.8°C water) has been analyzed by Brewer and Spencer (*op. cit.*) and by Brewer *et al.* (1965). In addition, Brewer *et al.* have analyzed a transitional water above the Discovery brine (69.2g/kg of Cl), and normal and slightly salt-enriched Red Sea Deep Water above this transitional water (22.54 and 22.68g/kg of Cl respectively); the 22.54 per mill water is essentially identical to

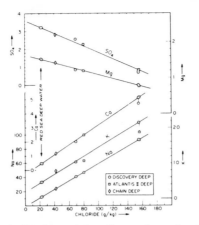

Fig. 1. Chloride variation diagram for major elements in the Red Sea brines.

normal ocean water of the same chlorinity.

In Fig. 1 the major element chemical data from these analyses are plotted on a chloride variation diagram. The linear relationships observed show that all intermediate salinity waters, including the Atlantis II 44°C water and the water in the Chain Deep, are formed by bulk mixing of the 156 per mill chloride brine and normal Red Sea Deep Water. The slight deviations observed for K in the Atlantis II brines are almost certainly analytical considering the conditions under which they were made (Brewer and Spencer, 1969). The Ca deviation of one Discovery brine sample represents the analysis by Brewer et al. (1965); the new Ca value by Brewer and Spencer (1969) is identical to their Ca value for Atlantis II brine.

The mixing lines for Mg and SO$_4$ in Fig. 1 have a special significance: they show that *the mixing relationship is linear between the brines and deep water of essentially the same salinity as that of present Red Sea Deep Water.* Concentration slopes for Ca, K, and Na vs. Cl in evaporation or dilution processes are close to those of the mixing curves in Fig. 1, so that mixing with evaporated Red Sea water would be difficult to detect with these components. However, the MgCl and SO$_4$Cl slopes for evaporation or dilution are approximately at right angles to the mixing curves observed, so that if the intermediate-salinity brines had formed by dilution with an evaporated Red Sea water of higher salinity than present bottom water, a significant convex curvature in the mixing lines would be observed. Consideration of these two curves indicates that if the brine dilution occurred once, with no further mixing, the diluting Red Sea water could not have differed by more than about one per mill in chlorinity from present Red Sea Deep Water. Alternatively, if continuous mixing has occurred, it has been fast enough to remove any trace of an earlier sea-water component with a different chlorinity from that of the present.

Fig. 2 shows the same type of diagram for the minor elements Sr and Br. The Sr analyses of Faure and Jones (1969), made by isotope dilution, are also shown; one of their intermediate composition Atlantis II samples has the same values as the Chain Deep sample. These data also show a simple bulk mixing relationship between sea water and the 156 per mill brine.

The only other element for which a range of samples is available is Mn, which has the same concentration in both the 44°C and 56°C brines of the Atlantis II Deep (Brewer and Spencer, 1969). Bischoff (1969) believes that the Mn concentration in the 44°C water is buffered by the presence of a solid phase. Also, the concentration of iron in the Atlantis 44°C water and in the Discovery Deep is much lower than in the Atlantis 56°C brine, so much so that precipitation of iron as well as dilution must have occurred during the mixing process. Other trace metals may show similar effects, though the various data are not in good agreement.

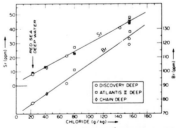

Fig. 2. Chloride variation diagram for strontium and bromine. Crossed symbols are Sr analyses made by isotope dilution (Faure and Jones, 1969).

The relationships in Figs. 1 and 2 show clearly that the Atlantis 44°C water, the Chain water, and the transitional Discovery Deep waters, are all formed by bulk mixing of the 156 per mill Cl brine with Red Sea Deep Water. In this process most of the brine components are conservative, but the concentrations of some trace metals are affected by chemical reactions and precipitation or solution.

Concentrations in the Original Brine

To what extent has the 155.5 per mill chloride brine in the Atlantis and Discovery Deeps been diluted with sea water mixing down from Red Sea Deep Water? One way to obtain an upper limit is by assuming that the original brine flowing into the depressions is saturated with e.g. NaCl. Addition of NaCl to the 155.5 per mill Cl brine in the laboratory resulted in saturation at 160.8g/kg of Cl at 25°C, corresponding to addition of 12.0g NaCl per kg of (unsaturated) brine (Craig, 1966). Solubility data indicate that NaCl is more soluble at 60°C than at 25°C by a factor of 1.023. If the observed brine is a mixture of sea water fluid and original brine fluid, the fraction of sea water (f_{SW}) is related to the concentrations of any element in the component fluids by:

$$f_{SW} = \frac{x_B - x_M}{x_B - x_{SW}} \quad (1)$$

where x is the mass fraction of any element and the subscripts are B for original, undiluted brine, SW for Red Sea Deep Water, and M for the observed mixture now in the brine depressions (Cl = 155.5g/kg). Thus if we assume a saturation chloride concentration in original brine at 60°C, of 164.5g/kg, and a concentration of 22.5g/kg in RSDW, an upper limit of about 6 per cent is obtained for f_{SW}. (A more accurate estimate could be obtained by evaporating some brine to dryness, and saturating another aliquot of brine with its own salt at 60°C.)

A better estimate can be obtained by using equation (1) with an element whose concentration in the brine is very small compared to sea water concentration; in such a case an upper limit for f_{SW} is obtained as x_B goes to zero in (1). The best element for this calculation is F which has concentrations of 1.6ppm in sea water of 40.6 per mill salinity (RSDW), and 0.0512ppm in the Discovery brine according to Brewer et al. (1965), and a concentration less than 0.02ppm in Atlantis brine according to Brewer and Spencer (1969). Then assuming $x_B = 0$, that is, that the original brine has *no* fluoride, and $x_M = 0.0512$ppm, we find a *maximum* value of $f_{SW} = 3.2$ per cent. For any finite concentration of F in the original brine, the fraction of sea water would be less. A similar calculation can be made using the NO_3 concentrations measured by Brewer et al., and gives an upper limit of 5.5 per cent.

The fraction of any element in the observed brine, derived from admixture with sea water, is given by:

$$f_i(SW) = x_{SW} f_{SW}/x_M \quad (2)$$

where x_M is the concentration of species i in the observed brine (i.e. mixture), and $f_i(SW)$ is the fraction of species i in the mixture which is derived from sea water. x_{SW} is the concentration of species i in Red Sea Deep Water of 40.6 per mill salinity. Using the maximum value of $f_{SW} = 3.2$ per cent calculated from the fluoride data, we obtain maximum values of $f_i(SW)$ for the individual species as shown in Table 1. In this calculation the sea water fraction for F is of course 100 per cent by definition.

The $f_i(SW)$ values are especially important in interpreting isotopic variations in the elements. For example Hartmann and Nielsen (1966) concluded from S^{34} measurements that the sulfate of the brine is derived from direct mixing with overlying Red Sea Deep Water. However, as shown in Table 1, only 12 per cent of the SO_4 in the observed brine can be so derived, and this is an upper limit which assumes that the original brine contains no fluoride at all. Therefore the RSDW is not a significant source of SO_4 by the process of direct mixing into the brine depressions.

Table 1 also shows that less than 4.4 per cent of the water and 3.0 per cent of the

Table 1 Maximum Values of $f_i(SW)$, the Fraction of Species i Which Can Be Derived from Red Sea Deep Water by Direct Mixing into the Brine Depressions, Calculated from Fluorine Mass Balance

Species	Brine Analyzed *	$f_i(SW)$ (%)
F	Disc.	100.0
SO_4	At.	12.0
I	Disc.	7.4
Mg	At.	6.3
H_2O	—	4.4
HCO_3	At.	3.0
Li	Disc.	2.7
B	Disc.	2.2
Br	At.	2.0
Cl	—	0.5

* Discovery brine: Brewer *et al.* (1965); Atlantis II brine: Brewer and Spencer (1969).

dissolved inorganic carbon of the brine can be derived from mixing with RSDW; these values are important for understanding the deuterium, oxygen-18, and C^{13} and C^{14} concentrations in the brine. The species "HCO_3" in Table 1 refers to total dissolved inorganic carbon, expressed as g/kg of bicarbonate ion; the concentrations used for HCO_3 are 0.143g/kg in the Atlantis brine (56°C water), measured by Weiss (1969), and an assumed concentration of 0.114g/kg in 19 per mill chlorinity sea water.

For all analyzed substances other than those listed in Table 1, $f_i(SW)$ is less than 1 per cent; the constituents in Table 1 are, of course, those with the highest concentration ratios x_{SW}/x_M. It should be emphasized that the tabulated values of $f_i(SW)$ are upper limits only. For example, the chloride concentrations (g/kg) which have been measured in the two high salinity brines are as follows:

Atlantis Deep	Discovery Deep	Reference
—	155.2	Brewer *et al.* (1965)
155.4 ± 0.1	155.5 ± 0.1	Craig (1966)
156.0	155.3	Brewer and Spencer (1969)

so that from the data of Brewer and Spencer we might conclude using equation (1), that Discovery brine has been diluted by addition of 0.5 per cent sea water to Atlantis brine. Because the Discovery brine is observed to be cooling at the bottom as well as the top (Pugh, 1967), it is widely believed that it represents an overflow from Atlantis Deep into an adjacent depression without direct influx of brine from below; the slightly oxidized nature of Discovery brine is consistent with this interpretation. Thus a slight dilution of the Discovery brine with sea water is plausible, and the value of 0.5 per cent is consistent with the results in Table 1. However, this effect should also be observable in sodium because of its high concentration in the brine, and the Na data measured to date (92.60g/kg in Atlantis brine, by Brewer and Spencer; 92.84 and 93.05g/kg in Discovery brine, by Brewer *et al.* and by Brewer and Spencer, respectively) do not agree with a chloride dilution in Discovery brine — in fact, they agree better with the chloride data reported by Craig (1966). Variations of the magnitude discussed here almost certainly reflect analytical and handling difficulties with such concentrated brines, and we must conclude that no significant differences in major element concentrations have yet been observed.

The upper limits for sea water contribution to the species listed in Table 1 depend on the F concentrations in the brines and RSDW — it would be important to analyze the Atlantis brine and the overlying deep water for fluorine in order to establish a more accurate upper limit for f_{SW}. Nevertheless, the present data show clearly that direct mixing with overlying sea water has not seriously affected the concentrations in the brines.

Origin of the Water

Deuterium-Oxygen-18 Salinity Relations

Fig. 3 shows the D-O^{18} isotopic relationships in waters of the Red Sea region (Craig, 1966). The δ values, in units of per mill are defined as

$$\delta = [(R/R_{SMOW}) - 1] \times 10^3$$

where R is the isotopic ratio HDO/H_2O or H_2O^{18}/H_2O^{16}, and $SMOW$ is a defined

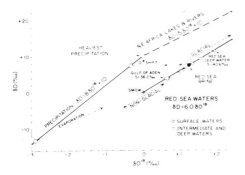

Fig. 3. Deuterium-oxygen-18 isotopic diagram for waters of the Red Sea region. SMOW is the standard Mean Ocean Water isotope standard; δ values are per mill variations relative to this standard.

Standard Mean Ocean Water with approximately the isotopic composition of average ocean water. The Red Sea waters whose compositions are plotted were collected on SIO Expedition Zephyrus in September, 1962; they extend over the salinity range from inflowing Gulf of Aden water at the south (S = 36.2 per mill) to Red Sea Deep Water (S = 40.6 per mill). Outside of the southern sill, in the Gulf of Aden and Indian Ocean, the waters are all less saline than the inflowing water of salinity 36.2 per mill, and have lower isotopic delta values. Within the Red Sea the D-O^{18} variations are linear with a slope of 6; this relationship is due to the evaporation-precipitation-inflow and outflow- and molecular exchange relationships (Craig and Gordon, 1965).

In glacial and non-glacial periods the isotopic composition and salinity of the ocean change slightly because of the formation and melting of continental ice sheets; the dashed lines in Fig. 3 show the maximum excursions of *SMOW* (marked by crosses) and corresponding isotopic trajectories assuming complete melting of continental ice during interglacial periods (Craig, *op. cit.*). Fig. 3 also shows the "precipitation locus" of slope 8 established for the isotopic variations in natural precipitation over the earth, and the trajectories of slope 5 followed by evaporating natural waters such as the North African lakes and rivers shown in the diagram. A complete discussion of these effects is given by Craig and Gordon (1965) and they are briefly reviewed by Craig (1966). Together with the Red Sea water locus, these lines establish the isotopic relationships in possible source waters for the brines.

The isotopic and chloride data measured on the brines are given in Table 2 (Craig, *op. cit.*), and plotted in Fig. 3; they plot directly on the locus of present-day Red Sea waters, approximately in the middle of the salinity range. The isotopic compositions of the Discovery and Atlantis II brines are identical within experimental precision and fit the observed Red Sea D-O^{18} relationship exactly. An extraordinary set of coincidences would be required to derive such water from fresh waters by evaporation from an initial water on the precipitation locus (lower open circle on that locus, with $\delta O^{18} = -2.9$, $\delta D = -13.2$), or by O^{18} exchange of a fresh water with carbonate minerals – the "isotope shift" observed in continental geothermal areas, which does not affect the deuterium concentration (Craig, 1963). That is, either process would require an exactly specified initial isotopic composition of source water (because the slopes are fixed), coupled with an exactly specified process intensity (evaporation or oxygen isotope exchange), to terminate the isotopic trajectory exactly at its intersection with the Red Sea water locus. The probability of two such coincidences occurring simultaneously in either the evaporation or isotope exchange process is sufficiently remote to eliminate both possibilities; we

Table 2 Isotopic Data (Relative to SMOW) and Chloride Concentrations (by Mohr Titration) in Red Sea Brine Samples

Brine Sample	Cl(g/kg)	δD(‰)	δO^{18}(‰)
Discovery Deep	155.4	7.6	1.22
(Discovery Station 5580)	155.6	7.3	1.22
Atlantis II Deep	155.3	7.4	1.18
(Atlantis II Station 544)	155.5	7.6	
Average values	155.5	7.5	1.21

must conclude that the brine water is derived from the Red Sea water itself.

The specific source water can be identified from the isotopic-salinity relationships shown in Fig. 4. The deuterium and oxygen-18 values (Table 2) correspond to Red Sea water with an initial salinity of 38.2 ± 0.2 per mill for the water which formed the brine. Water of this salinity is found only in the southern end of the Red Sea, extending down to the southern sill at 13°41′N in the Hanish Islands.

The seasonal fluctuation in salinity profile over the southern sill is shown in Fig. 5, from Thompson (1939). The upper section (September) shows the situation during the summer when NNW winds prevail; there is a surface outflow of high salinity water and a sub-surface inflow of low salinity water, with water of about 36.4 per mill salinity lying on the sill; occasional outward flow of higher salinity water at depth may occur during the flood tide. At the end of September the flow changes to the winter pattern shown in the lower diagram for May, the last month of the winter season. The winds are SE, the surface flow is inward with outward flow of deep water averaging about 40.3 per mill salinity over the sill. Similar observations for June were made by Neumann and Densmore (Neumann and McGill, 1962)

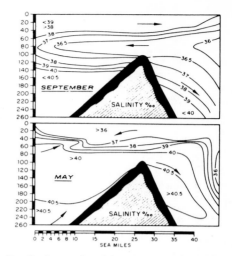

Fig. 5. Seasonal salinity variation with depth (meters) and distance (nautical miles) across the southern sill of the Red Sea (Thompson, 1939).

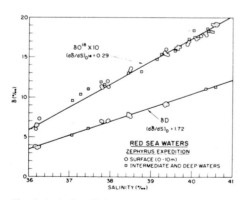

Fig. 4. Isotopic-salinity relationships for Red Sea waters. The deuterium and oxygen-18 isotopic delta values for the Red Sea Brine (+7.5 and +1.2) correspond to an initial salinity of 38.2 ± 0.2 per mill for the water which formed the brine.

who found 40.2 per mill water on the sill, as compared with Thompson's observation of 40.4 per mill. The water lying on the sill, averaged over an annual cycle, will therefore have a mean salinity very close to 38.2 per mill.

Northward into the Red Sea, water of 38.2 per mill can be found as far north as 19.5° at the end of winter in May (Thompson, op. cit.); the 38.2 per mill isohaline shoals northward from the sill (Fig. 5) and intersects the surface at this latitude. Farther north the water is more saline at all depths. Neumann and Densmore (op. cit.) observed 38.2 per mill water extending north to about 18.5°N latitude in June. In summer the 38.2 per mill isohaline moves south and occurs at two depths enclosing a core of lower salinity water; in September, 1962, at 15.5°N, I found the 38.2 per mill isohaline at 96m and 38 per mill water at the surface, with lower salinity water in between. At 16.5°N the 38.2 per mill isohaline occurred at 35m and 80m depth, enclosing less saline water; northward the depths converged and the isohaline vanished at about 17.5°N (Zephyrus Expedition hydrographic data report, Scripps Institution of Oceanography).

Thus in order to find a significant amount of 38.2 per mill average salinity water over a year cycle, either directly or as an alternation between higher and lower salinities, one is restricted to latitudes south of about 18°N at best, and more likely south of about 17°N. That is, source water of this salinity must ultimately flow northward some 500 to 900km, under the bottom of the Red Sea, to reach the brine depressions, the larger figure applying to flow all the way from the southern sill.

Evidence from Dissolved Gases

Concentrations of N_2, Ar, total dissolved carbonate (ΣCO_2), and C^{13} in the Atlantis and Discovery brines have been measured by Weiss (1969). In Table 3, the N_2, Ar and ΣCO_2 concentrations are given as cc(STP) of gas per kg of 38.2 per mill salinity water; that is, the measured concentrations (cc/kg of brine) are multiplied by 1.295, the ratio of water content in 38.2 per mill salinity normal sea water to the water content of the brine. The effect of salt addition or subtraction to the presumed source water of the brine is thus eliminated. The equilibrium concentrations in 38.2 per mill sea water, relative to air at a total pressure of 1 atmosphere including the saturation vapor pressure of H_2O, are also shown; N_2 and Ar values are from the experimental data of Douglas as treated by Weiss (*op. cit.*), while the ΣCO_2 value is a rough estimate from measurements on surface and deep waters in the Pacific, normalized for the salinity difference (Weiss and Craig, 1968).

The sea water salinity of 38.2 per mill used in Table 3 is the value obtained from the D-O^{18} isotopic data as the salinity of the source water of the brine. From the T-S diagram for Red Sea waters (Neumann and McGill, 1962) it is found that Red Sea water of 38.2 ± 0.1 per mill has a temperature of 28 ± 1°C, this temperature corresponding uniquely to that salinity value. Therefore, if the gases are conservative, their concentrations in the brine as predicted from the stable isotope data, should correspond to the saturation values in Red Sea water of T = 28°, S = 38.2 per mill. As shown in Table 3, the Ar values in such sea water and in Atlantis II brine are indeed found to be identical. (The temperature and salinity uncertainty ranges, ±1°C and 0.1 per mill, correspond to variations of only ±0.004 and 0.1 in Ar and N_2 concentrations, respectively.) Weiss (*op. cit.*) has developed this argument *ab initio* to derive minimum initial temperatures for sea water source fluid; he also shows that the Ar and N_2 concentrations in the brines are completely different from equilibrium solubilities in the brine. The Ar data thus provide striking confirmation of the D-O^{18} isotopic identification of source water for the Red Sea brine.

Discovery brine contains about 10 per cent excess Ar relative to Atlantis II brine;

Table 3 Dissolved Gases in Sea Water and the Red Sea Brines

Gas Concentrations Are Expressed as cc(STP)/kg of 38.2‰ Salinity Water, Calculated from Data of Weiss (1969). Δ Values Are Excess Concentrations Relative to Sea Water, δ Values Are Relative to PDB-1 Isotopic Standard

	Sea Water $S = 38.2‰$, $T = 28°C$	Atlantis II Brine	Discovery Brine
N_2	8.05	13.09	13.49
ΔN_2	–	5.04	5.44
Ar	0.2110	0.2110	0.2320
ΔAr	–	0.000	0.021
ΣCO_2 (ppm)	(125)	184.7	39.4
$\Delta \Sigma CO_2$ (ppm)	–	59.7	−85.6
δC^{13}(‰)	(−1)	−5.6	−16.8
$\delta C^{13}(\Delta CO_2)$(‰)	–	−15.2	+6.3

this is presumably excess radiogenic Ar[10] and, if so, it is consistent with a model of Discovery brine as a stagnant overflow from Atlantis Deep as discussed by Weiss (1969). The excess N_2 of about 5cc/kg in each brine is attributed by Weiss to decomposition of organic matter in sediments, as observed in other ground waters. Oxidation of NH_3 in this amount would consume about 7.5cc/kg of O_2, while the solubility of O_2 in the initial sea water would be about 4.1cc/kg, so that at least half of the excess N_2 would probably have to result from denitrification of nitrate or nitrite in sediments. (The observed nitrate depletion in the brines relative to Red Sea normal water (Brewer *et al.*, 1965) could produce only 3 per cent of the excess N_2.) N_2 and Ar concentrations in the brines are greater than in overlying Red Sea deep water, so the brine concentration cannot have been increased by direct mixing—nitrogen isotope measurements should show whether the excess N_2 is indeed derived from sediments.

Table 3 also shows that Atlantis II brine contains about 32 per cent excess CO_2 relative to sea water, while Discovery brine has only about 20 per cent of the Atlantis brine concentration, or about 30 per cent of the original sea water concentration; the brines also differ in isotopic composition from each other and from the estimated sea water value. Mean values of δC^{13} for the CO_2 added to or subtracted from the sea water to give the brine concentrations are also shown. The mean δ value of -15.2 per mill for excess CO_2 added to Atlantis II water is quite characteristic of carbon derived from a mixture of heavy carbonate and light organic matter, as observed in ground waters, or as a mean isotopic composition of particulate carbon in the oceans (e.g. whole foraminifera), while the mean δ value of 6.3 per mill for CO_2 removed from Discovery water is much more similar to that of carbonate precipitated inorganically from sea water. These differences are discussed in a later section.

Dissolved gas studies on brine samples are continuing and a detailed study is necessary for confirmation of these conclusions; however, the present data show very clearly that inert gas concentrations are also a powerful tracer for geothermal waters, and give strong support to the source water origin deduced from the isotopic water data.

Origin of the Added Salts

Although most components of the brine are greatly enriched relative to their sea water concentrations, certain elements such as Mg are present in much lower concentrations. It is convenient for the material balance calculations to classify the brine components in terms of their enrichment factors relative to the presumed source water of 38.2 per mill salinity normal sea water, as defined by the process of addition or subtraction of salts to or from the source water. The enrichment factor thus defined is

$$E_i = \frac{(C_i/C_w)_{\text{Brine}}}{(C_i C_w)_{SW}} \quad (3)$$

in which C_i and C_w are the concentrations of component i and of water (g/kg) and the subscript SW refers to normal sea water of 38.2 per mill salinity. In such a process E_w is of course unity; a value of $E > 1$ indicates a net addition of the component to the source water, while $E < 1$ indicates that the component has been removed from the original sea water. Enrichment factors for the various components are shown in Table 4, which also shows the concentration data used; these data refer to Atlantis II brine except when data are available only for Discovery brine.

In these and subsequent calculations the Na and Cl values given by Brewer and Spencer (1969) for Atlantis brine were adjusted slightly because their analysis shows an anion excess over neutrality by about 0.5 equivalent per cent. As noted earlier, their value of 156.03g/kg of Cl is higher than the value given by Craig (1966) for Atlantis brine and their own values for Discovery brine, and Craig (*op. cit.*) found both brines to have the same Cl content, 155.5g/kg. Also their Na value in Atlantis brine is *lower* than their value and that of Brewer *et al.* (1965) for Discovery brine;

Table 4 Enrichment Factors in the Red Sea Brine Relative to Normal Sea Water of 38.2‰ Salinity, Calculated for Addition or Subtraction of Salt

Component	Brine Analysis	Concentrations (g/kg) Brine	Concentrations (g/kg) Sea Water	Enrichment Factor
Pb	Atl. (1)	6.3×10^{-4}	3×10^{-8}	27,200
Mn	"	8.2×10^{-2}	4.2×10^{-6}	25,300
Fe	"	8.1×10^{-2}	$<2 \times 10^{-5}$	>5,200
Zn	"	5.4×10^{-3}	$<5 \times 10^{-6}$	>1,400
Ba	Atl. (2)	9×10^{-4}	16.7×10^{-6}	70
Cu	Atl. (1)	2.6×10^{-4}	5.5×10^{-6}	61
Ca	Atl. (1)	5.15	0.450	15
Si (3)	"	2.76×10^{-2}	2.4×10^{-3}	15
Na	"	92.85	11.75	10
Cl	"	155.5	21.13	9.5
Sr	"	4.8×10^{-2}	8.9×10^{-3}	7
K	"	1.87	0.423	5.7
Br	Atl. (1)	0.128	7.36×10^{-2}	2.2
B	Disc. (4)	7.8×10^{-3}	5.1×10^{-3}	2.0
Li	"	2.62×10^{-4}	2.07×10^{-4}	1.6
HCO_3^-	Atl. (5)	0.143	0.125	1.5
H_2O	Atl. (1)	742.50	961.65	1.0
Mg	Atl. (1)	0.764	1.413	0.70
I	Disc. (4)	3.0×10^{-5}	6.5×10^{-5}	0.60
SO_4	Atl. (1)	0.84	2.96	0.37
NO_3 (3)	Disc. (4)	4.4×10^{-5}	7.5×10^{-4}	0.08
F	"	5.12×10^{-5}	1.4×10^{-3}	0.05

Notes: (1) Brewer and Spencer (1969); values for Na and Cl slightly adjusted as described in text.
(2) Miller *et al.* (1966).
(3) Si and NO_3 sea water values are data for Red Sea Deep Water from reference (4), normalized to 38.2‰ salinity.
(4) Brewer *et al.* (1965). Miller *et al.* (1966) found $B = 11 \times 10^{-3}$ in Atlantis II brine, corresponding to an enrichment factor of 2.8.
(5) Brine value from Weiss (1969), sea water value as in Table 3, expressed as bicarbonate ion.

these differences account for the anionic imbalance. Accordingly the Cl and Na concentrations in Atlantis brine were taken as 155.5 and 92.85g/kg respectively as the most probable best values. This makes the total dissolved salt concentration in Atlantis brine 257.50g/kg as the sum of the analytical values (including total CO_2 calculated as HCO_3).

Table 4 shows that the observed enrichment factors fall roughly into four classes: (1) trace metals (Mn, Fe, etc.) with factors greater than 50, ranging up to 27,000; (2) normal components such as Na, K, Ca, and Sr with enrichments close to that of chloride — about 10 ± 5; (3) minor components only slightly enriched, with factors between 1 and 2; and (4) the components Mg, SO_4, F, and I which have factors less than one and have thus been subtracted from the original sea water. If Z_A and Z_S are the number of grams of salts respectively added to, and subtracted from, one kg of original sea water (38.2 per mill salinity, 38.5g/kg of total salts), then

$$(Z_A - Z_S) = 1{,}000 \left(\frac{C_{W\text{-}SW}}{C_{W\text{-}B}} - 1 \right) = 295.1 \text{g/kg}$$

is the net salt addition as given by the concentrations of water in the sea water and in the brine. The value of Z_S is given by

$$Z_S = \sum_s C_{i-SW} - 1.2951 \sum_s C_{i-B} \quad (4)$$

in which the summations are over the components with enrichment factors less than one (*subtracted* components: Mg, I, SO_4, F) in sea water and in brine. Thus the salt balance calculation gives

$Z_S = 2.30$ g/kg, subtracted salts
$Z_A = 297.4$ g/kg, added salts

as the amounts subtracted from, and added to one kg of 38.2 per mill salinity sea water, to form the brine.

Composition of the Added Salt

From the values of $(Z_A - Z_S)$ and Z_A for the brine, the mass fractions of the various components in the 297.4g of added salt per kg of initial sea water are given by

$$x_{i-A} = \frac{1.2951 C_{i-B} - C_{i-SW}}{297.4} \quad (5)$$

in which x_{i-A} is the mass fraction of component i in the added salt and the C_i are the concentrations in brine and 38.2 per mill salinity sea water. The calculated composition of the added salt, with the major cations expressed as chlorides, is given in Table 5, which shows that 99.8 per cent of the added salt consists of NaCl, KCl, and CaCl$_2$ with NaCl greatly predominant (93 per cent), and that the remaining components range from about 1 to 300ppm in concentration. The composition as a whole is quite normal for the halite-sylvite zones of evaporite deposits. Stewart (1963) has reviewed the chemistry of evaporite deposits in detail—trace element ranges in halite rocks of marine evaporites, given in his Table 24 (*op. cit.*) include: Sr, 52–180ppm; Si, 22–240ppm; Cu, 1–6.5ppm; Ni, 1.6–3.5ppm; Pb, 1–5ppm; and Ba, 1–3ppm. Bromine values in halite-sylvite rocks range from 140–3,300ppm (Table 28, *op. cit.*), and boron values, from two sets of data on halite rocks, range up to 50 and up to 379ppm (Table 27, *op. cit.*). All these values are quite comparable to the added salt composition in Table 5. Fe concentrations observed in halite-sylvite rocks range from 21–448ppm (Stewart, *op. cit.*, p. 37). Only Mn and Zn are considerably higher in the

Table 5   Composition of Salts Added to Normal Sea Water (S = 38.2‰) to Make the Red Sea Brine, and of Salts in Two Lousiana Brines from Salt Domes

	Red Sea Brine: Added Salts	Timbalier Bay, La. (1)	West Bay, La. (2)
Major Components (per cent)			
NaCl	92.7	89.9	81.2
KCl	1.3	0.3	0.8
CaCl$_2$	5.8	5.0	12.7
MgCl$_2$	–	2.85	2.09
SO$_4$	–	0.002	0.08
HCO$_3^-$	0.020	0.10	0.06
Minor Components (ppm)			
Sr	179	76	900
Br	310	591	1,964
B *	17	138	47
Si	112	83	37
Li *	0.45	38	95
F *	–	7.5	7.0
I *	–	131	90
Mn	357	7.6	150
Fe	353	59	550
Zn	23.5		25
Cu	1.11	0.14	2.0
Co	0.7		
Ni *	1.5	10.3	10.5
Pb	2.74		3.0
Ba	3.86	1,032	85

(1) Timbalier Bay oil field, Lafourche Parish, La.; chloride concentration 89.7g/kg of brine. Water analysis, from White *et al.* (1963, Table 12), has excess Cl equivalent to 1.7% of total salts over amount for neutrality.

(2) West Bay, Plaquemines Parish, La.; chloride concentration 124.0g/kg of brine. Water analysis, from White *et al.* (1963, Table 13), has excess Cl equivalent to 2.5% of total salts over amount for neutrality.

* Calculated from Discovery brine concentration data (*cf.* Table 3). Boron concentration would be 31ppm using Atlantis brine value of Miller *et al.* (1966).

added salt composition than in halite rocks, where observed values range up to 20ppm for Mn and about 3ppm for Zn; these two elements are concentrated in anhydrite and gypsum zones, and in salt clays.

A more direct comparison can be made with the salts found in brines collected around salt domes; Table 5 also shows the composition of the salts, recalculated to a water-free basis, in two such waters from Lousiana salt domes (White et al., 1963). The Timbalier Bay sample occurs in Pliocene sands at a depth of 1,800m with a temperature of 70°C; the well is on the north flank of a salt dome which was encountered at a depth of 2,370m. The West Bay brine is found at 2,500m in Miocene sandstone, only 60m from salt in the crest of the West Bay salt dome, which is reported to lack an anhydrite cap; the sample and bottom hole temperatures were 42°C and 87.5°C respectively. These two samples are probably the most representative analyses available of waters which have been in contact with salt domes, and as seen in Table 5, with the exception of Mn, the concentrations of all other elements bracket those in the added salts of the Red Sea brine. Some elements in the added salt are, however, conspicuously low — Mg, Li, F, I, Ba, and possibly B; reasons for this are discussed later on.

White (1965) has discussed the Ca/Cl and K/Na ratios in natural waters with respect to origin and mineral interaction; in a Ca/Cl vs. Cl plot (Fig. 1, op. cit.) the salt dome waters in Table 5 are more similar than any others to the Red Sea brine. The K/Na ratio of the brine is characteristic of the lowest temperatures (<100°C) for mineral-water interaction and is also most similar to the salt dome waters; while the brine has probably never equilibrated Na and K with the solid phases, the ratio at least indicates that it has not been subjected to high temperatures in the presence of feldspars and micas for any prolonged period.

In general, the composition of the added salt agrees very well with what we know about the composition expected for solution of salts from the halite-sylvite zone of evaporite and salt dome deposits; the high concentrations of trace elements in the brine demand only ppm concentrations in the original salt deposits, and, with the possible exception of Mn and Zn, are not higher than expected. Even Mn and Zn are not greatly enriched with respect to the salt dome waters in Table 5.

Isotopic Data on the Added Salts

The isotopic composition of lead in the Atlantis II brine has been measured by Delevaux et al. (1967), together with that of a series of Cenozoic ore leads from Saudi Arabia and Egypt. Although considerable variation in the ore leads is observed, the brine lead is very similar in both 207/204 and 208/204 ratios to lead in a Tertiary vein in Rabigh, Saudi Arabia, and to a series of galena samples in Miocene gypsum and lime sediments in Egypt (Um Gheig, Um Ans, and Bir Ranga deposits, Table 2, op. cit.). All of these leads are normal Tertiary J-type leads with isochron and 208/204 ages close to zero, quite different from Precambrian ore leads; in particular their isotopic composition is precisely what would be expected for lead from a Tertiary evaporite sequence.

The isotopic composition of strontium in the brine has been measured by Faure and Jones (1969), and found to be similar to that of strontium in marine carbonates, again consistent with derivation from marine sediments or from shells of marine organisms.

Subtraction of Salts from the Brine

Components in the Red Sea brine with enrichment factors less than 1 include Mg, SO_4, I, NO_3, and F. The amount of a component subtracted from the original sea water of 38.2 per mill salinity, in grams removed per kg of original sea water, is given by

$$Z_{i\text{-}S} = C_{i\text{-}SW} - 1.2951 C_{i\text{-}B} \qquad (6)$$

The amounts subtracted, and the percentage of each component which has been removed, from the original sea water are:

Mg = 0.424 g/kg   (30.0 per cent lost)

SO_4 = 1.872 g/kg   (63.2 per cent lost)

$I = 2.61 \times 10^{-5}$ g/kg (40.2 per cent lost)

$NO_3 = 6.9 \times 10^{-4}$ g/kg (92.4 per cent lost)

$F = 1.33 \times 10^{-3}$ g/kg (95.0 per cent lost)

for a total of $Z_S = 2.30$ g/kg removed. The SO_4/Mg ratio in the subtracted salt is 4.4, compared to the stoichiometric ratio in $MgSO_4$ of 4.0; the difference is within the analytical errors, so that the possibility that these components may actually have been removed as $MgSO_4$ must be considered. The F, I, and NO_3 brine concentrations were measured only in Discovery brine (Brewer *et al.*, 1965), which generally has lower trace metal concentrations than Atlantis brine, so that concentrations in Atlantis II brine may be significantly higher. However, Brewer and Spencer (1969) found that F in Atlantis brine was below their analytical limit of about 2×10^{-5} g/kg, while the concentration in Discovery brine was reported as 5×10^{-5} g/kg (Brewer *et al., op. cit.*), so F at least is not significantly higher in Atlantis brine.

It is worth noting that all these subtracted components occur in very low concentrations in halite-sylvite zones of evaporites (Stewart, 1963); fluorine in evaporites is found principally in sulfate deposits, while iodine is thought to be lost to the atmosphere during evaporation. Their low abundance, though not their subtraction from the source water, is thus consistent with the concentration pattern in the added salts. Fluorine and lithium occur in much higher concentrations in all so-called "volcanic" waters (White, 1965). The Li/Na ratio of 3×10^{-6}, and F/Cl of 3×10^{-7}, in Red Sea brine are the lowest ratios recorded for high salinity waters (cf. data in White *et al.*, 1963) and argue strongly against any significant magmatic or volcanic contributions to the brine.

Isotopic Composition of the Sulfate

The isotopic composition of sulfur in dissolved SO_4 ions has been measured by Hartmann and Nielsen (1966), and that of oxygen in the SO_4 ions by Longinelli and Craig (1967); the data can be summarized as follows:

	δS^{34} (per mill)	δO^{18} (per mill)
Red Sea Brine	20.3	7.3
Normal Red Sea Water	20.3	9.5

The S^{34} data refer to Atlantis II brine; O^{18} measurements were made on both Discovery and Atlantis II brines, with results of 7.5 and 7.2 respectively. O^{18} measurements on normal Red Sea water SO_4 gave $\delta = 9.3$ for Red Sea deep water, and $\delta = 9.6$ for SO_4 in surface inflow water from the Gulf of Aden. (Sulfur data refer to the meteoric sulfur standard, oxygen data to standard mean ocean water—*SMOW*.)

It has been shown in previous sections above that (1) the chloride and fluoride concentrations in the brine show that no significant fraction of the sulfate can be derived from direct mixing with overlying water (Table 1), and (2) that sulfate has actually been *removed* from the original water, 63 per cent of the original concentration having been lost. It is clear, therefore, that the δS^{34} values are simply those of the sulfate in the original sea water which formed the brine, and that SO_4 removal has taken place by a process which does not fractionate sulfur isotopes. The isotope data thus constitute additional evidence consistent with a brine origin from normal sea water.

The slight O^{18} decrease in the brines, relative to SO_4 in normal Red Sea water, probably represents an approach to isotopic equilibrium at the higher temperature of the brine, since the exchange time is probably less than 100 years at these temperatures (Longinelli and Craig, *op. cit.*); alternatively some oxygen isotope fractionation may have occurred in the removal process. Nevertheless, the S^{34} and O^{18} data together are more similar to values in sea water SO_4 than to anything else. Rafter (1967) has plotted S^{34} vs. O^{18} data from a wide variety of sulfate in natural barites, geothermal waters, and saline lakes; his diagram shows that the brine sulfate falls closer to sea water sulfate than to any other samples. Thus both the S and O isotopic labels indicate that the low sulfate concentration in the brine is a residual from the original sea water sulfate, most of which has been removed by a non-fractionating process for sulfur.

Kaplan *et al.* (1969) have studied sulfur isotope ratios in the brines and sediments in more detail. They find δS^{34} to be $+20 \pm 2$ per mill in sulfate in all the waters of both Atlantis and Discovery Deeps, with no significant differences between Atlantis 44° and 56° brines, Discovery brine, and Red Sea deep water overlying either brine; the isotopic data show no correlation with SO_4 concentration and individual variations appear to represent only analytical scatter. In Discovery Deep sediments, sulfides and elemental sulfur are very low in S^{34} ($\delta \sim -25$ per mill) and sulfate in interstitial water is about 7 per mill enriched in S^{34} relative to SO_4 in brine and sea water; this is clearly the result of *in situ* biogenic sulfate reduction in the sediments, as they point out. Biogenic sulfur and sulfides are also found in sediments outside the deeps and in the Atlantis II Deep central rise above the brine level. In all these areas the reduced sulfur is dominantly pyrite, with associated chalcopyrite, sphalerite, and elemental sulfur.

In Atlantis II Deep, sulfides and sulfur with $\delta S^{34} = +5$ per mill are found (range +3 to +11) (Hartmann and Nielsen, Kaplan *et al.*, *op. cit.*); sulfur and sulfides are dominated by sphalerite, and $CaSO_4$ is also found. Sulfate in interstitial water is slightly depleted in S^{34} relative to brine and sea water ($\delta S^{34} = 15$ to 21 per mill, averaging about 18), and anhydrite is about +23 per mill, consistent with precipitation from brine or seawater (Kaplan *et al.*, *op. cit.*). The origin of sulfur and sulfides in these sediments is obscure, especially as sulfur and sulfides of similar isotopic composition ($\delta S^{34} = +10$) were found by Kaplan *et al.* in Core 118K, 6km SE of the brine deeps and *below* normal biogenic sulfur and sulfides. In this core the dissolved sulfate in interstitial water is isotopically similar to that in Atlantis Deep cores ($\delta S^{34} = 16$ to 19 mill, averaging about 17.5), with the higher values (closest to sea water values) occurring in the upper region of light biogenic sulfur and sulfides.

The origin of the sulfide and sulfur in Atlantis Deep sediments, and the problem of their unique isotopic compositions, are obviously closely connected with the mechanism of sulfate depletion in the brine relative to its source water; these questions are considered in the next section.

Origin of Sulfide and Sulfur in Atlantis Deep Sediments

Two models have been proposed for sulfide origin and deposition in Atlantis Deep. Watson and Waterbury (1969) emphasize the difficulty of precipitating zinc sulfide and ferric hydroxide in the same environment, and postulate that precipitation of both takes place in the transition layer between the upper 44°C brine and normal Red Sea Bottom water, and that bacterial sulfate reduction is occurring in a shallow zone just above the 44°C brine. They suppose that just enough reduction takes place for ZnS precipitation so that ferrous iron diffuses on through this layer into the aerobic zone above where it is oxidized to the ferric state and precipitated. In order to account for the sulfide isotopic composition, δS^{34} about +5 per mill, they assume that a large enough fraction of the sulfate in the layer is reduced so that less than one enrichment stage is achieved. The sulfate in the reduction layer is assumed to be replaced continually by diffusion, so that it is never entirely used up; this is thus a steady-state model.

Later on, it is shown that a steady-state model of Atlantis Deep actually requires that essentially all of the Fe and Zn precipitation takes place within the 44°C convection layer. But one can readily show that the mass balance required by the W-W model cannot be achieved. Firstly, if we consider a closed system in which sulfate reduction is a Rayleigh process, with $\alpha = 1.040$ (the approximate isotopic separation factor for bacterial reduction), we find that starting with sea water sulfate ($\delta = +20$), 82 per cent of the sulfate has to be reduced in order for the *mean* δ value for all sulfide produced to be +5 per mill; the remaining 18 per cent of sulfate in the water then has $\delta S^{34} = +90$ per mill; this clearly has not occurred. In the proposed steady-state system, the sulfate is replaced by diffusion from both sides, and by the advective flux of new brine flowing into the depression. A simple diffusion-advection model (one-dimensional) with an advective flux of brine upward, and diffusion between the 44°C brine and sea water, gives an approximate solution for the mean value δ_J of sulfate at steady state:

$$\delta_J = \frac{\delta_{SW}[C_J(C_B - C_{SW}) - JC_{SW}] - \epsilon J(C_{SW} - C_J)}{[C_J(C_B - C_{SW}) - JC_{SW}] + 10^{-3}\epsilon J(C_{SW} - C_J)} \quad (7)$$

in which C is sulfate concentration and the subscripts SW, J, and B denote the values in Red Sea deep water, the thin sulfate-reduction layer (which can be assumed to be approximately uniform), and in the 56° brine. J is the sulfate reduction rate divided by the advective flux of brine into the deep, i.e. the removal rate in net brine-flow units. The factor ϵ is $10^3(\alpha - 1)/(1 + R)$, where R is the S^{34}/S^{32} absolute ratio = 0.044. In the steady-state replenishment model, the isotopic composition of precipitating sulfide at any time has to be given by:

$$\delta_S = [\delta_J - 10^3(\alpha - 1)](1/\alpha) \quad (8)$$

The relationships in (7) can be fixed as follows: the value of J must not be significantly greater than the flux of Zn into the reduction layer, or most of the iron would precipitate as FeS, rather than predominantly as oxidized iron as observed. The net flux of Zn into the layer (in brine flow units as used for J) is simply its concentration in the 56° brine, $= 5.4 \times 10^{-3}$ when the concentration units are g/kg. The corresponding sulfate reduction rate $J = 7.9 \times 10^{-3}$ as a maximum value. With this value of J and $\alpha = 1.040$, the following values of sulfate concentration in the reduction layer, and δ of sulfate in the layer and of precipitating sulfide, are found:

C_J (g/kg) =	0	0.006	1.00	3.20 (= C_{SW})
δ_J (per mill) =	59.4	45.2	20.6	20.3 (= δ_{SW})
δ_S (per mill) =	18.6	5.0	−18.7	−18.9

Values for $C_J = 0.006$ are those for which $\delta_S = +5$ per mill as actually observed; the other values indicate the limits on C_J and the range of variation which could be encountered. It is seen that the sulfate in the reduction layer must be as heavy as +45 per mill, and the sulfate concentration as low as 6 ppm, for this process to be operating at present. These values are quite impossible, as shown by the data of Kaplan et al. (1969) on sulfate concentration and δ values above Atlantis Deep. The essential point is that at steady-state the sulfate in the reduction layer has to be higher in S^{34} by one single-stage separation factor, as shown in the above values (this is true even if advection is omitted from the model). Thus the proposed steady-state process cannot account for the sulfide isotopic composition.

Kaplan et al. (1969) propose that the sulfate in the brine is derived from marine evaporites, some of which reacts with organic matter in shales to produce H_2S and perhaps elemental sulfur, removing sulfate from the brine. The metals are carried in as sulfides and precipitated in the Atlantis Deep, in some manner not described. The sulfide-sulfate fractionation of about 15 per mill is due either to a high-temperature isotopic equilibrium or to the difference in reduction rates of the two isotopes (α is observed to be about 1.022 for the relative rates at 18–50°C). They suppose that the similarly anomalous sulfides ($\delta = +10$) found at the base of core 118K, outside of the brine depressions, represent a thermal overflow of brine into that locality at some earlier time. Further, they conjecture that the water in the brine may originate from dehydration of gypsum to anhydrite (though such water would actually have a much different isotopic composition than is actually observed).

Arguments against the proposal of Kaplan et al. include the following:

1. The brine cannot have been in contact with shales at the high tempera-

tures (300–500°C) they require for equilibrium sulfur fractionation—an oxygen isotope shift in the water, due to exchange with silicates and carbonates, would surely occur (Craig, 1963, 1965).

2. Sulfate has been removed from the brine, not added from evaporites. If the removal had been accomplished by sulfate reduction, the remaining sulphate in the brine would now be enriched in S^{34} with respect to sea water sulfate. As shown previously, 63.2 per cent of the sulfate has been removed from the original sea water; the fraction of sulfate remaining is 0.368 which happens to be $1/e$, so that in the Rayleigh process calculation, the delta value of the present brine sulfate should simply be:

$$\delta_{SO4} = \delta_{SW} + 10^3(\alpha - 1)$$

if the original δ value is that of sea water sulfate (+20.3). Thus for $\alpha = 1.020$, the δ value of the present sulfate should be about +40, and for $\alpha =$ as low as 1.005 it should be +25. However, the sulfate in the brine is isotopically identical to that in sea water and has been removed by a non-fractionating process.

3. If reduction took place by means of organic carbon, δC^{13} of the dissolved carbon should be very low, about -30 per mill or less, whereas the Atlantis Brine dissolved carbonate has a δ value of -5.6 per mill (Weiss, 1969).

4. Sulfides and sulfur of about the same isotopic composition as Atlantis Deep sulfides, with $\delta S^{34} = +10$, occur outside the brine depressions, as found in core 118K (Kaplan et al., op. cit.); this sulfur is not biogenic and occurs as ZnS, FeS, and sulfur, at 390cm in the core, well below the biogenic sulfur ($\delta = -30$) at 110–175cm in the core. The existence of reduced sulfur so similar to that in Atlantis Deep sediments, some 6km outside of the area of the brines, shows that sulfides of this isotopic composition are not uniquely products of hot brine introduction.

Kaplan et al. suppose that these sulfides and sulfur represent a thermal overflow of hot brine into the area of core 118K at some earlier time. However, this is demonstrably not so, as shown by the detailed study of Deuser and Degens (1969) on carbonates in this core. They show that pteropods at the 390cm level are similar to those of the same age in sediments 160km away, and that there is no isotopic shift in O^{18} and C^{13} in the carbonate overgrowths such as is observed in pteropods and foraminifera of the same age in Discovery Deep, and in fossils in Atlantis Deep. The C^{14} stratigraphy, from the measurements by Ku et al. (1969), show that the biogenic sulfides at 110–175cm in core 118K correspond to the time of rising sea level, about 11,000 years ago, when the Red Sea freshened again, as shown by the O^{18} data on the fossils. The 390cm level in core 118K corresponds to a time more than 20,000 years ago when Pleistocene sea level lowering had caused the Red Sea to become hypersaline. Two points concerning the sediments in core 118K should be emphasized: (1) the fossils are similar in O^{18} concentration to those of a core much farther away from the brine deeps, with no O^{18} shift such as is shown by Atlantis and Discovery Deep fossils, and (2) near the core bottom where the S^{34} enriched sulfides occur, casts of original shells and secondary overgrowths on other shells have the same δO^{18} values, indicating equilibrium with water which had surrounded them (Deuser and Degens, op. cit.). The isotopic evidence thus indicates very strongly that (1) the sediments of core 118K have not been exposed to thermal brines at any levels and (2), as Deuser and Degens point out, all available present evidence indicates that no brine was present in the deeps before about 10,000 years ago, a time much later than the deposition of the S^{34} enriched sulfides at the bottom of core 118K.

From the oxygen and carbon isotope data presently available, we have to conclude that deposition of S^{34} enriched sulfides and sulfur at least as heavy as +10 per mill took place in Red Sea sediments as a normal, non-hydrothermal, process which could occur in the absence of any geothermal brines. The similarity of the sulfides and sulfur in Atlantis Deep and the bottom of core 118K ($\delta S^{34} = +3$ to +11 per mill), and of the dissolved interstitial sulfate (+15 to +20 per mill, averaging about +18 in both areas) implies very clearly that sulfide deposition in both areas occurred by similar processes which did not require, but also were not inhibited by, the presence of a geothermal brine. This in turn implies that sulfur introduction took place by an agency independent of the brine. The most plausible and simple explanation is that sulfur was introduced as H_2S from surrounding sediments into both these areas, precipitating sulfides and sulfur from geothermal brine in Atlantis Deep and from high salinity sea water in the Core 118K area.

The cyclic O^{18} and C^{13} correlation with Pleistocene glaciation and deglaciation, as measured by radiocarbon stratigraphy, is almost certainly a response to alternating hypersaline and freely circulating conditions in the Red Sea, as proposed by Deuser and Degens. Core 118K thus indicates the following sulfide cycle: during times of sea level lowering, corresponding to the bottom part of the core with large O^{18} enrichment, the bottom waters of the Red Sea were so saline that biogenic sulfate reduction could not occur. H_2S introduction from sediments into the high salinity waters resulted in precipitation of sulfides and sulfur with δS^{34} about +10 per mill. When sea level rose again during glacial melting, and normal circulation was re-established about 11,000 years ago, micro-organisms could again live in the accumulating sediments and biogenic sulfate reduction produced sulfides with the characteristic δS^{34} of −30 per mill in the 110–175cm level of the core.

The sudden change in isotopic composition of elemental sulfur at the 175cm level of core 118K (Kaplan *et al.*, *op. cit.*, Fig. 3) is strikingly parallel to the O^{18} and C^{13} depth profiles shown by Deuser and Degens (*op. cit.*, Fig. 2):

Depth (cm)	δO^{18} (per mill)	δS^{34} (per mill)
90	0.3	−31
100	0.6	−15
175	6.5	+8
390	5.5	+10

(the O^{18} values are those measured on the pteropod shells). Sulfides at the 175cm level are similar to those above, with $\delta S^{34} = -30$; at 390cm δS^{34} is +10, with no measurements in between. Deuser and Degens (*op. cit.*) point out that this transition layer consists of highly fractured and fragmented shells, indicative of turbulent sediment transport; a similar layer is observed at the same depth much farther south. The biogenic sulfides in this layer may therefore represent transported material; alternatively, biogenic reduction of sulfate and inorganic production of elemental sulfur may have persisted together for some time as the Red Sea freshened.

The slightly lower S^{34} content of dissolved interstitial sulfate in both Atlantis Deep sediments and core 118K, relative to sea water and brine, indicate that brine and sea water sulfate have not been involved in the sulfide precipitation process at all, but that some oxidation of lighter sulfides and sulfur has taken place *in situ*, after sulfide deposition, as suggested by Kaplan *et al*. The somewhat lighter sulfides in Atlantis Deep sediments, averaging +5 per mill vs. +10 per mill in core 118K, may indicate that the introduced H_2S, probably generated by diagenesis in underlying sediments, was lower in S^{34} than the sulfides formed from it, and that a larger fraction of the sulfur was precipitated in Atlantis Deep because of the much higher heavy metal content.

Still unexplained is the uniformly low dissolved sulfate concentration in the supposedly pre-brine levels of Discovery Deep sediments and in core 118K, of the order of 1g/kg, similar to that in the geothermal brines and Atlantis Deep sediments. The

Discovery Deep concentrations might be explained by diffusive interaction with overlying brine, but if the carbon and oxygen isotopic data correctly indicate that no brine has been present in core 118K, the low dissolved sulfate concentrations in this core are difficult to understand. Unfortunately no chlorinity determinations are available on the interstitial waters in Discovery Deep sediments and core 118K — without these data it is impossible to explain the sulfate concentrations and to be certain of the history of these sediments.

The Removal Process for the Subtracted Salts

The absence of S^{34} enrichment in the brine sulfate, and the absence also of a large C^{12} enrichment in the dissolved carbonate, indicate, as we have seen, that sulfate was not removed from the original water by reduction processes; nor do the sulfides in the sediments require any such origin. Similarly, it is difficult to account for the magnesium depletion in the brines by Mg fixation in dolomite or silicates such as chlorite: in such processes there would surely be oxygen isotope exchange between the water and the carbonates or silicates and an oxygen isotope shift would be observed in the water (Craig, 1963). Such exchange takes place in times of weeks for carbonates and feldspars at 100°C.

One process which could produce the observed depletion pattern of Mg, SO_4, I, and F, is that of membrane filtration, in which material is removed by selective filtration through clay minerals. White (1965) has recently reviewed the known and most probable characteristics of this process and has pointed out that Na^+, HCO_3^-, F^-, I, and B appear to be the most mobile components through such "membranes," followed by Cl^- and Mg^{++}; other components should be much less mobile. Neutral species and small, singly charged ions display the highest mobility, and White considers that I_2, H_3BO_3, and undissociated $NaHCO_3$ migration probably account for the mobility of these species.

Since it is known that an appreciable fraction of Mg and SO_4 exist as undissociated $MgSO_4$ in solutions as low in these components as sea water, selective loss of $MgSO_4$ by membrane filtration could account for the depletion of both these species in the stoichiometric proportions observed. About 10 per cent of the Mg ions in normal sea water are associated with SO_4 (Pytkowicz and Gates, 1968, and references therein) so that significant loss of uncharged $MgSO_4$ should occur in the filtration process. Sulfur isotope fractionation would probably be insignificant, because the sulfur atom is surrounded by oxygen atoms and probably has little interaction with the potential barriers involved in penetration of the membrane. The very modest boron enrichment in the brines can also be explained by a significant depletion in the filtration process subsequent to an initial higher enrichment.

The relative enrichment pattern in the bottom part of Table 4, is, with the exception of Li, therefore almost a classical pattern of expectation for the membrane filtration process. Lithium may actually be more highly enriched in the brine than indicated by the value of Brewer et al. (1965) used in Table 4; M. Peterson (personal communication, 1968) has measured concentrations of 4.5 and 4.0 ppm Li in Atlantis II and Discovery brines respectively, and 1.7 ppm in the 44°C brine layer of Atlantis Deep. These values are completely consistent with the mixing relations vs. chloride shown in Figs. 1 and 2, and, if correct, indicate that the enrichment factor for Li in the brines should be 28, rather than 1.6 as given in Table 4, making Li the most highly enriched element below the heavy metal group.

Salt and Water in Space and Time

Deuterium-O^{18} relationships in the waters of the Red Sea and the brines show, as we have seen, that water in the brines is isotopically identical to normal, present-day, Red Sea water of 38.2 per mill salinity. Although the data plotted in Fig. 3 show that the brine has about the median isotopic composition found in the Red Sea, it is important to note that this composition is by no means an *average* composition of Red Sea water; as discussed earlier, waters of equal or lower D and O^{18} concentrations ($S \leqslant 38.2$ mill) occur only in the southern part of the Red Sea, 400 to 800km south of the brine depressions in the region adjacent to and including the southern sill. There is no possibility that water of this composition can now be entering the recharge system on the flanks of the Red Sea adjacent to the brine depressions; all waters less than several hundred kilometers south of this latitude are already higher in D and O^{18} at all depths due to the progressive enrichment of inflowing water by evaporation and molecular exchange.

The southern sill at 13°41′N, is 850km south of the brine deeps, or just about 1,000km SE along the Red Sea axis. Water of the isotopic composition found in the brine corresponds closely to the mean water over the sill during an annual cycle (Fig. 5), and can be found in significant amounts up to about 430km SE of the brines, either as 38.2 per mill salinity water or as a virtual water which can be formed from direct mixing or annual cycling between more and less saline components. Such direct or seasonal mixtures are limited to depths of less than 200m within this region, as the water is always more saline than 40.2 per mill and higher in D and O^{18} below this depth. The sill depth itself is about 100m, and the 38.2 per mill isohaline lies about 40m above the sill in winter, and at levels of 60m above and 30m below the sill in summer, enclosing a core of lower salinity water. In summary, then, the brines occur about half-way along the length of the Red Sea, and water of similar isotopic composition can be found or formed in at most the southern one-third or one-fourth of the Red Sea, within a layer less than 200m below the surface.

Within this area Tertiary evaporite deposits are abundant both below the surface and as salt domes exposed on islands and along the shores (Girdler, 1958; Drake and Girdler, 1964; Heybroek, 1965). The seismic section measured on the southern sill itself by Drake and Girdler (*op. cit.*) shows some 2,500m of 3.7km/sec velocity material attributed to evaporites, lying between the crystalline basement below and some 250m of unconsolidated coral sand above—this section is shown in Fig. 6. Malone (discussion appended to paper of Girdler, 1958) has described the borehole on the island of Dahlak (15°N) which penetrated 2,000m of evaporites without ever leaving salt. Both the sill area and the flanks of the trough are extensively faulted and provide easy access for sea water to descend down into the salt.

In Fig. 6 some approximate isotherms have been marked in the section, assuming an average heat flow of 4×10^{-6} cal/cm² sec and conductivities as shown in the figure, and neglecting boundary effects. Since the sill depth falls off rapidly to the north and the brine deeps are some 2,000m below the surface, sea water can be heated to something like 100°C by the regional geothermal gradient which is more than sufficient to account for the observed brine temperatures, with no necessity for invoking any direct volcanic heating; these relatively low temperatures are consistent with the lack of an oxygen isotope shift in the water.

The minimum flow distance from source to brine depression is some 350km, for an origin at 18°N on the western flank of the Red Sea. The maximum distance is about

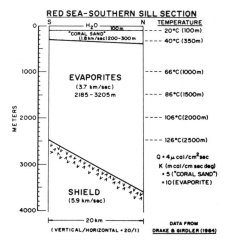

Fig. 6. Seismic section across the southern sill. Temperatures are calculated for the assumed heat flow and conductivity values shown, assuming no boundary effects.

1,000 km northwest from the sill. The driving force for the flow is the density difference between 2,000 m columns of essentially saturated brine and normal sea water, equivalent to about 40 atmospheres of pressure, so that the pressure gradient during the flow is of the order of 0.04 to 0.10 bars/km. At pressure gradients of 0.04 bars/km (corresponding to the 1,000 km distance), petroleum flow rates of 75 km/yr have been observed by Baker (1955) in fissure flow through rocks with fissures of about 0.25 cm by 300 m. Although viscosity and temperature effects are important, the flow velocity varied as the square of the fissure diameter, so that other things being equal, corresponding flow velocities of 1 km/yr could be attained in fissures of only 0.03 cm and a gradient of 0.04 bars/km. Although water is not petroleum, and such estimates are very crude, they indicate that even with fissure volumes of the order of 0.01 per cent of enclosing rock, there is no insuperable difficulty in carrying as much as 10^6 tons of brine per year even 1,000 km through a section 10 km by 300 m of rock with a travel time of about 1,000 years. Possibly the distances and velocities are less and the times somewhat longer, but the fact remains that unless water at some earlier time had the present isotopic composition of the brine at considerably higher latitudes, the source is at least 350 km to the south and more likely two to three times this distance, and the travel time must be significantly less than 10,000 years, according to the chronology of Ku et al. and Deuser and Degens. With a brine flux of this magnitude the salt addition to the Red Sea is only 10^{-5} of that added by the annual evaporation, so that even much greater fluxes would not be noticeable in the salt balance (Craig, 1966).

We must now ask about the history of the pattern of isotopic composition in Red Sea waters and in the ocean in general. A detailed consideration of the effects of the formation and melting of continental glaciers during the Pleistocene glaciation cycle, based on the new IGY information from Antarctica on ice thickness, indicates that average ocean water probably fluctuated between the limits marked for $SMOW$ by crosses in Fig. 3, between times of maximum ice formation and complete melting of continental ice (if complete melting in ice actually occurred)—(Craig, 1966). The indicated maximum range of δO^{18} in $SMOW$ is from about -0.5 per mill with complete melting of all present-day ice, to about $+1.0$ per mill during glacial maxima, which corresponds to a sea level lowering of about 200 m below the present level. The dashed lines marked "glacial" and "interglacial" in Fig. 3 are D-O^{18} Red Sea slopes originating from these points of maximum oscillation.

Curray (1960), in reviewing all the sea level chronologies based on radiocarbon dating, concludes that during the past 20,000 years of emergence from a glacial period, sea level has been rising at an average rate of about 8.6 m per 1,000 years. (This rate is equivalent to somewhat less than 1 per cent of the annual precipitation rate over the sea, so that the ocean is always about as well mixed isotopically as it is today.) Since the southern sill depth is about 100 m, assuming the topography has not changed, the Red Sea should have been isolated when sea level fell 200 m, and according to Curray's chronology, exchange of water with the Gulf of Aden should have been established again at $100/8.6 =$ approximately 11,600 years ago. This time is in striking agreement with the time of the 175 cm transition from very high δO^{18} values in $CaCO_3$ in Red Sea sediments to normal values, established by the radiocarbon measurements of Ku et al. (1969) on the Deuser-Degens cores as about 12,000 years ago, and is surely not a coincidence (Deuser and Degens, 1969, Fig. 2).

The question of the isotopic composition of the Red Sea water during its isolation from the open sea by sea level depression below the sill is difficult to discuss theoretically. Craig and Gordon (1965) have thoroughly discussed the evaporation-exchange isotope effects for the ocean, for isolated water bodies, and open systems, but the Red Sea is an awkward size—intermediate between an ocean which controls its own environment, and a small water body which responds to the atmospheric vapor characteristics imposed by the ocean. However, the carbonate data of Deuser and Degens indicate that during the times of isolation of the Red Sea, δO^{18} of the water (at least *some* water) increased to about +7 per mill, and from this empirical value, certain interesting points ensue. Most importantly, the Red Sea did not evaporate down as a closed body of water with no water inflow at all; if it had, evaporation would have ceased when the activity of water equalled the mean relative humidity, and the continuing molecular exchange of water vapor would have brought the water to approximate isotopic equilibrium with atmospheric vapor, so that the water would have been isotopically similar to local precipitation (Craig, 1966b). Since this is not in accord with observations, there must have been a continual input of precipitation and runoff, and, as the activity of the water decreased due to increasing salinity, the evaporation rate decreased until a new steady state for input-evaporation and isotopic composition was attained for the isolation period. The relevant equation describing the situation is obtained approximately by replacing the humidity h by h/a, where a is the steady state activity of water, in equation (32) of Craig and Gordon for their open-system model, as discussed by them (*op. cit.*, p. 65). In this way one can correlate the average humidity, activity, and isotopic composition of fresh water inflow and of atmospheric vapor, and one can see that the empirical value of +7 per mill for δO^{18} is not unreasonable for the water although there are too many variables hanging free for a detailed model. Perhaps more indicative are the observations of Craig *et al.* (1963) that water evaporating into marine vapor reached a steady-state composition of $\delta O^{18} = +6$ to +7 per mill, of Longinelli and Craig (1967) that open systems such as the Dead Sea and W. Coast saline lakes have δO^{18} values of about +5 per mill, and that the present lakes and rivers of E. Africa evaporate to δO^{18} values of +6 per mill (Craig, 1963) — these values are, of course, independent of the initial isotopic composition of the evaporating water. The subject is admittedly difficult, tedious, and not well-advanced, but, to sum up, enough is known at present to indicate rather clearly that (a) the Red Sea did not evaporate away, or reach a constant activity of water as an isolated body, and (b) the empirical value of +7 per mill for δO^{18} of the water, from the core carbonates, is a very reasonable figure for the value characteristic of periods when sea level fell below the sill and the Red Sea was cut off from the open ocean.

When sea level rose again due to melting of the ice sheets, the D/H and O^{18}/O^{16} ratios in sea water decreased because of the great depletion of these isotopes in precipitation stored in the ice sheets. At the time when sea level reached the sill, 100m below present level, *SMOW* would have had a δO^{18} value of about +0.5 per mill relative to the present zero value, due to melting of half the glacial increment of continental ice (Fig. 3). At this time, if oceanic circulation conditions were not drastically different from those of today, the entering water from the Gulf of Aden would have had a δO^{18} value of about +1.1 per mill, a salinity of about 37 per mill, and two opposing trends would have begun to affect the isotopic composition at the south end of the Red Sea: (1) entering water would have been slowly decreasing in δO^{18} from 1.1 to the present value of 0.6 per mill due to continued melting of ice, and (2) deep water to the north would have decreased in O^{18} and salinity due to the influx of sea water, and at some time later its density would have decreased to the point at which it could have begun to mix out over the sill at certain times. The Ku *et al.* chronology indicates that by 9,000 years ago, the water in the central section of the Red Sea had been essentially flushed to the isotopic composition and salinity it has today (Deuser and Degens, *op. cit.*, Figs. 2 and 5). Prior to this time, the proportion of deep water in the outflow over the sill would have been increasing as its density, salinity, and O^{18} concentration were decreasing by dilution so that it could mix with the near-surface waters. Thus the oxygen isotopic composition of the upper few hundred meters of water in the southern end of the Red Sea probably went through a highly damped, slight increase in δO^{18} relative to the surface inflow during the first few thousand years after the water flowed in, perhaps of the order of 1 or 2 per mill, and then declined to its present value of +1.2 per mill.

These considerations indicate that for a long period before the flow of brine into the deeps, and also since that emplacement began, there is no evidence that water in the immediate vicinity of the brine depressions could ever have been as low in deuterium and oxygen-18 as the water from which the brine has been made. The evidence from the oxygen-18 and radiocarbon stratigraphy in the Red Sea sediments therefore requires an origin of the brine far to the south, with the long flow trajectories of some 400 to 1,000km under the Red Sea floor before emergence into the deeps off Mecca.

The Glacial-Control Mechanism for Brine Flow

The isotopic evidence that the brine originates in near-surface waters at the south end of the Red Sea — on or adjacent to the southern sill — has the following consequence: during periods of continental glaciation when sea level falls below the sill and the level of the Red Sea drops, no water is available to form brine, and no flow will occur. As we have seen, the isotopic composition of the brine water indicates that it originates within the upper 200 — and probably upper 100 — m of the southern waters; with present evaporation rates the Red Sea would drop this much in 50–100 years after sea level falls below sill depths, and considerably more before the decreased evaporation rate could be balanced by precipitation and runoff input. The oxygen isotope shift evidence from the work of Deuser and Degens on carbonates, indicating that brine was not present in the depressions before about 10,000 years ago, is thus completely consistent with the glacial-control mechanism proposed here: brine could not begin to flow *until* sea level rose and water covered the area where the flow trajectory begins — some 100m below the present level of the Red Sea. The correspondence between time of brine emplacement and overflow of sea water across the sill, so strikingly shown by the Ku chronology, is thereby explained as a natural consequence of the location of the point of origin of the brine.

Two further consequences of the glacial-control mechanism are interesting. Brine may have flowed during interglacial periods previous to the present one, so that the deep sediment record may show one or more layers of heavy metal deposition *below* the present deposits, each separated from the others by a hiatus marking a glacial period when no brine flow occurred. Secondly, when the level of the Red Sea falls after sea level drops, the flowing brine has a greater hydrostatic head and can overflow the depressions until the aquifer is drained to the new level of the Red Sea — such overflow may have contributed very significantly to the heavy metal deposits. Thus it will be important to obtain long sediment cores from the brine depressions to determine if sizeable metal deposits have been formed below the last glacial hiatus.

A Steady-state Model of Atlantis II Deep

The recent surveys by Hunt and his co-workers indicate that Atlantis II Deep may contain as much as 3×10^9 tons of brine, or 7.5×10^8 tons of salts, in a volume roughly 3km $\times$ 6km $\times$ 150m. From the isotopic data on the water and the glacial-control chronology discussed in the last section, we can estimate that the hold-up or residence time of brine in the depression might be as long as 3,000 years, but not much longer. With these values, the lower limit for the flux of new brine into Atlantis II Deep is of the order of 10^6 metric tons of brine/yr, corresponding to 2.5×10^5 tons of salt/yr. (For comparison, evaporation contributes about 3.6×10^{10} tons of salt per year.) This flux of new brine flows from the 56° layer in the bottom of the deep, to the lower salinity 44° convective layer above, and is then mixed out to the Red Sea. Assuming that a rough approximation to steady-state has been reached, the dynamics and chemical characteristics of the system resulting from this steady input will now be discussed.

The bottom of Atlantis II Deep is filled with 135m of 56°C brine with a chlorinity of 155.5g/kg; the temperature and salinity are approximately constant with depth (Pugh, 1967). Above this layer, from 2,040–2,010m, is a 30m thick layer of 44°C, chlorinity 80g/kg, brine. Both layers appear to be uniformly mixed by active convection cells, and the transition between them is so sharp that no adequate data are available on the shape of the gradient. Above the 44°C layer there is a transition zone of about 30m separating it from normal Red Sea deep water with a salinity of 40.6 per mill. The structure of this transition zone is shown in Fig. 7, in which measurements of Na concentration and temperature, made by M. Peterson on the WHOI CHAIN 61 survey, are shown. (The Na data, which were volumetric, have been converted to g/kg by an approximate correction factor from the sea water and brine densities, assumed to vary as the reciprocal of density between these limits—possible deviations are too small to affect the shape of the gradient shown.) Measurements made by Peterson on Ca, Mg, and K (all as g/kg directly) show identical gradients, as would be expected from the relationships in Fig. 1 which show that all these elements are conservative. Sodium data are used here because the analytical scatter is much smaller than for the less abundant elements. The data in Fig. 1 are a compilation of all measurements on six different casts (Atlantis 715, 718, 722-1, 722-2, 726, and 727) and some of the scatter is certainly due to the difficulty of correlating depths within such a narrow interval on many different casts.

Mixing at the Brine-Sea Water Interface

In principle, it might be expected that mixing between the 44°C brine and overlying sea water would be a simple one-dimensional diffusion + advection process, in which molecular or turbulent diffusion operates vertically, and the influx of new brine into the 56° bottom layer is compensated by a net upward advective flow from the 44°C layer into normal sea water, i.e.

$$\kappa C'' = wC' \qquad (9)$$

in which κ is the diffusion coefficient, w the advective velocity, and the primes denote successive derivatives of concentration, C, with distance z (positive upwards). However, the Na gradient in Fig. 7 is exactly upside-down for such a process. The dashed line shown in the figure has been calculated from eqn. (9) with $\kappa/w = -4$m (advection down), for boundary conditions Na = 12.5g/kg at 1,980m and 46.9g/kg at 2,012m. That is, the linear gradient which would be observed for pure diffusion is displaced asymptotically downward from the upper boundary, so that in a one-dimensional model the advection must be downward. Since it is highly unlikely that sea water is being advected into the 44° brine from above we must conclude that the gradients in Fig. 7 reflect a two-dimensional mixing process in which a bottom current is flowing along Atlantic Deep above the 44° brine, removing the salts mixed into it by diffusion and (upward) advection. The profiles in Fig. 7 also characterize a two-dimensional process of this type in which horizontal flow has a quasi-parabolic velocity profile, falling off from the turbulent core of the flow towards the interface.

The temperature profile over Atlantis II Deep falls off more slowly than the salinity profile; Fig. 7 shows that temperatures at all depths in the transition region lie above the dashed line, which is calculated for the same one-dimensional mixing length used for the sodium-curve. If the salt diffusion is molecular, rather than turbulent, the thermal diffusivity in this system is about 100 times greater than the molecular diffusivity, so that a characteristic mixing length of the order of meters for salt would be of the order of hundreds of meters for temperature, and the temperature profile should be expected to be very close to linear unless

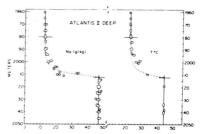

Fig. 7. Sodium and temperature profiles above the 44°C brine layer in Atlantis Deep, from data of M. Peterson (personal communication, 1968).

the horizontal advection rate is very great. More likely, the temperature profile represents an additional heat contribution by conduction from the surrounding sediments to the bottom current.

At the top of the transition zone (1,980m) salinity and temperature increase downward at a rate $dS/dT = 2$ per mill/°C, more than 5 times greater than the ratio of about 0.37 required for constant density; consequently the density increases rapidly, the stability is very high, and a convection cell does not develop above the 44° layer. However, above the hot sediments surrounding the brine depression there are probably rising convection currents in the normal bottom water, with downward flow of sea water above the brine itself and horizontal currents just above the interface. Diffusive mixing, probably turbulent but possibly molecular near the interface, takes place between the 44° layer and the horizontal bottom current, and in addition, in the steady-state model there is an advective flux out of the 44° layer equal to the influx of new brine into the 56°C bottom layer. A net brine influx of 10^6 tons/year would require an advective velocity of about 25×10^{-8} cm/sec out of the 56° layer (dimensions roughly 2×6km), and about 5×10^{-8} cm/sec out of the 44° layer (roughly 5×12km).

The 44°C brine layer is a mixture in almost equal proportions of 56° layer brine and normal Red Sea Bottom water, as demonstrated in Figs. 1 and 2, forming an independent convection cell which receives high-salinity brine from the 56° cell below, and mixes with the overlying sea water by diffusion. In this process there is a net advective flux of brine through the layer and the relative proportions of non-conservative brine salts are altered by chemical reactions, as discussed in the following section.

Steady-state Chemical Balance in the Atlantis II 44°C Brine Layer

The steady-state material balance for a component in the 44°C brine layer can be written as:

$$J_i = C_B - C_{44} - \phi_{SW}(C_{44} - C_{SW}) \quad (10)$$

in which C is the concentration of component i in 40.6 per mill salinity normal Red Sea bottom water (subscript SW), in 56°C Atlantis brine (subscript B), and in the 44° mixing cell (subscript 44). ϕ_{SW} is the total flux of sea water diffusing in from above, in units of the advection rate of new brine into the 56° layer, and J_i is the removal rate of component i from the 44° layer by any process other than diffusion and advection to the overlying sea water, also in 56°C brine-flux units. (The units of J_{Fe} are, e.g., (gFe/yr)/(kg 56° brine/yr) = g/kg when the concentration units are g/kg.) Thus, on the right-hand-side of (10) the first two terms are the advective flux of component i into, and out of, the 44° layer, respectively, and the last term is the net diffusive flux out to the sea water, all in units of the advective flux of new brine. The summation of the J_i is so small that the total advective flux from 44°C brine to sea water is essentially 1 in 56° brine-flux units, and ϕ_{SW} is also the total diffusive flux of material out to sea water.

For conservative elements (Figs. 1 and 2) $J_i = 0$, and ϕ_{SW} is obtained from the chloride concentrations (155.5, 80.0, and 22.5g/kg), as

$$\phi_{SW} = 1.3130 \quad (11)$$

in 56° brine-flux units. The steady-state chloride balance in the 44° cell (56° brine-flux units) is thus: 155.5g/kg advected in from below, 75.5g/kg as the net diffusion loss out to sea water, and 80.0g/kg advected out to sea water. For conservative elements ($J = 0$) we have

$$C_{44} = \frac{C_B + 1.313\ C_{SW}}{2.313} \quad (12)$$

as the steady-state concentration, and for trace elements of very low concentration in sea water, the steady-state concentration is simply:

$$C_{44} \rightarrow 0.432 C_B \quad (13)$$

as a criterion for $J = 0$. The principal elements for which J is clearly non-zero are Fe and Mn, and the various data on these elements are in good enough agreement to justify a detailed consideration.

Iron. The sediments in Atlantis II Deep contain precipitated iron as ferrous carbonate (siderite) and in the ferric form as hydrated iron oxides ranging from amorphous forms to goethite ($Fe_2O_3 \cdot H_2O$). Miller et al. (1966) proposed that iron oxides are precipitated by oxygen in the overlying sea water during pulses of brine, while Watson and Waterbury (1969), in their steady-state model for sulfide precipitation discussed in an earlier section, suppose that ferrous ions diffuse through their sulfate reduction layer and are continually oxidized and precipitated in the overlying oxygenated water as ferric hydroxide. On the other hand, Brewer and Spencer (1969) have proposed that iron is precipitated as $Fe(OH)_3$ at the 56°–44° interface.

Brewer and Spencer (op. cit.) report the Fe concentrations in Atlantis II brine as $C_B = 81$ppm (56° brine), $C_{44} = 0.20$ppm (44° brine). Peterson (personal communication) finds $C_B = 90$ppm, $C_{44} \approx 0.9$ppm (ranging from about 0.75 to 1.5); these measurements are similar enough that the calculations to follow are not seriously affected by the choice of data, and we shall use the Brewer and Spencer values. The concentration in normal sea water is less than 0.02ppm (Brewer et al., 1965), so that equation (13) shows that Fe is strongly non-conservative in the 44° layer. From equations (10) and (11) with the Brewer-Spencer data we find:

$$J_{Fe} = 80.5\text{ppm/kg}$$

as the removal rate of iron by precipitation *within* the 44° layer (56° brine-flux units). The flux of iron from 44° layer out to sea water by mixing, that is Fe which can be precipitated above the 44° interface, is thus only 0.5ppm/kg, so that essentially all the iron must be precipitated *within* the 44° layer.

Ferrous iron can be oxidized to the ferric state, and precipitated as ferric hydroxide, by the dissolved oxygen mixed in from overlying sea water:

$$Fe^{++} + \tfrac{1}{2}O_2 + 2H_2O + e^- \rightarrow Fe(OH)_3 + H^+ \quad (14)$$

We may assume that *all* the oxygen brought in by mixing is used to precipitate $Fe(OH)_3$. The dissolved O_2 concentration in Red Sea bottom water in this area is 2.5cc/kg (Neumann and McGill, 1962), and C_B and C_{44} will be zero, so from equation (10):

$$J_{O_2} = 3.28\text{cc/kg}$$

is the removal rate for O_2 by precipitation in the 44°C layer. Equation (14) shows that 1cc of dissolved O_2 corresponds to precipitation of 0.005g Fe, so that the maximum precipitation flux for Fe by reaction with dissolved oxygen is:

$$J_{Fe}(O_2) = 16.4\text{ppm/kg}$$

or only 20.4 per cent of the total removal rate of iron from the 44° layer. That is, 79.6 per cent of the precipitating iron must be removed without oxidation from dissolved O_2 mixed in from sea water. The total steady-state iron balance is as follows (fluxes in ppm/kg, fractions as percentage of the total Fe flux in and out of the 44°C layer):

In to 44° layer from 56° layer:
    81.0 (100 per cent)
Out to sea water by diffusion-advection:
    0.5 (0.6 per cent)
Precipitated as $Fe(OH)_3$ in 44° layer:
    16.4 (20.3 per cent)
Other precipitation in 44° layer:
    64.1 (79.1 per cent)

It is evident that another mechanism for Fe precipitation must be found, if the steady-state model is to work. Fe as well as Zn, Pb, and Cu, could be precipitated as sulfides in the 44° layer, by mixing in from sea water a small amount of reduced sulfur from biogenic sulfate reduction above the interface. But, as discussed in an earlier section, the sulfides should then have the biogenic δS^{34} values because only a small fraction of the sulfate can be reduced (since it does not show S^{34} enrichment); moreover, as Watson and Waterbury point out, FeS is not observed in the sediments, and it is highly unlikely that pyrite could precipitate directly in such a process.

At present, the most likely process seems to be precipitation of $FeCO_3$, since siderite is actually observed in the sediments, together with $MnCO_3$ (rhodochrosite). A precipitation flux of 64.1ppm/kg of iron as $FeCO_3$ requires $J_{HCO_3} = 70.0$ppm/kg as the removal flux of dissolved carbonate from the 44° layer. Using $C_B = 143$ppm for 56° brine (Weiss, 1969), and assuming $C_{sw} =$

135 ppm for Red Sea bottom water, as estimated in an earlier section, for the required value of J_{HCO_3} we find, from equation (10), $C_{44}(HCO_3) = 108$ ppm. This is to be compared with the value for $J = 0$ (no precipitation of carbonate): $C_{44}(HCO_3) = 138.5$ ppm. Therefore the $FeCO_3$ precipitation hypothesis can be tested directly, as a distinct minimum in the dissolved carbonate concentration should be observed in the 44° layer, relative to the brine and (actual) sea water concentrations, if this mechanism is responsible for the bulk of the iron precipitation.

Deuser and Degens (1969) measured a profile of C^{13} variations in dissolved carbonate through the brine; unfortunately the concentrations of carbonate were not measured. If we use their isotope data ($\delta_B = -4.2$, $\delta_{44} = -2$, $\delta_{SW} = +0.8$ per mill), then with $J_{HCO_3} = 70$, and with the isotopic analog to equation (10) we find $(\alpha - 1)$ for the precipitation process = 2.6 per mill. Weiss (1969) has shown that their brine data are probably affected by loss of CO_2 during sampling, as he finds lighter values for the brine ($\delta_B = -5.6$) as would be expected if their samples have lost CO_2. Using his value for δ_B and $\delta_{44} = -3.4$ per mill as an approximate correction, we obtain $(\alpha - 1) = +6.1$ per mill. Thus the isotopic fractionation factor for carbon probably lies somewhere between 1.0006 and 1.006, according to the Deuser-Degens isotope data and the estimated value for HCO_3 concentration in Red Sea bottom water, a range which is easily consistent with our present rather scanty knowledge of the expected factors (Thode et al., 1965; Deuser and Degens, 1967; Wendt, 1968).

Let us assume $\alpha = 1.004$ for carbon fractionation in the precipitation process (C^{13} concentrating in $FeCO_3$). Using the Deuser-Degens data for δ_B and δ_{SW}, we calculate for $J_{HCO_3} = 0$, $C_{44} = 138.5$ ppm, that $\delta_{44} = -1.4$ per mill. On the other hand, for $J_{HCO_3} = 70$, $C_{44} = 108$, we find the steady-state $\delta_{44} = -2.3$ per mill. That is, the difference is of the order of the scatter in their data, and one sees that the differences in carbonate *concentration* are a much more sensitive test of $FeCO_3$ precipitation than are the isotopic data. Of course the actual concentration effects have to be compared with actual measurements on Red Sea bottom water.

Much of the iron in Atlantis II sediments is present as amorphous iron oxides of various oxidation states, which is probably transforming to goethite; in addition siderite and pyrite are present in considerable amounts. The observations of Kaplan et al. (1969), that oxidation of sulfur and sulfides is taking place *in situ*, as demonstrated by local increases in SO_4 concentration with concomitant S^{32} enrichment, have been described earlier. Possibly $FeCO_3$ is being oxidized to ferric oxides by *in situ* mechanisms such as use of carbonate ion for oxidation of sulfur and H_2S,

$$CO_3^- + H_2S + H_2O = SO_4^- + CH_4 \quad (15)$$

with oxidation of ferrous iron by SO_2 produced as an intermediate. Also $FeCO_3$ may react with H_2S to form FeS, and then FeS_2 by interaction with elemental sulfur. It is evident from Kaplan's work that very complicated sulfur chemistry is going on in the sediments, involving all sulfur valences, so that it is difficult at present to decipher the various oxidation-reduction reactions affecting the ferrous-ferric balance.

Iron Recycling. If $FeCO_3$ is being precipitated in the 44° layer, some of the precipitate may settle into the 56° brine, redissolve, and be transported back to the 44° layer. In this case the observed HCO_3 and Fe concentrations in the 56°C brine layer are actually higher than those in new brine entering the reservoir, rather than being equal to the entering brine concentrations as assumed until now. Let us assume that the entering brine actually has lost or gained no dissolved carbonate relative to the original 38.2 per mill sea water with its (estimated) concentration of 125 ppm; then the entering brine concentration of HCO_3 is $125/1.295 = 96.5$ ppm, and the recycle flux of HCO_3 into the 56° brine layer is $143 - 96.5 = 46.5$ ppm/kg. With $J_{HCO_3} = 70$ ppm/kg in the 44° layer, this means that 46.5/70 or ⅔ of the precipitating carbonate dissolves in the 56° brine and recycles, and ⅓ is deposited directly in the sediments. Also ⅔ of the Fe is recycled into the 56° brine, so that of the 79.1 per cent of the Fe flux listed above as precipitating by other means, 52.5 per cent of the total flux is recycling and 26.6 per cent is depositing in the sediments. For these values, the new brine entering the reservoir has an Fe concentration only 47.5 per cent of that observed in the 56° layer, i.e., 38.5 ppm *vs.* 81 ppm in the 56° layer itself, representing the enrichment due to recycling. These processes have to be considered when chemical reactions occur in stacked convection cells, and one is comparing the trace element concentrations with those of other water types as a criterion of origin.

Manganese. A Mn-Ca variation diagram is shown in Fig. 8 for both brines, using the recent data of Peterson from the WHOI CHAIN 61 Survey; both elements were measured directly as mg/kg. The Discovery brine data indicate considerable scatter, but show essentially a mixing relationship between the depleted Discovery brine and sea water. In Atlantis Deep, however, the 44° brine Mn concentrations are about twice as high as the value of 37 ppm which would be expected for steady-state mixing between the 56° brine and sea water. Peterson's Mn data ($C_B = 86 \pm 4$, $C_{44} = 79 \pm 2$, and 56 ppm in Discovery brine) agree very closely with those of Brewer and Spencer (1969), (82, 82, and 54.6 ppm respectively), and the data for Atlantis II waters appear to be quite precise. The Ca data of Peterson on all brines are in exact agreement with those of Brewer and Spencer, but in Red Sea bottom water Peterson finds Ca = 432 ppm *vs.* the value of 471 ppm measured by Brewer *et al.* (1965) which is essentially the expected value from data on open ocean waters.

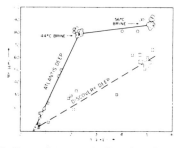

Fig. 8. Mn *vs.* Ca concentrations in waters of Atlantis and Discovery Deeps (data from M. Peterson, personal communication, 1968).

Using the Brewer-Spencer data in equation (10) we find $J_{Mn} = -108$ ppm/kg (Peterson's data give -97), indicating an *addition* of Mn to the 44° layer, from sources other than sea water or influx from the 56° brine. The steady-state balance is as follows (56° brine-flux units and percentages of total flux out to sea water):

In to 44° layer from 56° layer:
    82 (43 per cent)
In to 44° layer from other source:
    108 (57 per cent)
Out to sea water by diffusion-advection:
    190 (100 per cent)

The input flux to the 44° layer can conceivably come from direct solution of manganese minerals in contact with the 44° layer, or from recycling of Mn, as in the discussion of iron recycling, but in this case by recycling of Mn precipitated in the overlying sea water back into the 44° layer. If such recycling is occurring, then Mn should appear non-conservative in the profile through the Atlantis transition zone above the 44° layer: this profile, from Peterson's CHAIN 61 data, is shown in detail in Fig. 9, with Mn plotted both against Ca and against the (corrected) Na data, with all points from Peterson's 6 Atlantis casts with Mn values between 44° brine and sea water concentrations plotted. (Three Ca points, at Ca = 0.61, 0.80, 0.95, were reported with Ca concentrations higher than shown by a factor of 2; they have been renormalized for Ca from a Ca-Na plot.) The unenclosed solid points at the low concentration extremes are those for which Peterson reported a manganese upper limit only, i.e., Mn < 1 ppm. The lines in Fig. 9 are least squares fits to the enclosed points, and also include the mean 44° brine concentrations (Mn = 79 ppm, Ca = 2.30 g/kg, Na = 46.9 g/kg) as one point. The regression lines are:

$$Mn = 45.50 Ca - 24.44 \quad (S = 1.5)$$
$$Mn = 2.434 Na - 34.587 \quad (S = 1.3)$$

with Mn in ppm, Ca and Na in g/kg, and standard errors of Mn regression on Ca and Na as shown. Thus the relationship of Mn to both conservative elements is exceedingly linear, and there is absolutely no in-

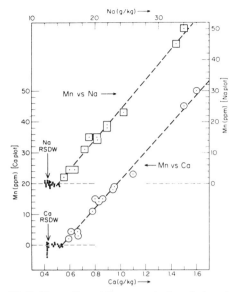

Fig. 9. Mn vs. Ca and Na concentrations in transition region above Atlantis Brine (data from M. Peterson, 1968). Na and Ca values in Red Sea Deep Water are shown by arrows; the unenclosed solid points show Na and Ca values measured for all waters with Mn reported as "less than one ppm." The lines are least squares fits, including the 44°C brine as one value, but excluding the Mn values for unenclosed points.

dication of manganese precipitation above the 44° brine-sea water interface.

Both the Mn-Ca and Mn-Na lines go to zero Mn at indicated concentrations well above sea water values, at Ca = 0.537g/kg and at Na = 14.20g/kg. Peterson's average values for Red Sea deep water were obtained by averaging his 12 analyses from these casts with Na less than 13g/kg. The values, with the standard error of the mean, are Ca = 0.432 ± 0.002g/kg and Na = 12.54 ± 0.08g/kg, so the offsets from the values at which Mn (= 0.004ppm in normal sea water) vanishes, are quite significant. As pointed out above, Peterson's Ca value is about 0.04g/kg lower than the most probable value for the deep water, but this difference is only a small fraction of the offset. The Mn offsets amount to 5 and 4ppm less than zero at the sea water Ca and Na values, which are very large differences compared to the Mn scatter, but the possibility of a systematic error which increases as the Mn concentration goes to zero cannot be discounted.

The question of the Mn offset at Red Sea deep water Ca and Na concentrations deserves serious attention because it bears so heavily on the possibility of Mn precipitation in the transition region and the possibility of manganese recycling from above. Nevertheless, it is extremely difficult to imagine a precipitation mechanism which would leave the Mn-Ca and Mn-Na relationships linear over the entire Mn range, and produce only a zero offset; moreover, the offset is probably not large enough to account for the amount of recycling required (108ppm/kg, or 57 per cent of the Mn mixed out to sea water). Thus at present it seems more likely that Mn is dissolving directly into the 44° layer from previously precipitated manganese in contact with the layer on its flanks. It should be noted that the Mn concentration in the 56° brine is probably significantly increased above the original value in inflowing brine by the same mechanism, and in fact the similarity of the concentrations in the 44° and 56° layers may indicate simply that both are saturated with manganese. Therefore, the enrichment factors for Mn (as well as for Fe if iron is recycled) shown in Table 4 may not reflect the actual enrichment in the flowing brine at all. This may also be true of the high values for lead and zinc, which may simply reflect recycling or re-solution.

Zinc, Copper, Lead, Cobalt. The differences between the data of Brewer and Spencer (1969) and those of Peterson on concentrations of these elements are so large that at present no meaningful statement even as to the sign of J can be made without choosing one set of data over the other. For concentrations in the 56° and 44° brine respectively (all ppm), Brewer and Spencer find: Zn = 5.4 and 0.15; Cu = 0.26 and 0.017; Pb = 0.63 and 0.009; Co = 0.16 and 8×10^{-4}, whereas Peterson's data indicate Zn = 7 (5 in Niskin bottles) and about 2.7; Cu = 0.3 (but 0.04 in Niskin bottles) and 0.3; Pb = 0.2 (0.10 in Niskin bottles) and 0.10; Co = 0.005 (but 0.03 in Niskin bottles) and 0.15. Similar discrepancies in lithium data have been noted earlier. The Niskin bottles contain no metal and may give better data for Cu and Pb, but other differences are difficult to explain. In any case, the Brewer and Spencer data indicate positive J values for all four metals with 93.5 per cent of the Zn flux from 56° brine to 44° brine, 85 per cent of the Cu flux, 97 per cent of the Pb flux, and 99 per cent of the Co flux, precipitated *within* the 44° layer. Peterson's data, however, show J approximately 0 for Zn (all removed by mixing), highly negative for Cu (input to 44° layer), and zero or negative for both Pb and Co. His data favor a conclusion that all these metals are conservative above the 56° brine, but any conclusion on the material balances must await a detailed study of the actual concentrations.

Discovery Deep

Element profiles for Discovery Deep, obtained by Peterson in his **CHAIN** 61 Survey, show a different pattern from that of Atlantis II Deep—from 1,980 to 2,010m they increase only slightly from sea water values, and then show an almost exactly linear slope to the brine interface at 2,065m. However, there are no samples between 2,040m, the depth at which the temperature has just increased to 36°C, and the brine interface, so that the question of the existence of a small 36° convection cell, implied by a few other measurements, cannot be answered. The temperature profile is non-linear and falls off more slowly from the brine interface than the concentrations. The profiles are roughly compatible with a model of a stagnant brine with no advective influx, but a more detailed set of measurements is required before this question can be answered. The intermediate brines in Discovery Deep, and the brine in Chain Deep, could not have formed by dilution with evaporated Red Sea water of much higher salinity than that of present Red Sea Deep Water, and remained stagnant since dilution, or the Mg and SO_4 mixing curves in Fig. 1 would be convex, as discussed in the first section of this paper. Thus mixing of these intermediate salinity brines with Red Sea Deep Water has probably been continuous and fairly rapid since these two depressions were filled.

Although the data plotted in Fig. 1 indicate a possibility that potassium concentrations might differ in the two deeps, Peterson's data show exactly similar values in Atlantis 56° brine and Discovery brine. All data show that Fe and Mn are much lower in Discovery brine than Atlantis 56° brine. Using the data of Brewer and Spencer (1969), for both brines, the Fe difference is 81ppm and the Mn difference is 27.4ppm; these differences correspond to removal of 88.5 and 30.4ppm of HCO_3 respectively, for a total of 119ppm of HCO_3 if the Fe-Mn depletion has occurred by precipitation of $FeCO_3$ and $MnCO_3$. The observed difference in HCO_3 concentration is 112ppm, in rather striking agreement with the carbonate precipitation model which thus provides a possible explanation for the carbonate difference.

Weiss (*op. cit.*) observed large differences in C^{13} concentration in the two brines; $\delta C^{13} = -5.6$ per mill in Atlantis 56° brine and -16.8 per mill in Discovery brine, and HCO_3 concentrations of 142.7 and 30.4ppm respectively. If a Rayleigh process has operated in Discovery Brine, with carbonate precipitation in a closed system, the fraction of carbonate remaining is 21.3 per cent, which requires a fractionation factor of 1.0073 for the process, roughly similar to the estimate for Atlantis II precipitation, and not an unreasonable value. The correspondence of depletion of both concentration of HCO_3 and C^{13} are difficult to explain by any mechanism requiring addition of light carbon, e.g., from organic matter, as the original carbonate has to be stripped out first. The closed system, carbonate precipitation model, is of course also consistent with a stagnant brine in Discovery Deep with no input of new brine.

Radiocarbon Measurements on Atlantis II Brine

Samples of Atlantis II brine and overlying sea water were collected, and the dissolved carbonate extracted for C^{14} measurements, by A. Jokela on the **CHAIN 61** Expedition. The water samples were collected in 400 liter Niskin-type bag samplers, kindly made available by W. Broecker of Lamont Geological Observatory, and the CO_2 was extracted in 55 gallon drums using the Lamont technique with two plastic KOH bubblers on each drum. The extracted carbonate was measured at the Tata Institute for Fundamental Research; initial results have been reported briefly by Craig and Lal (1968). An unopened KOH bubbler was also returned from the ship as a blank and processed for CO_2 in the laboratory; it contained no CO_2 within a detection limit of 0.2cc. Samples were collected at 1,000m depth, at 160m above the brine interface (approximately 1,850 in depth), and in the 56° brine itself. C^{13} measurements were made on the CH_4 used for counting, in order to normalize for isotopic fractionation.

Table 6 shows the results of the radiocarbon measurements. The samples were counted as CH_4 in Oeschger counters at 90cm pressure; the net counting rate for NBS Oxalic acid was 11.413 ± .085 (one sigma) cpm. δC^{14} is the per mill difference in counting rate between the sample and 0.95 × NBS Oxalic acid standard, and ΔC^{14} (per mill) is the difference corrected for the C^{13} variations (Lamont normalization):

$$\Delta C^{14} = \delta C^{14} - (2\delta C^{13} + 50)(1 + 10^{-3}\delta C^{14})$$

with all δ values in per mill. The errors shown are one standard deviation counting errors. Table 6 also lists the volumes of water from which CO_2 was extracted and the CO_2 yield expressed as ppm of HCO_3^-. The brine sample was extracted in two drums, and carbon from one first-stage KOH bubbler was measured separately in order to provide an independent check. Considerable fractionation occurs in the first bubbler stage, as shown by the δC^{13} value of this sample ($\alpha = 1.007$ from the C^{13} results and fractional yields), but the normalized C^{14} results are in excellent agreement.

The chlorinity of the collected brine sample was 152.34g/kg, *vs.* 155.5g/kg for the correct value, and it was observed that a small amount of sea water becomes trapped in the bag sampler when it is brought up unopened (A. Jokela, personal communication). The chlorinity difference indicates a 2 per cent contamination by sea water, and the mean ΔC^{14} value of the brine has been corrected for this, using ΔC^{14} of sea water = -18 per mill (average of the two measurements), an HCO_3 concentration in the brine of 143ppm (Weiss, 1969) and the previously assumed value of 135ppm for Red Sea deep water. The mean ΔC^{14} value of the brine, corrected in this way, is -866 per mill. The low yield of CO_2 from the brine (56 per cent, compared with the precise measurement by Weiss) reflects loss by gas evolution when the sample is brought to the surface, and the slight enrichment in C^{13} relative to a sealed sample (-4.8 per mill *vs.* -5.6 per mill) is in agreement with such an effect (Weiss, *op. cit.*).

Table 6 Radiocarbon Measurements on Red Sea Waters and Atlantis II Brine

TF*	Sample and Depth	Water Volume (liters)	ΣCO_2 as HCO_3 (ppm)	δC^{14}(‰)	δC^{13}(‰)	ΔC^{14}(‰)
690	Sea water—1,000m	120	104	39.3	−0.25	−12 ± 10
691	Sea water—160 m above brine, ~ 1,850m	197	98	28.1	+0.49	−24 ± 10
692	Brine: both KOH collectors on first drum, + second collector, second drum	(400)	(54)†	−842	−2.8	−849 ± 4
693	Brine: first KOH collector on second drum	(400)	(26)†	−846	−8.5	−851 ± 4
	Average brine, corrected for 2% sea water contamination	400	80	—	−4.8	−866 ± 4

* Tata Institute Radiocarbon Laboratory sample number.
†Yields of 9.6 and 4.7 liters STP of CO_2 obtained on samples 692 and 693 respectively.

A model for the interpretation of radiocarbon ages in flow systems with a reservoir has been developed previously (Craig, 1963) and has to be used for the Red Sea system in order to understand the measured values. In this model, there are two steady-state times: "t-time" and "τ-time," measuring the transit time for hydrodynamic or "pipe-line" flow from source, through aquifer, to the reservoir, and the residence time, or mean lifetime before removal, in the reservoir, respectively. Additionally, for short time periods of the order of the reservoir residence time, we must consider a third time, Δt, which measures the effect of the transient period before steady-state concentrations are attained. The general equation for radiocarbon concentration in the reservoir (i.e., the 56° brine) is:

$$\frac{R_T}{R_0} \frac{X}{X_0} = \frac{e^{-\lambda t}}{(1 + \lambda \tau)} (1 + T) \quad (16)$$

in which R_T is the C^{14}/C^{12} ratio in the reservoir, measured at some time *after* the reservoir has filled with fluid, R_0 is the C^{14}/C^{12} ratio in the source fluid beginning the hydrodynamic flow to the reservoir, t is the flow time in the aquifer, and τ is the residence time in the reservoir. The terms X and X_0 are the C^{12}/H_2O ratios in the reservoir and in the source material; using these ratios, rather than concentrations as given previously (Craig, 1963) eliminates the dilution effect of the added salts. λ is the radioactive decay constant, $= (1/8,267)(yr^{-1})$.

In this model, "dead" carbon can be added to the fluid at any point in the aquifer or in the reservoir. With $t = O$, the aquifer transport is very fast, and the radiocarbon decrease is entirely due to the time spent in the mixed thermal reservoir and to dead carbon addition to the reservoir. With $\tau = O$, the effects are only those of radioactive decay during aquifer flow (as if pieces of wood were being transported), and addition of dead carbon during flow. With both t and $\tau = O$, the radiocarbon decrease is due to dilution only. The residence time τ is the ratio of the amount of carbon (or water) in the reservoir to the flux of carbon (or water) from the reservoir (assuming carbon and water leave in reservoir proportions).

The transient term T decreases to zero as the time since reservoir filling increases; when it is insignificant, R_T becomes R_S, the steady-state C^{14}/C^{12} ratio in the reservoir, which no longer varies with time. Therefore $(R_T/R_S) = 1 + T$, and the function T measures the deviation from the steady-state radiocarbon activity.

The form of T depends on the model assumed for the transient stage while the reservoir is filling, because it is necessary to know the radiocarbon concentration at the end-point of the filling process, when outflow from the reservoir begins, in order to integrate the differential equation from which (16) is derived. In the model used here, we assume the fluid flows in continually and mixes within the reservoir (by convection) with no outflow until the reservoir is filled to its steady-state capacity. The aquifer flow time from the source is t, and the filling time is τ; thus from the initiation of the first flow into the aquifer to the time it is filled and outflow begins, the time $(t + \tau)$ has elapsed. At this time the C^{14}/C^{12} ratio in the reservoir is given by:

$$\frac{R_{(t+\tau)}}{R_0} \frac{X}{X_0} = \frac{e^{-\lambda t}}{\lambda \tau} (1 - e^{-\lambda \tau}) \quad (17)$$

and the reservoir begins to flush to steady-state (assuming constant input). Integrating from the C^{14}/C^{12} ratio in equation (17) to the value at some later time Δt, measured from the time when the reservoir is filled, the function of T in equation (16) becomes:

$$T = \frac{e^{-\frac{\Delta t}{\tau}(1+\lambda \tau)}[1 - e^{-\lambda \tau}(1 + \lambda \tau)]}{\lambda \tau} \quad (18)$$

and R_T is the ratio at any time $(t + \tau + \Delta t)$ since flow through the aquifer began, or at time Δt since the reservoir was filled, and approaches R_S as $\Delta t \rightarrow$ infinity.

Equations (16) and (18) are solved for t as a function of given values of τ and Δt, for observed or assumed values of the left-hand-side, which is given by:

$$\frac{R_T}{R_0} \frac{X}{X_0} = \frac{X}{X_0} \left[1 - \frac{(\Delta_0 - \Delta_B)}{(1,000 + \Delta_0)}\right] \quad (19)$$

(all delta values in per mill). Δ_B is the measured brine value in Table 6, $= -866$ per mill. The value of Δ_0 for the source water is assumed to be −50 per mill, according to present estimates of the activity in pre-nuclear, pre-industrial surface ocean waters. The values for Red Sea water in Table 6 have been increased by radiocarbon produced in nuclear weapon testing, as measurements made on Zephyrus Expedition (1962) gave values of −75 and −62 per mill in intermediate and deep waters (Craig, in preparation). The ratio (X/X_0) is the HCO_3^- enrichment factor, calculated in Table 4, assuming 125 ppm of HCO_3^- in the original 38.2 per mill salinity source water, = 1.48, and the value obtained in equation (19) is thus $1.48(0.414) = 0.209$.

Solutions to equation (16) are tabulated in Table 7, using 0.209 for the activity-dilution ratio. The final column shows the steady-state values of t as a function of τ, for $\Delta t = $ infinite, $T = 0$, and it is seen that in this system, in which the original carbon is assumed to have been increased by about 50 per cent by addition of dead carbon, the aquifer flow-time t is 12,940 years if the residence time in the reservoir goes to zero (this is the age which a block of wood with an original Δ value of −50 per mill would have for the present specific activity). For finite residence times, the time which can be allotted for aquifer flow decreases as the residence time increases, due to the holding period in the brine reservoir, and finally $t = 0$ for a maximum residence time $\tau = 31,390$ years. All t-τ combinations between these values are possible steady-state solutions to equation (16). These values (as well as the transient values) are not strongly affected by the value assumed for (X/X_0); if the original HCO_3^- concentration is assumed to be 110 ppm, which is probably a minimum estimate for 38.2 per mill salinity water, the range of t is 0–11,890 years for $\tau = 26,560$ to zero. For $(X/X_0) = 1$ (no dilution), t varies from 0 to 16,200 years, and τ from 50,360 to zero.

For finite residence times, the time which can be allotted to aquifer flow decreases as the transient period Δt increases, since maximum C^{14} depletion in the reservoir takes place at steady-state concentrations which are essentially achieved at $\Delta t \approx 2\tau$; thus the values of t for a given τ decrease across Table 7 as Δt goes from 0 to infinity. (This system does not require a period of 4 to 5 τ to reach steady-state concentrations because the C^{14}/C^{12} ratio at the time the reservoir is filled is already close to the final steady-state value; it can be shown that the ratio $R_{(t+\tau)}/R_S$ varies from 1 to 1 as τ varies from zero to infinity, with an intermediate maximum value at $(\lambda\tau) = 1.793$ for all systems. For C^{14} the maximum value is 1.56 for $\tau = 14,825$ years.)

In columns 2, 3, and 4 of Table 7 for

Table 7 Solutions to Equations (16) and (18) for the Observed (R/R_0) $(X/X_0) = 0.209$: Values of t (Aquifer Flow Time), for Given Values of τ (Residence Time) and Δt (Period Since Filling), All in Years. Values Above the Dashed Line Represent Solutions Consistent with the Ku et al. Deuser-Degens Chronology

	Values of t, for			
τ	$\Delta t = 0$	$\Delta t = 5,000$	$\Delta t = 12,000$	Steady-state
0	12,940	12,940	12,940	12,940
2,000	11,960	11,190	11,150	11,150
5,000	10,570	9,360	9,070	9,030
10,000	8,440	7,120	6,550	6,390
15,000	6,540	5,300	4,640	4,390
20,000	4,870	3,730	3,080	2,780
25,000	3,380	2,360	1,740	1,430
30,000	2,060	1,150	580	270

transient conditions, the total time involved since brine-flow began in the aquifer is $(t + \tau + \Delta t)$, time since brine began to flow into the depression is $(\tau + \Delta t)$, and the time since the depression filled to its steady-state level is Δt. It is evident that if the Ku et al.-Deuser-Degens chronology is correct, and brine-flow began during the past 12,000 years or so, solutions to the model are limited to the upper left quadrant of Table 7, from $\tau = 12,000$, $\Delta t = 0$, to $\tau = 0$, $\Delta t = 12,000$ years, above the dashed line in the table. However, these solutions require aquifer-flow times ranging from 8,400 to 12,900 years, and are thus generally not consistent with the isotopic composition of the water in the brine, which must date from not more than a few thousand years ago when the entering sea water approximately reached its present composition (say within 0.2 per mill for O^{18} and 1 per mill for D at most). Solutions for flow-times of a few thousand years are limited to residence times below the dashed line, greater than 15,000 years. We must therefore conclude that no solutions for the steady-state flow and dilution model are consistent with all the geochemical data now available.

It seems evident that in order to account for the extremely low radiocarbon activity in the brine, some isotope exchange with dead carbonate has to be assumed (carbonate, because of the relatively high C^{13}/C^{12} ratio in the brine). Exchange within the brine depression with the fossils deposited since brine flow apparently began, and recycling of carbonate would all add radiocarbon to the brine. On the other hand, exchange with very old carbonate during flow to the depressions would have the same effect in the model as increasing the numerical value of the C^{14} decay constant λ, by addition of a first-order rate constant to it. Thus in equation (16), if we set $(1/\lambda)$ equal to 4,133 years, twice the radioactive mean life, the flow times for $\tau = 0$ are all 6,470 years. For $\Delta t = 0$, $\tau = 10,000$, we obtain $t = 2,430$ years; for $\Delta t = 1,000$, $\tau = 5,000$, $t = 3,880$ years; for $\Delta t = 5,000$, $\tau = 2,000$, $t = 4,860$ years; etc. (The values of t become 50 per cent of those in Table 7 for $\tau = 0$, about 45 per cent for $\tau = 2,000$, about 38 per cent for $\tau = 5,000$, and about 27 per cent for $\tau = 10,000$ years.) An exchange-rate constant approximately equal to the radioactive decay constant therefore makes all the solutions above the dashed line in Table 7 consistent with the geochemical data, and requires that radiocarbon exchange out of the brine at an average rate about equal to the decay rate. This process is also consistent with the C^{13}/C^{12} ratio in the Atlantis II brine, which is about 6 per mill lower than that of sea water, in the direction in which dissolved carbon would shift if it approaches equilibrium with organically precipitated marine carbonates, especially with coralline material, which tends to be low in C^{13}.

Although more complicated processes can be designed to account for the low radiocarbon content of the brine, the exchange process with old carbonate is probably the simplest mechanism consistent with the present evidence on the origin of the brine. It is apparent that radiocarbon measurements on the Atlantis 44° brine and the Discovery brine would be extremely important in understanding the age and mixing relationships of these waters (e.g. the mixing rates in Atlantis Deep and the questions of the introduction time of, and possible influx into, Discovery brine). It is unfortunate that time could not be made available for sampling these two waters when Atlantis 56° brine was sampled, but hopefully future expeditions can afford the time.

Summary and Conclusions

The geochemical, chemical, and physical data on the Red Sea brines now available are consistent with a steady-state model of the Atlantis II Deep in which sea water circulates downward through evaporite sediments and flows northward, driven by the density difference between columns of brine and sea water, and heated by the local geothermal gradient which is quite sufficient to bring it to the observed temperature. The isotopic composition of the water, and the dissolved argon concentration, point to relatively warm, near surface waters, as the source of the brine. In particular, the deuterium and oxygen-18 concentrations indicate that the source water originates at least several hundred km south of the depressions, and probably about 800km south, near or on the southern sill, where water of 38.2 per mill salinity is abundant. The brine must flow at least 400km in a time of the order of a few thousand years, implying that fissure flow, probably in basalts along the central rift, is involved. The absence of an oxygen-18 shift in the water shows that temperatures are not much higher than 100°C at most.

The enrichment patterns of salts in the brine are consistent with an evaporitic origin, probably from the halite-sylvite zone, for most of the components. Certain constituents such as Mg, SO_4, I, and F have been depleted from the original sea water; this pattern and the almost stoichiometric loss of $MgSO_4$ may reflect loss by selective membrane filtration. Loss of NO_3 is probably due to nitrate reduction.

Flow of the present brine was probably initiated when sea level rose about 12,000 years ago to the level of the southern sill, and sea water became available on the sill and in the shallow regions adjacent to it. During glacial periods these areas were exposed and no supply of water was available at the faults which carry the water down. When the Red Sea is isolated during glacial periods, and its level falls, brine flow may temporarily increase because of the increased hydrostatic head, causing the brine to overflow its present level in the depression and possibly to precipitate large amounts of heavy metals by mixing with sea water. Several such cycles interrupted by periods of no flow during glacial times, may have occurred in the past.

At the present time, iron precipitation is occurring almost entirely in the 44°C layer of Atlantis Deep. Assuming a steady state, less than 1 per cent of the Fe entering this layer from the 56° layer is removed by mixing and the rest is precipitated. The supply of oxygen by mixing from overlying sea water is sufficient to precipitate only a small fraction of the iron being removed from this layer; most of the iron must be precipitating by another mechanism, possibly as ferrous carbonate. Magnanese is being supplied to the 44°C layer both from the 56° layer and from surrounding sediments, and there is no indication of Mn precipitation in this layer or in the overlying sea water. Iron recycling between the 44° and 56° layer may have increased the concentration of iron and other heavy metals in the 56° brine to steady-state levels considerably higher than in the inflowing new brine.

Sulfur and carbon isotope data show that the sulfides precipitating from the brine have not originated by sulfate reduction which has caused the observed sulfate depletion in the brine, nor are they precipitating at the 44° brine-sea water interface. In Atlantis Deep sulfides are probably precipitating because of introduction of H_2S from the surrounding sediments. A similar mechanism seems to have operated in the sediments outside of the brine area, during the period when the Red Sea was isolated by sea level lowering; at a later time when circulation was re-established, biogenic sulfate reduction began to produce sulfides and sulfur in these sediments.

At present there is no evidence requiring any contribution from volcanic or magmatic sources to the brines. The origin of Fe, Mn, and other trace metals, which show enrichment factors of the order of 10^2-10^4 (Table 4), remains unsettled. But proportions of these metals in the dissolved salts or added salts are quite comparable to those in the salts of salt-dome brines in non-volcanic areas (Table 5), and in most cases they are also similar to proportions in halite-sylvite rocks. Enrichments of these metals relative to major components can readily be accounted for by two processes:

(1) Water passing through evaporites will approach saturation with major components rather quickly, but will continue to extract trace metals which remain unsaturated.

(2) Recycling by steady-state precipitation in stacked convection cells can increase the concentration in the lowest layer by large factors, even orders of magnitude, relative to the actual fluid entering the

depressions. Thus the enrichment factors calculated for elements undergoing chemical reactions in such systems must be regarded as "apparent" enrichment factors only, until the original fluid can be sampled.

These and other concepts have been developed in considerable detail in order to provide a working framework for continued development and testing of our knowledge of the processes responsible for the origin and chemical history of the brine. Some of the present treatment is clearly speculative, some has the appearance of a tour-de-force treatment; but it is clearly important to determine whether or not a steady-state flow relationship has been reached or is being approached, most especially from the viewpoint of future exploitation of the brine. The arguments developed in the course of this work show that future expedition work should be aimed toward certain very detailed problems with important implications for understanding the processes operating in the brine areas:

1. Detailed temperature, salinity, dissolved oxygen, and trace element profiles across the brine-sea water transition regions should be made with great precision. Temperature-salinity profiles at various points above the brine can indicate the nature of the mixing processes involved, and oxygen, iron, manganese, SO_4 and other chemical profiles can delineate the precipitation processes, if any, occurring in this region.
2. Dissolved carbonate and C^{13} and C^{14} profiles should be measured from normal sea water continuously through the brines, because of the importance of carbonate chemistry and radiocarbon concentrations for many problems.
3. Chlorinity and sulfate measurements should be made simultaneously on interstitial waters both in the brine depressions and in normal sediments outside the depressions and correlated with S^{34}-C^{13}-O^{18}-C^{14} measurements on the sediments, in order to understand the very complicated history of sulfide-sulfate relationships in the area.

Acknowledgments

Many people helped immeasurably in the preparation of this paper; I am particularly grateful to A. Jokela for collecting the radiocarbon samples, to W. Broecker for providing the equipment, to D. Agrawal and S. Kusumgar of the Tata Institute for counting the samples, to John Hunt of Woods Hole for inviting the participation of A. Jokela and R. Weiss on the **CHAIN 61** Expedition, and to E. Degens and D. Ross for their assistance. R. Weiss provided critical and helpful discussion of many points as the paper was written, H. Oeschger provided sanctuary for the writing, and Mrs. Patricia Renner performed the difficult job of typing and assembly with great skill during my absence. Professor M. Bass read the manuscript and clarified many points. SIO Expedition Zephyrus was supported by NSF grant G-24479; participation of A. Jokela and R. Weiss on **CHAIN 61** was supported by ONR grant NONR-2216(23); preparation of the paper was supported by NSF grant GA-666.

References

Baker, W. J.: Flow in fissured formation. Proc. Fourth World Petroleum Cong., Section II/E, C. Columbo, Rome, 379–392 (1955).

Bischoff, J.: Red Sea geothermal brine deposits: their mineralogy, chemistry, and genesis. *In: Hot brines and recent heavy metal deposits in the Red Sea*, E. T. Degens and D. A. Ross (eds.). Springer-Verlag New York Inc., 368–401 (1969).

Brewer, P. G. and D. W. Spencer: A note on the chemical composition of the Red Sea brines. *In: Hot brines and recent heavy metal deposits in the Red Sea*. E. T. Degens and D. A. Ross (eds.). Springer-Verlag New York Inc., 174–179 (1969).

Brewer, P. G., J. P. Riley, and F. Culkin: The chemical composition of the hot salty water from the bottom of the Red Sea. Deep-Sea Res., **12**, 497 (1965).

Craig, H.: The isotopic geochemistry of water and carbon in geothermal areas. *In: Nuclear Geology on Geothermal Areas*, 1963 Spoleto Conference Proceedings, E. Tongiorgi (ed.). Consiglio Nazionale delle Richerche, Pisa, 17 (1963).

———: Isotopic composition and origin of the Red Sea and Salton Sea geothermal brines. Science, **134**, 1544 (1966).

———: Origin of the saline lakes in Victoria Land, Antarctica. Trans. Am. Geophys. Union, **47**, 112 (1966b).

——— and L. I. Gordon: Deuterium and oxygen-18 variations in the ocean and the marine atmosphere. *In: Stable Isotopes in Oceanographic Studies and Paleotemperatures*, 1965 Spoleto Conference Proceedings, E. Tongiorgi (ed.). Consiglio Nazionale delle Richerche, Pisa, 9 (1965).

Craig, H. and D. Lal: Radiocarbon age of the Red Sea Brine. Trans. Am. Geophys. Union, **49**, 193 (1968).

Craig, H., L. I. Gordon and Y. Horibe: Isotopic exchange effects in the evaporation of water. Jour. Geophys. Res., **68**, 5079 (1963).

Curray, J. R.: Sediments and history of Holocene transgression, continental shelf, northwest Gulf of Mexico. *In: Recent Sediments, Northwest Gulf of Mexico, 1951–1958*, Amer. Assoc. Petrol. Geologists, Tulsa, 221 (1960).

Delevaux, M. H., B. R. Doe, and G. F. Brown: Preliminary lead isotope investigations of brine from the Red Sea, galena from the kingdom of Saudi Arabia, and galena from United Arab Republic (Egypt). Earth and Planetary Science Letters, **3**, 139 (1967).

Deuser, W. G. and E. T. Degens: Carbon isotope fractionation in the system CO_2 (gas)-CO_2 (aqueous)-HCO_3^- (aqueous). Nature, **215**, 1033 (1967).

———, ———: O^{18}/O^{16} and C^{13}/C^{12} ratios of fossils from the hot brine deep area of the central Red Sea. *In: Hot brines and recent heavy metal deposits in the Red Sea*, E. T. Degens and D. A. Ross (eds.). Springer-Verlag New York Inc., 336–347 (1969).

Drake, C. L. and R. W. Girdler: A geophysical study of the Red Sea. Geophys. Jour., Roy. Astronom. Soc., **8**, 473 (1964).

Faure, G. and L. Jones: Anomalous strontium in the Red Sea Brines. *In: Hot brines and recent heavy metal deposits in the Red Sea*, E. T. Degens and D. A. Ross (eds.). Springer-Verlag New York Inc., 243–250 (1969).

Girdler, R. W.: The relationship of the Red Sea to the East African rift system. Quart. Jour. Geol. Soc., London, **114**, 79 (1958).

Hartmann, M. and H. Nielsen: Sulfur isotopes in the hot brine and sediment of Atlantis II Deep (Red Sea). Marine Geol., **4**, 305 (1966).

Heybroek, F.: The Red Sea Miocene evaporite basin. *In: Salt Basins around Africa*, Inst. Petroleum, London, 17 (1965).

Kaplan, I. R., R. E. Sweeney, and A. Nissenbaum: Sulfur isotope studies on Red Sea geothermal brines and sediments. *In: Hot brines and recent heavy metal deposits in the Red Sea*, E. T. Degens and D. A. Ross (eds.). Springer-Verlag New York Inc., 474–498 (1969).

Ku, T. L., D. L. Thurber, and G. G. Mathieu: Radiocarbon chronology of Red Sea sediments. *In: Hot brines and recent heavy metal deposits in the Red Sea*, E. T. Degens and D. A. Ross (eds.). Springer-Verlag New York Inc., 348–359 (1969).

Longinelli, A. and H. Craig: Oxygen-18 variations in sulfate ions in sea water and saline lakes. Science, **156**, 56 (1967).

Miller, A. R., C. D. Densmore, E. T. Degens, J. C. Hathaway, F. G. Manheim, P. F. McFarlin, R. Pocklington, and A. Jokela: Hot brines and recent iron deposits in deeps of the Red Sea. Geochim. et Cosmochim. Acta, **30**, 341 (1966).

Neumann, A. C. and D. A. McGill: Circulation of the Red Sea in early summer. Deep-Sea Res., **8**, 223 (1962).

Pugh, D. T.: Origin of hot brines in the Red Sea. Nature, **214**, 1003 (1967).

Pytkowicz, R. M. and R. Gates: Magnesium sulfate interactions in seawater from solubility measurements. Science, **161**, 690 (1968).

Rafter, T. A. and Y. Mizutani: Preliminary study of variations of oxygen and sulphur isotopes in natural sulphates. Nature, **216**, 1000 (1967).

Stewart, F. H.: Marine Evaporites. Chapter Y, Data of Geochemistry, U.S.G.S. Prof. Paper 440-Y, Washington, 1 (1963).

Thode, H. G., M. Shima, C. E. Rees, and K. V. Krishnamurty: Carbon-13 isotope effects in systems containing carbon dioxide, bicarbonate, carbonate, and metal ions. Canad. J. Chem., **43**, 582 (1965).

Thompson, E. F.: Chemical and physical investigations. The general hydrography of the Red Sea. John Murray Expedition 1933–34, Scientific Reports, **2**, 83; The exchange of water between the Red Sea and the Gulf of Aden over the "Sill." *ibid.*, **2**, 105 (1939).

Watson, S. W. and J. B. Waterbury: The sterile hot brines of the Red Sea. *In: Hot brines and recent heavy metal deposits in the Red Sea*, E. T. Degens and D. A. Ross (eds.). Springer-Verlag New York Inc., 272–281 (1969).

Weiss, R. F.: Dissolved argon, nitrogen and total carbonate in the Red Sea Brines. *In: Hot brines and recent heavy metal deposits in the Red Sea*, E. T. Degens and D. A. Ross (eds.). Springer-Verlag New York Inc., 254–260 (1969).

——— and H. Craig: Total carbonate and dissolved gases in equatorial Pacific waters. Trans. Am. Geophys. Union, **49**, 216 (1968).

Wendt, I.: Fractionation of carbon isotopes and its temperature dependence in the system CO_2-gas-CO_2 in solution and HCO_3-CO_2 in solution. Earth and Planetary Science Letters, **4**, 64 (1968).

White, D. E.: Saline waters of sedimentary rocks. Fluids in subsurface environments—a symposium. Am. Assoc. Petrol. Geol. Mem., **4**, 342 (1965).

White, D. E., J. D. Hem, and G. A. Waring: Chemical composition of subsurface waters. Chapter F, Data of Geochemistry, U.S.G.S. Prof. Paper 440-F, Washington, 1 (1963).

Copyright © 1970 by the Mineralogical Society of America

Reprinted from *Mineralog. Soc. Amer. Spec. Paper 3*, 291–306 (1970)

CHEMICAL EQUILIBRIA AND EVOLUTION OF CHLORIDE BRINES

Abraham Lerman

Canada Center for Inland Waters, Burlington, Ontario

Abstract

Dissolution of subsurface halite beds by ground water may result in the formation of chloride brines at rates which greatly depend on the physical characteristics of the environment, such as diffusivity in porous media, rate of water flow past the salt bed, and the presence of membranes impermeable to salt. For different sets of limiting conditions, estimated times from 30 to 150×10^6 years are required for the NaCl concentration in the subsurface brine (at 100 m above the salt bed) to attain the value of 90 percent saturation with respect to halite.

Sea water and ground waters of comparable ionic composition may evolve into chloride-rich brines, low in sulfate and carbonates, when the loss of H_2O from the brine is accomplished by: (1) relative enrichment in Ca such that the molar quotient $Ca^{2+}/(SO_4^{2-}+HCO_3^-)$ in the brines becomes greater than 1; (2) precipitation of $CaSO_4$ (and $CaCO_3$) mineral phases. After condition (1) has been fulfilled, process (2) results in continuous depletion of the brine in SO_4^{2-} (and carbonates) and further increase in the Ca^{2+} concentration. Data on 94 subsurface brines (5–290 g/liter dissolved solids) support a thesis of Ca enrichment at the expense of Mg in the brine, and an equilibrium precipitation of $CaSO_4$ minerals. The amount of gypsum or anhydrite formed from subsurface brines in such a process is very small relative to the total amount of sediment.

The composition of the chloride-type ground waters and brines of the Jordan-Dead Sea Valley is controlled by the history of halite dissolution and mixing with a brine of the Dead Sea-type. These brines could not have evolved from dilute ground waters by a simple loss of water and precipitation of calcium sulfate minerals. Thermodynamic models for calculation of the degree of saturation of chloride brines with respect to gypsum, anhydrite and halite produce results compatible with measurements and brine-mineral relationships in nature.

Introduction

The processes responsible for the formation of brines from more dilute solutions may be broadly classified in the following three major classes:
(I) processes in which water is being removed from the solution (deaquation), (II) processes in which solutions react with the surrounding rocks, and (III) mixing of brines of different composition and concentration. Within each class, processes of more restricted scope may be of either primary or ancillary significance in the formation of a brine. In the tabulation below the primary processes are designated by (p), and the ancillary processes by (a):

I. DEAQUATION
  1. Evaporation (p)
  2. Membrane filtration $(p?)$
II. SEDIMENT-BRINE REACTIONS
  1. Dissolution (p)
  2. Precipitation (a)
  3. Ion exchange (a)
  4. Biological activity (a)
III. MIXING (p)

Evaporation at the Earth's surface or from the groundwater table and, presumably, membrane filtration effectively removing H_2O from the brine are directly responsible for the increase in the concentrations of dissolved solids. Dissolution of salt beds and, possibly, reactions between HCl-containing solutions of deeper crustal origin and the surrounding rocks can also immediately result in large quantities of solids passing into solution. Mixing of a ground water with some highly saline brine may also be viewed as a primary process of the formation of a new brine.

Precipitation of mineral phases from a solution, ion exchange reactions between the solutions and sediments, and biological activity in the brine and sediments—all these may modify the concentrations of the individual components of a brine thereby contributing to a more or less pronounced change in its chemical nature. Such processes do not substantially increase the brine concentration and these are therefore treated as ancillary, even though their importance in shaping the final composition of the brine may be great.

Precipitation of halite in the course of evaporation of sea water is a well known example of a change in the brine composition (SO_4^{2-}/Cl^- ratio in the brine strongly increases in this process). Ion exchange, interpreted broadly, includes dolomitization reactions in the course of which the brine becomes enriched in calcium and depleted in magnesium. The main facet of the biological activity affecting the dissolved solids contents of brines is the activity of sulfate-reducing bacteria. The bacterial activity is considered responsible for the low sulfate concentrations in certain subsurface (Graf *et al.*, 1966, with a review of earlier literature) and surface brines.

Rates of Brine Formation

Stratigraphic information on a subsurface brine can in many cases tell a good deal about the age and origin of that brine, yet little can be inferred about the rate at which the brine formed. The picture is somewhat clearer with respect to many brine lakes of approximately known age, from which the minimum rates of the brine formation had been estimated (Langbein, 1961).

Considering the least clear first among the primary processes of the brine formation listed in the preceding section, mixing is likely to be a process of highly variable rates which depend on such factors as the flow rates and volumes of the brine reservoirs.

Membrane filtration, when effective, is a very slow process controlled by the water flow rates through poorly permeable beds. In experiments with dilute (under 3%) NaCl solutions effective hyperfiltration through various artificial membranes has been achieved at flow rates on the order of 5×10^{-5} to 5×10^{-4} cm^3 cm^{-2}·sec (Baldwin et al., 1969; Johnson and Harrison, 1969). These flow rates are equivalent to 15 to 150 m/cm^2·yr.

The rate of formation of a brine by surface evaporation and/or dissolution of saline minerals can vary greatly depending on the specific physical conditions of the environment, and any temporary interruptions or reversals in the process. However, limiting rates for these processes can be estimated for certain simplified physical situations, five such situations to be discussed in this section. It will be shown that relatively small changes in the physics of the environment result in great differences in the rates of increase in dissolved solids content of the brines forming by different processes. For the purpose of this illustration the five processes considered below apply to a brine column 100 m deep, containing initially NaCl at the concentration of 0.2 mole/1000 g H$_2$O (approximately 1% NaCl). The generality of the arguments is not lost by considering a pure NaCl solution rather than a solution containing different ionic species.

Surface evaporation. This is a relatively fast process: the mean global evaporation rate is 0.8–1 m/yr (Harbeck, 1955; Nace, 1967). In warm dry climates the rates are higher by a factor of two, although for a highly concentrated brine the rate of evaporation is lower, being proportional to the difference between the H$_2$O activity in the brine and the partial vapor pressure of H$_2$O in the air above it. At the mean rate of 1 m/yr, a 100-m deep column of initial solution (0.2 m NaCl) would be reduced to approximately 5 m in 95 years; at that stage the NaCl concentration in the brine (approx. 5.5 molal) would be 90 percent of its saturation value with respect to halite (6.1 molal) (Fig. 1).

Halite dissolution. In ground waters flowing at very slow rates and dissolving halite, the transport of the Na$^+$ and Cl$^-$ through the water column may be thought of as controlled by molecular diffusion and advection (i.e. flow) of the water mass. Diffusion due to a thermal gradient (the Soret effect) is negligible compared to the molecular diffusion driven by a chemical potential gradient: the Soret coefficients in ionic solutions are approximately three orders of magnitude lower than the molecular diffusion coefficients (Agar, 1959).

The diffusion coefficients of ionic species in water filled porous media are as a rule lower than the diffusion coefficients in bulk aqueous solutions. In the first approximation the value of a molecular diffusion coefficient in a porous medium is lower by a factor of $\sqrt{2}$, the factor arising from geometrical considerations of the diffusional free path in the pore space (Perkins and Johnston, 1963). Experimental determinations of the diffusion coefficients of univalent cations in "moderately crosslinked resins" give values 5 to 20 times lower than the corresponding values in free solutions (Helfferich, 1962, p. 309). The temperature, however, exerts an effect on the diffusion coefficients opposite to that of a porous medium: at 50°C the diffusion coefficients are by a factor of 1.5–3.7 higher than at 25°C (calculated as $D_{50}/D_{25} = \exp[-\Delta E(1/323 - 1/298)/R]$, for the activation energy of diffusion ΔE in the range 3–10 kcal/mole (Helfferich, 1962)). The concentration dependence of the diffusion coefficient is low for ionic species: the diffusion coefficient of NaCl (at 25°C) increases by less than 8 percent in the concentration range 0.2–5 molal, from 1.475×10^{-5} to 1.59×10^{-5} cm^2/sec (Robinson and Stokes, 1965, p. 515).

A constant value of the diffusion coefficient of NaCl in a "model" brine will be taken as 1×10^{-5} cm^2/sec, and no correction for the difference between the molar and molal concentration scales will be made.

The following four simple cases of a NaCl brine in contact with a halite bed will be considered (Fig. 1):

(i) An "infinitely high" brine column in a porous med-

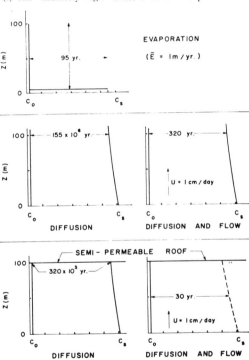

Fig. 1. Rate of increase in the NaCl concentration of 100 m thick brine layer under different conditions. C_o is initial concentration (0.2 molal NaCl); C_s—concentration at equilibrium with halite (6.1 molal). *Top*: concentration increase by evaporation at the mean evaporation rate $\bar{E} = 1$ m/yr. *Middle*: transport of dissolved NaCl from a halite bed at $z=0$ upwards by diffusion, and by diffusion and flow; flow rate $U = 1 \times 10^{-5}$ cm/sec (approx. 1 cm/day). *Bottom*: upward transport of dissolved NaCl as in the middle diagram, except for the semipermeable roof at $z=100$ which retains the salt within the brine layer.

ium above an "infinitely" thick halite bed. The initial NaCl concentration in the brine is $C_0 = 0.2$ m NaCl; the concentration at the brine-bed interface is maintained at the saturation value of halite $C_s = 6.1$ m NaCl. At the height $z = 100$ m above the brine-bed interface, the position of which in space is considered fixed ($z = 0$), the NaCl concentration (C_{100}) would increase as a function of time according to the relationship

$$C_{100} = C_0 + (C_s - C_0)\text{erfc}[z/2(Dt)^{1/2}] \quad (1)$$

where D is the diffusion coefficient of NaCl and t is time (Crank, 1956, p. 33; Carslaw and Jaeger, 1959, p. 63). From relationship (1), the NaCl concentration at 100 m above the interface comes to within 90 percent of the saturation value ($C_{100} = 5.5$ m NaCl) in approximately 155×10^6 years. The process is admittedly very slow when molecular diffusion is the sole transporting agent.

The time estimates from relationship (1), as well as (2), (3), and (5), discussed in the following paragraphs, were obtained by trial substitution of different values of t to give the value sought $C_{100} \approx 0.9 C_s$. The values of the error function (erf) and error function complement (erfc) were taken from the tables in Carslaw and Jaeger (1959, p. 485), and Abramowitz and Stegun (1964).

(ii) Under the conditions as in (i), the brine flows upwards in the z direction perpendicular to the halite bed. The flow rate is $U = 1 \times 10^{-5}$ cm/sec (approx. 3 m/yr). The equation describing the change in concentration at $z = 100$ m above the bed is:

$$C_{100} = C_0 + \tfrac{1}{2}(C_s - C_0)\left\{\text{erfc}\,\frac{z - Ut}{2(Dt)^{1/2}} + \exp(Uz/D)\right.$$
$$\left. \times \text{erfc}\,\frac{z + Ut}{2\sqrt{Dt}}\right\} \quad (2)$$

Relationship (2) is derivable from Carslaw and Jaeger (1959, p. 388) by substituting the following initial and boundary conditions: $C = C_s$ at $z = 0$, $t > 0$, and $C = C_0$ at $z > 0$, $t = 0$, which are the conditions of the problem discussed in the present section.

The NaCl concentration at 100 m above the brine comes to $C_{100} = 5.5$ molal in approximately 320 years. This is a relatively fast process, and the estimate of 320 years illustrates how a flow rate as low as 1 cm/day can accelerate the increase in the brine concentration by a factor of 5×10^5, compared with an estimate of 155×10^6 years obtained for the case of diffusion without the flow.

(iii) Under the conditions as in (i) a semi-permeable roof is located at 100 m above the brine-halite bed interface keeping all the NaCl diffusing upwards confined to within the 100 m column. This is a fairly realistic assumption, and such physical conditions are likely to be encountered in nature.

The initial and boundary conditions are:

$C_0 = C$ at $0 < z < l$, $t = 0$;
$C_s = C$ at $z = l$, $t > 0$;
$dC/dz = 0$ at $z = 0$.

The last condition of zero concentration gradient at the upper boundary defines the impermeability of the roof. It should be noted that for mathematical convenience the coordinates in this example have been changed by setting $z = 0$ at the upper boundary, and $z = l$ at the brine-bed interface. The distance between the boundaries is $l = 100$ m, as before. The appropriate equation for the concentration change at the upper boundary (C_{100}, at $z = 0$) as a function of time is (cf. Carslaw and Jager, 1959, p. 100):

$$C_{100} = C_s + \frac{4}{\pi}(C_s - C_0)\sum_{n=0}^{\infty}\frac{(-1)^{n+1}}{2n+1} \times \exp\{-Dt\pi^2(2n+1)^2/4l^2\} \quad (3)$$

The time required for C_{100} to attain the value of 5.5 m NaCl, or 90 percent saturation with respect to halite is approximately 320,000 years. An impermeable bed clearly has a pronounced effect on the rate of the brine formation as compared with a much longer time needed to attain the same concentration under the conditions of free diffusion through an upwards unlimited medium considered in case (i).

(iv) This case is a logical extension of case (iii): the roof at 100 m above the halite bed is a semipermeable membrane retaining all the salt but letting the water pass through. The brine flows upwards from the halite bed at the rate of 1×10^{-5} cm/sec. Under such conditions all the NaCl transported from the basal bed upwards is being retained within the 100 m thick brine column. A simple method of estimating the mean NaCl concentration within the brine column is first to assume that no semipermeable roof exists; second, the total amount of NaCl transported by diffusion and flow upwards may then be evaluated by integrating its concentration over the distance from the salt bed ($z = 0$) upwards ($z = \infty$): $\int_0^{\infty} C\,dz$, where the concentration C in the semiinfinite medium is given by relationship (2); third, the total amount of NaCl added to the brine column over a certain period of time, as obtained by integration of (2), may be divided into the volume 100 m $\times 1$ cm^2 to give a mean concentration within the 100 m column.

From (2), the total amount of salt added from the lower boundary into the semiinfinite medium (M, in g/cm^2) is

$$M = \int_0^{\infty}(C - C_0)dz = (C_s - C_0)\left\{\frac{Ut}{2} + \left(\frac{Ut}{2} + \frac{D}{U}\right)\right.$$
$$\left.\cdot \text{erf}\,\frac{U\sqrt{t}}{2\sqrt{D}} + \sqrt{\frac{Dt}{\pi}}\exp(-U^2t/4D)\right\} \quad (4)$$

and the mean concentration of NaCl ($\overline{C}$) in the 100-m layer can be found by dividing M into 10,000 cm and adding to the initial concentration (C_0 in g/cm^3 or moles/cm^3).

$$\overline{C} = C_0 + \frac{C_s - C_0}{10,000}\left\{\frac{Ut}{2} + \left(\frac{Ut}{2} + \frac{D}{U}\right) \times \right.$$
$$\left.\text{erf}\,\frac{U\sqrt{t}}{2\sqrt{D}} + \sqrt{\frac{Dt}{\pi}}\exp(-U^2t/4D)\right\} \quad (5)$$

Relationship (5) is physically meaningful only for those values of t which make the mean concentration not greater than the saturation value ($\overline{C} \leq C_s$).

Using, as before, vertical flow rate $U = 1 \times 10^{-5}$ cm/sec and diffusion coefficient $D = 1 \times 10^{-5}$ cm^2/sec, substitution of different values of t into relationship (5) gives, by trial and error, $t \simeq 30$ years as the length of time needed for the mean NaCl concentration to attain the 90 percent level of saturation with respect to halite. Here, as in case (ii), the vertical flow greatly accelerates the rate of brine formation.

It may be noted that at the flow rate of approximately 3.16 m/yr used in the preceding computation, the mean residence time of water in the column is approximately 30 years:

$$\frac{100 \text{ m}}{3.16 \text{ m/yr}} = 30 \text{ yr}$$

Thus it is clear that under the conditions chosen, the vertical flow, rather than diffusion, is the main mechanism of the salt transport. The scale length of the system is $D/U = 1$ cm, a very small number compared to the 100-m long brine column. For the diffusion to contribute significantly to the dispersion of solute in the presence of flow $U = 1$ cm/day, the diffusion coefficient should have been several orders of magnitude greater, making the scale length on the order of meters or tens of meters. Such high values of the diffusion coefficient, however, would fall in the range of turbulent, or eddy, diffusion, which is not a likely dispersion mechanism in a porous rock.

Concentration increase by flow and diffusion, and increase by mixing. Relationship (5) describing the increase in the solute concentration in the brine column above the salt bed may be simplified when long periods of time are considered. For large t the erf term on the right-hand side of (5) approaches 1 and the exponential term tends to 0; when the term D/U is small compared with $Ut/2$, as in the case discussed, the relationship may be written as

$$\overline{C} \cong C_0 + (C_s - C_0)Ut/10^4 \quad (6)$$

which shows that the rate of increase in the concentration in the brine column is a linear function of t. The relationship is valid for all those values of t as long as the condition $\overline{C} \leq C_s$ is met. Relationship (6) describes essentially a case of "piston flow" where the dilute brine of initial concentration C_0 is made to occupy progressively smaller fraction of the volume $(1 - Ut/10000)$ when it is being displaced due to the inflow of the concentrated brine (C_s) from below, such that the two brines do not mix in the process.

If, however, the concentrated brine (C_s) entering at the salt bed-brine interface undergoes complete mixing with the dilute (C_0) brine within the volume of 100 m $\times$ 1 cm^2 (*i.e.* the volume of the brine column is maintained constant by the removal of excess water through the semipermeable roof), then the concentration in the brine column increases as:

$$\overline{C} = C_0 + C_s Ut/10^4 \quad (7)$$

From (6) and (7) it follows that the rate of increase in concentration $(\partial \overline{C}/\partial t)$ is proportional to the concentration difference $\overline{C}_s - \overline{C}_0$ in the case of no mixing ("piston flow"), and to the concentration of the entering brine C_s in the case of complete mixing. The rates are very similar when the initial concentration C_0 is small compared with C_s:

$$(\partial C/\partial t)_{n.m.}/(\partial C/\partial t)_{m.} = (C_s - C_0)/C_s \cong 1 \quad (8)$$

where the subscripts n.m. and m. stand for "no mixing" and "mixing," respectively. When the concentration in the original brine (C_0) is not negligibly small compared with C_s, a "piston flow" will result in the formation of a brine at a lower rate than in the case of complete mixing:

$$(\partial \overline{C}/\partial t)_{n.m.}/(\partial \overline{C}/\partial t)_{m.} = (C_s - C_0)/C_s < 1 \quad (9)$$

In the cases discussed earlier, where the initial concentration value used $C_0 = 0.2$ was much lower than the value in the saturated brine $C_s = 6.1$, the times required for the brine column to attain saturation with respect to halite would be very similar (approximately 30 years) both for the complete mixing case and the case of "piston flow" without mixing.

Chloride Brines and Sulfate Depletion

A large class of brines, in part associated with oil fields and evaporite deposits, is characterized by the predominance of Cl$^-$ among the anions, and Na$^+$ and Ca^{2+} among the cations. The majority of the chloride brines are commonly thought of as having evolved from sea water and sea water brines entrapped in sediments (Graf et al., 1966; White, 1965; Bentor, 1969), although the origin of some brines can be traced, by the nature of their chemical composition, to the dissolution of rock salt (*e.g.*, Manheim and Bischoff, 1969). As most ground waters and sea water contain sulfate and carbonate in varying proportions in addition to chloride, straight deaquation of a brine would not normally produce a brine highly enriched in chloride. A number of ancillary processes effectively removing sulfate and carbonate species as the concentration of the brine builds up can, however, insure the ultimate low sulfate and carbonate content of a highly saline brine. The commonly invoked ancillary mechanisms are the bacterial sulfate reduction with the concomitant formation of HCO$_3^-$ due to the oxidation of organic matter (*e.g.*, Graf et al., 1966); removal of HCO$_3^-$ (balanced by Na$^+$) from the brines through subsurface clay membrane filters (Graf et al., 1966; White, 1965), its uptake in the regrading and reconstitution of clay minerals (Mackenzie and Garrels, 1966), and precipitation of CaCO$_3$.

It is the purpose of the present section to show that another mechanism—precipitation of calcium sulfate phases—can be an effective ancillary process which leads to the formation of chloride brines. Calcium sulfates (gypsum and anhydrite) are the only poorly soluble mineral phases which can control the concentration of sulfate in natural brines. If in the original brine the calcium concentration is lower than that of sulfate, then the precipitation of CaSO$_4$ minerals would remove essentially all the calcium from the brine, yet only part of the total sulfate equivalent to the amount of calcium would be removed. The concentration of the remaining sulfate in the brine may continue to increase as the brine looses water. Sea water and many fresh and brackish ground waters are characterized by Ca^{2+}/SO$_4^{2-}$ molar ratio lower than 1, and their progressive deaquation during continuous precipita-

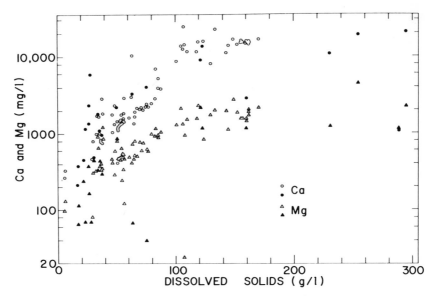

FIG. 2. Calcium and magnesium concentrations in subsurface brines plotted against the concentration of total dissolved solids. Data from White et al. (1963)—solid circles and triangles; Table 12, analyses 2–6; Table 13, analyses 1, 2, 3, 5, 6, 10, 12, 13; Table 15, analyses 3, 5, 9; Table 16, analyses 3, 7, 11, 14, 15. These are data from different localities, with the brine density values reported in the tables. From Bentor (1969)—open symbols; Appendix, analyses 1–18, 20–25, 27, 29–36, 38–46, 48–53, 55–78, 82, 83. The larger number of data on subsurface brines from Israel taken from Bentor (1969) as compared with a "worldwide" sample of White et al. (1963) is unlikely to introduce a lithological bias insofar as the Israeli brines occur in various types of sediment.

tion of gypsum would necessarily result in the sulfate enrichment of the brine. The importance of the Ca^{2+} concentration in determining whether the brine would evolve to either a chloride-sulfate or chloride brine has been recognized by Hutchinson (1957, p. 567–568).

When the removal of HCO_3^- and CO_3^{2-} from the brine is accomplished by $CaCO_3$ precipitation, a similar reasoning applies to the significance of the calcium concentration. Thus it is convenient to consider in brines the molar quotient $Ca^{2+}/(SO_4^{2-}+HCO_3^-)$ which is an indicator of whether a brine can become depleted in sulfate (and carbonates) by precipitation of $CaSO_4$ (and $CaCO_3$) mineral phases. If the quotient is smaller than 1, precipitation of gypsum would not deplete the brine of its sulfate. To achieve the latter the quotient value must be greater than 1. This means that either Ca^{2+} must be added to the brine or SO_4^{2-} and HCO_3^- must be removed from the brine, or both. The processes which may result in the calcium enrichment and sulfate (and carbonate) depletion have been briefly discussed in the introductory section.

To illustrate the possible significance of these processes in the increase of the $Ca^{2+}/(SO_4^{2-}+HCO_3^-)$ quotient values, 94 analyses of subsurface brines containing 5 to 290 g/liter dissolved solids, compiled by White et al. (1963) and Bentor (1969), will be considered. The concentrations of Ca^{2+}, Mg^{2+}, SO_4^{2-}, and HCO_3^- (in mg/liter) in those brines were plotted against the total dissolved solids concentration (in g/liter) as shown in Figures 2, 3.

The Ca and Mg data in Figure 2 show that in the brines of less than approximately 30 g/liter dissolved solids the concentrations of the two species are similar. In more concentrated brines, between 30 and 120 g/liter, the Ca^{2+} concentration increases more strongly than Mg. At higher brine concentrations the two alkaline earths slowly increase in approximately the same ratio.

If the relatively smaller increase in the Mg concentration in the brines, as compared with Ca (Fig. 2), were due to incorporation of Mg in silicate minerals then it would be reasonable to find some corroboration of the Mg trend in the HCO_3^- data, as this anion is the one which may be involved in the clay reconstitution reactions (Mackenzie and Garrels, 1966). However, in the absence of any clear trend in the HCO_3^- data (Fig. 3), as well as in the absence of details on the mineralogical and chemical composition of the rocks within which the brines occur, it is more plausible to account for the slower increase in Mg relative to Ca (Fig. 2) as being due to dolomitization reactions between brines and carbonate rocks.

The SO_4^{2-} and HCO_3^- data plotted in Figure 3 scatter greatly, with virtually no individual trends discernible. The diminution of the spread of the individual concentrations in the brines above 120 g/l dissolved solids might be due to the smaller number of samples representative of the high concentration range.

There is little in the sulfate and bicarbonate data of Figure 3 which may support a contention of a reciprocal relationship between the sulfate being reduced and bicarbonate being formed.

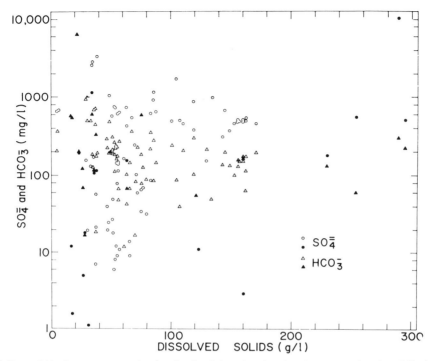

Fig. 3. Sulfate and bicarbonate concentrations in subsurface brines plotted against the concentration of total dissolved solids. Sources of data and symbols as in Fig. 2.

Two parameters were calculated from the analytical data: molar ratio of Ca^{2+} to the sum of SO_4^{2-} and HCO_3^- $[m_{Ca^{2+}}/(m_{SO_4^{2-}}+m_{HCO_3^-})]$, sensitive to the removal of $CaSO_4$ and $CaCO_3$ from the brine, as discussed earlier; and the molar fraction of Ca in the sum of Ca and Mg concentrations $[m_{Ca^{2+}}/(m_{Ca^{2+}}+m_{Mg^{2+}})]$, indicative of any amount of Ca-Mg exchange which might have taken place between the brines and sediments. The values of the $Ca^{2+}/(SO_4^{2-}+HCO_3^-)$ ratio for individual brines plotted in Figure 4 display a trend from the values of near and below 1 in more dilute brines to over 70 in the most concentrated brines.

The composition of the brines as given by the chemical analysis was averaged within successive intervals of 10–20 g/l dissolved solids, and the concentrations of Ca^{2+} and SO_4^{2-} were calculated for each of the "mean" brines at equilibrium with either gypsum or anhydrite at different temperatures. From the calculated concentrations new values of the ratio $Ca^{2+}/(SO_4^{2-}+HCO_3^-)$ were computed, and curves were drawn such as to smooth the calculated values (not shown in the figure). The curve for the anhydrite-brines equilibria at 75°C, shown dashed in Figure 4, fits the scatter of the raw data better than other curves tried for lower temperatures. The temperature of 75°C may be regarded as reasonable for many of the brines occurring at depths 1500–2500 m below the surface.

The concentrations of Ca^{2+} and SO_4^{2-} in brines at equilibrium with gypsum or anhydrite at different temperatures were calculated from the experimentally determined values (Marshall and Slusher, 1968; Marshall, 1967) of the solubility of calcium sulfates and dissociation of the $MgSO_4^0$ ion pair in chloride solutions of high ionic strength. The principles of the equilibrium solubility model and method of calculation are summarized in the next section.

The values of the $Ca^{2+}/(Ca^{2+}+Mg^{2+})$ ratio of the brines are shown in Figure 5. The physically meaningful limits of the quotient are 0 (no Ca^{2+}) and 1 (no Mg^{2+}). A gentle upward trend from the ratio value of approximately 0.5 to 0.85 is apparent in the plot of Figure 5. The dashed curve for the anhydrite-brines equilibria in Figure 4 was transferred onto Figure 5 using the calculated values of $m_{Ca^{2+}}$ and the values of $m_{Mg^{2+}}$ from the brine analyses. In this case the fit of the curve for 75°C is also better than for equilibria at lower temperatures.

The proximity of the calculated dashed curves for anhydrite-brines equilibria to the analytical data suggests that the brines above approximately 80 g/l dissolved solids are close to an equilibrium with anhydrite. In the lower concentration range, the points above the dashed curve in Figure 4 represent brines undersaturated with respect to anhydrite: it should be noted that if Ca^{2+} and SO_4^{2-} were added in equivalent amounts to those brines, the values of

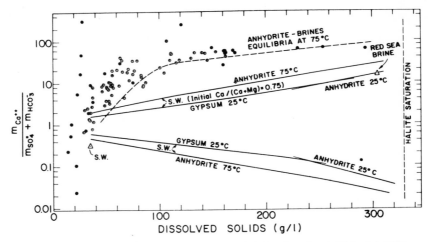

FIG. 4. Molar quotient $Ca^{2+}/(SO_4^{2-}+HCO_3^-)$ in subsurface brines. Solid circles—from data in White et al. (1963), open circles—from data in Bentor (1969). The quotient for the Red Sea brine shown for comparison with subsurface data [computed from analyses of Brewer et al. (1965) for depths 2105–2155 m. Discovery Station 5580]. Curves for equilibria between mineral phases and brines calculated as explained in the text. (S. W.-sea water).

the $Ca^{2+}/(SO_4^{2-}+HCO_3^-)$ quotient, all of which are greater than unity, would decrease falling closer to the dashed curve. In Figure 5, where the same brines are represented by the values of the quotient $Ca^{2+}/(Ca^{2+}+Mg^{2+})$ lesser than unity, addition of Ca^{2+} to the brines would displace the points upwards, closer to the equilibrium dashed curve.

The brines shown in Figures 4 and 5 as a group are likely to have evolved from the left to the right, i.e. from the lower to higher concentrations, rather than from right to left, by dilution of some more concentrated brines. Against the latter process it should be noted that simple dilution of a highly saline brine would not have affected the values of the quotients $Ca^{2+}/(SO_4^{2-}+HCO_3^-)$ and $Ca^{2+}/(Ca^{2+}+Mg^{2+})$. In the case, however, of a dilution process when the brine is maintained at equilibrium with either gypsum or anhydrite acting as a very large reservoir of Ca^{2+} and SO_4^{2-}, then the molar quotient $Ca^{2+}/(SO_4^{2-}+HCO_3^-)$ in the brine at low concentrations would approach the value of 1 from above, whereas the quotient $Ca^{2+}/(Ca^{2+}+Mg^{2+})$ would approach 1 from below. Neither of such trends is discernable in Figures 4 and 5.

At this stage two types of brines can be considered; one derived from sea water in which the quotient $Ca^{2+}/(Ca^{2+}+Mg^{2+})=0.15$, and the other from sea water in which part

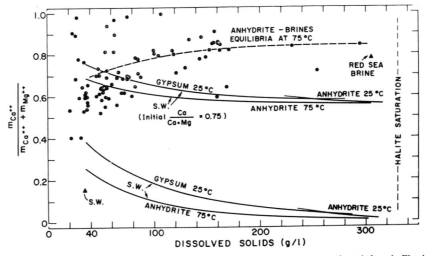

FIG. 5. Molar quotient $Ca^{2+}/(Ca^{2+}+Mg^{2+})$ in subsurface brines. Sources of data and symbols as in Fig. 4.

of Mg had been replaced by Ca to increase the quotient value to 0.75. An ordinary sea water is appreciably below saturation with respect to gypsum, and its equilibration with gypsum at 25°C, or with anhydrite at 75°C, raises the initial values of the two quotients significantly above the point for sea water, as shown in Figures 4 and 5. In the course of progressive deaquation of sea water, when equilibrium with gypsum or anhydrite is maintained, the ratios $Ca^{2+}/(SO_4^{2-}+HCO_3^-)$ and $Ca^{2+}/(Ca^{2+}+Mg^{2+})$ decrease at either 25 or 75°C, as shown by the calculated curves (lower set) for the two solid phases in Figures 4 and 5.

In the case of sea water with part of its Mg replaced by Ca, the initial value of the ratio $Ca^{2+}/(SO_4^{2-}+HCO_3^-)$ is greater than unity. Hence, in the course of deaquation of the brine, the ratio continually increases as Ca^{2+} and SO_4^{2-} are being removed from the solution and calcium sulfate minerals form at equilibrium, as shown by the upper set of solid curves in Figures 4 and 5.

The curves for the two types of sea water brines at equilibrium with calcium sulfate minerals were calculated by increasing the concentrations of the main dissolved species (Na^+, K^+, Mg^{2+}, Cl^-) in proportion to the amount of H_2O removed, and keeping the HCO_3^- concentration constant. The physical significance of the latter constraint is that every X moles of HCO_3^- removed in the process of deaquation of the brine are taken out as $0.5X$ moles of $CaCO_3$ and $0.5X$ moles of CO_2. For such brines the concentrations of Ca^{2+} and SO_4^{2-} at equilibrium with either gypsum or anhydrite were computed using the solubility data of Marshall (1967) and Marshall and Slusher (1968).

In Figure 4, the dashed curve identifying the anydrite-brines equilibria is positioned above a similar curve calculated for the Ca-enriched sea water brines, although the trends of the two curves are analogous. From the relative position of the two curves it may be concluded that a prerequisite for the formation of a chloride brine from sea water (or ground water of comparable ionic ratios) is an increase in the value of the molar quotient $Ca^{2+}/(SO_4^{2-}+HCO_3^-)$ to above 1. Dissolution of $CaSO_4$ minerals by a brine of initial ratio $Ca^{2+}/(SO_4^{2-}+HCO_3^-)<1$ cannot raise the value of the ratio above 1. However, the quotient value may be raised by the addition of Ca at the expense of Mg, as suggested by the data (Fig. 2). The position of the points plotted in Figure 4 also suggests that the process of the relative enrichment in Ca is probably not confined only to the early stages of the brine evolution, but it continues as the brine becomes more concentrated.

If Ca^{2+} and SO_4^{2-} are lost from the brines through precipitation, the gypsum or anhydrite forming in this process should be preserved in the sediment. However, the amounts of the sulfate minerals forming as the concentration of the brine increases are too small to affect the gross mineralogical composition of the sediment. The following example, taken from the data on which Figure 4 is based, illustrates this point: in the course of deaquation of the Ca-enriched sea water from 85 g/l to 165 g/l dissolved solids, the SO_4^{2-} concentration in the brine at equilibrium with anhydrite at 75°C decreases from approximately 0.015 to 0.011 moles/liter. The decrease is equivalent to precipitation of 0.004 moles $CaSO_4$, or 0.54 g anhydrite. Assuming that the porosity of the sediment is of the order of 10–20 percent, 1 liter of brine is distributed through a sediment column of 1 cm² in cross-section and 100–50 m in height. Addition of 0.5 g of a mineral to such a sediment column is obviously negligible.

Models of Gypsum, Anhydrite and Halite Solubility in Brines

In view of the importance of the chemical equilibria between brines and saline minerals to the evolution of the brine composition, some simplified models useful in the calculation of the state of saturation of a brine with respect to gypsum, anhydrite and halite will be considered in this section.

Gypsum and Anhydrite. Marshall and Slusher (1968) have shown that the solubility of gypsum and anhydrite in sea water concentrates and other chloride-sulfate type artificial brines may be satisfactorily predicted by a model which takes into account (i) the ionic strength of the solution, and (ii) concentration of the $MgSO_4^0$ ion pair in it. Marshall and Slusher (1968) have concluded that $MgSO_4^0$ is the most important ion pair which controls the solubility of the calcium sulfate minerals, whereas the existence of the $CaSO_4^0$ ion pair is only of minor consequences and it may in general be ignored. Those authors have defined a "practical" ionic solubility product for gypsum and anhydrite expressed in the form:

$$\log K_{sp} = \log(m_{Ca^{2+}} \times m_{SO_4^{2-}}) = \log K° + 8S\sqrt{I}/(1 + A\sqrt{I}) + BI - CI^2 \quad (10)$$

where $K°$ is the thermodynamic dissociation constant at the specified temperature, S is the Debye-Huckel limiting slope, and A, B and C and empirical constants dependent on temperature. From the definition of $\log K_{sp}$ in relationship (10) it follows that for gypsum at equilibrium with a brine the activity of H_2O is absorbed in the parameters B and C of the solubility equation. The values of $K°$, A, B, and C have been determined and tabulated by Marshall and Slusher (1968) for gypsum and anhydrite (and calcium sulfate hemihydrate) for a number of temperatures.

For the equilibria between brines and gypsum at 25°C, and between brines and anhydrite at 75°C, considered in the previous section, relationship (10) assumes the following forms:

Gypsum, 25°C:
$$\log K_{sp} = -4.373 + 8 \times 0.5080\sqrt{I}/(1 + 1.5\sqrt{I}) + 0.0194I - 0.0134I^2 \quad (11)$$

Anhydrite, 75°C:
$$\log K_{sp} = -4.884 + 8 \times 0.5645\sqrt{I}/(1 + 1.575\sqrt{I}) - 0.0201I - 0.0098I^2 \quad (12)$$

In order to calculate the activity product of the Ca^{2+} and SO_4^{2-} ions in a brine, or the concentrations of the two ions at equilibrium with either gypsum and anhydrite, it is necessary to know the amount of SO_4^{2-} complexed in the ion pair $MgSO_4°$. It can be shown (Marshall, 1967; Marshall and Slusher, 1968) that the concentration of the ion pair may be expressed as

$$(MgSO_4°) = K_{sp}(Mg^{2+})/[(Ca_{eq}^{2+})K_d + K_{sp}] \quad (13)$$

where the parentheses denote molal concentrations, Mg^{2+} is the total magnesium concentration in the brine, Ca_{eq}^{2+} is the calcium concentration at equilibrium with the given calcium sulfate phase, and K_d is the dissociation constant of the $MgSO_4°$ ion pair as determined by Marshall (1967). For the two temperatures considered in the previous section, K_d is given by the following relationships:

$$\log K_d = -2.399 + 8 \times 0.5080\sqrt{I}/(1 + \sqrt{I}) \quad (25°C) \quad (14)$$
$$\log K_d = -2.846 + 8 \times 0.5645\sqrt{I}/(1 + \sqrt{I}) \quad (75°C) \quad (15)$$

When a brine comes to an equilibrium with either gypsum or anhydrite, equivalent amounts of Ca^{2+} and SO_4^{2-} are either added to, or subtracted from, the brine, depending on whether it is undersaturated or supersaturated with respect to the given mineral. Therefore the ionic solubility product of a calcium sulfate mineral, as in relationship (10), may be written as

$$K_{sp} = (Ca_{eq}^{2+})(SO_{4eq}^{2-}) = (Ca_{or}^{2+} + x) \times (SO_{4or}^{2-} - MgSO_4° + x) \quad (16)$$

where the subscript or denotes the molal concentration in the original solution, and x is the number of moles of Ca^{2+} and SO_4^{2-} per 1000 g H_2O added or subtracted from the solution when it comes to an equilibrium with the given mineral. The quantity x may be either positive or negative. Rearrangement of (16) gives a more convenient form:

$$x^2 + x(Ca_{or}^{2+} + SO_{4or}^{2-} - MgSO_4°) + [Ca_{or}^{2+}(SO_{4or}^{2-} - MgSO_4°) - K_{sp}] = 0 \quad (17)$$

from which x can be determined when the remaining quantities are known.

In order to determine the concentration of Ca^{2+} and SO_4^{2-} in a brine at equilibrium with, for example, gypsum at 25°C, the ionic strength of the brine (I) was computed from the chemical analysis, and the quantities K_{sp}, K_d and ($MgSO_4°$) were determined using relationships (11), (13), and (14). In the first calculation of ($MgSO_4°$) the calcium concentration given by the chemical analysis (Ca_{or}^{2+}, in the present notation) was used instead of the unknown Ca_{eq}^{2+}. Next, equation (17) was solved for x and a new value of the ionic strength was computed as:

$$I' = I - 4MgSO_4° + 4x \quad (18)$$

Using the new value of I', equations (11), (13), (14), (17), and (18) were solved again, and the entire procedure was repeated three or four times until the new value of I' differed from the previous value by less than 0.5 percent. The values of x obtained by this procedure were algebraically added to the concentrations of Ca^{2+} and SO_4^{2-} given by the chemical analysis.

Halite. The solubility of halite markedly depends on the

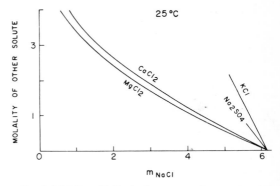

FIG. 6. Solubility of halite in the presence of some other electrolytes at 25°C. Data from averaged solubilities in Zdanovskii et al. (1953).

concentration and nature of other ionic solutes present. In Figure 6 are shown the effects on the solubility of halite due to some common dissolved components of natural brines. The solubilities shown in Figure 6 are at 25°C, although essentially the same picture holds at higher and lower temperatures, as the solubility of halite varies only little at temperatures below 100°C. Insofar as the solubilities of halite in the presence of either divalent metal chlorides or univalent metal chlorides (Fig. 6) are very different, it may be intuitively concluded that a satisfactory and sufficiently general model of the halite solubility in chloride brines should take into account the concentrations of the individual dissolved components of the brine. Such a model may in fact be based on the so-called Harned's rule (Harned and Owen, 1958, pp. 602–632; Robinson and Stokes, 1965) which describes the dependence of the mean activity coefficients of strong electrolytes in solutions also containing other dissociated solutes.

In Harned's rule notation the mean activity coefficient ($\gamma_\pm$) of a strongly dissociated electrolyte, such as NaCl, is given by a relationship of the following type:

$$\log \gamma_{\pm NaCl} = \log \gamma_{\pm NaCl(0)} + \alpha_{1i}\omega_i m_i + \alpha_{1j}\omega_j m_j + \cdots \quad (19)$$

where $\gamma_{\pm NaCl(0)}$ denotes the mean activity coefficient of NaCl in a pure solution containing only NaCl at the same ionic strength as the mixed solution; α_{1i}, α_{1j}, etc., are empirical parameters (Harned's rule coefficients); ω is the ionic strength factor ($\omega = 1$ for a NaCl-type electrolyte, $\omega = 3$ for MgCl$_2$-type); and m is the molality of the other electrolyte. The coefficients α_{12} are in general characteristic of the specific pair of solutes and they are independent of the presence of other solutes in solution. The relationship for the mean activity coefficient of a strong electrolyte as given by (19) is an approximation. A number of systems (of little geochemical interest, however) are known in which the mean activity coefficient is expressed as a power series in m, where additional terms of the type βm^2 are added to the right-hand side of (19).

From relationship (19) it is clear that the mean activity

coefficient of an electrolyte in a mixed solution can be calculated only for those ionic strengths for which the pure solutions exist and the values of $\gamma_{\pm 1(0)}$ are available. Thus, pure NaCl solution becomes saturated with respect to halite at near 6.1 molal NaCl ($I = 6.1$), whereas ionic strengths much higher than that may obtain in mixtures of NaCl and MgCl$_2$. This difficulty in obtaining the values of $\gamma_{\pm \mathrm{NaCl}(0)}$ for the ionic strengths in the supersaturation region of a pure NaCl solution is circumvented by the use of the mean activity coefficients of HCl which are simply related to the mean activity coefficients of such solutes as NaCl and KCl (Åkerlöf, 1937; Harned and Owen, 1958, p. 617):

$$\log (\gamma_{\pm \mathrm{NaCl}(0)}/\gamma_{\pm \mathrm{HCl}(0)}) = -0.088I \quad (20)$$

By rearrangement,

$$\log \gamma_{\pm \mathrm{NaCl}(0)} = \log \gamma_{\pm \mathrm{HCl}(0)} - 0.088I \quad (21)$$

The mean activity coefficients of HCl in aqueous solutions are available up to the ionic strength of 16 (Åkerlöf and Teare, 1937; Harned and Owen, 1958, p. 751), and from relationship (21) it is possible to calculate the mean activity coefficients of NaCl in its pure solutions of concentrations higher than the halite saturation. In this procedure it is assumed that relationship (21) holds for supersaturated NaCl solutions.

Before discussing in more detail the nature and methods of derivation of the α_{12} coefficients, their usefulness in the solution of the following common problems may be restated. For a chloride brine of a known composition the mean activity coefficient of NaCl may be determined by relationship (19) and then the value of the ionic activity product (IAP_{NaCl}) in the brine may be computed as follows:

$$IAP_{\mathrm{NaCl}} = a_{\mathrm{Na}^+} a_{\mathrm{Cl}^-} = m_{\mathrm{Na}^+} m_{\mathrm{Cl}^-} \gamma_{\pm \mathrm{NaCl}}^2 \quad (22)$$

Likewise, the NaCl concentration in a brine at equilibrium with halite may be determined from the basic relationship:

$$K_{\mathrm{hal}} = m_{\mathrm{Na}^+} m_{\mathrm{Cl}^-} \gamma_{\pm \mathrm{NaCl}}^2 \quad (23)$$

In the relationship (23) m denotes the molalities of the Na$^+$ and Cl$^-$ ions at equilibrium with halite. Specific examples of such a calculation by successive iterations have been described elsewhere (Lerman, 1967).

The coefficients α_{12} for various systems have been determined from the results of isopiestic measurements of vapor pressure over solutions and by the measurements of the individual ionic activities using ion-sensitive reversible electrodes (Lanier, 1965; Platford, 1968; Butler and Huston, 1967; Wu et al., 1968). A different method for the determination of α_{12} from the solubility data, such as those shown in Figure 6, has been described by Åkerlöf (1934). In this method the mean activity coefficient of NaCl in solutions containing NaCl and another solute at equilibrium with halite is determined for a number of different compositions using relationship (23), and the coefficient α_{12} is then calculated from each value of $\log \gamma_{\pm \mathrm{NaCl}}$ using relationship (19). When the variation between the different values of α obtained is not large, a mean value of α may be used for the entire range of ionic strength. Although this method is sensitive to the accuracy with which the composition of the solutions at equilibrium with halite has been determined, its advantage is that α_{12} values can be determined for very high ionic strengths.

Table 1 summarizes the value of α_{12} coefficients reported recently by some investigators, as well as the values calculated from solubility data and used in this work.

Although the differences between the values of α_{12} obtained by different methods are appreciable, it may be pointed out that if the values of $\alpha_{\mathrm{NaCl}-i}$ other than those given in the last column of Table 1 were used in the calculation of IAP_{NaCl} in the Dead Sea brines (Fig. 8), the agreement between the electrometric measurements and calculated results would not have been nearly as good. This in particular holds for the coefficient $\alpha_{\mathrm{NaCl}-\mathrm{MgCl}_2}$, as MgCl$_2$ is one of the main components of the Dead Sea brines.

For a chloride brine containing NaCl, MgCl$_2$, KCl, CaCl$_2$, and Na$_2$SO$_4$ as the main dissolved components (Lerman, 1967; Lerman and Shatkay, 1968), the activity product of the Na$^+$ and Cl$^-$ ions in the brine may be written, using the values of α_{12} from Table 1, relationships (19) and (21), in the following form:

$$\begin{aligned}\log IAP_{\mathrm{NaCl}} &= \log (m_{\mathrm{Na}^+} m_{\mathrm{Cl}^-}) + 2 \log \gamma_{\pm \mathrm{NaCl}} \\ &= \log [(m_{\mathrm{NaCl}} + 2 m_{\mathrm{Na}_2\mathrm{SO}_4}) \times \\ &\quad (m_{\mathrm{NaCl}} + m_{\mathrm{KCl}} + 2 m_{\mathrm{MgCl}_2} + 2 m_{\mathrm{CaCl}_2})] \\ &\quad + 2(\log \gamma_{\pm \mathrm{HCl}(0)} - 0.088I + 0.054 m_{\mathrm{MgCl}_2} - 0.0134 m_{\mathrm{KCl}} \\ &\quad + 0.0285 m_{\mathrm{CaCl}_2} - 0.075 m_{\mathrm{Na}_2\mathrm{SO}_4})\end{aligned} \quad (24)$$

In relationship (24) the value of $\gamma_{\pm \mathrm{HCl}(0)}$ is taken at the ionic strength I of the brine.

Setting $IAP_{\mathrm{NaCl}} = K_{\mathrm{hal}}$ in (24), the equation may be

TABLE 1. HARNED'S RULE COEFFICIENTS (α_{12}) AT 25°C FOR NaCl IN SOME SYSTEMS CONTAINING TWO ELECTROLYTES

	NaCl-MgCl$_2$				
I	1	2	3	4	5
1	0.0146	0.016	0.016	0.019	
3	0.0104	0.011	0.012	0.011	
5			−0.0022	0.008	
6	0.0060	0.009	−0.0004	0.007	
6.8–14.0					0.018

	NaCl–CaCl$_2$			
I	1	3	6	5
1	0.0040	0.010	0.0092	
3	0.0020	0.000	0.0018	
6	−0.0013	−0.004	−0.0016	
7			−0.0020	
6.9–12.9				0.0095

	NaCl-Na$_2$SO$_4$				NaCl-KCl	
I	1	4	7	5	I	8	5
1	−0.0605	−0.045	−0.0453		1	−0.0235	
3	−0.0537	−0.045	−0.0490		3	−0.0235	
6	−0.0500	−0.045			6	−0.0275	
6.7–7.7				−0.025	6.7–7.7		−0.0134

1. Lanier, 1965; 2. Platford, 1968, 3. Butler and Huston, 1967; 4. Wu et al., 1968; 5. This work; 6. Robinson and Bower, 1966; 7. Butler et al., 1967; 8. Robinson, 1961.

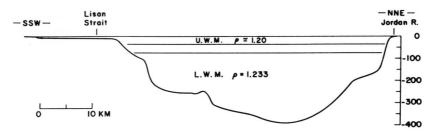

FIG. 7. Cross section of the Dead Sea basin showing the density stratification into the lighter upper water mass and denser lower water mass. Pronounced density gradient exists between the upper and lower water mass. Water depth scale on the right is in meters. Data from depth contour map in Neev and Emery (1967, p. 10).

used to compute the NaCl concentration in brines at equilibrium with halite. In this procedure the value of K_{hal} affects the value of m_{NaCl} obtained. Literature values of K_{hal} (some quoted in Lerman and Shatkay, 1968) differ by as much as 10 percent; therefore, depending on the value of K_{hal} taken, m_{NaCl} computed from (24) may vary by approximately 5 percent.

Application of relationship (24) to the evaluation of the ion activity products and the Na$^+$ and Cl$^-$ concentrations at equilibrium with halite will be demonstrated in the next section.

Chloride Brines of the Jordan— Dead Sea Rift Valley

The Jordan River—Dead Sea Rift Valley is a northern extension of the East African and Red Sea Rift Valley system. The Dead Sea and the saline ground waters in the Jordan—Dead Sea Valley constitute an excellent example of a suite of chloride brines in the range from dilute ground waters to brines precipitating halite. The more salient geological features of the Jordan-Dead Sea Rift Valley which bear on the composition of the ground brines are: (1) calcareous rocks forming the western escarpment of the rift valley; ground waters issuing along the western escarpment have a history of flow through the aquifers in calcareous rocks; (2) occurrence of numerous gypsum, anhydrite and halite beds, and salt structures in the older fill of the rift valley (Neev and Emery, 1967). The most pronounced and conspicuous salt structure is the partly exposed salt dome of Mount Sodom on the southwestern shores of the Dead Sea. The older gypsum, anhydrite and halite beds in the Dead Sea Valley sediments may be of local importance in the control of the composition of some of the saline springs as will be discussed in the following sections.

Present state of the Dead Sea brines. The cross-section of the Dead Sea in Figure 7 shows a pronounced density stratification of the water column. The stratification is apparently "permanent." The bottom sediment of the Dead Sea has been reported as made of halite, occurring below the 40-m water depth line. Thin discontinuous (0–40 cm) layer of detrital material, calcium sulfate and carbonate minerals, covers the halite on the bottom (Neev and Emery, 1967).

Measurements of the activity of the sodium and chloride ions in the Dead Sea brine samples taken at various depths throughout the water column (Lerman and Shatkay, 1968) have shown (Fig. 8) that the less saline upper water mass

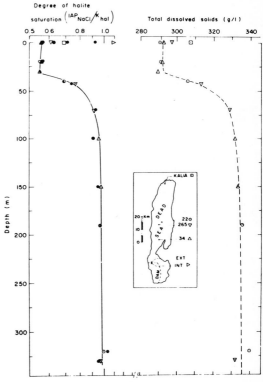

FIG. 8. Degree of brine saturation with respect to halite and total dissolved solids in the Dead Sea brine column. Open symbols refer to station numbers identified in the inset. Solid symbols —degree of saturation calculated from chemical analyses; open symbols—from electrode measurements in Lerman and Shatkay (1968). At station INT t.d.s. = 365 g/l.

(0–35 m) is appreciably undersaturated with respect to halite. The degree of saturation of the brine with respect to halite was expressed as the ratio

$$IAP_{NaCl}/K_{hal} = a_{Na^+}a_{Cl^-}/K_{hal} \quad (25)$$

where IAP_{NaCl} is the ionic activity product of the sodium and chloride ions in the brine as measured with a pair of sodium and chloride ion sensitive electrodes (details of the technique have been described by Lerman and Shatkay, 1968), and K_{hal} is the thermodynamic equilibrium product of the sodium and chloride ions in a solution at equilibrium with halite (at 25°C log $K_{hal} = 1.581_4$). The value of the $IAP/K = 1$ corresponds to the case of brine saturated with respect to halite. Values lower than 1 represent undersaturation, and values above 1, supersaturation.

It should be noted that the agreement between the measured and calculated values of the degree of halite saturation shown in Figure 8 does not depend on the value of K_{hal} taken: the same K_{hal} is divided into each measured and calculated value of IAP_{NaCl}, as in (25).

In the upper water mass of the Dead Sea the degree of saturation with respect to halite is in the range 0.55–0.7 (Fig. 8). The surface brine sample near saturation, labelled INT in the insert of Figure 8, was taken from a shallow section of the Dead Sea dammed off several years ago in order to increase the rate of evaporation of the brine utilized in an industrial process of carnallite recovery. At present halite forms on the bottom of the dammed section, and the slight supersaturation measured in that brine is compatible with the process observed.

The Dead Sea brine column below approximately 75 m depth is very close to saturation with respect to halite, as shown by the measured (and calculated) points close to the value $IAP/K = 1$ in Figure 8. The absence of any pronounced salinity gradient in the lower water mass (right-hand part of Fig. 8) suggests that it is relatively well mixed. The homogeneous salinity profile agrees well with nearly the same degree of halite saturation in the brine observed between 75 m and 330 m depth. The near saturation with respect to halite in the lower water is in good agreement with the reported occurrence of halite on the Dead Sea floor.

From the known chemical composition of the brines (Lerman and Shatkay, 1968) the values of the IAP_{NaCl} were computed using the thermodynamic model given by relationship (24).

The agreement between the measured and calculated values of the ionic activity product $IAP_{NaCl} = a_{Na^+}a_{Cl^-}$ plotted in Figure 8 (open symbols and solid dots) is as a whole very satisfactory, and it is a test of the validity of the model used in the computation of the IAP_{NaCl} values for the Dead Sea brines.

Historical Note on the Composition of the Dead Sea Brine. Antoine Laurent Lavoisier was the first to analyze the Dead Sea brine in 1778 (analysis cited in Herapath and Herapath, 1849). His results, however, reporting the density and concentration of the calcium and magnesium chloride (combined) and sodium chloride, are far off what may be considered even a crude approximation to the composition of the Dead Sea brine. Throughout the nineteenth century many chemists, geologists and explorers have analyzed the Dead Sea brine, some of the names retaining the historical

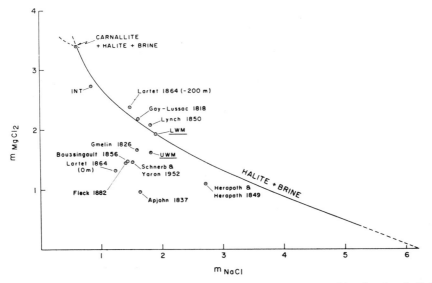

FIG. 9. Some 19th and 20th century results of the determination of the Dead Sea brine composition plotted as the NaCl and MgCl₂ molal concentrations. Analyses of Gay-Lussac, Gmelin and Apjohn cited in Herapath and Herapath (1849). Composition of the upper (UWM) and lower water mass (LWM) from analyses in Neev and Emery (1967). Derivation of the halite-brine and halite-carnallite-brine equilibrium curves discussed elsewhere (Lerman, 1967).

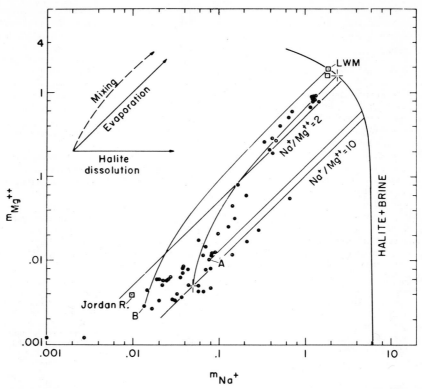

FIG. 10. Changes in the Na$^+$ and Mg^{2+} content of dilute brines due to dissolution of halite, evaporation or mixing with Dead Sea brines. Open circles: ground waters and saline springs from the southwestern Dead Sea Valley (analyses in Mazor et al., 1969).

memory and prominence even in modern times (Gmelin, Gay-Lussac). A number of analyses (not all) of the Dead Sea brine published in the 19th and 20th centuries are plotted in Figure 9 in terms of the two main components of the brine: NaCl and MgCl$_2$. All of the points plotted, unless specifically indicated, represent the surface water of the Dead Sea. The first chemical analyses confirming the density stratification of the Dead Sea were published by the French geologist Louis Lartet (1864), although the concentrations of the dissolved components in the deep water reported by him seem questionable. The analyses plotted in Figure 9 should be compared with the curve labelled Halite + Brine: the curve gives the composition of the brines at equilibrium with halite at 25°C, when the brines contain the major dissolved components (MgCl$_2$, KCl, CaCl$_2$, and Na$_2$SO$_4$) in the same proportions as the Dead Sea lower water mass. Thus the curve gives the NaCl molality at equilibrium with halite in the brines more dilute or more concentrated than the Dead Sea brine. The details of the computational procedure for the halite-brine equilibrium curve have been described elsewhere (Lerman, 1967). The position of the surface water analyses reported by Gay-Lussac in 1818 (Cited in Herapath and Herapath, 1849) and Lynch (1850) near the halite-brine equilibrium curve probably reflects analytical difficulties rather than the absence of density stratification in the early 19th century Dead Sea. The point INT shown in Figure 9 refers to the evaporating brine in the dammed section of the lake discussed earlier. Its position shows that the evaporation has progressed more than half-way to the brine in which carnallite (KMgCl$_3$·6H$_2$O) begins to form.

Modes of the brines formation. In an earlier section it was shown how the enrichment of the brine in Ca relative to Mg and subsequent precipitation of calcium sulfate minerals may result in the formation of a chloride-rich brine. The ground waters and springs in the Jordan-Dead Sea Valley, containing dissolved solids from approximately 0.2 g/liter and up, are characterized by the predominance of chloride over other anions (analyses in Bentor, 1969; Mazor and Mero, 1969; Mazor et al., 1969). As these waters occur within the Dead Sea drainage basin, their chemical composition may be viewed as bearing to a greater or lesser extent on the chemical composition of the Dead Sea water, or vice versa. The possible modes of evolution of the Dead Sea Valley brines, taken as a group, will be first considered in terms of the two main cations: Na$^+$ and Mg^{2+}. The NaCl and MgCl$_2$ content of a dilute brine plotted as a point in the coordinates representing the NaCl and MgCl$_2$ concentration would change along a path which depends on whether the brine is dissolving halite, undergoing evaporation, mixing with a more concentrated brine, or any combination of these processes. The chemical paths along which the brine may evolve are schematically shown in the upper left of Figure 10. Halite dissolution does not affect the Mg content of the brine, whereas its Na content increases, displacing the brine to the right in the coordinates of Figure 10. Evaporation of the brine increases the concentration of both Na^+ and Mg^{2+} but does not change the Na^+/Mg^{2+} molal ratio as long as the brine is undersaturated with respect to halite. In the log-log coordinates of Figure 10 this is represented by a straight line the slope of which is constant and determined by the initial Na^+/Mg^{2+} ratio. Mixing of the dilute brine with a more concentrated brine of a different Na^+/Mg^{2+} ratio is represented by a curvilinear plot; the curvature of the mixing line depends on the relative position of the two brines in the Na^+—Mg^{2+} coordinates, or, in other words, on the difference in the Na^+/Mg^{2+} ratio between the two brines.

Bearing these simple relationships in mind, near seventy ground waters and brines from the southwestern part of the Dead Sea Valley, the Jordan River, and the Dead Sea brine (lower and upper water mass) were plotted in Figure 10. The halite-brine equilibrium curve, shown in Figure 9, was transferred to the log-log coordinates of Figure 10. Two straight lines, labelled $Na^+/Mg^{2+}=2$ and Na^+/Mg^{2+} $=10$, were drawn to bracket two hypothetical cases of evaporation of dilute brines such that the values of the Na^+/Mg^{2+} ratio taken were close to the lower and upper limit recorded for the brines. Two curves representing mixing were drawn for the cases of (i) a dilute brine (point B) mixing with the lower water mass of the Dead Sea, and (ii) a brine similar to ocean water (lower cross in Figure 10; approximately 0.5 molal Na^+ and 0.05 molal Mg^{2+}) mixing with a hypothetical brine at equilibrium with halite, of composition only slightly different from the lower water mass (upper cross).

The scatter of the brine data, taken as a group, suggests that no single process could have been responsible for the shaping of their Na^+ and Mg^{2+} content. The brines in the lower concentrations range in Figure 10 suggest that dissolution of halite from the sediments and some deaquation are primarily responsible for the trend in the scatter from the lower left to the right. The higher concentrations are likely to be due to mixing of the more dilute with more concentrated brines in the subsurface, similar in their composition to the Dead Sea brines. The curvilinear trend

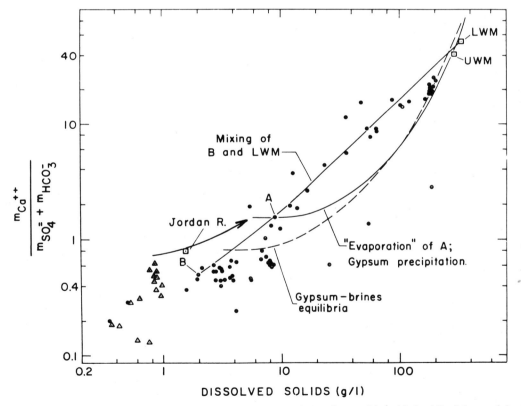

FIG. 11. Changes in the molar quotient $Ca^{2+}/(SO_4^{2-}+HCO_3^-)$ in dilute brines due to mixing with Dead Sea brines, and due to evaporation with concomitant precipitation of gypsum. Circles—Dead Sea Valley brines (analyses in Mazor et al., 1969); triangles—upper Jordan Valley, Lake Tiberias area (analyses in Mazor and Mero, 1969).

of the scatter of data in the higher concentration range supports this conclusion.

The mixing of the dilute and (hypothetical) subsurface brines similar to the Dead Sea brine, inferred from the analysis of the concentration of the two major components (NaCl and MgCl$_2$), is in good agreement with the conclusions of Gat (Gat et al., 1969), based on the isotopic composition of the brine water and the total salt content, regarding the contribution of the Dead Sea brines to the composition of the more saline spring waters (0.2–1 molal Na$^+$) plotted in Figure 10.

The mixing of the ground waters with the subsurface brines of the Dead Sea type is further strengthened by the distribution of the ratio Ca^{2+}/(SO$_4^{2-}$+HCO$_3^-$) in the brines shown in Figure 11. Although the scatter of the points is obviously large, mixing of the brine of the composition given by point B (Fig. 11) and the Dead Sea lower water mass results in a satisfactory fit to an appreciable number of different brines. The mixing hypothesis could be tested by choosing other dilute brine end members or mixing more than two brines in different proportions; then, depending on the brines chosen, a better or poorer agreement with groups of points plotted in Figure 11 can be obtained. Such attempts, however, to produce a better fit by mixing more than two different brines would be quite arbitrary.

The position of the gypsum-brines equilibria curve in Figure 11 shows that most of the brines are undersaturated with respect to gypsum at 25°C. Only one cluster of points, in the vicinity of 100 g/liter dissolved solids, is close to saturation with respect to gypsum.

The above considerations on the mixing of the more dilute brines and the Dead Sea-type water, shown in Figure 11, stress that the saline brines of the Dead Sea Valley do not represent an evolutionary sequence from dilute to highly concentrated chloride brines, as opposed to the case of the subsurface brines discussed in Figures 4, 5. If any of the more dilute brines, such as for example point A in Figure 11, characterized by its ratio Ca^{2+}/(SO$_4^{2-}$+HCO$_3^-$)>1, were deaquated while the brine remained at equilibrium with gypsum precipitating in the process, then the Ca^{2+}/(SO$_4^{2-}$+HCO$_3^-$) ratio of the brine would have changed along a path shown in Figure 11 which falls far from the points representing the Dead Sea Valley brines.

By analogy with the processes discussed in the section on subsurface chloride brines, the path of change in the brine composition in the course of gypsum precipitation, as shown in Figure 11, might represent the ancient course of events which led in the past to the formation of the Dead Sea-type brine. At present there are apparently no known significant sources of dilute brines characterized by the values of the ratio Ca^{2+}/(SO$_4^{2-}$+HCO$_3^-$)>1 in the immediate vicinity of the Dead Sea. However, some 100 miles north of the Dead Sea, in the Jordan rift valley in the vicinity of Lake Tiberias, several tens of ground water analyses (Mazor and Mero, 1969) are characterized by the ratio Ca^{2+}/(SO$_4^{2-}$+HCO$_3^-$)>1. Most of those ground waters contain less than 10 g/liter dissolved solids and, if shown in Figure 11, they would plot as a scatter above and below the line Ca^{2+}/(SO$_4^{2-}$+HCO$_3^-$)=1, in the vicinity of the arrow indicated in the figure. (In order to minimize cluttering of the graph by different symbols, only the very dilute brines from the limestone terrain of the northern Jordan rift valley were plotted as triangles in Figure 11; the more concentrated brines of the ratio Ca^{2+}/(SO$_4^{2-}$+HCO$_3^-$)>1 form a scatter essentially continuous with the triangles plotted).

Thus the conditions prerequisite for the development of chloride brines through the removal of sulfate by precipitation are also manifested in the Jordan Valley suite of relatively dilute ground waters.

Acknowledgments

A major part of the work reported in this paper was done at the Isotope Department, Weizmann Institute of Science, Rehovot, Israel. At various stages of the preparation of the paper I benefited from constructive discussions with Y. K. Bentor (The Hebrew University), H. P. Eugster (The Johns Hopkins University), R. M. Garrels (Scripps Institution of Oceanography), J. R. Gat (Weizmann Institute), R. F. Platford and Mary E. Thompson (Canada Center for Inland Waters).

References

Abramowitz, Milton, and I. A. Stegun (eds.) (1964) *Handbook of Mathematical Functions with Formulas, Graphs, and Mathematical Tables*. National Bureau of Standards, Washington, D. C.

Agar, J. N. (1959) Thermal diffusion and related effects in solutions of electrolytes. *In* W. J. Hamer, (ed.), *The Structure of Electrolytic Solutions*. Wiley, New York, 200–223.

Åkerlöf, Gösta (1934) The calculation of the composition of an aqueous solution saturated with an arbitrary number of highly soluble strong electrolytes. *J. Amer. Chem. Soc.* **56**, 1439–1443.

——— (1937) A study of the composition of the liquid phase in aqueous systems containing strong electrolytes of higher valence types as solid phases. *J. Phys. Chem.* **41**, 1053–1076.

———, and J. W. Teare (1937) Thermodynamics of concentrated aqueous solutions of hydrochloric acid. *J. Amer. Chem. Soc.* **59**, 1855–1868.

Baldwin, W. H., C. E. Higgins, and J. Csurny (1969) Preparation and hyperfiltration properties of detachable membranes. *Water Research Program, Bienn. Prog. Rep., Oak Ridge Nat. Lab.*, Oak Ridge, Tenn., Spec. Distr. ORNL-CF-69-5-41, 163–168.

Bentor, Y. K. (1969) On the evolution of subsurface brines in Israel. *Chem. Geol.* **4**, 83–110.

Boussingault, J. B. J. D. (1856) Recherches sur les variations que léau de la Mer Morte parait subir dans sa composition. *Ann. Chim. Phys.*, **48**, 129–170.

Brewer, P. G., J. P. Riley, and F. Culkin (1965) The chemical composition of the hot salty water from the bottom of the Red Sea. *Deep-Sea Res.* **12**, 497–503.

Butler, J. N., and R. Huston (1967) Activity coefficient measurements in aqueous NaCl-CaCl$_2$ and NaCl-MgCl$_2$ electrolytes using sodium amalgam electrodes. *J. Phys. Chem.* **71**, 4479–4485.

———, P. T. Hsu, and J. C. Synnott (1967) Activity coefficient measurements in aqueous sodium chloride-sodium sulfate electrolytes using sodium amalgam electrodes. *J. Phys. Chem.* **71**, 910–914.

CARSLAW, H. S., AND J. C. JAEGER (1959) *Conduction of Heat in Solids*, 2nd ed. Oxford Univ. Press, Oxford.

CRANK, J. (1956) *The Mathematics of Diffusion*. Oxford Univ. Press, Oxford.

FLECK, H. (1882) Constitution of the water of the Dead Sea. *J. Chem. Soc. (London)* **42**, 24.

GAT, J. R., E. MAZOR, AND Y. TZUR (1969) The stable isotope composition of mineral waters in the Jordan Rift Valley, Israel. *J. Hydrol.* **7**, 334–352.

GRAF, D. L., W. F. MEENTS, I. FRIEDMAN, AND N. F. SHIMP (1966) The origin of saline formation waters, III: calcium chloride waters. *Ill. State Geol. Surv. Circ.* **397**, 1–60.

HARBECK, G. E. (1955) The effect of salinity on evaporation *U. S. Geol. Surv. Prof. Pap.* **272-A**.

HARNED, H. S., AND B. B. OWEN (1958) *The Physical Chemistry of Electrolytic Solutions*, 3rd ed. Reinhold, New York.

HELFFERICH, FRIEDRICH (1962) *Ion Exchange*. McGraw-Hill, New York.

HERAPATH, T. J., AND W. HERAPATH (1849) On the waters of the Dead Sea. *J. Chem. Soc. (London)* **2**, 336–344.

HUTCHINSON, G. E. (1957) *Treatise on Limnology*, Vol. 1. John Wiley & Sons, New York.

JOHNSON, J. S., AND N. HARRISON (1969) Testing of membranes from other laboratories for hyperfiltration properties. *Water Research Program, Bienn. Prog. Rep., Oak Ridge Nat. Lab., Oak Ridge, Tenn., Spec. Distr.* ORNL-CF-69-5-41, 168–169.

LANGBEIN, W. B. (1961) Salinity and hydrology of closed lakes. *U. S. Geol. Surv. Prof. Pap.* **412**, 1–20.

LANIER, R. D. (1965) Activity coefficients of sodium chloride in aqueous three-component solutions by cation-sensitive glass electrode. *J. Phys. Chem.* **69**, 3992–3998.

LARTET, LOUIS (1864) Géologie. *In* Le Duc De Lyunes (1874) *Voyage d'Exploration a la Mer Morte, à Petra et sur la Rive Gauche du Jourdain*. Libr. Soc. Géogr., Paris, 278.

LERMAN, ABRAHAM (1967) Model of chemical evolution of a chloride lake—The Dead Sea. *Geochim. Cosmochim. Acta* **31**, 2309–2330.

———, AND A. SHATKAY (1968) Dead Sea brines: degree of halite saturation by electrode measurements. *Earth Planet. Sci. Lett.* **5**, 63–66.

LYNCH, W. F. (1850) *Narrative of the United States' Expedition to the River Jordan and the Dead Sea*, 2nd ed. Richard Bentley, London. [Also in *Amer. J. Sci., ser. 2*, **19**, 147–149 (1855)].

MACKENZIE, F. T., AND R. M. GARRELS (1966) Chemical mass balance between rivers and ocean. *Amer. J. Sci.* **204**, 507–525.

MANHEIM, F. T., AND J. L. BISCHOFF (1969) Geochemistry of pore waters from Shell Oil Company drill holes on the continental slope of the Northern Gulf of Mexico. *Chem. Geol.* **4**, 63–82.

MARSHALL, W. L. (1967) Aqueous systems at high temperature. XX. The dissociation constant and thermodynamic functions for magnesium sulfate to 200°. *J. Phys. Chem.* **71**, 3584–3588.

———, AND R. SLUSHER (1968) Aqueous systems at high temperature. XIX. Solubility to 200° of calcium sulfate and its hydrates in sea water and saline water concentrates, and temperature-concentration limits. *J. Chem. Eng. Data* **13**, 83–93.

MAZOR, EMANUEL, AND F. MERO (1969) Geochemical tracing of mineral and fresh water sources in the Lake Tiberias basin, Israel. *J. Hydrol.* **7**, 276–317.

———, E. ROSENTHAL, AND J. EKSTEIN (1969) Geochemical tracing of mineral water sources in the Southwestern Dead Sea Basin, Israel. *J. Hydrol.* **7**, 246–275.

NACE, R. L. (1967) Water resources: a global problem with local roots. *Envir. Sci. Technol.* **1**, 550–560.

NEEV, DAVID, AND K. O. EMERY (1967) The Dead Sea, depositional processes and environments of evaporites. *Israel Geol. Surv. Bull.* **41**, 1–147.

PERKINS, T. K., AND O. C. JOHNSTON (1963) A review of diffusion and dispersion in porous media. *J. Soc. Petroleum Eng.*, 70–84.

PLATFORD, R. F. (1968) Isopiestic measurements on the system water-sodium chloride-magnesium chloride at 25°. *J. Phys. Chem.* **72**, 4053–4057.

ROBINSON, R. A. (1961) Activity coefficients of sodium chloride and potassium chloride in mixed aqueous solutions at 25°C. *J. Phys. Chem.* **65**, 662–667.

———, AND V. E. BOWER (1966) Properties of aqueous mixtures of pure salts. Thermodynamics of the ternary system: water-sodium chloride-calcium chloride at 25°C. *J. Res. Nat. Bur. Stand.* **70A**, 313–318

———, AND R. H. STOKES (1965) *Electrolyte Solutions*, 2nd ed. (revised) Butterworths, London.

SCHNERB, J., AND F. YARON (1952) On the solubility of sodium chloride in Dead Sea brines. *Res. Counc. Israel Bull.* **2**, 197–198.

WHITE, D. E. (1965) Saline waters of sedimentary rocks. *In* A. Young and J. E. Galley (eds.), *Fluids in Subsurface Environments*. Amer. Assoc. Petroleum Geologists, Tulsa, 343–366.

———, J. D. HEM, AND G. A. WARING (1963) Chemical composition of subsurface waters. *U. S. Geol. Surv. Prof. Pap.* **440-F**.

WU, Y. C., R. M. RUSH, AND G. SCATCHARD (1968) Osmotic and activity coefficients for binary mixtures of sodium chloride, sodium sulfate, magnesium sulfate, and magnesium chloride in water at 25°. I. Isopiestic measurements on the four systems with common ions. *J. Phys. Chem.* **72**, 4048–4053.

ZDANOVSKII, A. B., E. I. LYAKHOVSKAYA, AND R. E. SHLEIMOVICH (1953) *Spravochnik eksperimental'nykh dannykh po rastvorimosti mnogo-komponentnykh vodno-solevykh sistem. Tom 1. Trekhkomponentnye sistemy.* [*Handbook of Solubility Data in Multicomponent Aqueous Salt Systems. Vol. 1. Three-component Systems*]. Goskhimizdat, Leningrad.

Author Citation Index

Abelson, P. H., 293
Abramowitz, M., 436
Ackley, C., 246
Adams, L. H., 44, 46
Agar, J. N., 436
Ahrens, L., 44
Åkerlöf, G., 436
Alberta Petroleum and Natural Gas
    Conservation Board, 310
Aldrich, L. T., 92
Aleinikov, G. I., 384
Alekseev, F. A., 220
Allee, W. C., 44
Allen, E. R., 226
Allen, E. T., 44, 158, 293, 383
Allen, H. A., 158
Alvarez, L. W., 206
American Public Health Association, 220
Anderson, C. C., 293
Anderson, D. L., 11
Anderson, D. W., 383
Anderson, E. M., 44
Anderson, E. T., 384
Armstrong, H. S., 92
Arrhenius, G., 136, 137
Arrhenius, S., 44
Association for Planning and Regional
    Reconstruction, 44
Aston, F. W., 44
Athale, R. N., 207
Atkins, W. R. G., 44
Ault, W. U., 92

Bainbridge, A. E., 206
Baker, W. J., 420

Baldwin, W. H., 436
Barnes, V. E., 45
Barrell, J., 44
Barth, T. F. W., 136, 383
Basharina, L., 347, 383
Bastin, E. S., 293
Basu, S. K., 92
Baumgart, J. L., 348
Baxter, G. P., 47
Bazan, F., 205, 207
Beerstecher, E., Jr., 294
Begemann, F., 205, 206
Bell, A. H., 221, 295, 310
Belousov, V. V., 347
Bemmelen, R. W. van, 294
Benfield, A. E., 44
Benioff, H., 44
Benson, C., 191
Bentor, Y. K., 386, 436
Berg, T. E., 272
Berger, W. R., 294
Berner, R. A., 5, 136
Berthon, L., 294
Betekhtin, A. G., 347
Bibron, R., 206, 207
Bidwell, R. M., 383
Bien, G. S., 136, 310
Billings, G. K., 271, 272
Birch, F., 44, 294
Biscayne, P. E., 136
Bischoff, J. L., 420, 437
Bjerrum, J., 111
Blackburn, P. E., 92
Blaedel, W. J., 383
Blifford, I. H., 246

Boato, G., 158, 159, 191, 220, 294, 383
Bodvarsson, G., 348, 383
Bolin, B., 191, 205
Booth, H. S., 383
Borchert, H., 62
Borovik-Romanova, F., 294
Botter, R., 191, 192
Boussingault, J. B. J. D., 436
Bowen, N. L., 44
Bower, V. E., 437
Boyce, I. S., 207
Brackett, F. S., 46
Bradley, W. F., 137
Braitsch, O., 62
Brancazio, P. J., 11
Brannock, W. W., 48, 296
Breuck, W. de, 142, 192
Brewer, P. G., 386, 420, 436
Brewster, D., 326
Bricker, O. P., 226, 267
Bright, H. A., 220
Broecker, W. S., 62
Brown, G. F., 421
Brown, H., 45, 92
Brown, H. T., 45
Brown, R. M., 206
Bruce, E. L., 45
Bucher, W. H., 45
Buddington, A. F., 45
Bullard, E. C., 45
Bullard, H. M., 220
Bullen, K. E., 45
Bunsen, R., 383
Burger, A. J., 92
Burton, E. F., 326
Butler, J. N., 436
Buttlar, H. von, 205, 206, 296

Cadle, R. D., 226
Callendar, G. S., 45
Cameron, A. G. W., 11
Cameron, E. N., 326
Cameron, J. F., 206, 207
Careri, G., 158
Carpenter, G. B., 327
Carsey, J. B., 45
Carslaw, H. S., 437
Case, L. C., 294
Chajec, W., 294
Chamberlin, R. T., 45, 348
Chamberlin, T. C., 45
Chao, E. C. T., 118
Chave, K. E., 62, 136, 220

Chebotarev, I. I., 294, 310
Chilingar, G. V., 271
Chivers, E., 384
Christ, C. L., 5, 267
Chu, C. C., 383
Clark, R. D., 220
Clark, S. P., Jr., 11
Clarke, F. W., 45, 118, 136, 294, 310
Clausen, H. B., 142, 143
Clayton, R. N., 220
Cleaues, E. T., 226
Cloud, P. E., 45
Coats, R. R., 294
Collins, W. H., 45
Contois, D. E., 136
Conway, E. J., 45, 118, 136
Cooper, L. H. N., 45
Cornog, R., 206
Correns, C. W., 136, 294
Cotton, C. A., 45
Coughlin, J. P., 92
Craig, H., 93, 142, 159, 191, 205, 206, 220, 294, 383, 420, 421
Crank, J., 437
Crawford, J. G., 45, 294, 310
Crooks, R. N., 246
Csurny, J., 436
Culkin, F., 386, 420, 436
Curloook, W., 92
Curray, J. R., 421
Currier, L. W., 45

Dakin, W. J., 45
Daly, R. A., 45
Damon, P. E., 92
Dansgaard, W., 142, 143, 191, 220
David, D. J., 383
Davidson, C. F., 92
Davidson, F. A., 45
Davy, H., 326
Davydov, Y. N., 207
Day, A. L., 44, 45, 158, 293, 383
Deffeyes, K. S., 136, 267
Degens, E. T., 220, 271, 386, 421
Deicha, G., 326, 327
Delevaux, M. H., 421
Delibrias, G., 207
Densmore, C. D., 386, 421
Deuser, W. G., 421
Devirts, A. L., 207
Dickson, F. W., 383
Dietz, R. S., 137
Dietzman, W. D., 220

Dobbin, C. E., 294
Dobkina, E. I., 207
Dobson, G. M. B., 45
Doe, B. R., 421
Dorsey, N. E., 327
Dott, R. H., 310
Drake, C. L., 421
Dreyer, R. M., 294
Dwyer, M. J., 62
Dzens-Litovsky, A. I., 294

Eaton, J. P., 92
Edmondson, W. T., 45
Edwards, M. L., 136
Egnér, H., 246
Ehhalt, D., 191
Einarsson, T., 383
Ekstein, J., 437
Elliott, W. C., 310
Ellis, A. J., 272, 348, 383
Elsasser, W. M., 45
Emerson, A. E., 44
Emery, K. O., 137, 294, 295, 437
Epstein, S., 149, 158, 159, 191, 220
Eriksson, E., 142, 192, 206, 246
Escombe, F., 45
Eskola, P., 45
Eugster, H. P., 383
Eulitz, G. W., 206
Evans, E. L., 93
Evans, J. W., 45
Ewart, A. J., 45

Facy, L., 192
Fairbairn, H. W., 137
Fairbridge, R. W., 11, 142
Fairchild, H. L., 45
Faltings, V., 206
Fanale, F. P., 11
Farrington, O. C., 45
Fash, R. H., 294
Faucher, J. A., 220
Faure, G., 421
Favajee, J. C. L., 137
Fenner, C. N., 45, 383
Fergusson, G. J., 206
Feth, J. H., 267
Fireman, E. L., 205
Fisher, E. M. R., 246
Fleck, H., 437
Fleischer, M., 47, 118
Fleming, R. H., 47, 48, 104, 246
Fleming, W. H., 93

Fletcher, N. H., 327
Foster, M. D., 45
Fournier, R. O., 272, 383
Friedman, I., 143, 149, 159, 192, 220, 272, 437
Fyfe, W. S., 294

Gail, F. W., 46
Garrels, R. M., 5, 62, 111, 137, 267, 294, 437
Gat, J. R., 437
Gates, R., 421
Gerhard, E. R., 246
Gibson, D. T., 46
Giese, A. C., 46
Giletti, B. J., 205, 207
Gilfillan, E. S., Jr., 192
Gilluly, J., 46
Ginter, R. L., 294, 310
Ginzburg, A. I., 347
Girdler, R. W., 421
Godfrey, A. E., 226
Goldberg, E. D., 137
Goldich, S. S., 92
Golding, R. M., 383
Goldschmidt, V. M., 11, 46, 111, 119, 236, 294
Goldsmith, J. R., 92
Gonfiantini, R., 142, 192
Gonsior, B., 206
Goranson, R. W., 46
Gordon, L. I., 142, 191, 421
Gorham, E., 137, 246
Goudie, A. J., 384
Gould, R. F., 5
Graf, D. L., 92, 220, 272, 437
Grechny, Ya. V., 327
Green, J., 92
Greer, F. E., 293
Griffin, G. M., 137
Griffin, J. J., 137
Grim, R. E., 137
Grønvold, F., 92
Grosse, A. V., 206
Grout, F. F., 119
Grubbs, D. K., 384
Grummitt, W. E., 206
Gruner, J. W., 46
Grushkin, G. G., 327
Gulbransen, E. A., 92
Gulyaeva, L. A., 294
Gunter, G., 46
Gustafson, P. E., 246
Gutenberg, B., 46
Gwinn, G. R., 45

Hagino, K., 384
Hague, A., 383
Hahn, W. C., Jr., 92
Halsted, R. E., 158
Hamilton, W. M., 348
Harbeck, G. E., 437
Harder, H., 384
Hardie, L. A., 220
Hare, C. E., 46, 295, 310
Harned, H. S., 437
Harris, J., 192, 220
Harrison, N., 437
Harteck, P., 206
Hartmann, M., 421
Harvey, H. W., 46, 111
Hathaway, J. C., 386, 421
Hawkins, M. E., 220
Haxel, O., 246
Healy, J., 384
Heck, E. T., 46, 294
Heezen, C. B., 137
Helfferich, F., 437
Hem, J. D., 267, 272, 294, 421, 437
Henderson, L. J., 46
Herapath, T. J., 437
Herapath, W., 437
Hess, H. H., 46
Heybroek, F., 421
Higgins, C. E., 436
Hillebrand, W. F., 220
Hillesund, S., 327
Hillig, W. B., 327
Himstedt, F., 294
Hinson, H. H., 293
Hitchon, B., 220, 271, 272
Hoffer, T. E., 327
Hoffman, J. I., 92, 220
Högbom, A. G., 46
Holden, F. R., 246
Holland, H. D., 62, 137
Holmes, A., 46, 95
Holser, W. T., 62
Hood, D. W., 5, 226
Hoover, W. H., 46
Horberg, L., 221
Horibe, Y., 142, 191, 421
Hoskins, H. A., 46, 294, 295, 310
Hostetler, P. B., 267
Hours, R. M., 207
Howard, C. D., 294
Hsu, P. T., 436
Hubbert, M. K., 46
Hudson, F. S., 294

Humphreys, W. J., 46
Hunt, J. M., 220
Hunter, J. G., 246
Hurley, P. M., 46, 137
Huston, R., 436
Hutchinson, G. E., 46, 92, 437

Ikeda, N., 294, 348
Ingerson, E., 46, 294
Ingram, M., 145
Ishizu, R., 294
Israel, G., 207
Ivanov, V. V., 347
Iwasaki, I., 384

Jaeger, J. C., 437
Jagger, T. A., 348
James, H. L., 92, 294
Jeffords, R. M., 294
Jeffreys, H., 46
Jensen, C. A., 46
Jensen, J., 294
Jessen, F. W., 310
Johns, W. D., 137
Johnsen, S. J., 142, 143
Johnson, J. S., 437
Johnson, M. W., 48, 104, 246
Johnston, E. S., 46
Johnston, J., 46
Johnston, O. C., 437
Johnston, W. M., 206
Jokela, A., 386, 421
Jones, H. S., 46
Jones, L., 421
Jones, O. W., 220
Junge, C. E., 137, 246

Kaiser, A. D., Jr., 138
Kalle, K., 246
Kalyuzhnyi, V. A., 327
Kamata, E., 226
Kamenetskaya, D. S., 327
Kanamori, N., 226, 236
Kanamori, S., 226, 236
Kaplan, I. R., 137, 421
Kato, K., 226
Kaufman, S., 205
Kaufman, W. J., 207
Kazmina, T. I., 294
Keenan, J. H., 46
Keevil, N. B., 46
Keily, H. J., 246
Keith, M. L., 221

Kelley, W. P., 310
Kellogg, H. H., 92
Kellogg, W. W., 226
Kennedy, G. C., 92, 384
Kennedy, V. C., 137
Kennedy, W. Q., 46
Keyes, F. G., 46
Khitarov, N. I., 348, 384
Kimura, K., 348
Kirshenbaum, I., 159
Kister, L. R., 136
Kitano, Y., 226, 272
Klinger, F. L., 92
Klingsberg, C., 92, 93
Klovan, J. E., 272
Knot, K., 191
Kobayashi, S., 236
Kokubu, N., 93
Kolesov, G. M., 206
Kontsova, V. V., 220
Korzhinskii, D. S., 348
Koyama, T., 226
Kramer, J. R., 11, 62, 111, 137
Krasintseva, V. V., 348
Krauskopf, K. B., 295, 348
Krishnamurty, K. V., 421
Krogh, A., 46, 47
Krueger, H. W., 92
Krynine, P. D., 93
Ku, T. L., 421
Kubaschewski, O., 93
Kuenen, P. H., 47, 137
Kuiper, G. P., 47, 93
Kulp, J. L., 92, 205
Kuroda, K., 236
Kuroda, P. K., 384
Kuwamoto, T., 236
Kuznetsova, G. A., 207

Ladd, H. S., 47
Lal, D., 206, 207, 421
Lamborn, R. E., 295
Landergren, S., 47, 295
Landsberg, H. E., 47
Lane, A. C., 47, 295
Langbein, W. B., 437
Langway, C. C., Jr., 142
Lanier, R. D., 437
Lartet, L., 437
Lasaga, A. C., 62
Latimer, W. M., 47
Lavrukhina, A. K., 206
Lawson, A. C., 47

Lazrus, A. L., 226
Legendre, R., 47
Leith, C. K., 47
Lerman, A., 386, 437
Lettau, H., 246
Levorsen, A. I., 310
Lewis, W. B., 383
Li, T. H., 62
Libby, W. F., 143, 205, 206, 207, 295, 296
Liddel, H. F., 384
Liebenberg, W. R., 93
Liebig, G. F., 310
Little, W. M., 327
Littlepage, J. L., 272
Livingstone, D. A., 11, 137, 226, 267
Lloyd, D. J., 327
Lloyd, E. T., 348
Lockhart, L. B., Jr., 246
Longinelli, A., 421
Lonsdale, K., 327
Lorius, C., 192
Lowenstam, H. A., 11, 47, 137
Lucia, F. J., 267
Lundell, G. E. F., 220
Lundgren, L. B., 206
Lusk, F. E., 295
Luyanas, V. Y., 205
Lyakhovskaya, E. I., 437
Lynch, W. F., 437

Macallum, A. B., 47
McCarter, R. S., 138, 310
McClendon, J. F., 47
McCrea, J. M., 158
McCue, J. B., 295, 310
Macelwane, J. B., 47
McFarlin, P. F., 386, 421
McGill, D. A., 421
McGrain, P., 295, 310
Macgregor, A. M., 47
Machta, L., 192
Mackenzie, F. T., 62, 137, 437
McKinney, C. R., 158
Macnamara, J., 93
Maere, X. de, 143, 192
Magill, P. L., 246
Mahon, W. A. J., 272, 384
Mair, J. A., 93
Malakhov, S. G., 207
Malinin, S. D., 348
Manheim, F. T., 386, 421, 437
Maricelli, J. J., 296
Markova, N. G., 207

Author Citation Index

Marshall, W. L., 437
Martell, E. A., 143, 206, 226
Martishchenko, L. G., 207
Mason, B. J., 327
Mathieu, G. G., 421
Mayeda, T., 93, 158, 159, 191, 220
Maynard, J. E., 47
Maynes, A. D., 93
Mazor, E., 437
Mead, W. J., 47
Meents, W. F., 220, 221, 272, 295, 310, 437
Meinzer, O. E., 47, 295
Menzel, D. H., 48
Merlivat, L., 192
Mero, F., 437
Merrill, G. P., 47
Meyer, G. H., 272
Miholić, S., 47
Miller, A. R., 386, 421
Miller, J. C., 295
Miller, J. P., 295
Mills, R. van A., 47, 295, 310
Milton, R. F., 384
Miyake, Y., 226
Mizutani, Y., 421
Moberly, R. M., Jr., 137
Moelwyn-Hughes, E. A., 384
Möller, F., 246
Moore, E. S., 47
Moore, R. C., 47
Moran, T., 327
Morey, G. W., 47
Morgan, J. J., 5
Morita, Y., 236
Morrow, M. B., 272
Moses, P. L., 221
Mossop, S. C., 327
Moum, J., 221
Mrazec, L., 310
Muan, A., 92
Murata, K. J., 48, 92, 296
Murray, J., 137

Naboko, S. I., 348, 384
Nace, R. L., 437
Nagel, J. F., 191
Naito, H., 236
Nakai, N., 226
Nash, L. K., 47
Neev, D., 437
Nekhorosheva, M. P., 207
Nel, L. T., 93
Nelson, B. W., 137

Neumann, A. C., 421
Newcombe, R. B., 47
Nichols, E. A., 221
Nicolaysen, L. O., 92
Nief, G., 191, 192
Nielsen, H., 421
Nier, A. O., 92, 144, 158
Niggli, P., 46
Nininger, H. H., 47
Nir, A., 207
Nissenbaum, A., 421
Nockolds, S. R., 295

Oana, S., 236
Odum, H. T., 11
Ogura, N., 142
Ohman, M. F., 295
Okabe, S., 236
Okuda, S., 272
Oliver, W. F., 326
Orr, W. L., 294, 295
Orville, P. M., 384
Osborn, E. F., 93
Osborne, F. F., 47
Ossaka, J., 272
Ostlund, G., 206
Ovchinnikov, A. M., 347, 348
Owen, B. B., 437
Ozawa, T., 384

Pan, K., 295
Panteleev, I. Y., 348
Papp, F., 295
Park, O., 44
Park, T., 44
Parks, G., 207
Patchett, J. E., 93
Patterson, C., 226
Payne, B. R., 206
Pearson, C. A., 220
Perkins, T. K., 437
Perry, E. C., Jr., 62
Pettersson, H., 47
Pettijohn, F. J., 47, 93
Picciotto, E., 142, 143, 192
Pidgeon, L. M., 92
Piip, B. I., 384
Pinson, W. H., 137
Piper, A. M., 47, 295
Platford, R. F., 437
Plummer, F. B., 310
Pocklington, R., 386, 421
Poldervaart, A., 92

Polzer, W. L., 267
Poole, J. H. J., 47
Powers, E. B., 47
Powers, M. C., 138
Price, P. H., 295, 310
Prikhid'ko, P. L., 327
Pugh, D. T., 421
Pytkowicz, R. M., 421

Quirke, T. T., 45

Rabinowitch, E. I., 47
Rafter, T. A., 421
Rama, 207
Ramdohr, P., 93
Rankama, K., 47, 295
Redfield, A. C., 192, 220
Reed, J., 384
Reed, W. E., 220
Rees, C. E., 421
Rees, O. W., 221, 295, 310
Reuter, J. H., 220
Revelle, R., 47, 138, 246, 295
Reznikov, M. I., 384
Richards, F. A., 138
Richter, C. F., 46
Riley, G. A., 47
Riley, J. P., 138, 386, 420, 436
Ringwood, A. E., 11
Ritchie, J. A., 384
Rittenberg, S. C., 137, 294, 295
Roberson, C. E., 267
Robinson, R. A., 437
Roedder, E., 327
Roether, W., 207
Rogers, C. G., 48
Rogers, G. S., 295, 310
Rogers, L. B., 246
Rolshausen, F. W., 310
Romanov, V. V., 206
Romo, L. A., 221
Ronov, A. B., 11
Rosenqvist, I. T., 221
Rosenstock, H. B., 246
Rosenthal, E., 437
Ross, C. P., 48
Ross, D. A., 386
Rossby, C. G., 246
Roth, E., 191, 192, 221
Rowe, J. J., 272, 383
Rowe, R. B., 326
Roy, R., 92, 93
Rubey, W. W., 11, 48, 62, 93, 138, 295, 310

Rublevskiy, V. P., 207
Rush, D. H., 220
Rush, R. M., 437
Russell, H. N., 48
Russell, K. L., 63
Russell, R. D., 93
Russell, W. L., 48, 295, 310
Rutherford, R. L., 48
Ryan, T., 246

Sahama, T. G., 47, 295
Salt, R. W., 327
Sandberg, C. H., 296
Sandell, E. B., 384
Sandoval, P., 206
Sargent, E. C., 310
Saukov, A. A., 348
Savchenko, V. P., 348
Sawyer, F. G., 295
Scarcia, G., 136
Scatchard, G., 437
Schafer, J. L., 383
Schairer, J. F., 44
Schmidt, K. P., 44
Schmölzer, A., 295
Schnerb, J., 437
Schoeller, H., 295
Schoen, B., 192, 220
Schoewe, W. H., 295, 310
Schopf, J. W., 63
Schröer, E., 295
Schuchert, C., 48
Schultz, P. W., 149
Schumann, G., 207, 246
Schwarzenbach, G., 111
Scofield, C. S., 295
Scriven, L. E., 327
Sederholm, J. J., 48, 119
Sevryugova, 192
Sewell, J. R., 384
Shallcross, F. V., 327
Shapiro, S. A., 295
Sharp, R., 191
Shatkay, A., 437
Shaw, D. R., 138, 271
Shcherbina, V. V., 348
Shelford, V. E., 48
Shepherd, E. S., 48, 295, 348
Shima, M., 421
Shimp, N. F., 220, 272, 437
Shleimovich, R. E., 437
Shumakov, V. I., 207
Siever, R., 63

Siffert, B., 267
Sillén, L. G., 63, 111, 138
Silnichenko, V. G., 384
Silverman, S. R., 158
Singh Pruthi, H., 48
Sitter, L. U. de, 294, 310
Sivertson, J. N., 384
Skinner, B. J., 2
Slichter, L. B., 48
Slusher, R., 437
Smith, C. L., 48
Smith, J. E., 295
Smith, W. C., 384
Smith-Johannsen, R., 327
Smolko, G. I., 295
Smyth, H. D., 158
Soller, R., 192
Sorby, H. C., 327
Soyfer, V. N., 206
Speer, M. G., 383
Spencer, D. W., 420
Spezia, G., 327
Spicer, H. C., 44
Spitzer, L., Jr., 93
Stahl, W., 206
Stanley, J. P., 45
Steacy, H. R., 93
Stearns, H. T., 48
Stearns, N. D., 48
Stefánnson, U., 138
Stegun, I. A., 436
Steiner, A., 220, 384
Sternling, C. V., 327
Stetson, H. T., 48
Stewart, F. H., 421
Stewart, N. G., 246
Stockman, L. P., 295
Stokes, R. H., 437
Strock, L. W., 295
Strutt, R. J. (Lord Rayleigh), 47, 295
Stumm, W., 5
Suess, E., 384
Suess, H. E., 48, 158, 206, 246
Sugawara, K., 226, 236
Sukharev, G. M., 295
Sverdrup, H. U., 48, 104, 246
Sweeney, R. E., 421
Synnott, J. C., 436

Tageeva, N. V., 295
Taggart, M. S., Jr., 138
Takahashi, T., 62
Taliaferro, N. L., 294

Tamers, M., 206, 207
Tan, F. C., 62
Tanaka, M., 236
Tarutani, T., 384
Taugourdeau, P., 327
Teare, J. W., 436
Teegarden, B. I., 206
Telegdi-Roth, K., 295
Telushkina, Y. L., 207
Terada, K., 236
Thatcher, L. L., 206
Thode, H. G., 93, 421
Thomas, H. C., 220
Thomas, J. W., 383
Thomas, W. H., 136
Thompson, E. F., 421
Thurber, D. L., 421
Tilbury, W. G., 221, 295, 310
Timm, B. C., 296
Timmermann, E., 48
Todd, D. K., 221
Togliatti, V., 142, 192
Tongiorgi, E., 142, 192
Torii, T., 272
Tovarova, I. I., 384
Truesdell, A. H., 272
Turekian, K. K., 11, 138, 143
Turnbull, D., 327
Twenhofel, W. H., 48
Twomey, S., 246
Tzeitlin, S. G., 296
Tzur, Y., 437

United States Department of Agriculture, 221
Urbain, M. P., 348, 349
Urey, H. C., 93, 111, 158, 296
Urry, W. D., 48
Utsumi, S., 384
Uvarof, 192

Valley, G., 48
Valyashko, M. G., 272
Van Hise, C. R., 48
Van Orstrand, C. E., 48
Van Sicklen, W. J., 327
Vaughan, T. W., 48
Vegard, L., 327
Verhoogen, J., 48
Vernadskii, V. I., 347
Villiers, J. W. L. de, 92
Vinogradov, A. P., 111, 205, 207, 296, 310, 347

Vlasova, N. K., 272
Vogel, J. C., 191

Waggaman, W. H., 45
Wahl, M. H., 158
Wahl, W., 48
Wakeel, S. K. El, 138
Walther, J., 48
Waring, G. A., 48, 267, 272, 296, 349, 421, 437
Washington, H. S., 118, 119
Waterbury, J. B., 421
Watson, S. W., 421
Wattenberg, H., 48
Watts, W. L., 296
Weaver, C. E., 138, 310
Weber, J. N., 221
Weeks, L. G., 48
Weis, P. L., 326, 327
Weiss, R. F., 421
Weissbart, J., 92
Weizsäcker, C. F. von, 111
Wells, R. C., 47, 295, 310
Wendt, I., 421
Werby, R. T., 137
Wetherill, G. W., 92
Weyl, P. K., 63, 267
White, D. E., 48, 159, 191, 220, 267, 272, 294, 296, 310, 349, 383, 384, 421, 437

Whitehouse, U. G., 138, 310
Whitney, J. D., 296
Wickman, F. E., 119
Wiik, B., 206
Wildt, R., 48
Wilson, A. T., 206
Wilson, J. T., 48
Wilson, S. H., 296, 349, 383, 384
Wlotzka, F., 384
Wolfgang, R. L., 206
Wu, Y. C., 437
Wyss, O., 272

Yamada, S., 236
Yaron, F., 437
Yarotskii, L. A., 347
Yefremova, G. P., 207
Yermakov, N. P., 327
Yokogama, Y., 348
Yoshioka, R., 226, 272
Young, S. W., 327

Zamchinsky, W. C., 149
Zdanovskii, A. B., 437
Zen, E-an, 221
Zeschke, G., 93
Zhavoronkov, 192
Zies, E. G., 44, 48, 296
ZoBell, C. E., 310
Zykova, A. S., 207

Subject Index

Acid ocean, 21–22, 98 (*see also* Ocean water)
Atlantis deep brine, 387, 390–391,
    393–395, 398–402, 412
  steady-state model, 407–409
Antarcticite, 270
Arsenic in precipitation, 224, 229
Atlantis deep sediments, 399–403, 409–414
Atmosphere
  chemical composition
    first stage, 65–77, 90
    second stage, 77–87, 90
    third stage, 87–90
    with time, 90
  dense primitive
    chemical effects, 21–24
    solution in melt, 18–21
  moderate primitive, gradual
    accumulation of ocean, 24–25
  present-day condition, 23
  strongly reducing, 64–65, 68, 70–72, 81

Biocarbonate ions, formation from carbon
    dioxide during weathering process,
    251–254
Body fluids, marine animals, 14–15
Brancazio, P. J., 10
Brine (oil field)
  alkaline earths (Mg^{2+}, Ca^{2+}, Sr^{2+}), 287
  alkalines (Li^+, Na^+, K^+), 282–283
  boron, 285
  carbon dioxide, 284–285
  chemical composition, 278, 280, 299
  combined nitrogen, 285–286
  dissolved salt, 281–282
  halogens (Cl^-, F^-, Br^-, I^-), 282

  hydrocarbon, 286
  hydrogen sulfide, 284
  increase in salt concentration, 423–425
  silica, 286–287
  sulfate, 283–284
Brine saturation
  carnallite, 433
  dissolved salt, 432–435
  halite, 432–434
Bromine in basal halites, 59

Calcite
  degree of saturation of seawater, 50–51
  solubility, 21–24, 102–103
Calcium chloride water, 270, 397 (*see also*
    Red Sea brine, depletion)
Calcium fluoride, solubility (100–400°C),
    373
Calcium sulfate (gypsum)
  formation
    in brine water, 426–427
    in seawater, 127–128
  solubility
    in chloride brine, 429–430
    in distilled water, 259
    in seawater, 54–55
    in sodium chloride solution, 55
Cameron, A. G. W., 10
Carbon dioxide
  atmosphere–ocean system
    equilibrium in seawater, 26–28
    pH and pCO_2 in seawater, 30–34
  in brine, 284–285
  continuous supply to atmosphere and
    ocean, 35–36

448

influence
    on composition of sediments (brucite), 29–33
    on leaching of rock (see Leaching, rock)
    on organisms, 28–29
    on weathering of rock (see Weathering of rock)
    inventory on earth's surface, 25–26
    in natural gases (see Gas, chemical composition)
    pCO_2 variation in past atmosphere, 33–35
    in primitive atmosphere, 22–26
    source of supply, 35–38
Carbon isotopes
    δC^{13} in limestone near brine, 217
    δC^{13}, δC^{14} in Red Sea brines, 393, 401–402, 414–418
Carbonate sediments
    formation
        in brine water, 426–429
        in Precambrian, 21–25, 30–35
        in Red Sea, 405
        in seawater, 10, 25–26, 55, 127, 129
    Sr/Ca ratio in carbonate fossils, 10
Chain deep brine, 387–388
Chloride brine
    in Dead Sea, 422
    in Jordan–Dead Sea rift valley, 432
    in oil field, 278
    in Red Sea, 387
Chloride ions in ancient seawater, 9
Chlorite in seawater, 127, 133
Climatic record in ice core, 140
Climatology of glacial cycle, 142
Cold origin theory, 9
Complex formation, 3, 107–109
Condensation of water, 1
Condensation temperature of water, 168
Condensation theory, 9
Connate water, 15, 275, 282–283, 288–291, 297
Craig, H., 140
Crust, chemical composition, 114–115
Cyclic salt, 22–24, 234
    noncyclic salt, 234
    semicyclic salt, 234

Dead Sea chloride brines
    chemical composition, 425–429, 432
        ancient brines, 433–434
        present brines, 432
    cross section of basin, 432

formation
    mode, 433–436
    rate, 422–425
sulfate depletion, 425–429 (see also Red Sea brine, depletion)
Debye–Hückel activity coefficient, 263
Deuterium (see also δD–δO^{18}; Isotopic fractionation; Tritium)
    measurement of D/H in water
        apparatus, 148
        conversion of water to H_2, 148–149
        conversion of water to HCl, 144
        gas pressure influence, 146
        H_3^+ ion correction, 145–147
        procedure, 144–149, 161
    natural occurrence
        in abalone shell, 156
        in Canyon Diablo meteorite, 157
        deep seawater (D/H–salinity relation), 140
        in gas in Yellowstone Park, 156
        in Greenland, 168
        in hydrosphere (H_2O^{16}, HDO^{16}, H_2O^{18}), 76, 161, 193
        in ocean water, 151–153, 155
        in organic matters, 156
        in river water, 154–155
        in snow, 154–155
        in surface seawater (D/H–salinity relation), 140, 159
δD–δO^{18} (see also Deuterium; Oxygen isotope; Tritium)
    brine waters
        in Alberta basin, 211, 213–214, 216
        in Gulf Coast, 211, 213–214, 216
        in Illinois basin, 210, 212, 214–215
        in Michigan basin, 210–212, 214–215
        in Red Sea, 390–393, 403–407
    connate water, 280
    core samples of Red Sea sediments, 402
    lake and river water, 152–153, 159
    lake water (East Africa), 159
    magmatic water, 280
    metamorphic water, 280
    meteoric water, 280
    natural water, 151–155, 159, 170–173, 183–188, 280
    along Nile River, 182
    ocean water, 151–152, 154–155, 159, 280
    precipitation
        in Africa, 181
        in Australia and New Zealand, 181
        in island stations (Apia, Barbados,

Subject Index

Bermuda, Canton, Christmas, Hawaii, St. Helena, Midway, Ship V, S. Tome, Wake), 177–180
  in Northern Hemisphere, 180
  in South America, 181
  in various stations of the world, 153, 170–173, 183–188
  snow and ice, 140, 166–169
Diabase, chemical composition, 79
Dipole moment of water, 1
Discovery deep brine, 387, 389–393, 399–403, 409–412
Dissociation constant of $MgSO_4^0$, 430
Dolomite, deposition, 55–56
Dry fallout in Japan, chemical composition, 231

Earth's formation
  precore stage, 79
  theory of birth, 97–99
    cold origin theory, 9
    condensation theory, 9
    reduction theory, 9
Enrichment coefficient (factor)
  atmospheric precipitation, 228–229
  Red Sea brine, 395
Environment, fitness for life, 15
Equilibrium (see also Phase diagram)
  acid–SO_4^{2-}–Cl^-–H_2O (pH 2–5), 352
  $Au(OH)_4^-$–H^+–Cl^-–$AuCl_4^-$–H_2O, 109
  carbonate in seawater, 102–103
  $CaSiO_3$–SO_2–H_2O–Ca^{2+}–SO_4^{2-}–SiO_2–H_2, 82
  $\langle C \rangle_c + 2(H_2)_g \rightleftharpoons (CH_4)_g$, 72
  $\langle C \rangle_c + 2(H_2O)_g \rightleftharpoons (CO_2)_g + 2(H_2)_g$, 72
  $(CO^2)_g + \frac{1}{2}(O_2)_g \rightleftharpoons (CO_2)_g$, 67
  $(CO_2)_g + 4(H_2)_g \rightleftharpoons (CH_4)_g + 2(H_2O)_g$, 67
  CO_2–CO, 17–18
  Co^{2+}–H_2O–H^+–$CoOOH(s)$, 109
  dissolved chemical species–sediment interaction, 3
  exchange equilibria (solution aluminosilicates), 101–102
  $FeOOH + 3H^+ \rightleftharpoons Fe^{2+} + 2H_2O$, 105
  Fe_3O_4–H_2O–Fe_2O_3–H_2, 82
  Fe_2SiO_4–SO_2–Fe_2O_3–FeS_2–SiO_2, 82
  gaseous volatile materials, 9
  $(H_2)_g + \frac{1}{2}(O_2)_g \rightleftharpoons (H_2O)_g$, 67
  $(H_2)_g + \frac{1}{2}(S_2)_g \rightleftharpoons (H_2S)_g$, 69
  H_2–CH_4–CO_2–NH_3, 72
  $(H_2S)_g + 2(H_2O)_g \rightleftharpoons (SO_2)_g + 3(H_2)_g$, 68, 80–81
  K-mica + H_2O + $H^+ \rightleftharpoons$ kaolinite + K^+, 101

  monmorillonite + $2H^+$ + $7H_2O \rightleftharpoons$ kaolinite + $8SiO_2$ + Ca^{2+}, 255
  olivine + $O_2 \rightleftharpoons$ pyroxene + magnesite, 77
  PbS–O_2–$PbSO_4$ (Pb–O–S), 86
  redox equilibrium
    $[BrO_3^-] / [Br^-]$, 108
    $[ClO_3^-] / [Cl^-]$, 108
    $[IO_3^-] / [I^-]$, 108
Evaporation of water, 1, 231
Evaporites
  in Dead Sea brines, 429–436
  marine
    brucite, 30–31
    chert, 60
    gypsum, 54–60
    halite, 55–60
  in Red Sea brines, 404
Evolution
  atmosphere, 12, 64
  seawater, 8–9, 12, 64
Excess volatiles (excess volatile material)
  chemical composition, 17–18, 68, 97
  comparison with gases from volcanoes, rocks, hot springs, 37–38, 68
  similarity with magmatic gases, 36–37

Fairbridge, R. W., 10
Fluid inclusion, 49, 312–313
Fossil water, 275 (see also Connate water)
Freezing temperature of liquid inclusion
  apparatus for measurement, 314–317
  operation
    calibration, precision, accuracy, 323–326
    determination of first melting temperature, 320–321
    determination of freezing temperature, 321–322
    freezing an inclusion, 318
    heat-exchange medium, 317
    lighting, 322–323
    problem of metastability, 318
    sample preparation, 317–318

Gas, chemical composition
  from basalt and diabase, 37
  from hot springs, geysers, fumaroles, 37
  from Kilauea, 37
  from obsidian andesite and granite, 37
  in Red Sea brines, 393
  in seawater, 393
Geochemical balance, 8, 15–16, 96–97, 112–113, 117–118
Ground water, chemical composition, 280

Harned's rule coefficient, 430–431
Heat flow in volcanic hot spring area, 277
Human activity, chemicals used
  for domestic purpose, 232
  for farming purpose, 232
  for industrial products, 232
Hydrogen pressure in atmosphere, escape of hydrogen, 70–73
Hydrosphere, 2
Hydrothermal ateration of rock, dissolution from rock
  ammonium, 366–368, 374
  arsenic, 375
  boron, 365–368, 371, 373–374
  calcium, 366
  cesium, 376
  chloride, 365–368, 370–372
  fluoride, 365–368, 372
  lithium, 366, 376
  magnesium, 377
  potassium, 366–368, 375
  rubidium, 376
  silicate, 365, 367–368, 377–378
  sodium, 366–368, 375
  sulfur, 365–366, 375

IAEA–WMO precipitation survey δD–δO^{18}, 161, 175–190
Igneous rock, average chemical composition, 279
Illite, formation in seawater, 127, 133
Industrial products in Japan, 231–232
Inert gas, 18
Ion pair, 3, 107–109
Iron, ferrous carbonate (siderite)
  in Atlantis II deep sediment, 409–410
  in Discovery deep sediment, 414
  formation in Precambrian, 15
  recycling in Red Sea water, 411–412
Isotopic fractionation (δD, δO^{18}) (see also Deuterium; Oxygen; Tritium)
  altitude, δD–δO^{18} variation for precipitation with, 190–191
  amount effect of rain, δD–δO^{18} variation for precipitation with, 169–170, 188–191
  equilibrium processes
    condensation, 163–164
    evaporation, 164–165
    Rayleigh condensation process, 164–165, 171
    sublimation, 164
    temperature effect, 165–166
    vapor–ice equilibrium, 165–166
  evaporation, δD–δO^{18} variation for precipitation with, 161, 172–174, 191
  monthly precipitation, 182–189
  nonequilibrium processes
    composition of evaporating ocean water, 174
    exchange, 174–175
    kinetic effect, 172–174
  seasonal variation in precipitation, 189
  unusual isotopic composition of precipitation, 169

Juvenile water, 113, 140, 276

Kramer, J. R., 9, 49, 102

Leaching, rock (see also Hydrothermal alteration of rock; Weathering of rock)
  reaction time, 367
  with thermal water, 361
Lead, in snow and ice, 224
Life, birth of, 24, 61
Lithogenic material, in atmospheric precipitation, 228–230

Magma, generation of local, 40–41
Magmatic gas, average chemical composition, 330
Magmatic water, 273, 280
Magnesian limestone, formation in Precambrian, 15
Manganese
  in Discovery and Atlantis deep waters, 412–413
  Mn–Ca concentration in Red Sea deep waters, 412
  Mn–Na concentration in Red Sea deep waters, 413
Marinogenic material, in atmospheric precipitation, 228–230
Membrane, filtration of chemical species, 403
Metamorphic water, 276, 280, 290–291
Meteoric water, 275, 351
Montmorillonite, formation in seawater, 127, 133

Nier-type mass spectrometer, 144, 149
Nitrate, limiting nutrient, 52

Ocean water
  chemical composition, 278, 280, 282–287

Subject Index

gradual accumulation, 24–25
isostatic balance with continental blocks, 41–43
mechanisms determining chemical composition, 9, 49, 95
primeval ocean, 8 (*see also* Acid ocean)
Organic matters, in natural waters, 4–5
Oxygen
  in atmosphere, 23, 103
    ancient, 24
    Precambrian, 15, 67, 77, 83–87
    present, 23–25
    time dependence, 88–89
  production and use, 86–89
  by photosynthesis, 87
Oxygen isotope (δO^{18}) (*see also* δD–δO^{18})
  in condensate with temperature, 166–167
  in Greenland ice cap stations, 141, 168–169
  in limestone lying close to brine, 217
  ocean water, with depth, 140, 174
  in precipitation
    monthly precipitation, 169, 182–189
    range of δO^{18} value, 185–186
Ozone, in upper atmosphere, 14

Parent rock, 97, 113–117
Period of passage, 116–117
pH
  CO_2, 28, 30
  marine animal, 29
  seawater, 28, 30, 99, 101–103
Phase diagram
  FeO–Fe_2O_3–SiO_2, 65–66, 69
  MgO–FeO–Fe_2O_3–SiO_2, 77–78
  Na_2O–Al_2O_3–SiO_2–H_2O, 257
  water, 1–2
Pliocene water
  Midway–Sunset area, California, 300–301
Pollution, natural water, 4–5, 224
Precipitation (atmospheric precipitation) (*see* Rainwater, chemical composition)
Primary rock, 97, 113–117
Primeval atmosphere, vapor pressure, 19–21
Primeval ocean, 8 (*see also* Atmosphere; Ocean water)
Pyrite, formation in seawater, 127–128

Radioactive substance, artificial, in ocean, 224 (*see also* Carbon isotopes; Tritium)

Rainwater, chemical composition
  calcium, 242–244
  chloride, 239–241
  Cl^-/Na^+, 241
  in Japan, 228–229
  potassium, 242–244
  sodium, 240–241
  sulfate, 244–246
  in United States, 240–243
Rayleigh process, 164–165
Red Sea brine (*see also* Atlantis deep brine; Discovery deep brine)
  chemical composition
    dissolved gas, 393
    heavy metals (Co, Cu, Mn, Ni, Pb, Zn), 395–396, 412–414
    major, 388–389, 395–396
    minor, 395–396
    potassium, 397
    seasonal variation, 392
    sodium, 408
    space and time variation, 403–407
  concentration of original brine, 389
  depletion (*see also* Removal of chemical species from seawater; Seawater, postdepositional modification)
    fluoride ions, 403
    iodide ions, 403
    magnesium ions, 403
    sulfate ions, 403 (*see also* Dead Sea chloride brines, sulfate depletion)
  glacial control for brine flow, 407
  isotopes
    δC^{13}, δC^{14}, 411, 415
    δD, 391–392, 405–407
    δO^{18}, 391–392, 402, 405–407
    δS^{34}, 398–402
  mixing at brine–seawater interface, 408–409
  origin and history, 387
  salt added to, 394–397
  southern sill section, 404
Reduction theory, 9
Removal of chemical species from seawater, chemical reaction and amount
  bicarbonate ions, 127–129
  calcium ions, 127–129
  magnesium ions, 127, 129, 133
  potassium ions, 127, 133
  silicate, 127, 132–133
  sodium ions, 127, 130, 133
  sulfate ions, 127–128

Residence time, 9, 50, 52–53, 245
  sodium, 121
  sulfur, 245–246
  water, 1
Rhodochrosite, formation in Red Sea sediments, 410, 414
River
  chemical composition
    in Japan, 229–231
    in world, 121, 224
  chemical mass balance between river and ocean, 120–133
  chemical species
    amount of, 117, 122–124 (see also Suspended material in river water)
    sources of, 231–234 (see also Weathering of rock)
Rockslide and landslide, forecast through water quality, 225
Rubey, W. W., 10

Seawater
  age, 113–117
  chemical composition, 27, 96–97, 106–107, 121, 280
    $Ca^{2+}-HCO_3^--SO_4^{2-}$ system, 54–59
    change with time, 23–24
    chemical equilibria between seawater and sediment, 50, 101–102
    chloride concentration in ancient seawater, 9
    constancy with time, 9, 52–53
    controlling factor, 50, 101–102
    equilibrium model and real system, 109–110
    in the past, 14–15, 33
  dissolved state of chemical species, 107
  geologic history, 12–13
  postdepositional modification (see also Red Sea brine, depletion; Removal of chemical species from seawater)
    barium ions, 307
    bicarbonate ions, 308
    bromide ions, 308–309
    calcium ions, 305–307
    chloride ions, 303
    iodide ions, 308–309
    magnesium ions, 305–306
    potassium ions, 303, 305
    sodium ions, 305
    strontium ions, 301, 307
    sulfate ions, 307–308
    total dissolved ions, 303
  siderite, deposition in earlier Precambrian, 61
  silica, high concentration in Precambrian, 60
  steady state, 52
Sediment, average composition, 97
Silica and silicate, formation in seawater, 127, 131–135
SMOW, 159, 161, 163, 168, 173, 209, 390, 398
Sodium cycle in hydrosphere, 9, 113
Sodium in igneous rocks and sediments, 113
Solid solution, 106–108
  (Ca, Sr) CO_3, 107–108
  (Fe, Co) OOH, 107–108
  (Mn, Pb) O_2, 107–108
Solubility
  brucite, 259
  calcite, 22, 259
  carnallite, 433
  gypsum, 259
  halite, 430–431, 433
  hydromagnesite, 259
  magnesite, 259
  sepiolite, 259
  silica
    amorphous silica, 259, 377
    cristobalite, 377
    quartz, 377
  water, in melt of granite, $CO_2-H_2O-K_2O-SiO_2$, 19–21
Solution chemistry, 4–5, 107 (see also Complex formation)
Subsurface water
  chemical composition, 297–298, 301–302
  diagenesis process, 270
Sulfate and sulfide
  isotope (δS^{34})
    in Red Sea brine, 398–402, 411
    in Red Sea sediment, 399–403
  origin, in Red Sea deep sediment, 399–400
  reduction of sulfate ions
    in polluted coastal area, 224
    in Red Sea brine, 400–401
    in river water, 125, 127, 233
  sulfur emitted to the atmosphere, 245
Suspended material in river water
  amount, 122
  type and reactivity, 123–124

Subject Index

Thermal spring water
  acid-chloride water, 252
  acid-sulfate water, 352
  chemical composition, 280, 292–293, 335–336, 342, 354–356
    in Japan, 231–232
  (principal) component of gas and ionic salt composition, 344
  (characteristic) environment, 344
  estimate of underground temperature, 270
  formation process, 344–345
  from graywacke and sediments, 359
  metamorphic process, 345
  neutral chloride water, 351–352
  in New Zealand, 353–356
  origin of initial water, 344
  in reducing environment, 337–338, 342–343
  secondary phenomena and processes, 345
  type, 344
  in upper oxidizing zone, 335–336
Tritium
  distribution
    in atmospheric precipitation, 201, 204
    in atmospheric water vapor, 201
    in groundwater, 199
    in Moscow water, 202
    in nature, 193–194, 202
    in rain at Ottawa in 1952–1963, 195
    in surface water, 202
    in troposphere in 1965–1968, 203
  formation, 193–194
    cosmic and nuclear origin, 195
    rate, 194–195
  method for determination
    counting apparatus, 197–198, 200
    enrichment, 196
    hydrocarbon preparation, 196–197

Uniformitarianism, 15

Volatiles (volatile matter), 8–9, 17, 97
  escape during magma crystallization, 38–39
  escape through hot springs, 39–40
  excess volatiles, 17–18, 38, 65, 68
Volcanic gas
  in Hawaiian volcanoes, 38, 68, 78
  in Kilauea, 37–38
  in Kilauea lava, 329

  in Showa volcano, 330
  from volcanoes, rocks, and hot springs, 37–38
Volcanic (hot spring) water, chemical composition, 279–280
  alkaline earths (Ca, Mg, Sr), 287
  alkalines (Li, Na, K), 282–283
  boron, 285
  bromide ions, 282
  carbon dioxide, 284–285
  chloride ions, 282
  combined nitrogen, 285–286
  dissolved salt, 281–282
  fluoride ions, 282
  hydrocarbon, 286
  hydrogen sulfide, 284
  iodide ions, 282
  silica, 286–287
  sulfate ions, 283–284

Water
  condensation, 1
  dipole moment, 1
  distribution in earth's surface, 1–2
  evaporation, 1
  movement of, and chemical materials, 3
  organic matters, 4–5
  pollution, 4–5
  precipitation, 1–2
  residence time, 1
  separation, 1
  solubility in a melt of granite, 19–21
  transportation, 1–2
  unique nature
    potential for dissolving matter, 1
    potential for fractionating elements and isotopes, 1
    three phases, 1
  waste material, 2
Weathering of rock
  chemical reaction, 251–254
  dissolution
    calcium, 251–254
    calcium carbonate, 251–254
    magnesium, 251–254
    potassium, 251–254
    silica, 253–254
    sodium, 251–254
  formation
    kaolinite from biotite, 251–254
    kaolinite from plagioclase, 251–254

montmorillonite from plagioclase, 251–254
source material, 251–254
total mass of eroded rocks and deposited sediments, 16–17, 22–24
weathered product, 22–23, 251–254

weathering, 15–17, 22–23, 113, 117, 251
weathering process, 251–254
White, D. E., 270
WHOI CHAIN 61 Survey (Red Sea), 408, 412, 414–415